Intelligent Systems Reference Library

Volume 134

The aim of this series is to publish a Reference Library, including novel advances and developments in all aspects of Intelligent Systems in an easily accessible and well structured form. The series includes reference works, handbooks, compendia, textbooks, well-structured monographs, dictionaries, and encyclopedias. It contains well integrated knowledge and current information in the field of Intelligent Systems. The series covers the theory, applications, and design methods of Intelligent Systems. Virtually all disciplines such as engineering, computer science, avionics, business, e-commerce, environment, healthcare, physics and life science are included.

More information about this series at http://www.springer.com/series/8578

Colette Faucher
Editor

Advances in Culturally-Aware Intelligent Systems and in Cross-Cultural Psychological Studies

Editor
Colette Faucher
LIP6 (Laboratoire d'Informatique de Paris 6)
UPMC (Université Pierre et Marie Curie)
Paris
France

ISSN 1868-4394 ISSN 1868-4408 (electronic)
Intelligent Systems Reference Library
ISBN 978-3-319-88365-6 ISBN 978-3-319-67024-9 (eBook)
https://doi.org/10.1007/978-3-319-67024-9

Softcover reprint of the hardcover 1st edition 2017

Printed on acid-free paper

This Springer imprint is published by Springer Nature
The registered company is Springer International Publishing AG
The registered company address is: Gewerbestrasse 11, 6330 Cham, Switzerland

For Nick, my son and my pride, with my unconditional love, admiration and respect.

—Colette Faucher

Preface

In this era of globalization, people from different countries and cultures have the opportunity to interact directly or indirectly in a great diversity of situations. Despite differences in their way of thinking and reasoning, their behaviors, their values, lifestyles, customs and habits, languages, religions, in a word, their cultures, they must be able to collaborate on projects, to understand each other's views, to communicate in such a way that they do not offend each other, to anticipate the effects of their respective actions on others, and so on. It is then of primary importance to understand how culture affects people's mental activities, such as perception, interpretation, reasoning, emotion, and people's behavior, in order to anticipate possible misunderstandings due to differences in handling the same situation, and to try and solve them.

Artificial intelligence, and more specifically, the field of Intelligent Systems design, aims at building systems that mimic the behavior of human beings in order to complete tasks more efficiently than the latter could by themselves. Consequently, in the last decade, experts and scholars in the field of Intelligent Systems have been tackling the notion of cultural awareness. A Culturally-Aware Intelligent System can be defined as a system, where culture-related or, more generally, sociocultural information is modeled and used in designing its human–machine interface or intervenes in the task carried out by this system, be it reasoning, simulation, or any other task involving cultural knowledge.

The first part of the book is devoted to the presentation of some Culturally-Aware Intelligent Systems, devised in the field of artificial intelligence.

The two following parts intend to be a source of inspiration for building modelizations of culture and of its influence on the human mind and behavior, to be used in new Culturally-Aware Intelligent Systems. They, respectively, deal with the results of experiments carried out in two fields that study culture and its influence on the human mind's functions: Cultural Neuroscience and Cross-Cultural Psychology. Cultural Neuroscience is a sub-domain of both Cognitive Neuroscience and Cultural Psychology and Cross-Cultural Psychology is a sub-domain of Cognitive Psychology, that is the reason why, in Chap. 1 of Volume 1, entitled "Introduction," before the part devoted to the Culturally-Aware

Intelligent Systems, the goals that characterize each discipline and the tools used to conduct experiments in each field will be recalled.

Here is a brief outline of each chapter of the three parts.

Overview of the Chapters

Part I Culturally-Aware Intelligent Systems

Chapter 2: Culturally-Aware HCI Systems

R. Heimgärtner

Culture strongly influences human–computer interaction (HCI), since the end user is always operating within a certain cultural context. Chapter 2 describes a Culturally-Aware HCI System, in the context of automotive navigation, which culturally adapts its interaction with the end user over time.

Chapter 3: Building Time-Affordable Cultural Ontologies Using an Emic Approach

J. Petit, J.C. Boisson and F. Rousseau

Culturally-Aware Intelligent Systems often need to have a representation of a given culture. Most of the time, this representation is subjective because it is based on an etic approach, aiming at discovering cultural universals from an outsider perspective. However, this clearly fails at capturing a culture's specificities. Chapter 3 presents an emic approach that consists in building cultural knowledge in the form of ontologies, whose goal is to identify from an insider view, concepts, and behaviors that constitute typical traits of the concerned culture.

Chapter 4: Teaching an Australian Aboriginal Knowledge

C. Kutay

Teaching cultural competency is an important purpose of Culturally-Aware Intelligent Systems. Chapter 4 presents gaming environments involving intelligent agents modeling cultural rituals, values, and emotional responses to support the learning of cultural competency. It describes the development of cultural knowledge sharing processes to allow students to experience the conflicts felt by Aboriginal Australians within the mainstream culture.

Chapter 5: Culturally-Aware Healthcare Systems

L. Yin, T. Bickmore

Culturally-Aware Intelligent Systems often use conversational agents representing a specific culture that matches the user's one. Moreover, the messages sent through these agents, targeted and tailored for this culture, increase their impact on users belonging to that culture. Chapter 5 presents such a system, whose aim is to improve the efficacy of health care on patients.

Chapter 6: Combining a Data-driven and a Theory-Based Approaches to Generate Culture-Dependent Behaviours for Virtual Characters
B. Lugrin, J. Frommel and E. André
To incorporate culture into Intelligent Systems, for example, to generate culture-dependent behaviors in conversational agents or any virtual agents, two approaches are commonly used: theory-based approaches that build computational models based on cultural theories to predict culture-dependent behaviors, and data-driven approaches that rely on multimodal recordings of existing cultures. Chapter 6 presents a hybrid approach combining a data-driven and a theory-based approaches to generate culture-dependent behaviors for virtual characters.

Chapter 7: Mental Activity and Culture: The Elusive Real World
G.J. Hofstede
Chapter 7 deals with the design of social agents in Culturally-Aware Intelligent Systems, like the conversational agents seen in Chap. 5. Social agents have a mental activity in a social world. Mental activity includes three steps—perceive, interpret, select action—that can result in many differences between agents from different cultures. The social world—in the form of generic sociological theory—and these differences—in the form of Cross-Cultural Theory—can be used for designing these agents. The chapter gives examples from recent literature that can serve as points of departure for further work.

Chapter 8: Affective Body Movements (for Robots) Across Cultures
M. Rehm
Body movements in the human being express affective information but may be interpreted differently depending on the culture of the interpreter. The purpose of Chap. 8 is to both generate and interpret body movements in robots by using a methodological approach that takes into account the cultural background of both the developer and the user during the development process.

Chapter 9: Modeling Cultural and Personality Biases in Decision-Making
E. Hudlicka
The process of decision-making is well-known for including cultural, personality, and affective biases. Chapter 9 describes a method for modeling multiple decision biases resulting from cultural effects, personality traits, and affective states, within the context of a symbolic cognitive-affective agent architecture: the MAMID methodology and architecture.

A great deal of Culturally-Aware Intelligent Systems is made for military and security purposes, because military situations typically involve people from different cultures. Chapters 10 and 11 deal with the notion of culture in such contexts.

Chapter 10: Considering the Needs and Culture of the Local Population in Contemporary Military Intervention Simulations: an Agent-Based Modeling Approach

J.Y. Bergier, C. Faucher

Chapter 10 presents SICOMORES, a simulation system, based on a multi-agent architecture representing the local population in the context of an asymmetric conflict, which simulates the effects of military actions of influence on the population. It takes into account the way culture influences these types of actions, treated like cognitive processes.

Chapter 11: Simple Culture-Informed Models of the Adversary

P.K. Davis

Cognitive models of the adversary are useful in a variety of domains, including national security analysis, but good cognitive models must avoid mirror imaging, which implies recognizing ways in which the adversary's reasoning may be affected by history, culture, personalities, and imperfect information, as well as by objective circumstances. Chapter 11 describes a series of research efforts over three decades to build such cognitive models.

Part II Cultural Neuroscience

Chapter 12: Cultural Neuroscience

R.T. Bjornsdottir, N.O. Rule

Chapter 12 provides an overview of the research in the burgeoning field of Cultural Neuroscience that results from the convergence of Cultural Psychology and Cognitive Neuroscience, and outlines the history of the field and its origins. This specific field encompasses a wide variety of research and provides a unique lens through which to study cultural differences. Notably, research in this field has provided evidence of subtle and nuanced differences across cultures, where behavioral evidence alone could not, demonstrating the importance of the neuroscientific approach.

Chapter 13: Cultural Neuroscience and the Military: Applications, Perspectives, Controversies

K. Trochowska

Chapter 13 shows the evidence of the culture-brain nexus and its numerous implications for human mind functioning in a variety of domains, reviews the existing solutions and projects that leading military institutions are already conducting in the cognitive field, and, in light of the newest findings of Cultural Neuroscience, proposes new potential solutions and enhancements for the design and conduct of military training and of non-kinetic aspects of military operations.

Part III Cross-Cultural Psychology

Chapter 14: Cross-Cultural Dimensions, Metaphors and Paradoxes: An Exploratory Comparative Analysis

M.J. Gannon, P. Deb

Chapter 14 compares the three most popular ways of describing and analyzing cross-cultural similarities, differences, and areas of ambiguity: dimensions like in Hofstede's works, according to an etic approach, cultural metaphors according to an emic approach, and paradoxes again according to an etic approach. These paradigms are actually complementary in the global understanding of a given culture, and one can go back from one level to a previous one that is then better understood (« feedback loops »).

If the paradigm of dimensions has been largely exploited in AI Systems, we think it would be beneficial to researchers in this field to develop works on operationalizing the concepts of cultural metaphors and paradoxes as well as the notion of feedback loops.

Chapter 15: A Model of Culture-Based Communication

B. Martinovski

Both humans and virtual agents interact with intercultural environments and need to behave appropriately according to the environment. Chapter 15 proposes a dynamic modular model of culture-based communication, which reflects intercultural communication processes and can be used in the design of life-like training scenarios where culture is defined as a semiotic process and a system, which builds upon self and other identities and which is sustained and modified through communication and cognitive-emotive mechanisms such as reciprocal adaptation, interactive alignment, and appraisal.

Chapter 16: Dynamic Decision-Making Across Cultures

C.D. Güss, E. Teta

Chapter 16 studies the decision-making process in complex, uncertain, and dynamic situations, whereas decision-making research had so far focused on simple choices.

The chapter discusses a methodology especially suited for the study of dynamic decision-making and then discusses new empirical research on how culture influences dynamic decision. Such findings contribute to a more comprehensive theory of decision-making and allow for a better understanding of decision-making conflicts. Finally, applications of these findings are discussed and can be utilized in cultural competency training programs or international work teams.

Chapter 17: When Beliefs and Logic Contradict: Issues of Values, Religion and Culture

V. Cavojova

In real debates, we often do not think about the validity of the arguments from the strictly logical point of view and we often disagree even before we hear the particular argument: This is the *confirmation bias*. Chapter 17 deals with the confirmation bias in reasoning about controversial issues (in this case abortions), and it examines the effect of cultural parameters, like values (pro-life, pro-choice, neutral) and religious and political affiliations, on syllogistic reasoning. This chapter shows that when beliefs and evidence clash, it is often belief that wins. It is no surprise that people untrained in critical, scientific thinking resort to beliefs as their compass in navigating through the vast ocean of conflicting information (obtained from scientific research) and conflicting values (such as the rights of children vs. the rights of their mothers) that are contemporary, globalized human societies.

Chapter 18: Social Influence and Intercultural Differences
L. Rodrigues, J. Blondé and F. Girandola
Chapter 18 studies the effects of cross-cultural differences (individualistic vs. collectivistic cultures) on cognitive dissonance, social influence, and persuasion. It shows that intra-individual processes, such as the reduction of dissonance and the processing of persuasive information, are regulated by cultural orientations and cultural aspects of the self (independent vs. interdependent self-construal). Considering these cross-cultural effects, new avenues of research open up on change and resistance to change in many fields such as health, environment, consumption, and radicalization.

Chapter 19: The Influence of Emotion and Culture on Language—Representation and Processing
D.M. Basnight-Brown, J. Altarriba
Research focused on the study of emotion, specifically how it is mentally represented in the human memory system, is of great importance within the study of cognition. Chapter 19 examines the factors that make emotion words unique, as compared to other word types (e.g., concrete and abstract words) that have traditionally been of interest. This chapter also describes the factors that influence how those who know and use more than one language process *and* express emotion, and the role that language selection plays on the level of emotion that is activated and displayed. Finally, cross-cultural differences in emotion are examined, primarily as they relate to differences in individualistic and collectivistic contexts.

Chapter 20: Creating a Culture of Innovation
A. Markman
In the context of Chap. 20, the term culture is not limited to nations or ethnic groups, but is broadened to include any group that influences the individual's behavior, actually any organization. This chapter describes a culture of *innovation*, that is, the process of generating and implementing practical new ideas, where key factors are the need to favor innovation over efficiency, to tolerate failure, and to have the flow of information and ideas circulate between people.

To operationalize this broadened view of the notion of culture, we think it would be worth designing new concepts defining a culture and studying the way they can be used to model the effects on the reasoning and behavior of the group of people that adopts it.

Chapter 21: The Wonder of Reason at the Psychological Roots of Violence
M. Maldonato
The last chapter, Chap. 21, stands apart from the other chapters and deals with the role of culture at the scale of mankind from a psychoanalytical and philosophical point of view.

Aggression, violence, and destructiveness have been part of human nature since its origins. Their roots can be traced back to the unconscious and an elaboration of mourning that uses division in order to save oneself from anguish and guilt, attributing all good to one's own object of love and all evil to an external enemy—just as it happens a stranger, considered dangerous and an enemy, is the object of anguish not because he really is dangerous, but because onto him the internal enemy is projected. Chapter 21 seeks to show how this permanent psychic tension derives from the meeting of opposing, heterogeneous, and unpredictable forces and movements which can be neutralized but are never canceled out. The balance between instinct and rationality can be lost at any time, and, on an individual or collective level, it can degenerate into pure violence. But, if life expresses itself through biological functions of a very high complexity, it also does so through history and culture. A sense of guilt can be elaborated in order to build better civilization and allow for the development of life protected from the worst excesses of violence.

Paris, France Colette Faucher

Contents

Contributors

Jeanette Altarriba Department of Psychology, Social Science 399, University at Albany, State University of New York, Albany, NY, USA

Elisabeth André Human Centered Multimedia, Augsburg University, Augsburg, Germany

Dana M. Basnight-Brown United States International University - Africa, Nairobi, Kenya

Jean-Yves Bergier Paris, France

Timothy Bickmore College of Computer and Information Science, Northeastern University, Boston, MA, USA

R. Thora Bjornsdottir Department of Psychology, University of Toronto, Toronto, ON, Canada

Jérôme Blondé LPS, Aix Marseille University, Aix-en-Provence, France

Jean-Charles Boisson CASH Team, CReSTIC Laboratory (EA 3804), University of Reims Champagne-Ardenne, Reims, France

Vladimíra Čavojová Centre of Social and Psychological Sciences SAS, Bratislava, Slovak Republic

Paul K. Davis Pardee RAND Graduate School, Santa Monica, CA, USA

Palash Deb Strategic Management Group, K-503, New Academic Block, Indian Institute of Management Calcutta, Joka, Kolkata, India

C. Dominik Güss Department of Psychology, University of North Florida, Jacksonville, USA

Colette Faucher LIP6 (Laboratoire d'Informatique de Paris 6), UPMC (Université Pierre et Marie Curie), Paris, France

Julian Frommel Institute of Media Informatics, Ulm University, Ulm, Germany

Martin J. Gannon University of Maryland and Cal State San Marcos, San Marcos, CA, USA

Fabien Girandola LPS, Aix Marseille University, Aix-en-Provence, France

Rüdiger Heimgärtner Intercultural User Interface Consulting (IUIC), Undorf, Germany

Gert Jan Hofstede Department of Social Sciences, Applied Information Technology Group & SiLiCo, Wageningen University, Wageningen, The Netherlands

Eva Hudlicka Psychometrix Associates & College of Information and Computer Sciences, University of Massachusetts-Amherst, Amherst, MA, USA

Cat Kutay Faculty of Engineering and Information Technology, The University of Technology, Sydney, Australia

Birgit Lugrin Human-Computer Interaction, University of Wuerzburg, Würzburg, Germany

Mauro Maldonato Naples, Italy

Arthur B. Markman Department of Psychology, University of Texas, Austin, TX, USA

Bilyana Martinovski Department of Computer and Systems Sciences (DSV), Stockholm University, Stockholm, Sweden

Jean Petit Capgemini Technology Services, Suresnes, France

Matthias Rehm Aalborg U Robotics, Technical Faculty of IT and Design, Aalborg University, Aalborg, Denmark

Lionel Rodrigues LPS, Aix Marseille University, Aix-en-Provence, France

Francis Rousseaux MODECO Team, CReSTIC Laboratory (EA 3804), University of Reims Champagne-Ardenne, Reims, France

Nicholas O. Rule Department of Psychology, University of Toronto, Toronto, ON, Canada

Elizabeth Teta Department of Psychology, University of North Florida, Jacksonville, FL, USA

Kamila Trochowska State Security Institute, National Security Faculty, War Studies University, Warsaw, Poland

Langxuan Yin Jersey, NJ, USA

Chapter 1
Introduction

Colette Faucher

Abstract In this introductory chapter, we would like to clarify the purpose and the tools used in two disciplines, Cross-Cultural Psychology and Cultural Neuroscience, as in Part II and Part III of this book, will be presented studies conducted respectively in both research domains. We begin to describe Cognitive Psychology as Cross-Cultural Psychology is one of its subfield, then Cognitive Neuroscience as well as Cultural Psychology from which Cultural Neuroscience originated.

Keywords Cognitive Psychology · Cross-Cultural Psychology
Cognitive Neuroscience · Cultural Psychology · Cultural Neuroscience

1.1 Cognitive Psychology

Cognitive Psychology studies the main psychological functions of the human being. The general notion of cognition subsumes the set of the functions that psychology has dealt with since its origins: sensation, perception, learning, memory, reasoning, as well as all the activities that are involved in the production and understanding of language. In this sense, one can characterize it as "the set of processes whose function is to produce and to use knowledge". It also concerns the formulation of hypothesis about the way knowledge is organized within the human memory (individual symbolic entities or networks) and the architecture that links all the components.

Cognitive Psychology postulates that one can infer representations, structures and mental processes from the study of behavior. Contrary to behaviorism, it makes it clear that psychology is the study of the mind, not of the behavior. Moreover, introspection is not seen as a reliable enough way to explore the mental.

C. Faucher (✉)
LIP6 (Laboratoire d'Informatique de Paris 6), UPMC (Université Pierre et Marie Curie), Bureau 26:00, 503, 4, Place Jussieu, 75005 Paris, France
e-mail: Colette.Faucher@lip6.fr

C. Faucher (ed.), *Advances in Culturally-Aware Intelligent Systems and in Cross-Cultural Psychological Studies*, Intelligent Systems Reference Library 134, https://doi.org/10.1007/978-3-319-67024-9_1

What characterizes Cognitive Psychology is a general goal: to establish a conception of human cognition inspired by the concepts provided by the information processing theory. For this purpose, it is necessary to account for the mental functions that apply to information in general and more specifically to information that is likely to become knowledge. Several steps are distinguished: the initial input of perceptive information, its transformation and its storing in memory, its organization and its evolution within this memory and finally its retrieval in view to later use, in the context of new situations implying, for example, the search for a solution to an unusual problem.

Cognitive Psychology preferentially uses experimentation and behavioral measures that study in particular reaction time or the time necessary to an operation (time to complete the task, time to read exhibition), the accuracy of the response (for instance, the rate of right or wrong answers), or even cognitive eye tracking or physiological data (functional imaging, etc.). Computer modeling also plays an important role.

Some scholars devote their research to the cognitive architecture of the mind. One then finds experiments intended to elucidate the different «modules» that take charge of the main functions of cognition. These modules do not necessarily correspond to specific cerebral units, but rather to functional entities that may mobilize a variety of distinct cerebral structures. For example, one can refer to the distinction between working memory and long-term memory or semantic memory and episodic memory. Cognitive Psychology also works with the concept of association.

Other researchers aim at describing the strategies that individuals implement to deal with daily life tasks, problem solving tasks, decision-making or even professional tasks (medical diagnosis, air traffic control and so on).

The cognitive activity is seen as the interface between two sets of entities: representations (informational structures «written» in memory) and processes of treatment applicable to these representations (activation, comparison, combination, transformation).

An idea that has been prevailing for a long time is that cognition deals mostly with higher level processes, which has sometimes led some scholars to posit a gap between «perception» and «cognition». However, this idea has since been largely abandoned and all forms of information processing (including at its lowest level) have been progressively gathered together under the generic term «cognition».

It is not by chance that the expansion of the cognitive concepts in the psychology of the last quarter of the twentieth century coincides with the taking into account of the concept of cognition by other disciplines, first and foremost by the ones concerned at that time with the simulation of the functions of the human mind, like Artificial Intelligence. This approach postulated that it is possible to build artificial systems able to contain knowledge, to manipulate its components, and to apply to it computational mechanisms allowing for the production of reasoning or decision-making. Therefore, cognitive mechanisms similar to those supported by biological systems were supposed to be transposable to other supports and result in similar outputs than the ones from human intelligence. This ambition to «mimic» cognitive functions and to have them performed by non-biological systems allowed scholars to reveal that eminently cognitive concepts like «representation» or «intelligence» had a validity far beyond psychology. Actually, those concepts had not been borrowed or transferred from a scientific field to another one, but the acknowledging of those concepts was possible at a higher level of abstraction and generality than for each of the concerned disciplines. This situation has forced the researchers to elaborate description languages compatible with both the concepts used to account for natural cognition and those used to account for artificial intelligence. Some of these efforts resulted in symbolic models, others in connectionist or hybrid models.

1.2 Psychology and Culture

For about 60 years, Cognitive Psychology has been developing without paying a lot of attention to Culture. Indeed, we tend to suppose that our ways of being, thinking and acting are «natural» and «universal». This postulate was at the core of the scientific approach of the psychologists. It is only around the sixties that more and more researchers began to question that postulate and to conduct researches within several distinct cultures.

Three trends then appeared within Cognitive Psychology concerning the notion of Culture:

- **Cross-cultural Psychology** where researchers observe differences between individuals coming from different cultural groups. At this level, cultural similarities or differences are observed separately in each group.
- **Intercultural Psychology** where scholars study phenomena related to the contact between the members coming from different cultures. The main concept is the one of *acculturation*, which refers to all the changes (psychological, cultural, etc.) induced by contacts between people from different cultural backgrounds.
- **Cultural Psychology**, which supports a theoretical approach differing from both Cross-Cultural and Intercultural Psychology. First, Intercultural Psychology tends to view Culture as a set of factors or conditions external to the individual, whereas Cultural Psychology considers Culture as being essentially internal to the person. Second, whereas Intercultural Psychology usually adopts a universalist point of view on the nature and unity of mankind, Cultural Psychology tends to favor relativism and the idea that culture and mind participate in the constitution of each other. These distinctions are important as they lead to opposite viewpoints on the question of the universality of the cognitive processes.

1.3 Cognitive Neuroscience

The research domain covered by Cognitive Neuroscience is fairly recent. It dates back to the end of the 1970s where Michael Gazzaniga and George Miller introduced this expression.

The aim of Cognitive Neuroscience is to identify cerebral processes at the source of our mental capacities and to assess, in neurobiological terms, the plausibility of the models proposed in Cognitive Science. How can the functioning of our brain underlie our immaterial mind? To perceive, to recognize, to decide, to act, to memorize, to speak, to reason, to be aware of, to pay attention to… What cerebral organization, what neural processes, what neural coding can explain the main cognitive capacities in the human being and the animal? Cognitive Neuroscience is at the interface between Neuroscience, Psychology, Computer Neuroscience, Cognitive Science and Philosophy. Researchers in this area often combine paradigms from Cognitive Psychology with approaches that are specific to the study of the brain and neural functioning. The spectacular technical evolution of the last 30 years has allowed for a huge rise of Cognitive Neuroscience.

Our knowledge about the relations between the brain and cognitive capacities has long depended on Neurology and Neuropsychology. The deterioration of some patients' cognitive functions allowed to match a given function with a specific cerebral area.

The contribution of cerebral imaging has been tremendous: while our knowledge depended on the study of patients whose cerebral functioning was impaired by trauma, it became possible to visualize the brain activity of a healthy subject while he was performing a specific cognitive task.

The techniques used to this purpose are Positron Emission Tomography (PET) (which requires the injection of a weakly radioactive tracer to get a picture of the brain) and Magnetic Resonance Imaging (MRI) (that is a totally non-invasive method). Such techniques produce anatomic pictures of the brain in two or three dimensions, but mostly pictures of the brain in activity during the execution of a task (fMRI, Functional Magnetic Resonance Imaging).

fMRI does not directly measure neural activity, but it reveals the active cerebral areas by measuring blood flow, since blood brings to the brain the elements necessary for its activation, especially oxygen. While a subject performs a given task, a larger volume of oxygenated blood flows towards the active cerebral areas and the BOLD (Blood Oxygenation Level-Dependent) signal reflects these flow modifications.

However, if fRMI is useful to visualize the set of cerebral structures that become active during a cognitive task, it does not provide any temporal information about the order in which they are activated. For that purpose, it is necessary to make use of Electroencephalography (EEG) or Magnetoencephalography (MEG), that register the synchronous activity of several thousands neurons by recording an electric or magnetic signal. However, it remains very difficult to determine the localization of the neuron populations that are at the source of the signal.

Researchers solve that problem by combining the «temporal resolution» of EEG and the «spatial resolution» of RMI and then merging the information provided by means of both techniques, but this double approach is quite complex and is not used very often.

One can then understand how the extraordinary rise of Cognitive Neuroscience is closely linked to the considerable technical progress that revolutionized Neuroscience, but those techniques still cannot answer the fundamental question that Cognitive Neuroscience asks: what is the nature of neural representations in our brain? How can the functioning of neuron populations organized in cerebral networks explain our capacity to perceive the world that surrounds us, to interact with it, to reason, to anticipate, to understand others?

The notion of «neural representation» is at the heart of Cognitive Neuroscience as the notion of «mental representation» is at the heart of Cognitive Psychology. For some researchers, the representation of a stimulus is «distributed» and implies the activation of a very large set of neurons—several millions, possibly billions—none of which is really specific to this stimulus. For others, this representation implies only a few very specific neurons and one then refers to a «scattered» representation. The most extreme hypothesis (called the hypothesis of the «grand-mother» cell) postulates that a unique neuron is activated only if the precise stimulus for which it has been specializing occurs (for example, the neuron devoted to the neighbor's dog).

All the techniques we mentioned are not able to provide a definitive answer to the partisans of the distributed and the scattered approaches. This question requires the study of the coding of information at the level of the individual neuron and the recording of the activity of unitary neurons is an invasive method since one or several electrodes must enter the brain. However, some experiments support the hypothesis of the scattered representation.

In any cognitive task, an individual's decisions depend on both their internal motivations and the consequences of their decisions on the complex social network to which they belong. To decode the emotions and the intentions of others from their facial expressions and the direction of their gaze is essential to establish harmonious social relationships.

Cognitive Neuroscience has made great progress to pinpoint the neural substrates on which emotions like fear, disgust, and anger depend. It is still necessary to be able to interpret this emotion or this gaze. An autistic child knows how to recognize the direction of the other's gaze, but he does not know how to interpret it to deduce the other's interest or desire.

Interacting is also to anticipate the consequences of others' actions and to use others' mistakes to adapt one's own behavior. Which neural coding is at the source of such capacities?

For a few years, very ingenious protocols having two or three monkeys interacting have been allowing to obtain very interesting data. Researchers got to show that a monkey knew how to use the mistakes of another monkey that he observed to adapt his own behavior. They highlighted, in the frontal lobe, a neural population which codes in a very specific way the mistakes made by the observed monkey!

Other neurons even anticipate the consequence (reward or not) of the decisions of the observed monkey. These studies require the complex conditioning of several monkeys and numerous controls, but Cognitive Neuroscience is developing in the domain of social interactions and will allow to characterize the underlying neuronal processes and to better understand the disfunctioning of the social behavior associated to certain pathologies.

Cognitive Neuroscience however should reach one of its limits, as, in the human being, beyond his relationships with the other, it would need to take into account dimensions like the individuals' history or cultural identity.

1.4 Cultural Neuroscience

Human beings are physiologically constituted in the same way across the world. So far, one assumed that memory, intelligence, brain functions worked exactly the same way for all human beings. By considering that Culture and the human mind are inevitably linked and inseparable, Cultural Psychology questions this assumption.

Traditionally, experimental psychologists assume that the human mind consists of the same components and follows the same general rules of functioning whatever the Culture, like the central unit of a computer. However, in the framework of Cultural Psychology, several reasons lead to think that very often the functioning of the human mind is not independent from Culture and the understanding of cognitive phenomena will remain incomplete as long as one will not take into account the cultural context within which they take place. Contrary to a computer, the human brain changes and evolves across time, depending on our experiences. Given that our experiences influence our mind, and that cultures differ as regards the types of experiences they may bring, some researchers attempted to illustrate the idea that Culture and the human mind shape each other. One can for instance mention Shobu Kitayama's works at the University of Michigan. Kitayama uses MRI to show that the areas of the prefrontal cortex known for being involved in the processing of information related to the Self are activated when one asks Chinese people or Westerners to judge information related to the Self. However, an interesting cultural difference is observed when the participants have to make similar judgments towards their mother. In that case, the same cerebral area is activated but only in the Chinese participants. This result is consistent with the idea of collectivism and interdependent Self in the Chinese culture, contrary to individualism and the concept of independent Self in the Westerners.

Cultural Neuroscience, stemming from both Cognitive Neuroscience and Cultural Psychology, is concerned with the relations between the values and the cultural practices on the one hand and the functioning of the brain on the other hand and will probably develop more and more in the coming years. One can expect that future research will focus on the study of the mechanisms allowing to explain the influence of Culture on human cognitions and behavior.

Most of the studies in Cognitive Psychology that underlie AI Culturally-Aware Intelligent Systems come from Cross-Cultural Psychology as the latter model the influence of Culture within one given culture and by using the metaphor of the brain functioning as an information processing computer unit. To our knowledge, very few not to say no AI Culturally-Aware Intelligent Systems are based on Intercultural Psychology. Finally, the postulate that characterizes Cultural Psychology, which is the mutual influence of Culture and the human mind, viewed through the lens of Cognitive Neuroscience, has not yet been used as a basis for AI Culturally-Aware Intelligent Systems. By describing experiments in the approaches to Culture that one can find in Cross-Cultural Psychology and Cultural Neuroscience respectively, we hope to inspire new research in AI and Culture that will result in more realistic models of the influence of Culture on the human mind and behavior.

References

1. Fabre-Thorpe, M., Thorpe, S.: «NEUROSCIENCES COGNITIVES», *Encyclopædia Universalis*, online, seen on 23 June 2017, http://www.universalis.fr/encyclopedie/neurosciences-cognitives/
2. Denis, M.: «PSYCHOLOGIE COGNITIVE», *Encyclopædia Universalis*, online, seen on 23 June 2017, http://www.universalis.fr/encyclopedie/psychologie-cognitive/
3. Markus, H.R., Kitayama, S.: Culture and the self: implications for cognition, emotion, and motivation. Psychol. Rev. **98**, 224–253 (1991)
4. Anderson, J.R.: Cognitive Psychology and its Implications, 7e édn. Worth Publishers, New York (2009)
5. Bayne, T., Cleeremans, A., Wilken, P.: The Oxford Companion to Consciousness. Oxford University Press, Oxford (2009)
6. Dehaene, S.: Vers une science de la vie mentale. Fayard, Paris (2006)
7. Denis dir, M.: *La Psychologie cognitive.* La Maison des sciences de l'homme, Paris (2012)
8. De Vega, M., Glenberg, A.M., Graesser, A.C.: Symbols and Embodiement: Debates on Meaning and Cognition. Oxford University Press, Oxford (2008)
9. Johnson-Laird, P.N.: How we Reason. Oxford University Press, Oxford (2006)
10. Kahneman, D.: Thinking, Fast and Slow, Farrar. Straus & Giroux, New York (2011)
11. Lemaire, P.: *Psychologie Cognitive*, 2e édn. De Boeck université, Bruxelles (2006)
12. Le Ny, J.-F.: Comment l'esprit produit du sens: notions et résultats des sciences cognitives. Odile Jacob, Paris (2005)
13. Richard, F.: *Les Activités mentales*, 4e édn. Armand Colin, Paris (2005)
14. Ceballos, M.: Simulating the effects of acculturation and return migration on the maternal and infant health of Mexican immigrants in the United States: a research note. Demography **48**(2), 425–436 (2011)

Part I
Culturally-Aware Intelligent Systems

Chapter 2
Culturally-Aware HCI Systems

Rüdiger Heimgärtner

Abstract Culture influences human–computer interaction (HCI) heavily, since the end-user is always operating within a certain cultural context. First, cultural and informational factors jointly influence the look and feel of interactive systems, for example, widget position or information density. In addition, each individual develops a specific culture (eating style, walking style, etc.)—that is, their own characteristics, behavior, attitudes, and values. Consequently, individual adaptivity is sometimes a key factor in covering the disparate needs of culturally but uniquely imprinted end-users; this may involve such tasks as reducing the workload by recognizing the individual expectations of each end-user. This improves usability, shortens training units, and improves universal access. For Culturally-Aware HCI systems socio-cultural information is used and modeled in the design and application of the human–computer interface (HCI) of such systems. In this chapter, we describe a Culturally-Aware HCI system in the context of automotive navigation that culturally adapts its interaction with the end-user over time. We analyze the way in which culture influences reasoning and the way the users feel and behave in HCI in order to establish a model for automatically adapting HCI to users using a Culturally-Aware adaptive HCI system. Fundamental theoretical reflections are presented and exemplified, and design and functioning are thus described in both theory and practice.

Keywords Culturally influenced HCI model · Model · Culture
Architecture · Culturally-Aware HCI systems · System · HCI
Demonstrator · Cultural interaction indicator · Cultural dimension
HCI dimension · Structural equation model (SEM) · Culturally
adaptive HCI architecture (CAHCI) · Principle of culturally adaptive
HCI · Intercultural user interface design (IUID) · UID
Intercultural HCI design

R. Heimgärtner (✉)
Intercultural User Interface Consulting (IUIC), Lindenstraße 9, 93152 Undorf, Germany
e-mail: ruediger.heimgaertner@iuic.de
URL: http://www.iuic.de

C. Faucher (ed.), *Advances in Culturally-Aware Intelligent Systems and in Cross-Cultural Psychological Studies*, Intelligent Systems Reference Library 134, https://doi.org/10.1007/978-3-319-67024-9_2

2.1 Approach to Culturally-Aware HCI Systems

Culturally-Aware systems can be defined as systems where culture-related or, more generally, socio-cultural information is modeled and used to design HCI, or such information intervenes in the task carried out by this system, whether it is during reasoning, simulation, or any other task involving cultural knowledge. This information must be encapsulated in a cultural model in the Culturally-Aware HCI system in order to effect the right HCI adaptations to be made in a cultural context. In the following, a method for creating Culturally-Aware HCI systems is presented.

2.1.1 Intercultural HCI Design

One method for finding differences between cultures is analyzing critical interaction situations between humans [1]. Honold [2] introduced this approach to HCI analysis by considering the occurrence of critical interaction situations in problematic user interfaces and system functionality as well as considering systems as (artificial) agents with their own culturally imprinted behavior caused by the developer's culture. The user's internal model of the system is imprinted by their culture, by their expectations of the system's properties, and by their interaction experience with the system.

After deriving intercultural factors from cultural dimensions that describe the behavior of members of cultures [3], for example, [4] empirically showed the direct influence of cultural markers on the performance of users interacting with the system and showed the connection between culture and usability. In addition, [5] investigated the influence of culture on usability. In his conclusion, [5] stated that "Individualism/Collectivism is connected to and has an effect on usability" ([5], p. 17): Hofstede's individualism index [6] is significantly connected to the user's attitude toward satisfaction and their attitude toward product usability.

Röse [7] suggested a "method of culture-oriented design" (p. 103) (MCD) that integrates aspects from new concepts of culture-oriented HCI design and knowledge about cultural differences into existing concepts of HCI design. Relevant cultural variables for intercultural HCI design have to be determined and specified analytically through a literature review and requirements analysis. The values of the variables represent culture-dependent variations that can be found at all levels of HCI localization (surface, functionality, and interaction) [7] and can be exploited in intercultural user interface design (IUID) [8]. However, areas strongly influenced by culture do not come to the surface directly—only behavior is visible on the surface, which is imprinted by cultural aspects over time; only the user's behavior itself yields deep insights about the user's cultural imprint [7]. Therefore, one of the most promising methods for obtaining cultural differences in HCI is observing, analyzing, and evaluating user–system interaction. Qualitative and quantitative empirical analyses must show if the results of studying HCI correspond to cultural models. Finally, the values of the cultural variables need to be considered to develop guidelines for intercultural HCI design and intercultural usability engineering [8].

2.1.2 *Determination of Cultural Interaction Indicators and Formation of HCI Dimensions*

A first step toward a theory of culturally influenced HCI is to develop a set of cultural interaction indicators (CIIs), which establishes the basis of a model for describing cultural differences in the interaction behavior of the user by representing the relationship between (the values of) cultural dimensions and (the values of) the dimensions of user interaction behavior; that is, (the values of) "human computer interaction dimensions" (HCIDs) such as information speed, information density, interaction speed, and interaction frequency [9]. They represent the characteristics of HCI by describing the HCI style of the user, that is, the method of information processing and the interaction style exhibited by the user. Frequency, density, order, and structure are particularly affected during information processing; frequency and speed are affected by user–system interaction. The quality of information processing and interaction is nourished by effectiveness and efficiency. To be able to measure these parameters, the specifics of the HCI dimensions must be as concrete and exact as possible. They can be represented by quantitative variables to build a basic measuring apparatus, from which they can then be connected to cultural dimensions, thereby forming empirical hypotheses. Some of these quantitative variables are explained later in this chapter.

The most common approach to quantifying the influences of culture on HCI is to perform qualitative and personal studies. Although this process is relatively controllable, it is very expensive and time consuming. Conversely, asking many users online is a relatively quantitative and less controllable process. The advantages of both approaches can be combined to solve this dilemma: many users can be asked to respond to special use cases on their PCs and the resulting qualitative data can then be collected quantitatively. The "Intercultural Interaction Analysis" tool (IIA tool) was developed by [10], based on [11] and [12], to automatically obtain qualitative data regarding cultural differences in HCI and to find metrics which are adequate for measuring cross-cultural HCI.

2.1.3 *Creating a Model for the Relationship Between Cultural Dimensions and HCI Dimensions*

The collected quantitative data can be analyzed using statistical methods (such as ANOVA, the Kruskal–Wallis test, post hoc tests, etc.) to reveal the cultural differences in HCI (e.g., using potential CIIs). Structural equation modeling (SEM) serves to verify the postulated relationships between cultural dimensions' and HCIDs' values to confirm or to modify (or even to identify) relationships (e.g., in combination with factor analysis). From this, usability metrics [13] can be derived that are empirically valuable in terms of measuring quantitative variables in culturally influenced HCI. Thereby, HCIDs are connected to cultural models to construe a model that explains the relationship between HCI and culture, which, in turn, must also be empirically validated.

2.1.4 *The Approach to Culturally-Aware HCI Systems in a Nutshell*

The first task in obtaining a Culturally-Aware HCI system is to consider socio-cultural information in the design and application of such systems. Thereby, the influence of culture on the reasoning as well as the way the users feel and behave in HCI is analyzed (Sect. 2.2). Moreover, in order to adapt HCI in Culturally-Aware systems to the users, a cultural model has to be established that can be used to generate the adaptation rules for the required contexts of use (Sect. 2.3). This model uses cultural metrics to connect the adaptation rules in the system to the environment or obtain the environmental information for the system's model, leading to a generic framework and architecture for Culturally-Aware HCI systems (Sect. 2.4). The theoretical reflections are then empirically exemplified by a Culturally-Aware mobile driver navigation system (Sect. 2.5). Finally, the implications for Culturally-Aware HCI systems are summarized and recommendations given (Sect. 2.6).

2.2 Gathering Empirical Data

In order to adapt Culturally-Aware HCI systems to a user's cultural needs, first the differences in the cultural needs of the users—and hence the cultural differences in HCI at all levels of the HCI localization (surface, functionality, and interaction)—need to be investigated. Thus, areas such as the presentation of information (e.g., color, time and date format, icons, font size), language (e.g., font, direction of writing, naming), dialog design (e.g., menu structure and complexity, dialog form, layout, widget positions), and interaction design (e.g., navigation concept, system structure, interaction path, interaction speed) are considered [14]. Hall [15] found differences in communication speed between cultures, which also implies differences in information speed ("duration of information presentation"), information density ("number of parallel pieces of information during information presentation"), and information frequency ("number of information presentations per time unit").

2.2.1 Empirical Evidence: Cultural Interaction Indicators

Using a literature review and analytical reasoning, 118 *potentially culturally sensitive parameters have been identified* [16], implemented in the IIA tool, and applied by measuring the interaction behavior of test persons in relation to their culture using the IIA tool. Two online studies were conducted, in 2006 and 2007, respectively, with almost 15,000 speakers of Chinese (C), German (G), and English (E) located around the world using the IIA tool. Almost 1000 complete and valid data sets are available for evaluation. The test persons had to do short test tasks (mainly concerning driver navigation), where their interaction behavior was recorded during their working. From the 118 potential variables, 18 showed significant differences and which therefore can be called CIIs. These indicators represent significant differences in user interaction due to the different cultural backgrounds of the users. F expresses the ratio of explained to unexplained variance: a high value of F indicates a high probability that the mean values of two samples are different. The level of statistical significance is referenced with asterisks (*$p < 0.05$, **$p < 0.01$).

- *Opentaskbeforetest* (F(2,94) = 3.234*) is a metric variable which represents the number of open tasks in the working environment (i.e., running applications and icons in the Windows™ task bar) before the test session with the IIA data collection tool began.
- *Messagedistance* (F(2,94) = 6.625**) denotes the temporal distance of sequentially showing the maneuver advice messages in the maneuver guidance test task.
- *Infopresentationduration* (F(2,94) = 4.595*) represents the time the maneuver advice message is visible on the screen.
- *Mg.speed* (F(2,94) = 8.665**) indicates the driving speed of the simulated car in the maneuver guidance test task.
- *Mg.mouseclicks* (F(2,94) = 3.627*) specifies the number of mouse clicks during the maneuver guidance test task.
- Similarly, *uv.mouseclicks* (F(2,94) = 4.274*) counts the mouse clicks in the test task "uncertainty avoidance."
- *YesCounter* (F(2,94) = 4.012*) contains the number of acknowledged systems messages by the user during the whole test session.
- *Infohierarchy.number* (F(2,94) = 3.422*) indicates the number of entries in list boxes, group boxes, or combo boxes specified by the test persons.
- *Interactionexactness.duration* (F(2,94) = 3.892*) measures the duration of the abstract test task of "clicking dots away."
- *MouseMove_norm* (F(2,94) = 4.473*) contains the number of mouse moves during the whole test session divided by the duration of the whole test session (indicated by the suffix "_norm").
- *Break10s_norm* (F(2,94) = 5.150**) represents the number of interaction breaks with the mouse greater than 10 s divided by the duration of the whole test session.

- *Sem* (F(2,94) = 3.398**) measures the number of semantic events triggered by the user during the whole test session (e.g., the number of initiations of functions).
- *MoveAgent_norm* (F(2,94) = 44.204**) specifies the number of temporary moves of Microsoft's avatar "Merlin" into the middle of the screen every 30 s (which disappears again after 5 s) divided by the duration of the whole test session. If this indicator is low, the user switched off the agent quickly (as German- and English-speaking users did, in contrast to Chinese-speaking users).
- *Number of Chars* ($\chi^2(2) = 14.593$**) contains the number of characters entered by the user during the maneuver guidance and map display test tasks when answering open questions.
- *MouseLeftDown_norm* ($\chi^2(2) = 6.053$*) counts the number of clicks with the left mouse button divided by the duration of the whole test session.
- *MaxOpenTasks* ($\chi^2(2) = 10.061$**) is similar to *OpenTasksBeforeTest* explained above, but measures the maximal number of open tasks during the test session.
- *NoCounter* ($\chi^2(2) = 20.696$**) contains the number of refused system messages by the user during the test tasks which is similar to the variable *RefuseMessage* ($\chi^2 = 13.864$**), which measures the same but during the whole test session.
- *Break1 ms* ($\chi^2(2) = 23.430$**) represents the number of interaction breaks with the mouse greater than one millisecond, which effectively measures the speed of mouse movements made by the user.

The significant differences in the CIIs can also be seen when applying the IIA data analysis tool to plot "cultural HCI fingerprints" (in the style of [17]), which represent the cultural differences in HCI with respect to several variables for HCI design that depend on the cultural background of the potential target group of users (Fig. 2.1).

The data analysis showed that there are correlations between the interaction of Chinese and German users with a computer system (HCI) and their cultural background [9] concerning layout (more complex vs. simpler), information density (higher vs. lower), personalization (higher vs. lower), language (symbols vs. characters), interaction speed (higher vs. lower), and interaction frequency (higher vs. lower).

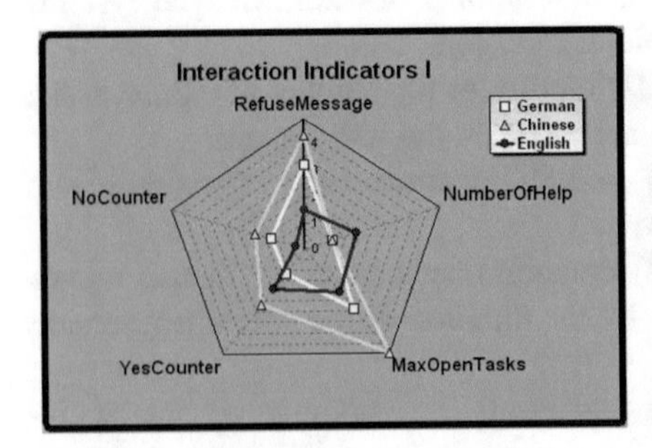

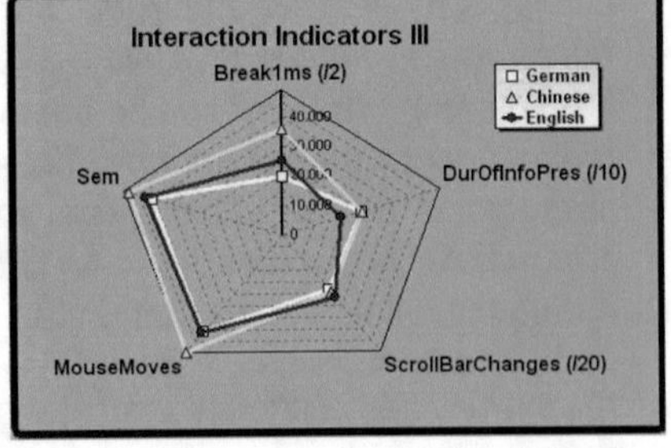

Fig. 2.1 Cultural HCI fingerprints (different values of the CIIs according to test languages) plot by the IIA data analysis tool

2.2.2 *Possible Relationship Between Cultural Dimensions and HCI Dimensions*

The cultural influence on HCI design can be represented by the relationship between the values of cultural dimensions and the values of the variables relevant for HCI design. Several basic assumptions were derived from the work of [18] regarding the connection between cultural dimensions and HCI dimensions [9]. The empirical hypotheses primarily concern quite basic user behavior, described with the following cultural dimensions: time orientation, density of information networks, communication speed, and action chains (sequential actions). Accordingly, it is reasonable to assume that HCI dimensions such as information speed (distribution speed and appearance frequency of information units), information density (number of and distance between information units), or information structure (order of information units) stand in relation to the culturally different basic behavior patterns of the users. If this is the case, the differences that [18] discovered also imply differences in information speed ("duration of the information presentation"), information density ("number of pieces of information presented in parallel"), and information frequency ("number of presentations of information units per time unit"). Interaction style should therefore also be affected. Table 2.1 shows some of these assumed connections.

For instance, based on the action chain dimension of [18], it can be assumed that German users' responses to questions are more linear (i.e., answered consecutively) in comparison to Chinese users. Furthermore, due to the high task orientation, the number of dialog steps taken until the completion of the task could be lower for German users. In addition, it can be assumed for German users that the number of interactions (such as the usage number for optional functions and help functions or adjusting colors, etc.) is higher because of their desire to work very exactly. However, the number of mouse movements or mouse clicks by German users should turn out to be lower than Chinese users due to the higher uncertainty avoidance and strong task orientation of German users [6, 19]). For these reasons, an interaction step (and thereby the complete test duration) might last longer for German than for Chinese users. Moreover, to save face [20], it is expected that Chinese users possibly do not click the help button as often as German users do. Finally, the speed of mouse movements should be lower for German users in accordance with higher uncertainty avoidance, lower communication speed, and low relationship orientation.

Table 2.1 Possible relationship between cultural dimensions and HCI dimensions

HCI dimension	Examples of specifics for HCI dimensions	CIIs	Cultural dimension
Information frequency (IF)	Number of words, sentences, propositions, or dialogs per minute)	Message distance, number of pieces of information per time unit	Relationship versus task orientation, individualistic versus collectivistic orientation, uncertainty avoidance, action chains, network density
Information density (ID)	Number of pieces of information presented simultaneously, distance of pieces of information to each other (e.g., images, words, sentences, dialogs)	Number of points of interest (POIs)	Relationship versus task orientation, individualistic versus collectivistic orientation, uncertainty avoidance, network density
Information /processing parallelism (IP)	Sequential or parallel presentation or reception of information units and information arrangement or order (e.g., widget positions, image–text distribution)	Maximal open tasks, refused system messages, time to disable virtual agents	Relationship versus task orientation, individualistic versus collectivistic orientation, all time-relevant cultural dimensions (such as uncertainty avoidance, action chains, time orientation)
Interaction speed (INS)	Clicking mouse buttons, length of mouse track per second, speed of entering chars	Mouse interaction breaks less than 1 ms	Relationship versus task orientation, individualistic versus collectivistic orientation, all time-relevant cultural dimensions
Interaction frequency (INF)	Overall mouse clicks, mouse moves, number of function, or help initiations per session	Number of left mouse button presses, number of mouse moving events	Relationship versus task orientation, individualistic versus collectivistic orientation, all time-relevant cultural dimensions

2.3 Toward a Model of Culturally Influenced HCI

The analysis of the collected data from Chinese-, German-, and English-speaking users showed that there are correlations between the users' interaction behavior with the system and the users' culture. There are CIIs in HCI which depend on the culture of the user (and which partly apply independent of the meaning of the application). The found CIIs can be applied by analyzing user interaction in order to describe the users' needs regarding HCI in terms of the users' culture; thereby, an explanatory model of culturally influenced HCI as well as a usability measurement system [13] for culturally influenced HCI has to be derived that is very valuable empirically.

2.3.1 *Analytical Evidence: Structural Equation Model*

In order to identify the correctness of the postulated relations between the cultural dimensions' and HCI dimensions' values, a structural equation model (SEM) can be employed to compare variances [21]. Confirmative factor analysis or regression analysis can support this process of modeling and explanation finding.

The primary objective is to reveal the connections between the interaction indicators and their cultural causes. The HCI dimensions can be represented on the right side of the structural equation models and the cultural variables on the left side connected with the suspected connections (parameter–variable combinations). A theory is then best explained if the left and right parts of the modeled structural equations correspond to the modeled variables; that is, the explanatory model is better if more variances in the empirical data can statistically be explained by the structural equation model. The structural equation is modeled by adding or removing variables or relations in order to improve the quality of the explanation. Figure 2.2 shows the sub-model of one side of the complete model, which arose from literature studies, posited hypotheses, and the empirical results of the described studies.

At the moment, the cultural shaping of the user is not connected with the HCI dimensions using cultural dimensions, but only using the variable "nationality." The effect of "nationality" on interaction fault (1.00) and information density (0.80) is not insignificant, even if the influence of the user age (−1.97) is even stronger. However, sex (0.18) and PC experience (2.08) obviously have far less effect on information density in HCI than nationality (89.05). Thus, using this structural equation model, it turns out that nationality has considerably more influence on information speed, information density, and interaction faults than age (36.13), PC

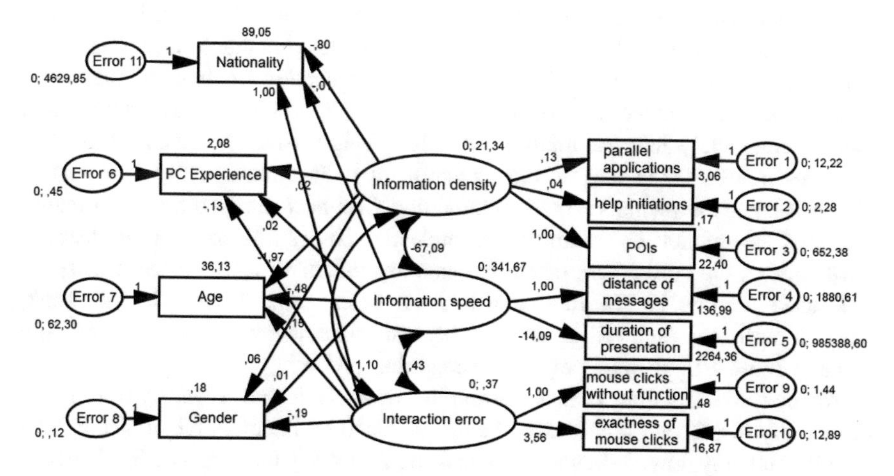

Fig. 2.2 Model for the explanation of the relations between HCI dimensions and CIIs on the *right* side and HCI dimensions and nationality or disturbance variables on the *left* side

experience, or sex—even if age represents a significant interference factor [though not on interaction faults (0.15)]. Furthermore, interaction faults, which are immediately connected with the interaction medium "mouse," are considerably less influenced by those disturbance variables than information density (−1.97) and information speed (0.48). The main emphasis of the evaluation of the data, therefore, was on log files generated by measuring the interaction with the mouse device.

Nevertheless, the model shows the relations between some specifics for HCI dimensions, for example, advertisement parameters as part of information density and information frequency or interaction precision as a part of interaction faults. Other combinations of CIIs can also be assigned to the HCI dimensions. Further modeling must show, however, which combinations provide the highest explanation quality. Furthermore, the other side of the structural equation model, where the cultural dimensions must be inserted and connected, also must be investigated in detail.

2.3.2 *Discussion*

Despite probable objections, it is not trivial to derive general guidelines for intercultural HCI design from the results of the studies. The results of the qualitative studies for interaction analysis also have a doubtful character, since very dynamic phenomena (such as interaction speed or information frequency) cannot be observed and recognized by people without the support of special tools like the IIA tool. Therefore, the qualitative studies that were carried out in parallel with the described quantitative studies offered no useful design recommendations for intercultural interaction design in the field of HCI.

In addition, an enormous amount of interpretation is necessary (even in quantitative studies) to achieve plausible, reliable, and valid results from which valid conclusions can be derived. Furthermore, it is not trivial to recognize differences in the interaction behavior that are not culturally dependent but that have, for example, demographic causes (e.g., different information reception or another interaction style due to age differences). Therefore, it is extremely problematic to bring cultural models completely into accordance with HCI design. Not all possible disturbing variables can be taken into account because of cultural complexity. The results obtained by the explanatory model containing CIIs necessarily differ from reality (because no model by definition completely covers all aspects of reality). Furthermore, the correctness of the explanatory model varies with the number of CIIs used for one HCID. In this sense, the explanation strength is still weak, because until now each HCID has only been substantiated by a few CIIs and only some of those CIIs display very high separation power.

Another consideration is that the amount of analysis and SEM, as well as the availability of relevant sets of data, being produced by intercultural studies in HCI is still relatively low. Likewise, the task of modeling the connections between cultural-, informational-, and interaction-related variables as a structural equation model using CIIs is not completely soluble at the moment due to too few or missing data, indicating the need for further urgent empirical data collections. As a result, only individual parts (sub-models) of the definite structural equation model are in the foreground for the moment. Hence, much work is still required to complete an adequate explanation model for cultural HCI.

Nevertheless, reciprocally confirming aspects attest the high reliability and criteria validity of the statistical results of the two studies: there is a high discrimination rate of over 80% using CIIs to classify users as Chinese (C) and German (G) as well as a high accordance of the HCIDs and the CIIs found by applying different statistical methods [9]. Therefore, the results found in the studies lead to the conviction that it is justified, reasonable, and encouraged to use CIIs in intercultural HCI research to develop an explanatory model of culturally influenced HCI, which is a necessary component in Culturally-Aware HCI systems.

In addition, it has been proven empirically that the interaction of the user with the system is influenced not only by cultural parameters, such as nationality, mother tongue, country of birth, etc., but also by other parameters such as experience or age [9]. Therefore, it is difficult to separate cultural influences from experience because experience is culturally imprinted too (depending on the definition of the terms). To escape these difficulties at least partly, corresponding measures were used (e.g., sensible rating of samples, clear up data sets, keep disturbing variables constant).

2.4 Framework for Culturally-Aware HCI Systems

In the following, the considerations so far will be integrated into a framework for Culturally-Aware HCI systems. First, the principle of culturally adaptive HCI systems derived from the findings of the model concerning the relationship between culture and HCI is presented. From this, a culture-adaptive interface agent architecture is developed, which in turn can be implemented in a demonstrator in order to prove that its functionality is empirically usable in Culturally-Aware HCI systems.

2.4.1 Principle of Culturally Adaptive HCI Systems (CAHCI)

The principle of CAHCI by [9] (Fig. 2.3) is in accordance with the ideas of [22], [23], and [24] and represents a feedback control system that allows the deduction of values of cultural dimensions by analyzing monitored user interaction behavior and retrieving associated cultural parameters stored in a database format (both during the design phase and runtime).

Suitable aspects for cross-cultural user interface design (parameters for cultural adaptation) can be derived herewith that allow the adaptation of both the "look" (appearance) and "feel" (behavior) of HCI according to the cultural needs of the

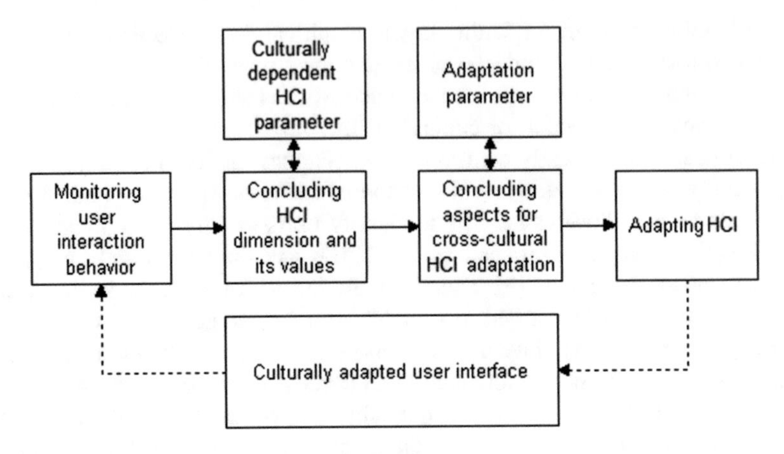

Fig. 2.3 Principle of culturally adaptive HCI systems

user. Adaptation parameters are directly visible cultural variables concerning "look" that are immediately visible (such as color, font, and menu position). The "feel" of HCI is affected by adaptation parameters (direct hidden cultural variables) that are perceivable over time, such as menu structure, usage of scroll bars, information presentation speed, or frequency of messages.

The system has to analyze the interaction behavior of the user to discover interaction patterns and to behave similarly to the user. This means gaining the acceptance of the dialog partner through unobtrusive imitation of the behavior of the dialog partner so that the cognitive models of the system and the user are on the same wavelength. Reconciling the cognitive models of the system and the user ensures a certain basic acceptance as well as a fundamental benevolence of the user with regard to the system (cf. the principle of charity according to [25]). This increases the possibility that the user will buy a device from the same company again, provided that its usability, functionality, and politeness are sufficient enough.

The system monitors and records the user's interaction behavior with the system. It then analyses this data using cultural interaction criteria to determine the cultural characteristics of the user. Finally, the system adapts the HCI according to the cultural preferences of the user, employing HCI design guidelines for intercultural interface design after asking the user or automatically if expectance conformity is not hurt or an emergency situation forces it to do so. The basic principle of each adaptivity consists in observing the behavior of the user with the system, generating a user model with the system, and automatically adapting the system to the user [26].

As a result, no cultural dimension will be used to relate the user–system interaction behavior to a certain culture. Only the interaction behavior itself will be classified according to the informational dimensions, whose specifics depend on the cultural background and imprint of the user. Hence, it is not necessary to classify the user as belonging to a certain culture, but to a certain interaction behavior from which the cultural settings the user presumably prefers are known. For instance, if the user interacts very frequently and quickly with the system, it can be assumed that either the user is very experienced or they belong to a cultural group which is highly relationship oriented, such as that of China. By understanding the default values of the variables of the informational dimensions determined for different cultures in the design phase, the system can compare those values with those actually initiated by the user currently interacting with the system. The best matching patterns allow the system to deduce the cultural adaptation parameters with which to adapt the HCI with the highest probability of coping with the user's cultural needs.

2.4.2 Culturally Adaptive HCI (CAHCI) Architecture

The principle of culturally adaptive HCI systems can be implemented in a culturally adaptive HCI architecture. A culturally different user employs the device. The system monitors and records the interaction patterns. The system classifies the interaction patterns into interaction classes using its knowledge about culturally dependent variables (CIIs). The principle works if the interaction classes are built up according to the culturally different users. After recognition of the culturally imprinted interaction pattern (CIP), the device should be able to adapt to the interaction preferences of the user and, if defined in the design phase by the determined guidelines for intercultural HCI design, to the preferences of the user regarding surface, functionality and interaction. According to the results of the reflected model and the architecture, additional adaptation can be made for the preferences of the user that emerge during runtime.

The culturally adaptive HCI architecture consists of several subagents, each of which fulfills a special sub-task (Fig. 2.4).

The HCI monitoring agent serves to record the interactive user behavior with the system to retrieve localized data, which is stored in a database communicating with the database agent. By specifying several profound use cases, the relationship between the use case and the cultural dimension has been extracted empirically (as shown in Sect. 2). From this, implications could be made concerning the intercultural parameters (cf. CIIs). The CIIs can be stored in a lookup table [23] and parts of them can be retrieved to adapt the graphical user interface (GUI) or the speech user interface (SUI). Thereby, existing possibilities for tracing and logging user interactions and in general determining and showing the user's behavior with the system can be used to recognize the cultural variables and their values within the localization process. However, the general preparation of the system for many localized configurations must also be considered to fulfill the concept of internationalization. The interaction agent deduces the interaction patterns of the user with the system. In combination with the learning agent, the patterns can be analyzed and recognized over time according to the identified user to facilitate acquisition of an adequate model of them, which also contains their cultural characteristics. The

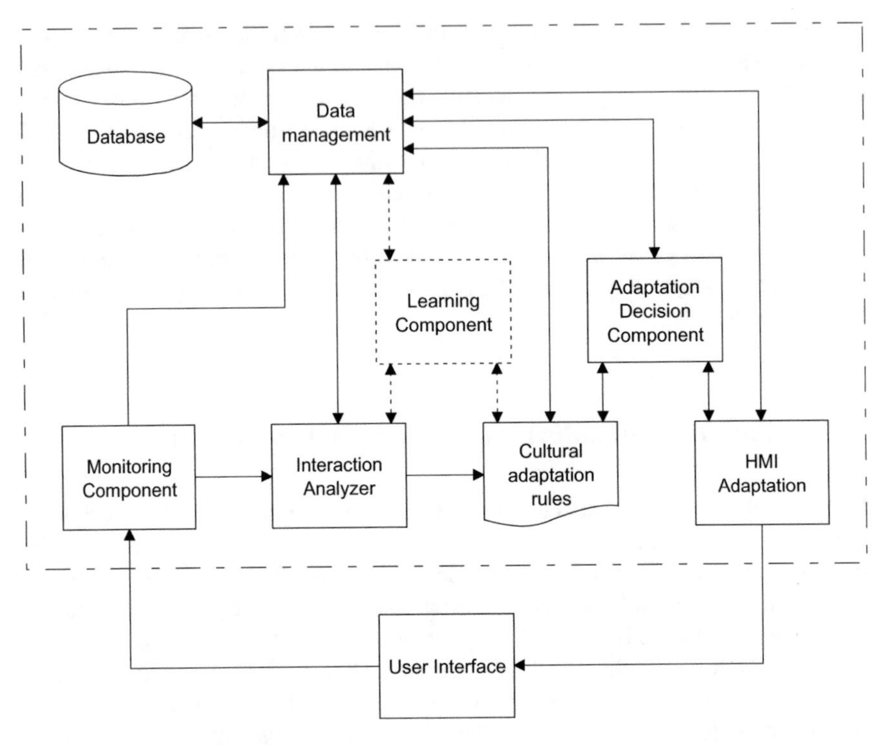

Fig. 2.4 Culturally adaptive HCI architecture

adaptation agent retrieves the recognized cultural characteristics from the database agent and adapts the HCI according to the connected HCI aspects for the desired cultural needs of the user.

2.4.3 *Exemplification: Culturally-Aware HCI Systems in the Automotive Context*

Today, in the automotive context adaptivity in driver information and assistance systems is necessary because the functional and informational complexity of infotainment systems (a mixture of information and entertaining systems) can no longer be handled solely by the driver without employing adaptivity [27]:

- It is hard for the driver to handle the functional and informational complexity of such systems in extreme driving situations: the mental workload which is caused by all possible senses (i.e., resulting from visible, audible and haptic information) simply exceeds the mental capacity of the driver.
- The mental workload should be maintained within acceptable limits in dangerous driving situations if the system adapts the information flow for the user

automatically. Due to this, adaptivity must also take external input sources into account (for instance from pre-crash sensors).
- The output modality has to be adapted automatically for at best the least workload (for example by using different displays).

It is also necessary to *culturally adapt* driver information and assistance systems because:

- The user preferences must be considered and covered, which depend on the cultural background of the user.
- The cultural background of the driver also determines behavior in certain (especially dangerous) driving situations.
- There are many different groups of drivers that exhibit their own "culture" (for example interaction behavior), whether this is regarding groups at an international level or at a national level (such as social, ethnic, or driver groups).
- The local market for cars has increasingly become a worldwide market. Future infotainment systems will have to handle the demands of various drivers and various cultures. This aspect can only be covered within a single system if this system is adaptable and configurable.

Hence, it is necessary to build Culturally-Aware HCI systems because of urgent application cases. Enhanced algorithms are needed to enable the system to automatically and correctly adapt itself to the culturally imprinted needs of the user to bring the mental model of the system in line with the user's mental model.

2.5 Culturally-Aware HCI Demonstrator

The significant statistically discriminating cultural interaction indicators identified by the studies mentioned above motivated the author to demonstrate that they also work in a real environment. By means of a demonstrator, some important cultural variables, as well as the CAHCI principle (cf. Fig. 2.3), were exemplified to support the following argumentation: If the CAHCI demonstrator is capable of classifying the user according to their interaction with the system and correctly according to their cultural characteristics, then there is empirical proof that parts of the interrelationship between the HCI dimensions relevant for HCI design and the cultural dimensions are correct.

A demonstrator regarding culturally (adaptive) HCI requires the following properties to show the correctness of the theorized models and guidelines as well as the CAHCI principle :

- Parameterization with intercultural properties for different cultural groups;
- Recognition of the specifics of the cultural variables through user monitoring;
- Automatic adaptation of the HCI to the cultural needs of the user (i.e., CAHCI).

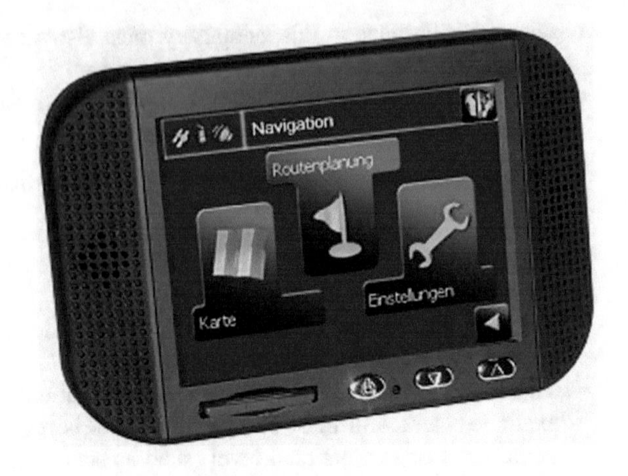

Fig. 2.5 The CAHCI demonstrator based on a mobile driver navigation system

CAHCI functionality has, therefore, been integrated in a portable navigation system called the "CAHCI demonstrator" (Fig. 2.5).

2.5.1 Setup, Runtime, and Using the CAHCI Demonstrator

Some of the variables for cultural adaptation that best categorize the HCI style of people from different cultures have been implemented in the CAHCI demonstrator in order to show that the CAHCI principle works in a real system—and not only statistically (cf. Sect. 2.2.1).

The following aspects can be covered and adapted within the CAHCI demonstrator:

- Color scheme of the map display;
- How often a voice output will be repeated automatically (NIT);
- Speed of the voice output (TPI);
- Number of displayed road names (~POI);
- Number of buttons and configuration possibilities in menus (~POI);
- Number of POIs (POI);
- Language.

The CAHCI demonstrator consists of three modules according to the three main parts of the principle of culturally adaptive HCI: the monitoring module, analysis module, and adaptation module. Table 2.2 gives an overview regarding some aspects of the possible adaptation levels and parameters within the CAHCI demonstrator for the adaptation of the HCI according to preliminary Chinese user expectations obtained through a qualitative survey with 20 Chinese students in Hangzhou.

Only five adaptation levels (none of which are fuzzy) have been implemented into the demonstrator due to cost restrictions and the system performance

Table 2.2 Adaptation levels provided by the CAHCI demonstrator depending on the C-value

Adaptation level	Number of POI (MD) [POI]	Highway color (MD)	Route color (MD)	Number of announcements (MG) [NIT]	Voice speed (MG) (Words per second) [DIP]	C-value (%)
1	40	Blue	Light blue	1	50	0–20
2	80	Light blue	Light violet	2	100	20–40
3	120	Turquoise	Violet	3	150	40–60
4	160	Light turquoise	Light red	4	200	60–80
5	200	Green	Red	5	250	80–100

limitations of the embedded technology. The C-value is the cultural index, which expresses the assumption strength that the user is a Chinese user (calculated by the system) as a percentage; that is, if the C-value is 100, the system has recognized a Chinese user; if the C-value is 0, the system has recognized a German user. MD indicates that the adapted aspect is mostly relevant for the use case of presenting information on a map display. MG indicates the "maneuver guidance use case."

Depending on the C-value, the look and feel of the demonstrator will change according to the adaptation levels presented in Table 2.2. (S) means "standard," that is, the default settings for German users are used, and (A) means that the HCI is adapted to Chinese users (for instance, a status bar with icons is displayed in the bottom right instead of in the top left of the screen). According to the results of the qualitative survey (cf. Sect. 2.2), Chinese users prefer the most important information to be in the top left (vs. the least important information in the bottom right). If the system recognizes that the user behaves like a Chinese user, it adapts the HCI to the Chinese settings according to the content of the C-value (Fig. 2.6). Figure 2.7 shows the differences in the appearance of the map display of the navigation system for German and Chinese settings.

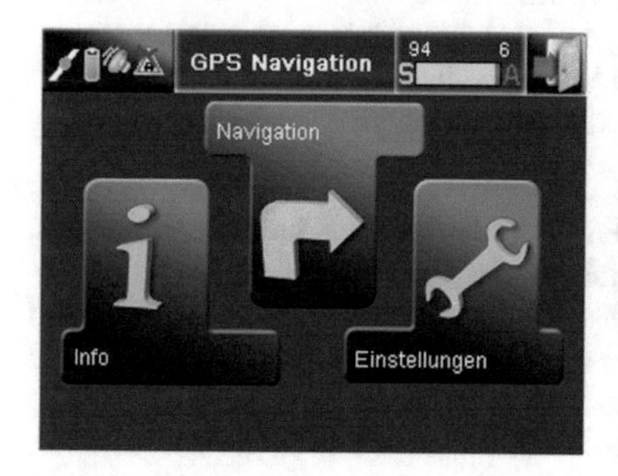

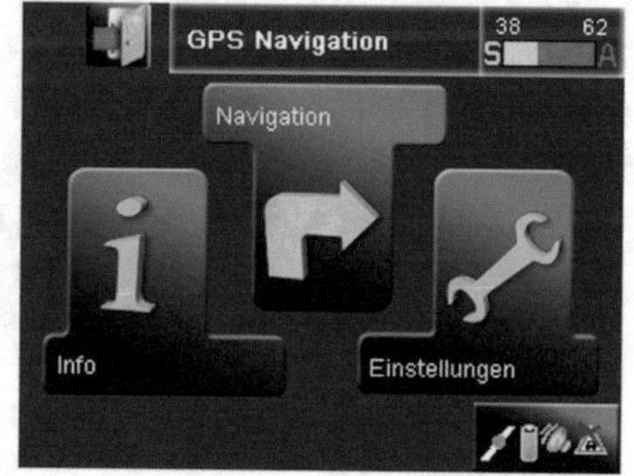

Fig. 2.6 The initial screen of the CAHCI demonstrator (standard (*left*) for German users and adapted (*right*) to Chinese users)

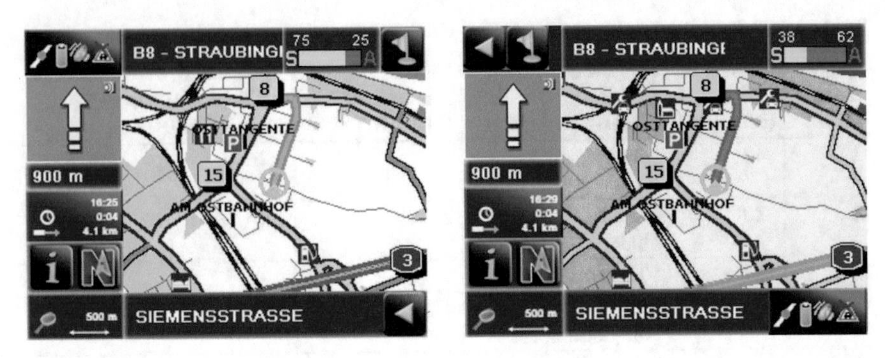

Fig. 2.7 Differences in map display according to user interaction behavior [*left*: German setting (adaptation level is 2, C-value = 25); *right*: Chinese setting (adaptation level is 4, C-value = 62)]

Many more elements could be considered and adapted in intercultural HCI design regarding icon and symbol design, layout, language, text size, format, units, street names, etc. [28]. However, to localize and internationalize driver navigation systems in general remains a task for software developers and HCI designers in the industry.

2.5.2 Evaluating the CAHCI Principle

The functional test presented in this section served to prove the classification correctness and the proper functionality of the CAHCI principle, that is, that the basic principles of cultural adaptivity (monitoring, analyzing, adapting) work.

Twenty-five (13 Chinese and 12 German) users were asked to complete several test tasks as quickly as possible. Questions could be directed to the test leader, if necessary. For evaluation at the time, two groups from different cultures with similar conditions regarding use case, education, profession, age, and gender were built. Interaction data that emerged during the interaction of the users with the CAHCI demonstrator were recorded by the logging module of the demonstrator. The data sets were analyzed using the IIA data evaluation module using a neural network, as described in [10] . The results of the analysis led to the following statements:

- The number of total entries in the log file (representing the amount of user interaction), error clicks, and mouse moves classified very well.
- The more frequent the interaction breaks greater than 10 s, the less experienced or trained the user is at handling the application—a higher cognitive processing time of the user could also explain this—or the user is Chinese.
- The longer the test duration, the more exact or less experienced the test person is.
- The more interaction breaks <1 ms (equal to the number of scrolls <1 ms, i.e., moves with the finger on the touch screen, a.k.a. “mouse moves” (MM), or

"mouse movement speed" (MMS)), the less experienced the user is at dealing with use cases of driver navigation systems, or the user is very hasty or Chinese.
- Cross-validated interaction breaks (<1 ms and >10 s) classify up to 72% of the users correctly to their cultural background (Chinese or German). Including the third parameter ("test duration") within a linear discriminant analysis, the classification rate reached 74%.

Results showed that interaction differences exist between Chinese and German users. The averaged values of the culturally different groups tend to always be in one direction, which indicates a trend regarding CIPs. In addition, from the 25 data sets, nine were analyzable by linear discriminant analysis to calculate the statistical classification power of the CIIs used in the CAHCI demonstrator. Surely, this result reflects the small sample size and depends on several statistical settings (such as how many and which variables are in the set to which the linear discriminant analysis is applied or what including and excluding statistical limits are set). However, one-way ANOVA showed that the CIIs work very well: their discrimination power is high, and hence their weight within the adaptivity algorithm implemented in the CAHCI demonstrator using production rules is also high. Table 2.3 lists a ranking of the excellence of the CIIs used in the CAHCI demonstrator (green-marked).

Out of the implemented cultural indicators in the CAHCI demonstrator, two CIIs ("Mouse Up" and "Scrolls <1 ms") classified (cross-validated) 80% of the users correctly to their cultural background (Chinese or German). "Mouse Up" represents the measuring variable "mouse clicks" (MC). "Scrolls <1 ms" measures the number of interaction breaks less than 1 ms, representing the MM from which the MMS can be derived over time. Additionally, it can be proven that the CAHCI demonstrator classifies the nationality of the test user correctly to the interaction behavior of the test user. For example, the CIIs Nr_Of_Scrolls_Shorter_Than_1 ms, Breaks_Greater_Than_10 s, and Nr_Of_Scrolls_Over_1 ms, which represent interaction breaks, using a touch screen, and mouse up und counter entries, classify the respective nationality up to 80% correctly for the interaction of the test user.

There are classification quotes significantly over 50%, which proves that the results have not been found randomly, but support themselves mutually. Hence, these results prove that the CAHCI principle works not only statistically, but also within a real system exemplified by a mobile driver navigation system.

The obtained results using the CAHCI demonstrator justify the direction of research in this work, which supports further studies that will increase the exactness and the completeness of the results as well as the discriminatory power and separation effect of the CIIs.

Table 2.3 CIIs used in the CAHCI demonstrator

Cultural Interaction Indicator	*Weight in %*	*F-Value*	*Significance*	*Homogeneity of variances [h]*	*Weight [0;1]*	*Ranking*
Nr of IO Breaks Over 10s_NORM	100	9.06	0.006	0.539	1.000000000	1
Mouse Up_NORM	59	5.353	0.03	0.867	0.5908388521	2
Counter Entries_NORM	49	4.478	0.045	0.947	0.4942604857	3
Total Entries_NORM	48	4.382	0.048	0.82	0.483664459	4
Mouse Down_NORM	46	4.141	0.054	0.722	0.4570640177	5
Nr of IO Breaks Over 1s_NORM	41	3.69	0.067	0.571	0.407284768	6
Mouse Up_Error Click_NORM	40	3.629	0.069	0.111	0.4005518764	7
Mouse Down_Error Click_NORM	40	3.619	0.07	0.053	0.3994481236	8
Nr of Error Clicks_all_NORM	40	3.619	0.07	0.53	0.3994481236	9
Normal Entries_NORM	39	3.519	0.073	0.769	0.3884105960	10
Nr of IO Breaks Shorter than 1ms_NORM	34	3.091	0.092	0.053	0.341169977	11
Nr of Mouse Clicks_all_NORM	27	2.436	0.132	0.512	0.2688741722	12
Keyboard Button_NORM	22	2.007	0.17	0.674	0.2215231788	13
No Button_NORM	16	1.472	0.237	0.969	0.1624724062	14
Nr of Mouse Clicks_Since Start Driving_NORM	11	0.976	0.333	0.643	0.1077262693	15
Whole Scrolling Time_NORM	7	0.617	0.44	0.098	0.0681015453	16
Average Scrolling Time_NORM	6	0.54	0.47	0.113	0.0596026490	17
Duration of Test in Min	3	0.263	0.613	0.047	0.029028697	18
Nr of Scrolls_NORM	2	0.189	0.668	0.49	0.0208609272	19
Nr of IO Breaks Shorter than 1s_NORM	2	0.158	0.694	0.332	0.017439293	20

Legend:

Best variable with best significance and very high F-value	100	9.06	p=0.006	h=0.047	Ref.: 1.00 = 100% of F_{max} (9.06)	1
Very good variable with significance p < 0.05 and very high F-value	48–59 (100)	4.382–9.06	p < 0.05			
Good variable with significance p < 0.1 and high F-value	34-46	3.091-4.141	p < 0.1			
Bad variable without significance and very low F-value	2–27	0.158–2.436	p > 0.2			
Worst variable with worst significance and very low F-value	2	0.158	p=0.694	h=0.969	0.017439293	20

2.5.3 Enhancing the CAHCI Demonstrator

Cultural adaptivity does not only concern the look and feel of the user interface, but also the interaction devices as well as the number and the type of system functions [14] that can be changed dynamically according to user preferences and the usage context [29]. Thus, designing an appropriate system according to the user in the design phase helps to avoid the problems arising from adaptivity. For instance, it is problematic that automatic adaptation (adaptivity) depends on maximum data when observing new users: the system needs more data in order to be able to release information about the user as well as to be able to infer the characteristics of the user regarding information presentation, interaction, and dialogs. Furthermore, the knowledge gathered about the user can be *misleading* or simply false. Hence, the reliability of assumptions can be a problem: the behavior of the system has to be in accordance with the beliefs of the user to prevent unexpected situations for the user. In addition, legal restrictions have to be taken into account, as only the effects of user actions are allowed to be permanently stored, but not the log files of the personalized sessions themselves.

Additionally, there are many open questions that have to be addressed very carefully: How many dynamic changes are optimal for and will be accepted by the user? When does a "hidden" adaptation occur? How can this be prevented? How much does the user trust the adaptive system? Adaptivity may not surprise the user but must be in accordance with the mental model of the user [30]. Additionally, there are culturally dependent questions which have to be answered. For example, what cultural aspects must be adapted? Which of them can be adapted automatically? Additional technical problems include when to stop behavior analysis and start adapting ("the bootstrapping problem") (one example for a possible solution to the bootstrapping problem can be found in [31]).

As long as no solution is available that can achieve meaningful adaptations from minimum data automatically, it remains necessary to investigate standard parameters and their values very early in the design phase, and long before runtime, in order to integrate them into the system. Therefore, it is necessary that the system already has corresponding user knowledge (standard parameters) before the user's first contact with the system occurs. Before using the system for the first time, it must be adjusted, for example, to the nationality of the user (which indicates the main affiliation of the user to a cultural group) and the corresponding cultural parameters can be set simultaneously as standard parameters for the desired country. Furthermore, the adaptive system obtains the adequate characteristics of the user more quickly at runtime, because there is "more time" to collect the culture-specific data for the user, since a basic adaptation to the most important user preferences has already been performed before runtime (by putting the standard parameters into the system).

The near-term objective is to enhance the tool for "cross-cultural HCI analysis" by applying enhanced techniques using statistical and data mining methods and semantic processing to extract cultural variables and their values as well as guidelines for cross-cultural HCI design in a more automatic way. The mid-term objective is to analyze and evaluate the test data in more detail to generate several algorithms for adaptivity based on neural networks as well as structural equation models to prove basic theoretical cultural interaction models. In the long-term view, the best discriminating algorithms for adaptivity will be transformed and implemented in Culturally-Aware HCI systems to be evaluated qualitatively using intercultural usability tests with users of different cultures and users under mental stress.

2.6 Implications for Culturally-Aware HCI Systems

The findings so far indicate some recommendations for designing Culturally-Aware HCI systems. CIIs from the culturally influenced HCI model serve as basis for the quantitatively derivation of adaptation rules concerning HCI according to the culturally imprinted interaction patterns of the user with the system at runtime to ensure cultural adaptivity of HCI (CAHCI) in Culturally-Aware HCI systems.

2.6.1 Quantitative Apparatus of Recognition

The specifics (values) of intercultural variables can be determined purely quantitatively through analysis of the interaction of the user with the system (considering only the interaction tracing log file of the system). Hence, it is possible to determine the culturally imprinted characteristics of the user by analyzing the interaction of the user with the system. Furthermore, the greater the cultural distance, the greater the difference in the kind of interactions of humans with a system (computer, machine, navigation system, etc.). It has been statistically proven that there are significant cultural differences in the interaction behavior of the user with the system using the IIA tool. The combination of cultural differences represented by CIIs form CIPs according to the cultural imprint of the user. There are different patterns of interaction in HCI (composed of combinations of CIIs) that are culturally significant depending on the cultural imprint of the user; that is, it has been statistically and empirically proven that the interaction of the user with the system depends on the user's cultural background. Furthermore, the cultural interaction differences of the users with the system have been identified quantitatively and not qualitatively (by using interaction times or the number of interactions): they can be statistically identified and measured by a computer system (using the IIA tool and the CAHCI demonstrator). Cultural interaction differences in HCI can be recognized and measured quantitatively by a computer system (although only by monitoring and analyzing the interaction of the user with the system quantitatively, resulting in the adequate culturally dependent specifics for the intercultural variables). For this purpose, it is suggested to deploy the principle of culturally adaptive HCI systems (cf. Sect. 2.4.1) as well as the culturally adaptive HCI architecture (cf. Sect. 2.4.2) in Culturally-Aware HCI systems.

2.6.2 Deploying the Culturally Influenced HCI Model

The interaction of the user with the system in HCI is influenced by static aspects (preferences), which are present due to the cultural shaping of the user and their experience with the system, and due to dynamic aspects depending on the situation. Therefore, the type of user–computer interaction within HCI must be adaptable in such a way that the system can do justice to the interaction requirements of the user (i.e., the specifics of the HCI dimensions).

The empirical results obtained through the described study partly confirm the relationships theorized in the literature by showing that there are metrics composed of CIIs, which are adequate for measuring culturally influenced HCI as a basic property of Culturally-Aware HCI systems. The values of the CIIs revealed interesting tendencies in user interaction behavior (i.e., HCI style or HCI characteristics) related to the cultural imprint of the user. Therefore, it should be possible to complete and optimize the explanatory model of culturally dependent variables for HCI design using the methods of factor analysis and SEMs by revising the relationship between user interaction and user culture.

The design of Culturally-Aware HCI systems can continue to profit as long as the presented culturally influenced HCI model is developed and validated. The ideas presented in this chapter represent a reasonable step toward an explanatory model of culturally influenced HCI. As this process continues, the connections between HCI and culture will become clearer and comprehensive in the end, even if much work still remains (for instance, improving the separation power of the CIIs or the explanation strength of the model of culturally influenced HCI).

2.6.3 Design Recommendations

With regards to additional parameters for adjusting HCI according to the cultural needs of the user to provide for and tend to adequate slots for the CIIs presented in this work, the following aspects should at least be considered very carefully when designing new architectures of *Culturally-Aware* HCI systems or extending existing systems in addition to explicit formation principles when designing adaptive HCI systems for vehicles in the Automotive context, cf. [28, 30, 32–36]:

- The number of information units presented simultaneously (e.g., POIs in the map display should be about a half in number for German than for Chinese users. Number, duration, and frequency of information units presented sequentially (e.g., (system) messages or maneuver advice in maneuver guidance should be lower in number for German than for Chinese users).
- When designing information systems, consider that the frequency and usage speed of interaction devices (e.g., a touchscreen, hard keys, or a mouse) are almost twice as high and fast but less exact for Chinese than for German users

Furthermore, some results of this work can be expected to be valid for HCI design in general, because there are culturally sensitive variables that can be used to measure cultural differences in HCI simply by counting certain interaction events without the necessity of knowing the semantic relations to the application. Such indicators include the number of MM, breaks in the MM ($\neg IN_{MM}$), MMS, MC, interaction breaks in general ($\neg IN$), and the number of acknowledging system messages (AM) or refusing system messages (RM). Surely, all these indicators can also be connected semantically to the use cases of applications running on the system. However, simply counting such events related to the session duration from users of one culture and comparing them to those of users of another culture is obviously sufficient to indicate differences in the interaction behavior of culturally different users. Furthermore, the values of the CIIs change in a similar way even if different use cases and test tasks are applied. Hence, those CIIs can be called "general cultural interaction indicators" (GCIIs) and that can be applied in Culturally-Aware HCI systems in general.

Using methods of artificial intelligence may help to fulfill the steps to achieving cultural adaptivity, which in turn will broaden universal access in the application of the following culturally adaptive HCI principles (cf. Sect. 2.4):

- Learning the differences in the interaction of the users of different cultures.
- Classifying interaction patterns according to culture.
- Determining the user preferences according to culture.
- Adapting HCI according to user preferences.
- Learning user preferences by observing HCI over time.
- Integrating knowledge from observation into the system's user model.

Finally, to improve system intercultural usability, it is necessary to internationalize and to localize intercultural variables; that is, to consider such variables within the process of product design [5].

2.7 Conclusions and Outlook

Culturally-Aware HCI systems need a cultural model that allows them to automatically derive adaptation rules to adapt the system's HCI to the culturally imprinted user's needs (cf. Sect. 2.1). Statistically valid and significant results from two empirical studies confirmed (cf. Sect. 2.2) that *special combinations of cultural interaction indicators (CIIs)*, are *statistically discriminating enough* to enable computer systems to detect different culturally influenced interaction patterns (CIPs) automatically and to relate users to a certain culture behavior according to the theorized principle of culturally adaptive HCI.

Reflections have been made to generate a structural equation model (SEM) of the relationship between HCI dimensions and cultural dimensions (cf. Sect. 2.3). For example, the higher the relationship orientation (e.g., toward collectivism), the higher information density, information speed, information frequency, interaction frequency, and interaction speed are and vice versa. However, further research showed that no cultural dimension has to be used in the first place to relate the interaction behavior of the user with the system to a certain culture. Only the interaction behavior itself will be classified according to the HCI dimensions, whose specifics depend on the cultural background and imprint of the user. Therefore, it is not necessary to classify the user to a certain culture, but to a certain interaction behavior from which the cultural settings the user presumably prefers are known.

According to the CAHCI Architecture (cf. Sect. 2.4), by knowing the default values of the variables of the HCI dimensions determined for different cultures in the design phase, the system can compare those values with those actually initiated by the user currently interacting with the system. The best matching patterns allow the system to deduce the cultural adaptation parameters and adapt the HCI with the highest probability of coping with the user's cultural needs.

Even though the results of the quantitative studies conducted up to now have primarily concerned and demonstrated the cultural differences in HCI related to use cases in driver navigation systems for the Automotive context exemplified by the CAHCI demonstrator (cf. Sect. 2.5), they serve to offer some confirmed facts and a basis for providing at least some general recommendations (even if no guidelines are available) for the design of intercultural user interfaces in Culturally-Aware HCI systems (cf. Sect. 2.6).

It is up to future studies to derive scientifically sound and practically relevant design guidelines from explanatory models for culturally influenced HCI. Furthermore, they must reveal, for instance, the degree of *acceptance of cultural adaptivity* of the user as well as the degree to which the user's mental workload is affected using cultural adaptivity in Culturally-Aware HCI systems.

Acknowledgements I would like to thank everyone who supported the cultural adaptivity project, which was the basis for the reflections on Culturally-Aware HCI systems presented in this chapter.

References

1. Thomas, A.: Psychologie interkulturellen Handelns. Hogrefe, Göttingen, Seattle (1996)
2. Honold, P.: Interkulturelles usability engineering: Eine Untersuchung zu kulturellen Einflüssen auf die Gestaltung und Nutzung technischer Produkte. Als Ms. gedr. ed. Vol. 647. VDI Verl, Düsseldorf. 343 (2000)
3. Hofstede, G.H., Hofstede, G.J., Minkov, M.: Cultures and Organizations: Software of the Mind, 3rd edn. McGraw-Hill, Maidenhead (2010)
4. Badre, A, Barber, W.: Culturabilty: the merging of culture and usability. In: Proceedings of the 4th Conference on Human Factors and the Web. AT and T Labs, Basking Ridge, NJ, USA (1998)
5. Vöhringer-Kuhnt, T.: The Influence of Culture on Usability. Freie Universität Berlin (2002)
6. Hofstede, G, Hofstede, J.G.: Cultures and Organizations: Software of the Mind, 2nd edn. McGraw-Hill, New York, USA (2005)

7. Röse, K.: Methodik zur Gestaltung interkultureller Mensch-Maschine-Systeme in der Produktionstechnik, vol. 5. Univ. 244, Kaiserslautern (2002)
8. Heimgärtner, R.: Intercultural user interface design. In: Blashki, K, Isaias, P. (eds.) Emerging Research and Trends in Interactivity and the Human-Computer Interface, pp. 1–33. Information Science Reference (an imprint of IGI Global), Hershey (2014)
9. Heimgärtner, R.: Cultural differences in human computer interaction: results from two online surveys. In: Oßwald, A. (ed.) Open Innovation, pp. 145–158. UVK, Konstanz (2007)
10. Heimgärtner, R.: A Tool for getting cultural differences in HCI. In: Asai, K. (ed.) Human Computer Interaction: New Developments, pp. 343–368. InTech, Rijeka (2008)
11. Shneiderman, B.: User Interface Design. MIT Press, Munich (2002)
12. Dix, A., et al.: Human-Computer Interaction. Prentice Hall, London (1993)
13. Nielsen, J., *Usability metrics.* nngroup, 2001
14. Röse, K., Zühlke, D.: Culture-oriented design: developers' knowledge gaps in this area. In: 8th IFAC/IFIPS/IFORS/IEA Symposium on Analysis, Design, and Evaluation of Human-Machine Systems, Preprints, Kassel, Germany, 18–20 Sept 2001 (2001)
15. Hall, E.T.: Beyond Culture. Anchor Books, New York (1976)
16. Heimgärtner, R.: Research in progress: towards cross-cultural adaptive human-machine-interaction in automotive navigation systems. In: Day, D., del Galdo, E.M. (ed.) Proceedings of the Seventh International Workshop on Internationalisation of Products and Systems (IWIPS 2005), pp. 97–111. Grafisch Centrum Amsterdam, The Netherlands, Amsterdam (2005b)
17. Smith, A., Chang, Y.: Quantifying Hofstede and developing cultural fingerprints for website acceptability. In: V.R.K.H.P.C.J.D.D. Evers (ed.) Proceedings of the Fifth International Workshop on Internationalisation of Products and Systems, IWIPS 2003, Germany, Berlin, pp. 89–102. University of Kaiserslautern, Kaiserslautern, 17–19 July 2003
18. Hall, E.T.: The Silent Language. Doubleday, New York (1959)
19. Halpin, A.W, Winer, B.J.: A factorial study of the leader behavior descriptions. In: Stogdill, R.M., Coons, A.E. (eds.) Leader Behavior: Its Description and Measurement, pp. 39–51. Bureau of Business Research, Ohio State University, Columbus, OH (1957)
20. Victor, D.A.: International Business Communication, 7th edn. Harper Collins, New York, NY, 280 (1998)
21. Bortz, J., Döring, N.: Forschungsmethoden und Evaluation: Für Human- und Sozialwissenschaftler, 4., überarb. Aufl. ed. Springer, Berlin. 897 (2006)
22. Stephanidis, C., Savidis, A.: Universal access in the information society: methods, tools, and interaction technologies. Univ. Access Inf. Soc. **1**(1), 40–55 (2001)
23. Baumgartner, V.-J.: A Practical Set of Cultural Dimensions for Global User-Interface Analysis and Design. Fachhochschule Joanneum, Graz, Austria (2003)
24. Leuchter, S., Urbas, L.: Useware engineering mit kognitiven Architekturen. Automatisierungstechnische Praxis **46**(9), 68–72 (2004)
25. Davidson, D.: Inquiries into Truth and Interpretation. Clarendon Press, Oxford [u.a.]. 292 (1984)
26. Mandl, T., Schudnagis, M., Womser-Hacker, C.: A framework for dynamic adaptation in information systems. In: Stephandis, C. (ed.) Human Computer Interaction: Theory and Practice. Proceedings of 10th International Conference on Human-Computer Interaction, pp. 425–429. Springer, Heidelberg, 22–27 June 2003 (2003)
27. Heimgärtner, R., et al.: Towards cultural adaptability to broaden universal access in future interfaces of driver information systems. In: Universal Access in Human-Computer Interaction. Ambient Interaction, 4th International Conference on Universal Access in Human-Computer Interaction, UAHCI 2007 Held as Part of HCI International 2007 Beijing, China, 22–27 July 2007 Proceedings, Part II, Stephanidis, C. (ed.). Springer, Heidelberg. pp. 383–392 (2007)
28. Röse, K., Heimgärtner, R.: *Kulturell adaptive Informationsvermittlung im Kraftfahrzeug: 12 Uhr in München – Stadtverkehr, 18 Uhr in Shanghai – Stadtverkehr.* i-com, 7 (3 Usability und Ästhetik), pp. 9–13 (2008)

29. Heimgärtner, R., Holzinger, A.: Towards cross-cultural adaptive driver navigation systems. In: Holzinger, A., Weidmann, K.-H. (eds.) Empowering Software Quality: How can Usability Engineering Reach these Goals? 1st Usability Symposium, pp. 53–68, HCI&UE Workgroup, Vienna, Austria, 8 Nov 2005. Austrian Computer Society, Vienna (2005)
30. Kobsa, A.: User modeling in dialog systems: potentials and hazards. AI & Soc. **4**(3), 214–231 (1990)
31. Heimgärtner, R.: Cultural Differences in Human-Computer Interaction. Paperback B: Einband - flex. (Paperback) ed. Vol. 1. 2012: Oldenbourg Verlag. XVIII, 325 S. - 24,0 x 17,0 cm
32. Kolrep, H., et al.: Mobile Anwendungen im Kraftfahrzeug - Mensch-Maschine-Interaktion und Akzeptanz, in Informatik 2003 - Innovative Informatikanwendungen, Band 2, Beiträge der 33. Jahrestagung der Gesellschaft für Informatik e.V. (GI), 29. September - 2. Oktober 2003 in Frankfurt am Main, D.K. R., et al. (eds.), pp. 386–391. GI, Bonn (2003)
33. Malinowski, U., Obermaier, A.: Adaptivität und Benutzermodellierung - Was ist ein adäquates Modell? In: Kobsa, A.e.a. (eds.) Tagungsband KI-93-Workshop "Adaptivität und Benutzermodellierung interaktiven Softwaresystemen", pp. 61–68. Univ.-Verl. Konstanz, Konstanz (1993)
34. Kobsa, A., Wahlster, W., Carberry, S.: User models in dialog systems: [based on UM86, the first International Workshop on User Modeling, held in Maria Laach, Germany]. Springer, Berlin (471) (1989)
35. Dieterich, H., et al.: State of the art in adaptive user interfaces. In: Schneider-Hufschmidt, M., Kuehme, T., Malinowski, U. (eds.) Adaptive User Interfaces Principles and Practice, pp. 13–48. Elsevier Science Publisher B.V., North-Holland (1993)
36. Shneiderman, B., Plaisant, C.: Designing the User Interface: Strategies for Effective Human-Computer Interaction, 4th edn. Pearson, Boston, Mass (652) (2005)

Author Biography

Dr. Heimgärtner earned undergraduate degrees in Psychology and Computer Science. He did graduate work in Information science, Linguistics, Philosophy, and Religion Studies and obtained his Ph.D. in Information Science from the University of Regensburg. Dr. Rüdiger Heimgärtner is a specialist of cultural differences in HCI and has worked in software and HCI projects at Siemens AG and Continental AG. He is the founder and managing director of the company Intercultural User Interface Consulting (IUIC) and has provided training and consultation for Intercultural User Interface Design (IUID). The areas of interest of Dr. Heimgärtner include machine learning, automated reasoning, psychology of judgment, and mathematical psychology. One of his main contributions to information science, usability engineering, and HCI design is detailing IUID in articles, proceedings, and books, e.g. as author of the first German speaking book on IUID. Dr. Rüdiger Heimgärtner is iNTACS certified ASPICE assessor with more than 10 years of experience in quality and process management in international software and HCI projects. He is member of the International Usability and User Experience Qualification Board (UXQB) as well as of the working team of ambient assisted living (AAL) at the German Institute for Standardization (DIN).

Chapter 3
Building Emic-Based Cultural Mediations to Support Artificial Cultural Awareness

Jean Petit, Jean-Charles Boisson and Francis Rousseaux

Abstract Recently, studies about culturally-intelligent systems have arisen to manage digitized cultural diversity. The current systems possess an artificial awareness of cultures by mediating them through representations. Coming from an etic approach, these universal representations facilitate the mediation of different cultures but limit their understanding and thus, prevent the development of an higher degree of awareness. In this research, we propose a methodology to construct artificial cultural awareness from emic-based representations. We tested the latter through an experiment on the domain of 'abortion' with the Pro-Choice and Pro-Life communities.

Keywords Culturally-Aware systems · Culturally-intelligent systems
Artificial cultural awareness · Prototypical cultural models · Cultural ontologies · Cultural mediations

3.1 Introduction

Since the 2000s, with the rapidly expanding web, computer systems are more exposed than humans to culture diversity. To deal with this diversity, it is essential for these systems to develop cultural awareness. In the literature, there are many

J. Petit (✉)
Capgemini Technology Services, 7 Rue Frederic Clavel, 92287 Suresnes, France
e-mail: jean.petit@capgemini.com

J.-C. Boisson
CASH Team, CReSTIC Laboratory (EA 3804), University of Reims Champagne-Ardenne, Reims, France
e-mail: jean-charles.boisson@univ-reims.fr

F. Rousseaux
MODECO Team, CReSTIC Laboratory (EA 3804), University of Reims Champagne-Ardenne, Reims, France
e-mail: francis.rousseaux@univ-reims.fr

C. Faucher (ed.), *Advances in Culturally-Aware Intelligent Systems and in Cross-Cultural Psychological Studies*, Intelligent Systems Reference Library 134, https://doi.org/10.1007/978-3-319-67024-9_3

systems called 'Culturally-Aware'. Blanchard et al. [1] define Culturally-Aware systems as "any system where culture-related [knowledge has] some impact on its design, runtime or internal processes, structures, and/or objectives". This definition encompasses three kinds of systems: those managing cultural data, those enculturated and those culturally-intelligent. Cultural data management systems are produced for domains and activities related to culture. Their objective is to retrieve and structure cultural data to facilitate their access. Such system can be found in the electronic collection of the Human Relations Area Files[1] (HRAF) called eHRAF World Culture where cultural data is stored in a database and classified through two indexes: Outline of Cultural Materials (OCM) and Outline of World Cultures (OWM). The enculturated systems are designed according to cultural specifics to meet the needs of particular cultural groups [1]. The term enculturation refers to the process by which a group is transferring cultural elements at the conception of a child. Applied to computer systems, it defines the intentional process by which designers include cultural artifacts in the systems' components. This process can be achieved automatically by relying on formal cultural knowledge and using rules to transpose the knowledge into enculturating actions. An example of enculturated system is Rehm's [2] conversational agents which implement "culture-specific emblematic gestures". Culturally-intelligent systems are systems that provide the right enculturation by managing various cultural contexts. They produce culturally-relevant decisions driven by a form of artificial cultural awareness. This awareness comes from the mediation of cultural representations. Intercultural educational/collaboration systems are kinds of culturally-intelligent systems. For instance, Johnson [3] presents a system to develop intercultural competences based on immersive simulations. The cultural representations are based on the situated culture methodology which allows finding relevant cultural information for American soldiers going in mission to Afghanistan.

Cultural awareness requires on one hand to be aware of one's own culture as well as the culture of the others [4] and to be "conscious of similarities and differences among cultural groups" [5]. In other words, to be Culturally-Aware, one needs to possess representations of his own culture and others, but also has to produce culturally-relevant mediations between these representations. As such a system possessing an artificial cultural awareness should be able to manage both aspects. Therefore, calling Culturally-Aware the cultural data management systems and manually enculturated systems might be confusing or inappropriate because the cultural awareness remains within the experts designing them. As for the culturally-intelligent systems, to our knowledge the cultural mediations are implicitly embedded in the systems through the cultural representations. As a consequence, the mediations are not explicitly produced by the system and thus, the artificial cultural awareness can only be considered as partial.

Currently, many culturally-intelligent systems happen to be designed with illegitimate conceptualisation of culture based on folk approaches as well as cultural

[1]http://hraf.yale.edu.

representations coming from an 'etic' approach [6]. "Folk approaches stem from subjective, personal descriptions and perceptions of cultural contexts which are used to represent cultural features" [6]. An etic approach has for objective to discover cultural universals from an outsider perspective. In contrast, an 'emic' approach aims to identify from an insider view singular concepts and behaviors which constitute cultural specifics.

Cultural representations coming from the etic approach are intrinsically intercultural because they are supposedly universals. These representations can be used to describe and mediate several cultures. Therefore, they fit the requirements to develop artificial cultural awareness. That is why as the main etic cultural representation Hofstede's value system of national culture [7, 8] is popular in the works related to Culturally-Aware systems [9–14]. However, these etic representations of cultures come with to a granularity problem. Because they are built on few shared features, they are too coarse-grained and thus limit the understanding of culture by the systems. Therefore, it leads to a bottleneck in the development of higher performance Culturally-Aware systems. As observed by Mohammed and Permanand [6], these systems strive for finer-grained representations based on legitimate conceptualisations of culture.

To overcome this limitation, artificial cultural awareness has to emerge from legitimate (conceptually sound) and emic-based cultural representations. Besides being hardly available, such representations are naturally heterogeneous. As a consequence, they are not adapted for cross-cultural mediation. That is why to our knowledge, there is not yet any attempts to develop artificial cultural awareness based on such cultural representations.

In this research, we design a methodology to develop artificial cultural awareness based on Prototypical Cultural Models (PCMs) [15] and Ontology Mediation. PCMs are emic-based legitimate cultural conceptualisations coming from Cognitive Anthropology. As for Ontology Mediation, it addresses the heterogeneity problem by dealing with conceptualisation level mismatches between local ontologies to produce a single global ontology. Because culture is often bound with language, it is necessary to specify that our methodology was designed with a single language in mind: English.

Our paper is composed of theoretical and practical parts. We start the theoretical part by introducing the PCMs and a methodology to build them. Then, we explain their formalisation into cultural ontologies. We end by presenting cultural mediations as the result of aligning cultural ontologies. The practical part consists of an example to develop artificial cultural awareness. We describe our process and the experiment, and analyse the results. We close this chapter with a short conclusion.

3.2 Prototypical Cultural Models (PCMS)

The development of artificial cultural awareness is composed of two parts. The cultural representations are mandatory to become aware of the cultures. The cultural mediations are necessary to become aware of the similarities and differences between these representations. In this section we address the first part through the introduction of PCMs. We start by explaining the underlying cognitive theory of culture. Then, we describe mental models as the building blocks of PCMs. We finish by presenting the emic methodology leading to their acquisition.

3.2.1 Cognitive Theory of Culture

To date there is no general consensus on what culture is exactly. Kroeber and Kluckhohn [16] identified not least than 164 definitions. Spencer [17] aggregated many quotes from researchers describing culture that he tried to classify. We mostly rely on his work to present what culture is and introduce the cognitive theory of culture.

1. Culture is learned and shared.
2. Culture is socially-constructed, thus is about social groups [17]. But, "culture is as much [a] psychological construct as it is a social construct" [18] as it ultimately resides in individuals' mind. "Culture is always both socially and psychologically distributed in a group, and so the delineation of a culture's features will always be fuzzy" [17]. Culture is not uniformly shared by the whole but consensually shared. Intracultural variations is the manifestation of its irregular distribution [19]. "Culture can be differentiated from both universal [inherited] human nature and unique [inherited and learned] individual personality" [17]. But, culture consists in "both universal (etic) and distinctive (emic) elements" [17].
3. "Culture is manifested at different layers of depth" [17]. Here, Spencer uses quotes from Schein [20, 21] who describes manifestations of culture through artifacts and values. Artifacts such as behavior or art are observable but hardly understandable. In contrast, the visibility of values depends on the level of awareness. Hidden values are considered as basic assumptions and represent those either taken for granted, invisible or pre-conscious. These different levels of awareness of cultural values refer to their explicitness/tacitness.
4. "Culture affects behavior and interpretations of behavior" [17]. It means that both the behavior and its interpretation are dependent. The visible behavior is encoded with invisible meaning that outsiders cannot decipher properly. Therefore, only people sharing a similar understanding (thus be- longing to the same group) can correctly interpret it.
5. "Culture influences biological processes" [17]. By shaping individuals' behavior, culture impacts eating habits, physical preferences, etc. As an illustration, Ferraro [22] describes Westerner people having a psychosomatic culturally-

induced reaction (vomiting) after hearing that they just ate freshly killed rattlesnakes.

6. "The various parts of a culture are all, to some degree, interrelated" [17]. Cultures "tend to be integrated systems with a number of interconnected parts, so that a change in one part of the culture is likely to bring about changes in other parts" [22]. As such it is "subject to gradual change" [17]. Culture is relatively stable or perceived as invariant [23] for the short or mid-term. However, the interrelated parts of culture evolve and may bring changes on the long-term.

For D'Andrade [24], there is now an agreement in Cognitive Anthropology that culture is about symbols, concepts and ultimately meanings. According to Bennardo and De Munck [15], this consensus was already present among sociologists and socio-anthropologists [16, 25, 26]. The cognitive theory of culture situates culture in the mind as a system of learned and shared knowledge [27]. As such, "buildings, behaviors, movies, prehistoric artifacts, and anything else "out there" [disqualify] from being culture" [15]. What essentially constitutes culture is socially rooted consensual knowledge.

3.2.2 *Mental Models*

Mental models were introduced by Craik [28] in 1943. Jones [29] describes a mental model as "a simplified representation of reality that allows people to interact with the world". Mental models are partial and imperfect abstractions, representing what is considered to be true. They potentially contain errors [30] up to contradictory propositions [31].

Mental models conceptualize abstract or concrete objects through signs, mental images or verbal descriptions [32]. Because they are composed of knowledge structures, mental models are able to interpret external data [29, 33–35]. Jones observed that "people tend to filter new information according to its congruence or otherwise with their existing understandings" [29]. This "tendency to search for and use information that supports one's beliefs" [36] is a phenomenon known as confirmation bias.

Mental models are dynamic. New knowledge is built upon prior knowledge [37] either by reasoning or by learning. Reasoning is achieved through the computational structure of mental model [38]. Learning emerges from the experiences coming from the interpretation of data. The knowledge created is either assimilated or accommodated [39]. The accommodation refers to the indirect integration of the new knowledge after adapting the mental model structure while the assimilation is the direct integration of the new knowledge without any change occurring.

"The maximum size of a mental model is thought to be determined by the capacity of working memory, the mental workbench on which people store information temporarily while thinking about it" [40]. The number of concurrent distinctions while processing information was set by Miller [41] to 7 plus or minus to.

Mental models range from idiosyncratic to collective [15]. Idiosyncratic mental models are shaped by personal experiences while collective ones are modeled by a "social process of elaboration, communication and dissemination of knowledge systems" [42].

Cultural models are particular collective mental models. They are defined as presupposed taken-for-granted and distributed mental models socially constructed and intersubjectively shared by a social group [43–45]. As such they have special properties, they are mature [46], tacit [47] and durable [48, 49].

'Prototypical cultural models' (PCMs) are the consensual representation of similar cultural models. They constitute virtual "culturally standard or "prototypical" model[s] that [are] shared but not held by [anyone]" [15]. Because they come from inside a social group, their elicitation fundamentally requires an emic approach.

3.2.3 Methodology for Building PCMS

To our knowledge, building PCMs is mostly based on the ethnographers' experiences. The most used consists in (1) selecting a sample based on features indicative of social link, then (2) eliciting individuals' domain to (3) discover the cultural domain which (4) facilitates individuals' mental model elicitation used to (5) build the prototypical cultural model through consensus analysis [50]. The general methodology can be summarized in three steps: ethnographic sampling, individuals' mental model elicitation and cultural consensus analysis.

3.2.3.1 Ethnographic Sampling

The ethnographic sampling step is based on the idea that cultural models are socially-constructed. It aims to capture a representative number of individuals likely to share the same collective models. This task is generally achieved through the identification of a community, a set of individuals with long-term, strong, direct, intense, frequent and positive relations [51]. A community can be identified through shared socially-related criteria such as genders, religions, jobs or areas - working places [52], towns [53] or regions [54].

3.2.3.2 Individuals' Mental Model Elicitation

Knowledge is personal and roots deeply in the subconscious of one self in a tacit state [55]. In order to elicit knowledge, it has to become object of thought [56]. Knowledge elicitation enables to explicit tacit internal knowledge structures.

The purpose of this second step is to elicit for each individual constitutive of the sample their mental model (their knowledge about a domain). Because we are interested in the construction of a PCM, the elicitation will focus on a cultural model and thus the knowledge about a cultural domain.

Jones et al. [29] distinguish two categories of knowledge elicitation: direct and indirect. In the first category, knowledge is directly elicited by the individual possessing the knowledge whereas in the second, knowledge emerges from the analysis of data collected from the individual.

The elicitation of domain knowledge is composed of two elements: concepts and relations. The specifics of one's own bias are embedded in this structure and manifest through their particular organisation. DeChurch and Mesmer-Magnus [57], after a survey about mental model elicitation, distinguish two key approaches. In the first one, predefined categories of cognitive content (concepts) are provided to respondents to link them. In the second approach, knowledge from respondents is induced without using predefined categories. These two approaches are consistent with the structured and open-ended implementations of the Conceptual Content Cognitive Map (3CM) method [58]. The goal of this interview-based method is to spatially elicit the cognitive map of a participant by having him write terms associated to a domain and their relations. In the open-ended implementation, the participant has to build the map from scratch whereas in the structured implementation he is constrained to use a provided list of terms.

An example of direct domain knowledge elicitation can be found in Vuillot et al.'s [54] study of farmers' ways of farming and ways of thinking. They record individual's mental models through the open-ended implementation of the 3CM using the Cmap Tool software.[2] Another direct elicitation is presented in Freeman. This time the elicitation of the concepts and relations are distinct. 900 statements are elicited for success and failure through free-listing. They are quantified and sorted. The top 50 are kept for both. The relations between the remaining terms are produced using the pile-sorting technique where participants are asked to assess how similar the terms are.

3.2.3.3 Cultural Consensus Analysis

The cultural consensus analysis is the last step to build a PCM. As a theory, the consensus analysis enables the operationalization of culture [15]. Cultural Consensus Theory (CCT) "formalizes the insight that agreement among [individuals] is a function of the extent to which each knows the culturally defined 'truth'" [60]. In other words, the degree of sharedness of some knowledge is representative of its cultural dimension ranging from personal to universal when it is shared respectively by none to mankind.

[2]http://cmap.ihmc.us.

CCT also "refers to a family of models that enable researchers to learn about [individuals'] shared cultural knowledge" [61]. They are based on two fundamental axioms [15]: culture consists of a pool of information that is shared and it is distributed. Typically, cultural consensus analysis uses questionnaires or structured interviews to obtain answers. Then, the respondents' answers are statistical analysed to determine the consensually shared information. Depending on the form of the elicited knowledge, either formal or informal cultural consensus models are used [62, 63]. However, simple aggregations, majority or averaging responses across respondents also constitute reasonable cultural estimates [64].

These models work generally on three restrictive assumptions [65]. First, for the sake of knowledge consistency [64], the sample has to share a single culture. Second, the elicitation of the individuals' knowledge should be done independently to avoid the apparition of shared knowledge emerging from collective interactions. Third, the knowledge should be homogeneous, that is about a common domain and on a similar level of difficulty [64]. However, a model extension produced by Batchelder and Romney [66] copes up with the heterogeneity issue.

As a method, cultural consensus analysis provides a way: (1) to determine whether observed variability in knowledge is cultural, (2) to measure how much cultural competence each individual possesses and (3) to ascertain the culturally correct knowledge [67].

1. Eigenvalues are derived from the factor analysis of the agreement matrix. The agreement matrix contains the agreement between each individual analysed. Eigenvalues are measures indicative of the knowledge variability across the sample. A general rule is that there is a single culture when the ratio is 3 or greater [64]. Gatewood and Lowe [68] set this value to 3.5.
2. The estimated knowledge, competencies or loadings of each participant refer to the degree of similarity between one individual's and the overall' knowledge [15]. An individual with a high loading possesses knowledge representative of the culture.
3. The answer key is the result of the cultural consensus analysis. It consists in the consensual and considered-true knowledge [15].

Applied to the individuals' elicited cultural model, the cultural consensus analysis produces an answer key which embodies the Prototypical Cultural Model.

Relying on the cognitive theory of culture, PCMs provide legitimate and relatively fine-grained representations of cultures. In fact, the granularity of the mental models produced in the second step determines the granularity of the final PCM. However as such, they cannot be used for the development of cultural awareness because computers systems are not yet able to make sense of them. To be understandable, they have to be formalized.

3.3 From Prototypical Cultural Models to Cultural Ontologies

As they are, PCMs are informal. To become machine-readable cultural representations usable by any systems they have to be represented with a formal language. In addition, the systems can try to build cultural mediations only if the cultural representations are formalized. Because PCMs constitute prototypical cultural conceptualizations, they can naturally be embedded into ontologies which are defined as explicit and formal specifications of shared conceptualizations [69–71].

3.3.1 Formal Ontologies

Ontologies are composed of conceptual structures. Their principal components are labels, concepts, relations and axioms. Axioms are rules associated to the relations used to enable reasoning. Their formal specification enables their interoperability, re-usability, understandability and machine-readability. Ontologies range from highly informal (expressed loosely in natural language) to rigorously formal (meticulously defined terms with formal semantics) [72].

The Resource Description Framework (RDF) is the main formal language as it supports the development of the semantic web. RDF is based on entities (resource, property, value) which constitute triples of the form (subject, predicate, object). Resources are concepts described thanks to an Uniform Resource Identifier (URI). Properties can be attributes or any other kind of relations which are themselves concepts. Values are literals pointing either to a symbol or another resource. The common syntax to formalize RDF is XML, called RDF/XML. Ontologies written in RDF are interpretable by machines through SPARQL Protocol and RDF Query Language (SPARQL).

3.3.2 Framework to Produce Ontologies

Similarly to PCMs, methodologies to create ontologies are mostly based on experience [73]. The METHONTOLOGY is a proven framework describing the general steps to build an ontology [74]. Common steps are composed of specification, conceptualization, formalization, implementation and evaluation.

The specification consists in planning the production and exploitation of an ontology. At a minimum, it defines its primary purpose, level, granularity and scope. These specifications mainly constrain the conceptualization. Typically, the conceptualization step is carried out by a group of domain experts. The goal is to discover the significant concepts and associated relations [71]. The formalization step expresses the conceptualization with formal languages. It is often manually supervised by knowledge engineers or with the support of a software like Protégé.[3] Mapping techniques can also be used to automatically transpose informal to formal knowledge [75]. The implementation step addresses the technical and practicable aspects associated with the usage of an ontology by a computer system. The evaluation step validates each step according to the specifications.

3.4 Formal Cultural Mediation

While trying to bridge various local ontologies, many kinds of mismatches are generated. Klein [76] provides a classification of these mismatches mainly based on Visser's work [77]. The main distinction is between the language and ontology. Language level mismatches are caused by the language used to formalize the ontology. Ontology level mismatches are due to the represented knowledge itself, the differences appearing in the conceptualization and explication. Conceptualization mismatches come from the bias intrinsic to conceptual structures constituting the ontologies while explication mismatches "result from explicit choices of the modeler about the style of modeling" [76].

Building cultural mediations consists in identifying continuities and discontinuities between heterogeneous cultural ontologies and thus, it is about dealing with conceptualization mismatches. In our case, the cultural dimension of these mismatches is guaranteed by the PCMs. Therefore, our idea is to use ontology mediation which addresses conceptualization level mismatches, to produce cultural mediations.

Eventually, for culturally-intelligent systems to make use of these mediations, they have to be expressed in a formal language.

3.4.1 Ontology Mediation

Ontology mediation includes many tasks such as ontology mapping, alignment, articulation, merging, integrating and combining. Ontology alignment aims at bridging several ontologies into mutual agreement, making them consistent and coherent [78]. Ontology combining refers to using several ontologies for a task in which their mutual relations are relevant [79]. In ontology alignment and combining, ontologies themselves are not modified contrary to ontology merging and integrating. Ontology merging or integrating has for purpose to capture all knowledge from the source ontologies into a single new ontology. Whether it is ontology alignment, combining, merging or integrating, all these operations are supported by ontology matching techniques.

[3]Available at: http://protege.stanford.edu/.

3.4.2 Ontology Matching

Ontology matching is the specific task of finding correspondences between entities in two ontologies. Matching techniques are often divided between syntactic and semantic. "In syntactic matching, the labels and sometimes the syntactical structure of the [ontologies are] matched and typically some similarity coefficient [0, 1] is obtained, which indicates the similarity between two concepts. Semantic matching computes a set-based relation between the [concepts], taking into account the meaning of each [concept]" [80]. Euzenat [81] goes further with a more detailed classification of concrete matching techniques.

Among the matching techniques presented on Fig. 3.1, four kinds are particularly adapted to address conceptual mismatches because they are related to structural aspects of ontologies. They are based on: formal resources, taxonomies, graphs and models.

The matching results constitute mappings which relate similar concepts or relations from different source ontologies to each other [79, 82]. Given two ontologies O_1 and O_2, a mapping is generally represented [83, 84] as a 5-tuple (i, e_1, e_2, r, s) where i is a unique identifier for the correspondence, e_1, e_2 are the matching elements, with $e_1 \in Q(O_1)$ and $e2 \in Q(O_2)$ where Q denotes ontology constructs, r is a logical relation and s a score. The score is generally indicating the probability of the matching result.

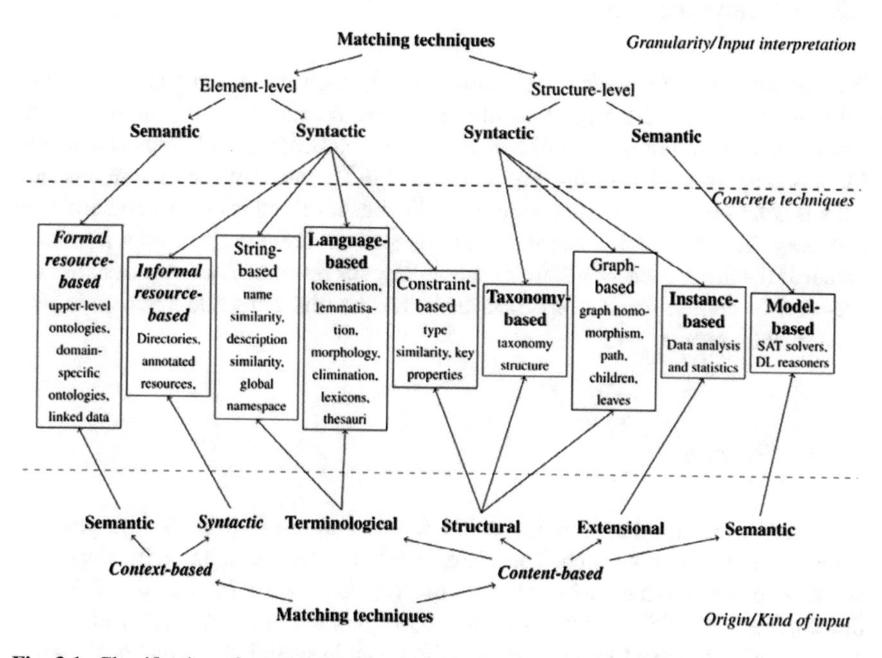

Fig. 3.1 Classification of current matching techniques (*source* Euzenat [81])

3.4.3 Creating Cultural Mappings

The two main parts of any mapping process are mapping discovery/evaluation and mapping representation/storage [80, 85]. Mappings are discovered by matching several ontologies and generally manually reviewed by an expert to be accepted. Then, the mappings found are formally represented and stored. It can constitute an intermediate ontology defined as an articulation ontology [86]. The MAFRA (MApping FRAmework for distributed ontologies) [87] called the latter the Semantic Bridging Ontology. In the end, the articulation ontology produced from local ontologies may be used as input to merge them into a single global ontology or as a mean to align them when needed [88].

Following this process, building formal cultural mediations is within reach. Using as inputs both the cultural ontologies and appropriate matching techniques, it is possible to discover, evaluate, represent and store culturally-relevant mappings.

Once a system has access to formal cultural mediations between cultural ontologies, it gains access to an artificial awareness of these cultures through their representations. Using the mediations, a Culturally-Aware system can produce culturally-relevant decisions to become culturally-intelligent. One form of this intelligence could be to adapt its own enculturation according to its former design cultural bias and the cultural background of its users.

3.5 Demonstration

We presented in a theoretical fashion the development of an emic-based artificial cultural awareness. The awareness of the cultures coming from the cultural representations are supported by cultural ontologies embedding Prototypical Cultural Models. The awareness of the similarities and differences between the representations is achieved by the formal cultural mediations resulting from conceptualization matching. To give some consistency to our methodology, we will now provide a practical demonstration. We start by presenting our process. Then, we describe the use-case we choose for our experiment. We end by observing and commenting the experimental results.

3.5.1 Process

The process we develop is rather crude as our purpose is to provide a practical demonstration of our research. Therefore, to ease the understanding of the latter, we decided to remind the main steps of the methodology with the Fig. 3.2. The development of artificial cultural awareness begins with the conception of PCMs. Their creation is driven by three steps: ethnographic sampling, individuals' mental

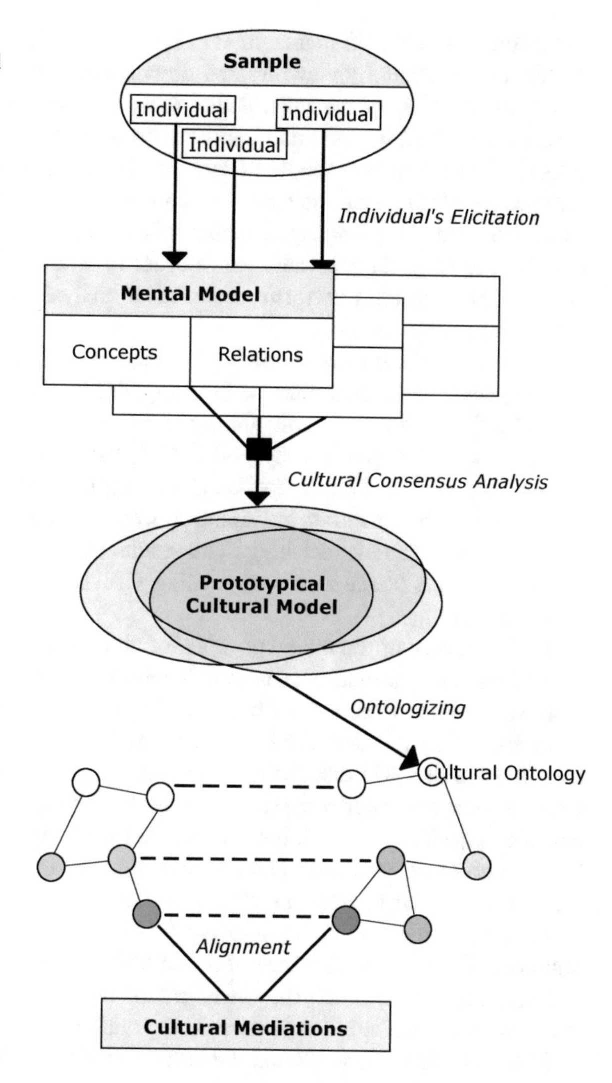

Fig. 3.2 Main steps of our methodology to build artificial cultural awareness

model elicitation and cultural consensus analysis. For our process, to compose the samples we assumed the role of an ethnographer. The criteria used to select the individuals were based on both a domain of study and on shared social features.

For the elicitation of the individuals' mental model, we developed an indirect elicitation process. We followed the idea that the mental models can be extracted from textual productions [89]. The process was based on two steps: data collection and data analysis. The data were retrieved directly from the web. Textual data collection was achieved thanks to HTTRACK[4] which is a tool that can mirror the content of a website by crawling and downloading its files.

For the analysis of the individuals' data, we used text-mining to find the concepts (symbolized by nominals) and relations (in the form of pairs of nominals) associated to the mental model of individuals. Mining textual data generally starts by a pre-processing task. Its goal is to improve the quality of the data by removing noise and by adding language specific information. The process ends by using an algorithm to mine the relevant data in order to produce the desired result. Most of the time the algorithm performs a statistical analysis which is achieved by quantifying the textual data.

The texts of the collected documents were extracted using Apache Tika.[5] Then, the language was identified with LangDetect [90] and English texts kept. OpenNLP[6] enabled us to split the latter in sentences and filtered the duplicates. Then, we used the Stanford CoreNLP API[7] for natural language processing operations: tokenization, Part of Speech (PoS) tagging and lemmatization. Eventually, nominals which constitute the main concepts of conceptualizations were found through simple pattern matching. The results of the preprocessing were stored as annotations in a 'serial data store' using GATE[8] (General Architecture for Text Engineering).

The discovery of the main concepts for an individual was basically achieved by correlating the importance of nominals with their number of occurrences. To this end, we quantified the lemmatized and lower-cased form of nominals and sorted them according to their number of occurrences. To find important relations we followed Harris' [91] distributional linguistic hypothesis. It considers that 'close' words appear in similar context. Therefore, by relying on co-occurrence statistics, it becomes possible to determine the proximity of two words. The co-occurrences between nominals was determined with a window size of 5. Then for each nominal, we measured their relatedness with others through the Jaccard formula:

$\mathrm{Jaccard}(n_1, n_2) = O(n_1 \cap n_2)/(O(n_1) + O(n_2) - O(n_1 \cap n_2))$, with n_1 and n_2 two nominals, $O(n)$ the occurrences of a nominal and $O(n_1 \cap n_2)$ the co-occurrences.

The higher is the score, the closer are the nominals. By ranking the relations of every nominal, we could thus determine which ones were the most important. To build the PCMs we first defined the cultural domains. To determine these domains we aggregated the rank of the significant concepts for each individual. According to their distribution in their respective samples, those relevant were filtered through a 'sharedness' threshold. Given a number, the top nominals were kept to constitute the cultural domains.

[4]http://www.httrack.com/.

[5]https://tika.apache.org/.

[6]https://opennlp.apache.org/.

[7]http://stanfordnlp.github.io/CoreNLP/.

[8]https://gate.ac.uk/.

The partial structure of the PCMs was produced using a nominal as a seed. This seed was used to initiate the discovery of shared relations. For each individual we retrieved a fixed number of best relations associated to the seed and belonging to the cultural domain. Shared relations were found by aggregating them and ensuring that their number was representative of the majority. Finally, we manually evaluated and labeled the relations and only retained those of interest. In particular, we focused on lexico-semantic relations. Apart from common transitive relations (meronymy, causality, …), we payed special attention to hypernym/hyponym lexico-semantic relations as they are known as the backbone of conceptualizations. The nominals constitutive of those relations were used as new seeds. We repeated this operation to a maximum of 7 times in accordance with the relative size of the PCMs. We cleaned the PCMs by removing the logical duplicates based on transitivity. For example, from the three relations (animal, hypernym, mammal), (mammal, hypernym, rabbit) and (animal, hypernym, rabbit) the last relation was deleted. We also ensured that the concepts were useful for the structures. This task was achieved by representing the PCMs as semantic graphs with Gephi.[9] Then we used the topology filter to keep only the nodes/concepts having a minimum of two edges/relations.

After having created the PCMs, they have to be formalized in order to be machine-readable. This is the next step toward the development of our artificial cultural awareness.

The formalization of the PCM in the RDF/OWL formal languages was done with Apache Jena.[10] The nominals belonging to the cultural domains were transposed as labels representative of classes. We did not manage polysemy: one nominal for one class with a single label. Then, the hypernym/hyponym relations were transposed as *rdfs:superClassOf/rdfs:subClassOf* relations, transitive relations as http://www.w3.org/2002/07/owl#TransitiveProperty and the remaining ones as *rdfs:seeAlso*.

Now that we dispose of formal cultural ontologies, we need to produce cultural mediations. This task constitutes the last step for computer systems to develop an artificial awareness about cultures through their representations. For the mediation of the cultural ontologies, we built our own conceptualization matcher. The semantic matching between two concepts was achieved by comparing the set of labels of every concept associated to them through transitive relations. It led to a score between 0 and 1. The scores close to 0 should indicate that the concepts are completely different while those close to 1 mean that they are similar. The alignment of two ontologies produced a number of mappings composed of their own identifier, the resources and their respective ontology, the kind of relation and the score. The formal cultural mediations were finally built by reviewing these mappings.

[9]https://gephi.org/.

[10]https://jena.apache.org/documentation/ontology/.

3.5.2 Experiment

For our experiment we chose a controverted topic, composed of two major and distinct communities with point of views drastically different. We are talking about 'abortion', those advocating the right for women to freely choose and those willing to protect 'unborn babies', that is the 'Pro-choice' and 'Pro-life' communities. The goal of this use case is neither to participate to the debate nor to judge a particular position. We aim to develop an artificial cultural awareness that can grasp the cultural similarities and differences.

We constituted two samples with our individuals being websites managed by identified Pro-Life and Pro-Choice communities. Considering websites as individuals may not be the best choice to carry out our experiments. However, this decision was driven by the necessity of being able to collect for a consequent number of 'individuals' large amount of textual data written in the same language and about a single domain.

For our experiment, we did not produce complete PCMs as the time required is too consequent and because the objective of the latter was to illustrate our methodology. Therefore, we decided to build partial PCMs centered around the conception of 'abortion'.

The samples we constituted contained respectively 3 and 5 individuals for the Pro-Choice and Pro-Life communities. We collected the data of each individual. Then, we preprocessed the data (additional information is available in Table 3.1).

The cultural domains were defined arbitrarily by aggregating the top 10000 nominals of every individual and by filtering those whose score/vote obtained the majority. The structures of the PCMs were produced using the term abortion as a seed and the top 50 associated relations. The consensually-shared relations were evaluated and the relations identified and tagged. The new concepts we encountered were used as new seeds. The result of the first iteration is presented in Table 3.2.

The resulting PCMs were verified, cleaned and formalized (jointly represented on Fig. 3.3).

Table 3.1 Amount of useful textual data for each individual

Sample	Individual	Number of sentences
Pro-choice	http://www.prochoiceactionnetwork-canada.org/	17,682
Pro-choice	http://www.prochoiceamerica.org	30,110
Pro-choice	http://www.prochoiceforum.org.uk/	18,788
Pro-life	http://www.nrlc.org/	136,028
Pro-life	http://www.priestsforlife.org/	235,694
Pro-life	http://www.standtrue.com/	15,885
Pro-life	http://studentsforlife.org/	19,629
Pro-life	http://www.lifenews.com/	76,929

Table 3.2 Cultural relations of interest associated with the seed 'abortion' with manually labeled classes (*h* hypernym, *t* transitive, *r* related; *1* direct, *2* reverse)

Community	Source	Destination	Relation
Shared	Abortion	Risk	t1
		Decision	h2
		Reason	t1
		Issue	h2
		Woman	r
		Choice	h2
		Legal_abortion	h1
		Pregnancy	R
		Right	h2
Pro-life		Mother	r
		Birth	r
		Baby	r
		Murder	h2
Pro-choice		Health	h2
		Surgical_abortion	h1
		Birth_control	h2
		Procedure	h2
		Contraception	h2
		Fetus	r
		Illegal_abortion	h1

The cultural ontologies produced were aligned. The final result is the set of mappings which corresponds to the cultural mediations between the Pro-Choice and Pro-Life communities.

3.5.3 Results

The Pro-Choice and Pro-Life ontologies shared 48 concepts. On those 48, equivalent mappings were produced for 37 of them. For example, 'medical_procedure' was found as an identical concept. The cultural ontologies possessed 77% of conceptual similarities. This result was worst than what Fig. 3.3 suggested. We sorted the concepts according to the lowest scores to observe the major cultural differences.

The ranking on Table 3.3 seems to confirm that the matching scores are culturally-relevant. Concepts with a score close to 1 indicate that both communities share similar conception. This is the case for 'birth'. When the scores are near 0.5, cultural dissimilarities appear. For example, some aspects of the 'abortion' are mutually present for the Pro-Choice and Pro-Life communities: 'procedure', 'decision' or 'pregnancy'. However, only the Pro-Life community considers that

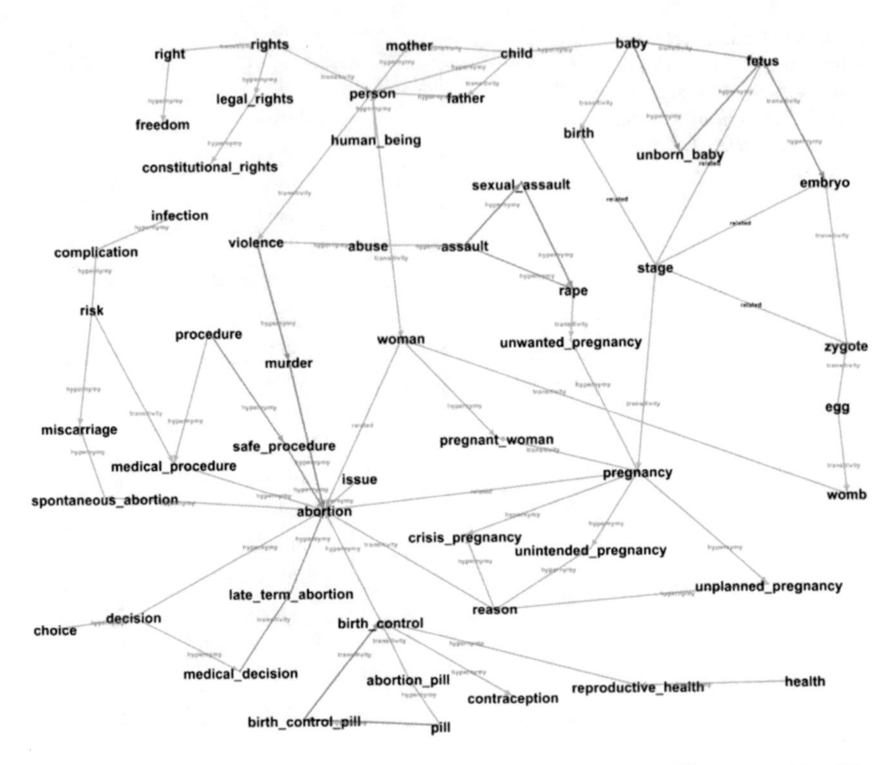

Fig. 3.3 Prototypical cultural models for both the pro-choice and pro-life communities (*blue* shared, *green* Pro-Life, *red* Pro-Choice)

Table 3.3 Concepts with the lowest matching score

Concept	Equivalence score
Embryo	0.2
Fetus	0.3
Abortion	0.55
Spontaneous_abortion	0.57
Miscarriage	0.58
Birth_control	0.8
Contraception	0.83
Rape	0.88
Baby	0.9
Unwanted_pregnancy	0.91
Birth	0.91

destroying an embryo is like murdering somebody. Finally, when the scores fall towards 0, it means that there is at least one fundamental cultural difference.

Based on this analysis, we can infer that the main disagreement between the two communities is not about the ‘abortion’ but their respective conceptions of an 'embryo' and a 'fetus'. This interpretation seems consistent with our own judgments about the core cultural divergences of these two communities.

To obtain practical cultural mediations to develop an artificial cultural awareness, we use a threshold of 0.75 to distinguish appropriately the concepts to produce valid cultural mediations. Below this threshold, the conceptions are disjoint while over this score they are equal.

This experiment tends to demonstrate that the scores obtained through cultural ontology matching are good cultural indicators and are able to produce cultural mediations.

3.6 Conclusion

In this paper we proposed a methodology to build artificial cultural awareness based on a legitimate conceptualization of culture as well as an emic approach. It started with the creation of PCMs and their formalization into cultural ontologies. It ended with the production of relevant cultural mediations resulting from the ontology alignment. We designed a process driven by this methodology for our demonstration. The experimental results tended to confirm the validity of our methodology.

Nevertheless, the effective use of cultural mediations by culturally-intelligent systems remains to be studied: "How these cultural mediations can drive the adaptation of these systems?"

In addition, culture and language are correlated. Therefore, the multilingual dimension needs to be included in our methodology.

References

1. Blanchard, E.G., Mizoguchi, R., Lajoie, S.P.: Structuring the cultural domain with an upper ontology of culture. The Handbook of Research on Culturally-Aware Information Technology: Perspectives and Models, pp. 179–212 (2010)
2. Rehm, M., Nakano, Y., André, E., Nishida, T., Bee, N., Endrass, B., Wissner, M., Lipi, A.A., Huang, H.-H.: From observation to simulation: generating culture-specific behavior for interactive systems. AI & Soc. **24**(3), 267–280 (2009)
3. Johnson, W.L.: Using immersive simulations to develop intercultural competence. In: Culture and Computing, pp. 1–15. Springer, Berlin (2010)
4. Tomalin, B., Stempleski, S.: Cultural Awareness. Oxford University Press, Oxford (2013)
5. National Center for Cultural Competence (2004)
6. Mohammed, P., Mohan, P.: Breakthroughs and challenges in culturally-aware technology enhanced learning. In: Proceedings of Workshop on Culturally-aware Technology Enhanced Learning in Conjuction with EC-TEL 2013, Paphos, Cyprus, 17 September 2013
7. Hofstede, G.H., Hofstede, G.: Culture's Consequences: Comparing Values, Behaviors, Institutions and Organizations Across Nations. Sage, Thousand Oaks (2001)

8. Hofstede, G., Bond, M.H.: Hofstede's culture dimensions an independent validation using rokeach's value survey. J. Cross Cult. Psychol. **15**(4), 417–433 (1984)
9. Khashman, N., Large, A.: Investigating the design of arabic web interfaces using hofstede's cultural dimensions: a case study of government web portals. In: Proceedings of the Annual Conference of CAIS/Actes du congrès annuel de l'ACSI (2013)
10. Mascarenhas, S., Paiva, A.: Creating virtual synthetic cultures for intercultural training. In: Third International Workshop on Culturally-Aware Tutoring Systems (CATS2010), p. 25. Building Time-Affordable Cultural Ontologies using an Emic Approach 19 (2010)
11. Reinecke, K., Bernstein, A.: Tell me where you've lived, and i'll tell you what you like: adapting interfaces to cultural preferences. In: International Conference on User Modeling, Adaptation, and Personalization, pp. 185–196. Springer, Berlin (2009)
12. Marcus, A., Gould, E.W.: Crosscurrents: cultural dimensions and global web user-interface design. Interactions **7**(4), 32–46 (2000)
13. Chandramouli, K., Stewart, C., Brailsford, T., Izquierdo, E.: Cae-l: an ontology modeling cultural behaviour in adaptive education. In: Semantic Media Adaptation and Personalization, 2008. SMAP'08. Third International Workshop on, pp. 183–188, IEEE (2008)
14. Mohammed, P., Blanchard, E.G.: Leveraging comparisons between cultural frameworks: preliminary investigations of the mauoc ontological ecology. In: Sixth International Workshop on Culturally-Aware Tutoring Systems (CATS2015), p. 1 (2015)
15. Bennardo, G., De Munck, V.C.: Cultural Models: Genesis, Methods, and Experiences. Oxford University Press, Oxford (2014)
16. Kroeber, A.L., Parsons, T.: The concepts of culture and of social system. Am. Sociol. Rev. **23** (5), 582–583 (1958)
17. Spencer-Oatey, H., Franklin, P.: What is culture. A compilation of quotations. GlobalPAD Core Concepts (2012)
18. Matsumoto, D.R.: Cultural Influences on Research Methods and Statistics. Brooks/Cole Publishing Company, USA (1994)
19. Pelto, P.J., Pelto, G.H.: intra-cultural diversity: some theoretical issues. Am. Ethnol. **2**(1), 1–18 (1975)
20. Schein, E.H.: Coming to a new awareness of organizational culture. Sloan Manag. Rev. **25**(2), 3–16 (1984)
21. Schein, E.: Organizational culture. Am. Psychol. (1990)
22. Ferraro, G.: The Cultural Dimension of International Business, 3rd edn (1998)
23. Mathews, H.F.: Uncovering cultural models of gender from accounts of folktales. In: Finding Culture in Talk, pp. 105–155. Springer, Berlin (2005)
24. D'andrade, R.: A cognitivist's view of the units debate in cultural anthropology. Cross Cult Res **35**(2), 242–257 (2001)
25. Bartlett, F.C.: Remembering: An Experimental and Social Study. Cambridge University, Cambridge (1932)
26. Turner, V.W.: The Forest of Symbols: Aspects of Ndembu Ritual, vol. 101. Cornell University Press, New York (1967)
27. Goodenough, W.H.: Culture, Language, and Society. Benjamin-Cummings Pub Co, San Francisco (1981)
28. Craik, K.: The Nature of Explanation (1943)
29. Jones, N., Ross, H., Lynam, T., Perez, P., Leitch, A.: Mental models: an interdisciplinary synthesis of theory and methods. Ecol. Soc. **16**(1) (2011)
30. Moray, N.: Models of models of… mental models. In: Perspectives on the Human Controller, pp. 271–285 (1997)
31. Byrne, R.M.: The rational imagination (2005)
32. Forrester, J.: W. Industrial Dynamics, vol. 1, no. 961, pp. 1–464. MITPress, Cambridge Mass (1961)
33. Chi, M.T.: Three types of conceptual change: belief revision, mental model transformation, and categorical shift. In: International Handbook of Research on Conceptual Change, pp. 61–82 (2008)

34. Johnson-Laird, P.N.: Mental Models: Towards a Cognitive Science of Language, Inference, and Consciousness, No. 6. Harvard University Press, Cambridge (1983)
35. Gentner, D., Gentner, D.R.: Flowing waters or teeming crowds: mental models of electricity. Tech. Rep. DTIC Document (1982)
36. Bank, T.W.: Thinking with mental models (2015)
37. Wiig, K.: People-focused knowledge management. How effective decision making leads to corporate success (2004)
38. Rutherford, A., Wilson, J.R.: Models of mental models: an ergonomist psychologist dialogue. In: Selected papers of the 8th Interdisciplinary Workshop on Informatics and Psychology: Mental Models and Human-Computer Interaction 2, pp. 39–58. North-Holland Publishing Co., New York (1989)
39. Piaget, J., Cook, M.: The Origins of Intelligence in Children, vol. 8. International Universities Press, New York (1952)
40. Doyle, J.K.: Measuring change in mental models of dynamic systems: an exploratory study. Ph.D. thesis, Department of Economics, Siena College (1998)
41. Miller, G.A.: The magical number seven, plus or minus two: some limits on our capacity for processing information. Psychol. Rev. **63**(2), 81 (1956)
42. Wagner, W., Hayes, N.: Everyday Discourse and Common Sense: The Theory of Social Representations. Palgrave Macmillan, UK (2005)
43. Gee, J.P.: Video games and embodiment. Games Cult. **3**(3–4), 253–263 (2008)
44. Holland, D., Quinn, N.: Cultural Models in Language and Thought. Cambridge University Press, Cambridge (1987)
45. D'Andrade, R.: A folk model of the mind (1987)
46. Mathevet, R., Etienne, M., Lynam, T., Calvet, C.: Water management in the camargue biosphere reserve: insights from comparative mental models analysis. Ecol. Soc. **16**(1) (2011)
47. Eraut, M.: Non-formal learning and tacit knowledge in professional work. Br. J. Educ. Psychol. **70**(1), 113–136 (2000)
48. Berninger, K., Kneeshaw, D., Messier, C.: The role of cultural models in local perceptions of sfm–differences and similarities of interest groups from three boreal regions. J. Environ. Manage. **90**(2), 740–751 (2009)
49. Mathieu, J.E., Heffner, T.S., Goodwin, G.F., Cannon-Bowers, J.A., Salas, E.: Scaling the quality of teammates' mental models: equifinality and normative comparisons. J. Organ. Behav. **26**(1), 37–56 (2005)
50. Stone-Jovicich, S., Lynam, T., Leitch, A., Jones, N.: Using consensus analysis to assess mental models about water use and management in the crocodile river catchment, South Africa. Ecol. Soc. **16**(1) (2011)
51. Wasserman, S., Faust, K.: Social Network Analysis: Methods and Applications, vol. 8. Cambridge university press, Cambridge (1994)
52. Mathieu, J.E., Rapp, T.L., Maynard, M.T., Mangos, P.M.: Interactive effects of team and task shared mental models as related to air traffic controllers' collective efficacy and effectiveness. Hum. Perform. **23**(1), 22–40 (2009)
53. Young, J.C.: A model of illness treatment decisions in a tarascan town. Am. Ethnol. **7**(1), 106–131 (1980)
54. Vuillot, C., Coron, N., Calatayud, F., Sirami, C., Mathevet, R., Gibon, A.: Ways of farming and ways of thinking: do farmers' mental models of the landscape relate to their land management practices? Ecol. Soc. **21**(1), 1–23 (2016)
55. Polanyi, M.: Personal Knowledge: Towards a Post-Critical Philosophy. University of Chicago Press, Chicago (1958)
56. Alexander, P.A., Schallert, D.L., Hare, V.C.: Coming to terms: how researchers in learning and literacy talk about knowledge. Rev Edu Res **61**(3), 315–343 (1991)
57. DeChurch, L.A., Mesmer-Magnus, J.R.: Measuring shared team mental models: a meta-analysis (2010)

58. Kearney, A.R., Kaplan, S.: Toward a methodology for the measurement of knowledge structures of ordinary people the conceptual content cognitive map (3 cm). Environ. Behav. **29**(5), 579–617 (1997)
59. Freeman, H., Romney, A., Ferreira-Pinto, J., Klein, R., Smith, T.: Guatemalan and us concepts of success and failure. Hum. Organ. **40**(2), 140–145 (1981)
60. Kempton, W., Boster, J.S., Hartley, J.A.: Environmental Values in American Culture. MIT Press, Cambridge (1996)
61. Oravecz, Z., Vandekerckhove, J., Batchelder, W.H.: Bayesian cultural consensus theory. Field Methods **26**(3), 207–222 (2014)
62. Romney, A.K., Weller, S.C., Batchelder, W.H.: Culture as consensus: a theory of culture and informant accuracy. Am. Anthropol. **88**(2), 313–338 (1986)
63. Romney, A.K., Batchelder, W.H., Weller, S.C.: Recent applications of cultural consensus theory. Am. Behav. Sci. **31**(2), 163–177 (1987)
64. Weller, S.C.: Cultural consensus theory: applications and frequently asked questions. Field Methods **19**(4), 339–368 (2007)
65. Garro, L.C.: Remembering what one knows and the construction of the past: a comparison of cultural consensus theory and cultural schema theory. Ethos **28**(3), 275–319 (2000)
66. Batchelder, W.H., Romney, A.K.: Test theory without an answer key. Psychometrika **53**(1), 71–92 (1988)
67. Borgatti, S.P., Halgin, D.S.: 10 consensus analysis. A Companion to Cognitive Anthropology, p. 171 (2011)
68. Gatewood, J.B., Lowe, J.W.: Employee Perceptions of Credit Unions: Implications for Member Profitability. Filene Research Institute, Madison, WI (2008)
69. Gruber, T.R.: Toward principles for the design of ontologies used for knowledge sharing? Int. J. Hum Comput. Stud. **43**(5–6), 907–928 (1995)
70. Borst, W.N.: Construction of Engineering Ontologies for Knowledge Sharing and Reuse. Universiteit Twente, Enschede (1997)
71. Studer, R., Benjamins, V.R., Fensel, D.: Knowledge engineering: principles and methods. Data Knowl. Eng. **25**(1–2), 161–197 (1998)
72. Uschold, M., Gruninger, M.: Ontologies: principles, methods and applications. Knowl. Eng. Rev. **11**(02), 93–136 (1996)
73. Gòmez-Pérez, A., Benjamins, R.: Overview of knowledge sharing and reuse components: ontologies and problem-solving methods. IJCAI and the Scandinavian AI Societies. CEUR Workshop Proceedings (1999)
74. Fernàndez-Lòpez, M., Gòmez-Pérez, A., Juristo, N.: Methontology: From Ontological Art Towards Ontological Engineering (1997)
75. Pennacchiotti, M., Pantel, P.: Ontologizing semantic relations. In: Proceedings of the 21st International Conference on Computational Linguistics and the 44th annual meeting of the Association for Computational Linguistics, pp. 793–800, Association for Computational Linguistics (2006)
76. Klein, M.: Combining and relating ontologies: an analysis of problems and solutions. In IJCAI-2001 Workshop on ontologies and information sharing, pp. 53–62, USA (2001)
77. Visser, P.R., Jones, D.M., Bench-Capon, T.J., Shave, M.J.: Assessing heterogeneity by classifying ontology mismatches In: Proceedings of the FOIS, vol. 98 (1998)
78. Noy, N.F., Musen, M.A., et al.: Algorithm and tool for automated ontology merging and alignment. In: Proceedings of the 17th National Conference on Artificial Intelligence (AAAI-00). Available as SMI technical report SMI-2000–0831 (2000)
79. Su, X., Gulla, J.A.: Semantic enrichment for ontology mapping. In: International Conference on Application of Natural Language to Information Systems, pp. 217–228. Springer, Berlin (2004)
80. De Bruijn, J., Ehrig, M., Feier, C., Martìn-Recuerda, F., Scharffe, F., Weiten, M.: Ontology mediation, merging and aligning. Semantic web technologies, pp. 95–113 (2006)
81. Shvaiko, P., Euzenat, J.: Ontology matching: state of the art and future challenges. IEEE Trans. Knowl. Data Eng. **25**(1), 158–176 (2013)

82. Calvanese, D., De Giacomo, G., Lenzerini, M.: Ontology of integration and integration of ontologies. Description Logics **49**(10–19), 30 (2001)
83. Meilicke, C., Stuckenschmidt, H., Tamilin, A.: Repairing ontology mappings. In: AAAI, vol. 3, p. 6 (2007)
84. Calì, A., Lukasiewicz, T., Predoiu, L., Stuckenschmidt, H.: Tightly coupled probabilistic description logic programs for the semantic web. In: Journal on Data Semantics, pp. 95–130. Springer, Berlin (2009)
85. Granitzer, M., Sabol, V., Onn, K.W., Lukose, D., Tochtermann, K.: Ontology alignment—a survey with focus on visually supported semi-automatic techniques. Future Internet **2**(3), 238–258 (2010)
86. Kalfoglou, Y., Schorlemmer, M.: Ontology mapping: the state of the art. Knowl. Eng. Rev. **18** (01), 1–31 (2003)
87. Maedche, A., Motik, B., Silva, N., Volz, R.: Mafra—a mapping framework for distributed ontologies. In: International Conference on Knowledge Engineering and Knowledge Management, pp. 235–250. Springer, Berlin (2002)
88. Su, X., Gulla, J.A.: An information retrieval approach to ontology mapping. Data Knowl. Eng. **58**(1), 47–69 (2006)
89. Carley, K., Palmquist, M.: Extracting, representing, and analyzing mental models. Soc. Forces, 601–636 (1992)
90. Shuyo, N.: Language detection library for java, vol. 7, p. 2016. Retrieved July 2010
91. Harris, Z.S.: Distributional structure. Word **10**(2–3), 146–162 (1954)

Chapter 4
Teaching an Australian Aboriginal Knowledge Sharing Process

Cat Kutay

Abstract Experiential learning of other cultures not only provides knowledge of the protocols and values of a different culture, but also enables the learner to realize there are such differences. It is this awareness that enables us to better understand our own culture and how we communicate within and between cultures. We are using intelligent agents modelling cultural rituals, values and emotional responses within gaming environments to support the learning of cultural competency. In this chapter, we describe the development of cultural knowledge sharing processes. Starting with information sharing, in class role play and recorded material, we are expanding the interactions and scripting options to allow students to experience the conflicts felt by Aboriginal Australians within the mainstream culture. We analyse the different teaching methods and the suitability of the material.

Keywords Indigenous knowledge · Serious cultural games
Aboriginal languages

4.1 Introduction

Australia is home to the longest running cultural system in the world. The Aboriginal people have lived here 68–100,000 years [1], trading with neighbours in Indonesia and the Pacific, while developing a culture that prioritises relationships between humans and the environment, and minimizing human impact. Partly due to the highly irregular climate and hence food supply, in Australia a lot of time is spent travelling to reduce the burden on each region of a clan's land. This has also led to the development of a highly mobile and shared technology system.

C. Kutay (✉)
Faculty of Engineering and Information Technology, The University of Technology, PO Box 123 Broadway, Sydney 2007, Australia
e-mail: Cat.Kutay@uts.edu.au

C. Faucher (ed.), *Advances in Culturally-Aware Intelligent Systems and in Cross-Cultural Psychological Studies*, Intelligent Systems Reference Library 134, https://doi.org/10.1007/978-3-319-67024-9_4

Despite the effort of the Europeans who have settled in Australia since 1788, the Aboriginal culture has retained much of its traditional forms in the urban, regional and remote communities of Australia.

Hence all Australians are living in a community that is home to two very different cultures. The actor Jack Charles recently said on television [2] that Australia has a unique form of racism towards its Aboriginal people. It is unique both in the world, in that the cultural difference is so vast that misunderstanding and malpractice abound, and unique in Australia in that the racism expressed against the Aboriginal people is both more pervasive and more complex than other forms of racism in Australia.

To provide cultural teaching, we are looking to traditional oral methods. In Australian and North American Aboriginal groups, the oral tradition is a skill learnt and passed on as a discipline, involving repetition, praise and critique [3] to train the young in this method of history retention. Yet the importance of Aboriginal oral memories in terms of retaining a true history of Aboriginal collective identity and knowledge is generally denigrated in the non-Aboriginal view, as oral records are perceived as coloured by personal experience [4], and in constant flux.

This chapter is about attempts to help non-Aboriginal people understand the culture of the Australian landscape, as well as understanding the oral knowledge sharing process. We describe the development that led to the use of games for this project. We began with face-to-face workshops training professionals who were going to work with communities, while developing web services for Aboriginal communities to share knowledge relating to their own language and culture. This involved long community consultation over access and interface design. But most importantly were the understanding of protocols around knowledge sharing and the tools that would support this occurring in a more seamless manner.

4.2 Cultural Training

Much of Aboriginal traditional storytelling presented as Dreamtime stories, which are about a time that is neither past nor present but for all time. They incorporate experience from the past presented with relevance to the present audience [5]. While they are about a spiritual realm where Kangaroos are people and people are kangaroos, they are also stories of Law, how to behave, how to look after country. They are told from when a person is a child until when they learn to hunt and grow food, with more information added as they understand more. The knowledge is given as required, as the person's awareness grows, but also the story carries many levels so that a variety of listeners can learn from them [6].

The general approach we took to developing an understanding of Aboriginal culture was to both replace the current deficit model with an understanding of the wealth of the Aboriginal culture, as well as provide this opportunity to educate students on the variety of human values and aspirations.

Three focuses of cultural training have emerged and provide different development routes and strategies. Firstly to train the professionals who go out to communities. Secondly to provide support for Aboriginal people to learn and share their cultures amongst themselves. Thirdly to provide all Australians with an education in the local cultures.

All projects are focused on the holistic knowledge sharing practices in the community and how to emulate and support these through technology. A second offshoot of this work is the increased understanding of new ways of knowledge sharing, especially tacit knowledge sharing, that has emerged from this work, which we will mention in the next section. Hence, we first need to explain the knowledge sharing processes as used traditionally by Aboriginal people to understand how this procedure may be represented online (see studies in [7–9] for other work in this area).

4.2.1 Traditional Knowledge Sharing

The oral tradition that is practised and perfected by Aboriginal people provides a way to remember extensive details of the landscape and the maintenance of the food supply, while only teaching knowledge that is correct for the moment in a highly variable climate.

As an example we consider the story of how the kangaroo got its tail [10]. This is not a story told because Aboriginal people believe that sometime long ago a wombat was running around, got speared in the rear and the spear grew into a tail. The story is a theme, a 'textbook', on the care and preservation of the kangaroo.

When a child hears an elder start to tell about 'how the kangaroo got his tail', they know they are entering a lesson about the kangaroo, much as our children know they are entering a maths lesson when they are told to get out their maths textbook.

The story also covers many levels to engage audience with different skills. The moral tales and human relations are more for children, the practical details are comprehended better when people are familiar with the landscape, and the spiritual knowledge is unclear to those without prior understanding [11]. This understanding of the levels of engagement with oral knowledge links this research with the issues that arise in tacit knowledge sharing, which is the workplace knowledge of components and procedures, often shared only in oral form [12].

A story will be started by the elder at whatever point they know, or the one that is relevant to the listeners. The story they will tell is only the part that they have authority to tell others, and this helps prevent false information, or gossip, being inserted into a story.

The story in its entirety tells of the travel of the hunters, where they find seeds, fruit and water, and, by the position of the stars, how you can tell the season, along a route that the people will travel often in their search for kangaroo. The story tells where the kangaroo get good food, where they breed and how to look after them.

The people can recite the stories as they move through the country. Like the memory palace developed by Cicero, where an orator will place topics around their mental room so that they have a location in their head to trigger the memory of the point they want to make, so the storytellers of the Aboriginal community have a memory aid for where the food is through locational triggers, including the position of the stars in each season. Also the layout of the stars can be related to land tracks, providing a further guide for travellers. Stars are also used as a representation of kinship relations as a teaching aid for the community.

Also it is these storytellers who are responsible for the well-being of each totem, and when a community wants to hunt for kangaroo, they will go and ask permission of the elder responsible for kangaroo in the area, who will know if there are enough to allow hunting at that juncture.

Learning to hunt is however very difficult. The clan cannot afford for young hunters to go out and be unsuccessful in throwing a spear and frighten away the whole herd. Aboriginal people are in general uncomfortable with trial and error training. This has been an issue in schools in Australia, where a Socratic style is used which is alienating and confronting to Aboriginal students.

Hence the corroboree for the kangaroo provides an immersive and practical training. Here dance, music, storytelling and drama merge as a teaching method where a story if performed in front of people gathered from surrounding regions, sharing their part of the knowledge of the kangaroo, its breeding places, where the feed is good, etc.

To prepare for a corroboree performance, the elders sit together and discuss the desired format: the context of the corroboree; what is significant in the present situation for the people; what is relevant to the user's desires within the environment. They evaluate the community's goals and beliefs. This involves considering the themes that need to be covered for learning about the present context, which links to the audience's goals. They interpret and interact with this environment. This develops the cohesion of the narrative, what will be presented for the social and creative linkage of information. So they decide what will be performed.

Once this corroboree has started, individuals contribute stories or songs relating to how their own knowledge fits into the previous narrative and so select the stories that are shared. The rendition of these stories will be through various performances [13].

The corroboree stories say that in the Dreamtime, men and kangaroos changed into each other regularly and in a corroboree the elder who was responsible for kangaroo knowledge will indeed become a kangaroo. He will show how the kangaroo smells the wind, and reacts to sounds.

There are stories of historical films of Aboriginal hunters, one showing an older hunter standing up to throw a spear towards a herd of kangaroos. There was a medium sized male off to the right of the mob. The man throws his spear. The kangaroo was grazing and lifted its head, it smelled the wind, then leapt to the right, right into the path of the spear. To learn a skill like that you need many lessons with real kangaroos, or as near to this as you can get. The corroboree provides this opportunity.

4.2.2 *Analysis*

We now consider the significant features of Indigenous knowledge as expressed through stories and performances. The difference between Indigenous and non-Indigenous knowledge is summarised by Nakata [14] and these differences can be used to understand the mode of knowledge sharing [15]. These points are expanded here to consider the particular aspects that require encoding to support online transmission of this knowledge:

- **Thematic structure**. Knowledge is developed around a theme of morality or subsistence. The stories in this sequence are linked together in a simple framework (for example, a Dreamtime story) which is used for children and gradually over time embellished with further detail, while retaining the same basic story line.
- **Story-path or Songline**. Knowledge is generally presented in a synchronous and repetitive manner to enable memorising in a transient (oral) form. It is the landscape and the season of telling that provides this order to the story or song.
- **Contextual**. Knowledge cannot be presented without the context of the theme and the story-path which requires both the whole contextual framework and those who have a right to tell the story being included in the teaching/learning process.
- **Seasonal**. Time may cycle as the story moves across the landscape, describing the area as it is when you travel through at a walking pace.
- **Continuity of time**. Past and present knowledge are continually repeated, with knowledge placed in a larger context according to location and theme rather than historical time of occurrence.
- **Spiritual**. Stories move between western concepts of physical reality and spiritual or intuitive descriptions with complete fluidity. The Past or Dreamtime is continually existing in the Present.
- **Collaborative**. Knowledge is learnt from many people, not just your immediate superiors. Other people in the same role in other groups, also contribute to that knowledge telling.
- **Depth development**. Knowledge can only be shared with those already equipped with the requisite background knowledge to understand and memorise the new knowledge.
- **Access rights**. Knowledge is often not shared between certain people. Those who have different roles in society do not share these responsibilities to those outside their 'story' or knowledge context.
- **Knowledge is about intuition as well as observation**. One of the processes used in Indigenous knowledge sharing is the verification of intuition through comparison with others experience.

When considering how a knowledge story is formed, the first important factor is the theme. This may be related to morality or survival information. Then within the story there are four dimensions of knowledge sharing. First there are the story-paths or the thread of the story, which provides the sequence. These are generated along geographical paths. Then there is the dimension of time, which is cyclic in terms of seasons and continuous in terms of dreamtime events that still occur; or changed in terms of events long past that are now different. Thirdly at each location along the story-path there are the stories inserted by individuals with that knowledge, at various depths and providing a lattice of linking ideas, and each explaining some aspect of the theme or some aspect of the environment, that occurred or occurs at that location.

The final aspect is the verification. While teaching the community the elders are also checking and updating their knowledge. This arises from the observational and often intuitive nature of the knowledge gathering, and the need to update to the frequent and irregular changes in climate.

At this last level, complexity arises in providing the detailed enumeration of the various sub-themes and topics (physical, spiritual and social) that may be used to assist the linking of material in a combined story-path that retains its cohesiveness and veracity. While these stories have a strong sequence, it is geographical rather than time based, and a time base may not even be used when describing events at one location.

In a story as part of a knowledge theme, different sub topics are chosen, or the ecosystem is divided into subjects from the environment, such as water, trees, and winds and the responsibility for these assigned to different people through their kinship role. However, events also may be described along the story-path as unfolding through the seasons, so that the progression through the seasons matches a person's travel across the land, and at each time and location the type of sustenance that is available will be described.

The use of time in 'the Dreamtime' is a description that enables the two situations of continuity with the past, and discontinuity to be described. People talk of how a resource has been maintained 'since the Dreamtime'. Alternatively, at one location, variation in events may be described, such as 'in the Dreamtime our ancestors gathered shell fish here and ate them along the waters edge, see here are the middens' on the Barrup Peninsula (in the Pilbara region of Western Australia) located in the hills now many meters above sea level. Or the fishing boundaries in the water off Millingimbi community, which follow the line of ridges under the surface, which would have been exposed 15,000 years ago in the last ice age [16].

Finally there is another aspect to these stories, which is the extent to which the subject is physical, emotional or spiritual. There is no differentiation made between these in the Aboriginal view, hence these cannot be classified as dimensions. It is the 'physical' nature or the strong reality of this spiritual view as well as its emotional hold which has caused most dissent, misunderstanding and harm in negotiations with non-Aboriginal people. For example landholders have claimed a strong 'spiritual' link to their land over a few generations, attempting to mirror or match statements used in land claims by Aboriginal people [17]. The understanding of spirituality and how it is part of the knowledge development process is not understood by a culture that denigrates intuition and any emotional understanding of what effects an outcome.

In particular, it is important in the process of providing online web based services or Indigenous knowledge, to avoid any idea of assigning a western view of true or false to the stories. Such oral records have been considered unsuitable for historical or legal evidence [18–20], however they are vital to the preservation of Aboriginal history and identity [4]. Using an Aboriginal methodology requires a holistic approach. In terms of understanding the environment and knowledge about people and relationships, this means we need to understand that inanimate objects are sentient beings and part of the moral code of the landscape [21].

4.2.3 Knowledge Sharing

Tacit knowledge sharing is an important area of study in western systems, and the experience from indigenous communities has much to offer in understanding the process. In particular the concern is the 'verification' or reliability of knowledge shared orally. The model in Fig. 4.1 is used by Zaman et al. [22] to develop an

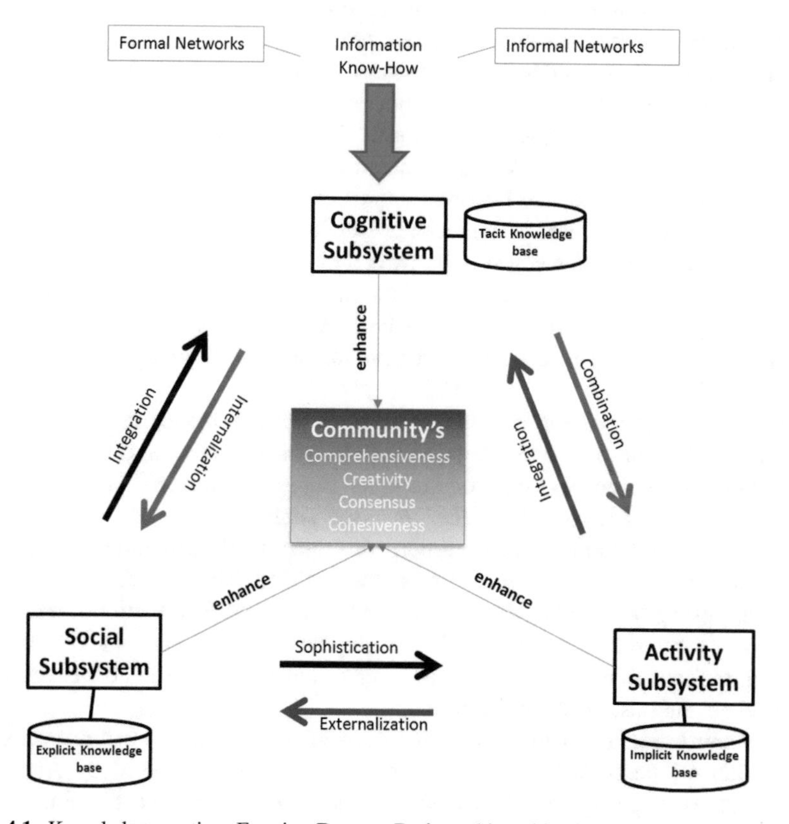

Fig. 4.1 Knowledge creation: Framing Buayan–Bario paddy cultivation example [22]

Indigenous Knowledge Government Framework (IKGF) and such studies of Indigenous cultures has strengthened our understanding of this process [23], in particular how knowledge is embedded in artefacts created by humans and socially constructed.

In Australia research has also shown there is an interlinking framework of holistic aspects that effect the governance and management of knowledge. Using the corroboree model above we have found many features of tacit knowledge sharing to be relevant, but the process is also multi-levelled [12]. The importance of features such as the right to know, the right to tell, and the relevance of knowledge to the specific setting result in a highly flexible sharing system where the governance is complex and tacit in itself.

The model of tacit knowledge sharing developed in this work is described in Sect. 4.4.8 as protocols to be followed in our design.

4.3 Training Professionals

The first application of Indigenous culture sharing was for training professionals who were conducting research and those providing engineering support for communities. There has been much misunderstanding and much harm done with the aim of progressing or integrating communities into mainstream Australian culture and economy. The role of those professionals is to act as gatekeepers for further researchers who wish to work with communities.

The first aspect is the holistic nature of knowledge, which is hard for non-Aboriginal researchers to appreciate. Each story told by a community member does not constitute knowledge on its own. A story requires understanding of the environmental and spiritual context and the thematic intent for it to be understood. This has caused problems when non-Aboriginal scientists seek knowledge from elders relating to care of the land and have not wished to engage with the whole story [24, 25].

In training engineers for work in communities, whether urban, rural or remote, we present the holistic nature of the project development such as the following table (Table 4.1) which covers some of the main issues experienced. We will discuss these in relation to language projects later (see Table 4.2).

Winschiers-Theophilus et al. report on their work providing training for researchers to prepare to work with communities [26]. They explain the process is about exploring the bonding aspects of community-researcher relationship and how this is key to the success of community development research projects. They run role playing sessions to emphasize the need to carry out research that is beneficial to the researchers and also to the community.

In running similar sessions we ensure our students understand and respect social and cultural aspects such as the kinship system, the method used by Aboriginal communities to specify the relationship between all people in the community, across clans, languages, etc. Through this system people are linked to their totem, and the responsibility for the preservation of that totem. The people of a certain totem will be

Table 4.1 Aspect for community engagement

Overall aspect	Project-based aspect	Possible obstacle
Governance	Control of land and resources	Access by all families to infrastructure
Existing resources to build on	Material suitable for work	Not possible to transplant projects between communities
Personnel or champions	Relationship to researchers	Researcher will not be accepted without a champion
History of occupation	Previous experience with resources e.g., IT	Lack of trust or competence in working with new systems
Cultural requirements	Men's and women's business	Need to work with stable portion of population
Relation to surrounding lands	Consultation requirements	Project may need to be delayed to reach agreement
Community responsibility	Relation to community plan	Time pressure on staff

distributed across Australia, and share responsibility for maintaining that animal, plant or resources and also will be responsible for maintaining the story that carried the knowledge for this to occur.

Another main use of the kinship system is for marriage. By specifying the group into which a person may marry to those who are most distant genetically, the process of procreation is maintained in a mobile and disperse group with maximum avoidance of inheriting recessive genetic diseases.

Working with the engineers who are going into communities, we are training people who are traders in knowledge. The kinships system, although it varies across languages, was used to assist any trader who enters new country, as they can be fitted into the local system through their kinship elsewhere. The value for external researchers is that when people come to the community, to trade in goods or knowledge, they can quickly find who are the people they should be talking to about business, who they can go to for advice, who to be cheeky to, and who they should not talk to as this crosses marriage boundaries. This includes a forbidden communication path between in-laws, a boundary that is commonly felt in many societies.

A face-to-face kinship training workshop has been recorded and is now used to train university students in various disciplines on the significance of the kinship grouping in community relations. Simple flash games are used to emulate aspects of the workshop (see Fig. 4.2). Related stories from community members are also collected and used as exemplars.

Throughout the online workshop the workshop instructor and stories from community members provide examples of how misunderstandings and assumptions affected their life and those around them. Similar histories by other Aboriginal people can be added at these junctures in the form of a playlist. The initial workshop provides a structure for the cultural training session, while the opportunity to interleave new stories helps younger people to gain a greater variety of experience in the history of this cultural clash.

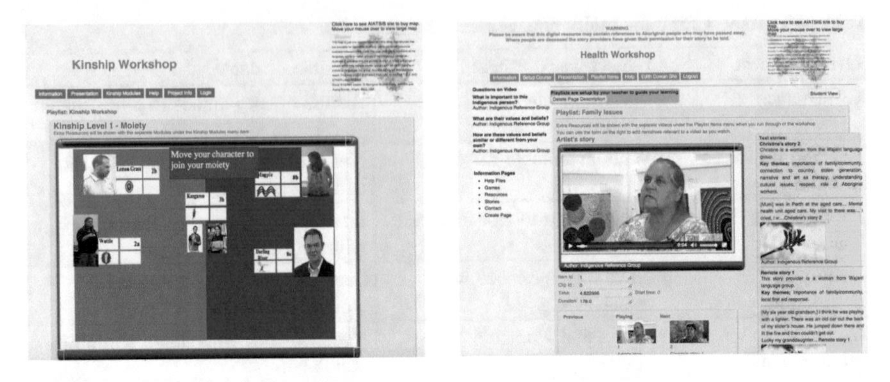

Fig. 4.2 Flash game for kinship site and health site showing playlist of stories

As the next step the community stories were provided as a playlist with option for the adding of further comments and videos. This was used to develop a site for teaching health professional.

Another similar project is the Bringing Them Home Oral History Project which includes many stories from the Stolen Generations online. The history of the campaign for recognition of the Stolen Generation was one of storytelling and illustrates the power of this oral process. When children were removed from their families, the parents and children affected all had stories, and feelings of guilt as they were often told they neglected their children. Social and geographical barriers prevented them from meeting to combine their stories, and it was only with the advent of the Link Up network to search family records that their stories also linked up, and that the real historical reasons and consequences of this period in Australian history became known outside Aboriginal communities [18, 27].

This Stolen Generations site can be used by external learning environments which link similar stories by common themes, such as place or time, and thus deliver to the user a combined series of stories that provide a greater depth of understanding of the issues expressed by the authors.

4.4 Developing Knowledge Sharing Interfaces

To enable the process of knowledge sharing we first consider how to use technology to help communities share their knowledge amongst themselves. This process is in its infancy, so we present here a few different approaches used, but show the similarity in the issues dealt with.

The application of this work is on language sharing to aid in revitalisation of the many languages of New South Wales (NSW). However we will also compare to similar studies for language reclamation for the Malaysian Penan and the use of knowledge sharing websites.

For Aboriginal languages, while much knowledge has been preserved, such as language recordings, there has been less provision of resources for Aboriginal people to control their own knowledge storage and distribution. Hence we need to apply knowledge about Indigenous knowledge sharing in how language resources can be represented in systems of ICT. Any system is required to provide a way for Indigenous people to first share their own knowledge amongst themselves so that they can control how it is shared externally.

The introduction of ICT into communities has been slow. The take up has been affected by the political, economic, and social conditions, and the historical focus on mobile technology. Recent focus has been on providing mobile apps for knowledge sharing such as language learning, however this has to be backed up with more robust and data-intensive application to store, analyse and serve the date, both on web services and through community computer based workshops [26, 28].

The process of setting up projects in community and the lack of cultural understanding in this process have led to the development of a website What Works [29] funded by the Jumbunna Indigenous House of Learning at the University of Technology in Sydney. This project is to campaign for proper protocols and procedures to be followed in funding and running projects.

4.4.1 Community Engagement Model

Zaman et al. and Hunter argue to design appropriate ICT tools for indigenous knowledge management, the information technology professionals carrying out the study need to understand the holistic nature of indigenous knowledge management and so model and formalise this within any project for technology design and approaches [30, 28]. This approach has led to a layered system of analysis [31] that incorporates the various interrelated components of the knowledge system (Capital, IK Governance, Activity, Knowledge Management, Data Repository and Community Engagement, all linked to the External Environment layer). This model was developed in the design of a botanical knowledge system [22].

The model considers the network of responsibilities and considerations required to run projects with communities as discussed in the design of the training modules in Sect. 4.3 above. The design team then used the design process for developing a knowledge management tool for the Oroo' language of the Penan. They ran workshops to develop a classificatory system for language words [28] (see Fig. 4.3).

The aspects shown in Table 4.2 that were extracted from Indigenous projects in Australia show components that are interlocking, rather than layers, but there are many similar features. Using the table provided in Sect. 4.3 above, we give an example of how issues effect the implementation of language reclamation projects. Then we consider how we worked on such a project to develop the present interface for language teaching (see Table 4.2).

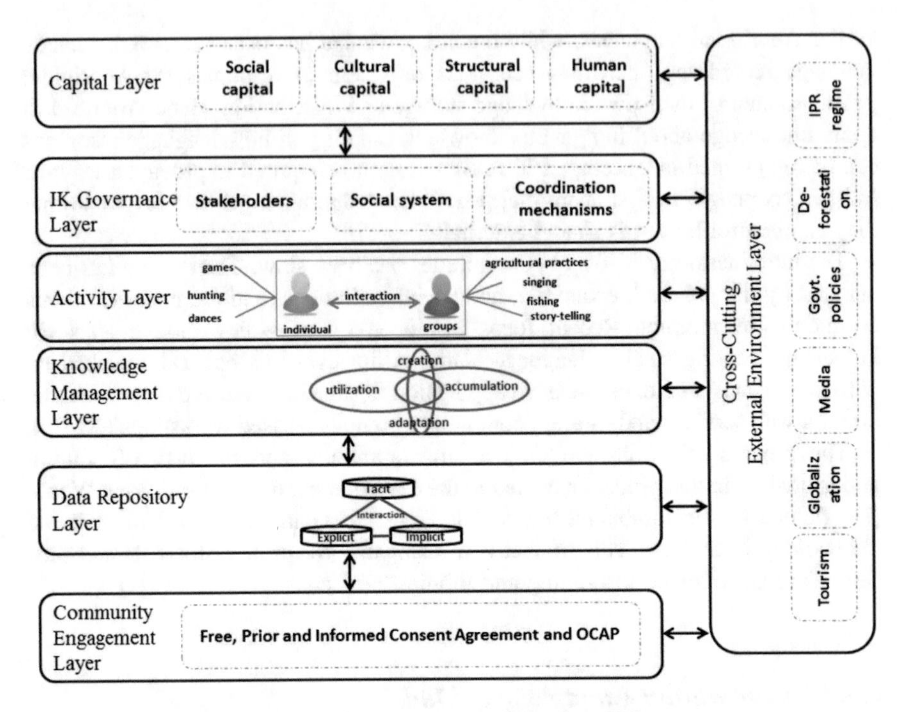

Fig. 4.3 The logical architecture view of a layered IKG system [31]

Table 4.2 Community engagement in relation to language reclamation

Overall aspect	Language reclamation and teaching
Governance	Different dialects will present with stronger spokespeople. Access to local computer centres may be controlled by single families making it hard for general access
Existing resources to build on	Some languages have archival recordings, some still have speakers. The variety of resources requires a flexible web service to collate these
Personnel or champions	The presence of community members with computer skills as well as language skills is necessary to engage other community members
History of occupation	If many people were stolen or imprisoned for speaking their language the memory makes progress difficult
Cultural requirements	Those who can speak the language may not be able to talk to each other due to kinship requirements. Negotiating how the language is reclaimed is slower
Relation to surrounding lands	Negotiating with other dialects in terms of how the different languages are incorporated or separated is important
Community responsibility	The community needs to develop the staff and local computer resources to maintain the material on the web themselves

The Penan model relates strongly to community governance and ensures that any project fits within this model, where the language model cuts across communities as an online and mobile model. At the same time we are also dealing with protocols for suitable consultation and negotiation with communities.

4.4.2 Previous Work on Interface Design

A study by George et al. [32] of urban Aboriginal people used Hofstede's [33] cultural model to analyse websites and provide a method of classifying salient features. They stated that cultural schema must be supported within a context before the culture can be conveyed. In our case the schema is the linkage of knowledge through story, the ability for community to contribute to develop the knowledge, and the levels of access to knowledge. This emphasis on the multiple layers of knowledge representation within the culture [34] is also reflected in work with the Penan in Malaysia discussed above.

Further research is needed for human computer interaction with different cultures. Workshops run with the Penan found that the older community members have different schemas for language classification to the younger members, which will make the development of a suitable interface complex, or requiring adaptation. Similarly workshops run with Aboriginal language speakers have shown that there are many different design needs for the representation of language online for different communities.

Another project enables the sharing of the 'alternate' Arandic sign language used in Central Australia, in various contexts by people who also use spoken language [35]. This required extensive community consultation on how the words are delineated and constructed, as well as how the signing should be authentically represented in an online environment.

The complex process of designing language repositories is repeated with every new project, as the communities deal with a variety of different environmental and social factors that provide a unique system of knowledge and understanding. We now apply the methods developed above and the patterns extracted to new situations for Aboriginal culture sharing and so establish the features of each specific site or module developed. In the final section we look at how this process developed into the creation of storytelling games.

4.4.3 Interface Design Model

A study was run on the way Aboriginal people viewed the online environment and how they could use this for knowledge sharing [6, 36]. In providing interfaces for community use, we are attempting to relate to existing knowledge sharing practises to reduce the cognitive load of the community members engaging with the system to learn what is a new language after years of denial of access.

Support for the users' external cognition arises from the interaction between internal and external representations when performing tasks that reduce the user's cognitive effort through the use of external representations, without reducing the information provided. This provided a model that both developed on the different layers of needs and the different design concepts listed by Rogers and Scaife [37], who developed guidelines for studying users interactions with different kinds of graphical representations (including diagrams, animations, and virtual reality) when carrying out cognitive tasks.

We used their properties and design dimensions to determine which kinds and combinations of graphical, audio and linkage representations would be effective for supporting different activities. The matrix of affordances provided the semantics of the interface pattern language (see Fig. 4.4). Pirolli and Card [38] conceptualized searching for and making sense of information, using concepts borrowed from evolution, biology, and anthropology together with classical information processing theory called the information foraging food-theory (IFT). They describe searching strategies in terms of making correct decisions on where to search next, influenced by the presence or absence of "scent."

We provide a conceptualisation or representation of searching, in this case for language information, from the perspective of Indigenous learning within the corroboree setting, where the re-enactment of the real environment assists the user in the construction of their knowledge.

Using the levels of an ecological framework [39] as levels of analysis on the left, we mapped this to the pattern attributes of content, context and cohesion of knowledge within the site. These were then mapped to the interface analysis

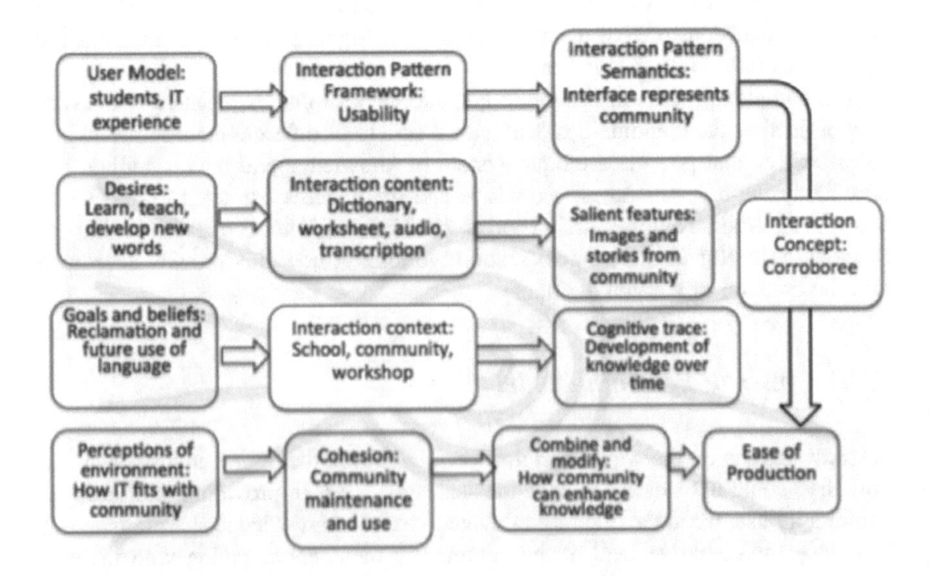

Fig. 4.4 Patterns for language site interactions [36]

techniques of semantics, salient features and cognitive trace, plus the final step of combining and modifying information on the site to form knowledge. In particular the ease of production of this knowledge as a sharable resource was the focus of the analysis framework.

The framework was then used to assist and evaluate the development of a language learning platform described in the next section. Here we introduce the process used to enact this pattern language on the web.

When working with knowledge artefacts, there will be no 'elders' or over-arching knowledge holders online to tie the information together into knowledge to be understood. In effect an online system provides isolated media packets from which the user has to draw sense. The interface design framework we provide here has been developed around this need to design tools for information selection (thematic content), the interface format (context), and information linkage (cohesion) to create a knowledge repository.

The aim of the visual design of knowledge learning web sites is to reduce the cognitive load in learning through:

1. Representations within real world, or juxtapositions that can represent processes in real life narratives. This provides the narrative for the user to follow.
2. Visual representations of temporal, cultural and spatial boundaries such as dialect, speaker, learning environment (worksheet, dictionary, etc.) that provide constraints and affordances to assist the learning enquiry, and so select the resultant artefacts.
3. Artefacts found for further enquiry, such as audio example, and the authority of different annotations available on these artefacts. This will vary with the thematic structure of the activity.
4. Graphical elements which provide affordances or constrain the kinds of inferences that can be made about the relevance of the search artefacts and relevance to their focus audience, including the level of language used in the document. This provides a context for the user's search activity.

We provide an example for the language sites discussed below, where the aim is to support both teachers who are searching for related material to present to students, and the students doing with their own searches to collate knowledge.

4.4.4 *Implementation for Language Learning Sites*

The design concepts that we developed from the studies described above were used to derive the component systems in the language learning sites we developed. These language sites [40] are based on a python system that provides easy access to the data for mobile apps through the application programming interface (api) and the ability to parse and analyse data uploaded to the site. Using a RESTful interface provides increased ease of access to the data [41].

The language sites have three main context components. Firstly the dictionary, usually developed by a linguist and transferred to a database. Then archival material in the form of audio and (possibly) transcripts are uploaded and linked to the dictionary words where possible as time-aligned text. These audio files are often in the form of songs which emphasise the imagery of the subject matter through sound [42]. Therefore much of the wealth of the material lies in the complete tape rather than any segmentation into word or phrase examples. Thirdly teachers can edit and save wordlists and use these to develop worksheets (see Fig. 4.5) which are exercises based on a topic or wordlist.

These worksheets are based on the Accelerate Second Language Acquisition method (ASLA) developed by Stephen Neyooxet Greymorning, an Arapaho teacher from Montana. This method is used at Muurrbay Aboriginal Language and Culture Cooperative for language teaching. This provides a context for the learning, where a wordlist is used each week and the learning exercises are based around this list and include actions (e.g. 'the girl touched the tree' is linked to a picture of this action).

Finally cohesion is supplied by software that links all parts of the web site, such as archival material, time-aligned text transcripts, further community contributed recordings, and text material.

The site has four functionalities to support the design proposed above. Firstly the data structure is highly interactive, allowing community to add resources which are then updated as part of the site information without requiring extensive tagging.

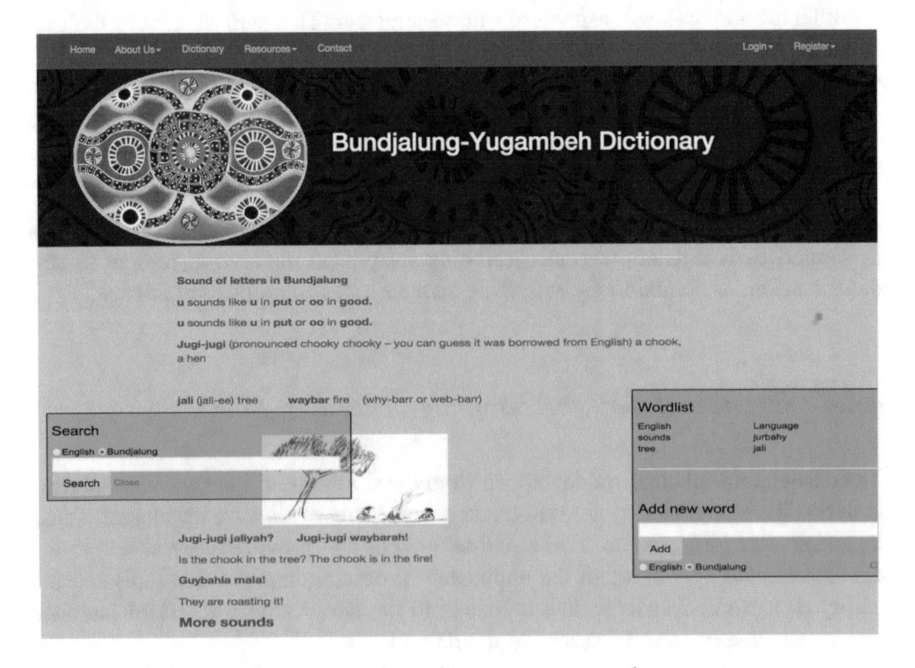

Fig. 4.5 Language site showing worksheet with popup support tools

Secondly there are functions that enable example sentences to be linked to ample recordings of their constituent phrase or word. Hence when linguist's transcripts include recordings that contain a searched phrase, these will be automatically linked to the words in the example. In the dictionary learners can search for words, and will be provided with example sentences linked to the word. The interface to time aligned segments of tapes is being developed to retain the full context as an option for the user.

Thirdly there are various JavaScript functions that provide support for users. Data hidden on the page can be accessed by JavaScript to verify the status and access of the user (e.g., whether teacher or learner) and then support can be tailored to their needs. For instance if a sound file exists for a group of words or a single word, it is shown as a link for listening. While working on a lesson resources are provided to search for a word (student), or creating a word list (teacher), and learners have access to the wordlist to help them complete lessons.

Finally the site provides an api for mobile users to access words and wordlists for weekly exercises, based on worksheets. This enables teachers to set up wordlists for users to practise on their mobile. The new mobile app being developed also relies much more on swiping to move around between different views of words (e.g., full text, word list, examples, memory game, etc.) to enable the user to change context while retaining their context.

Also another benefit of using python system is we have linked in the python natural language toolkit [43] to parse example sentence from linguists, and also those entered as answers by users. This provides functionality on the site for teachers to set interactive questions and some options for flexibility in the answers. However it is acknowledged that Aboriginal language structure is highly fluid, using markers to distinguish parts of speech and so sentence comparison can be difficult.

The site has been set up for some NSW languages and a separate interface used for each one to provide for material specific to that group, for instance in the Sydney Aboriginal community identity is often disputed so much of the site explains the groups genealogy and heritage. This issue arises as the population was decimated early after the English arrived, through disease and deliberate hunting down of clans. Also the language has few archival materials and a limited wordlist. The site supports learning language by providing recent recordings, and places an emphasis on linking wiki-style pages on local history and genealogy.

The web service code on github [44] is continually updated with a single system for the whole set of sites supporting the different functionality options. In this way the functionality and resources can be shared across sites with little resources, while retaining the different cultural needs of the different language groups.

4.4.5 Interface Design

Using the model in Sect. 4.4.3, we consider here one workshop where the language site was developed. We had with us the dictionary to be used on the site which included many example sentences, and audio archival tapes, as well as a language speaker present. The language centre has a series of images on various topics that we could use to move around to make scenes and activities. There was some transcription of tapes, but that was not in a searchable order. As mentioned we are using the ASLAN process of learning and the workshop was partly to consider how to support this online.

The main proposal was to provide weekly wordlists on mobile and the website on worksheets. The aim is to assist teachers to produce these and students to use and share them in their learning (ease of production).

We looked at the semantics of the website, how we are to create meaning in the language when many students and teachers have limited vocabulary. By understanding how users interact with the site we can assist them in this meaning making.

The interaction patterns with teachers working from existing offline resources showed that they would search the dictionary for word, and then seek their own dialect first from the options. As a second option they used neighbouring dialects. They would then ask others for the pronunciation, or seek an audio version. Then they would check a usage example, also in the dictionary. Finally they would give a changed example that related to their world.

The interaction content they used was the dictionary and each other. They also were keen to listen to the old tapes to hear 'how the language was supposed to be pronounced' but they did not see this as a resource to take apart or to make relevant to any particular topic. Hence the audio tapes do not provide a salient feature for specific language learning. We hope that with transcription this may change.

We also had some images that are shared between many language groups at the centre and teachers also produced ones they had made. There was an emphasis on the need to have images to point to and make the language more active.

The cognitive trace through the material was the theme or word they had chosen and were following up. The conversation may start with 'what is that word' and an attempt to say a half-remembered word, or it may be 'what was [a person] talking about the other day' or 'how do we say this'. The last format came more when planning a lesson, not so much out of community interest. Also the emphasis was to get the version for their own dialect if possible.

Then the content was combined to provide example sentences with images and sound to provide exercises the students could do where they spoke and listened to each other. When reproduced online this was seen as the need to share graphics, link to audio to help practice and for community to upload new audio when needed.

The cognitive load for a learner was considered under the four aspects above. The learners and teachers (who are also learning) are focused on reconstructing the language as close as possible to the last spoken form.

1. Representations within real world, or juxtapositions that can represent processes in real life narratives.
   The worksheet system is set up with editing tools, shared images and parsing support to assist the teacher creating the sheet and the learner following the material to link unfamiliar words with their example sentences and audio. The links are done within the sheet where possible or in sliding windows that follow down the page. This was to replace the verbal questioning used in the workshop.
2. Visual representations of temporal and spatial constraint that provide constraints and affordances to assist the learning enquiry, and select the resultant artefacts.
   The learner is provided with colour coded dialect options, and a map to show where the dialects are located, as they may wish to select for and use nearby words if no local one is available. They can also select to see matches that are exact, or from related words.
3. Artefacts found for further enquiry, such as audio example, and the authority of different annotations available on these artefacts. This will vary with the thematic structure of the activity.
   Audio examples are chosen by the dialect option used by the teacher or learner. Also the gender of the speaker can be chosen to obtain a speech tone that is easier to emulate. This option is not used yet as we only have a single speaker for most examples. At present we have an option to listen to both male and female if available.
   To assist in constructing wordlists we are including categories for most dictionary words which provides another search option.
4. Graphical elements which provide affordances or constrain the kinds of inferences that can be made about the relevance of the search artefacts and relevance to their focus audience.
   The dialects are shown in order of locality to the user. Also audio examples are sought as a word group and provided as a link to that group, before individual words, as the sound will change in context. This system has an added benefit that we are now constructing a separate website based on a single dialect. We will need to integrate the examples uploaded by this community with the examples from the original site with all dialects combined.

4.4.6 Grammar of Knowledge Sharing

In languages it is grammar that provides the cohesion in the transmission of knowledge. To provide more than just a system of information sharing we aim to provide a knowledge cohesion system that is respectful of the culture being shared.

The first aspect of the Aboriginal language that was instrumental in initiating the revitalisation process in NSW was the naming of place. This arises from the strong cultural tie to land and the fact that languages are used to name the land and create ties between people and land. This links to the way that Aboriginal knowledge is remembered and re-expressed as a story in place. For instance Langton [45] notes that through the cyclic nature of the kinship system, a person's mother's mother and father's father will be the same moiety, and hence will often relate in the same manner to the same country, which reinforces this link to place to the grandchild. Starting with this relationship mode we look at how to support cultural knowledge.

For thousands of years, Indigenous people have been sharing knowledge through oral means on how to live in and maintain both themselves and their physical and social environment. While much of this knowledge is now recorded, and some is available on the Internet, the online framework for this knowledge is highly unstructured. Research is being done on providing ontologies and frameworks that will provide online learning spaces for this knowledge, especially while retaining the oral format [6].

The conception of an oral storytelling grammar is to support the sharing of Aboriginal knowledge online while respecting the cultural representation of knowledge. It is recognised that Aboriginal people have avoided colonisation in many aspects of their culture while living within the mainstream (e.g. [46]) and wish to retain alternative means of living and knowing.

Online repositories of stories, supported by the cultural grammar, are becoming a learning tool for those within the culture, as well as those outside the culture, to increase their understanding. It also enables Aboriginal trainers to access resources from the community to provide a broader cultural training.

4.4.7 Syntax

The Indigenous Australian story telling is a communal form of oral history designed to fit the inheritance structure of a nearly non-hierarchical society. While social status is granted to people based on their skills and experience, this authority is shared with others of equal skill in other areas, as well as those with the same kinship and hence the same social and environmental responsibilities.

Aboriginal people use the group story telling process to select the stories that are valuable and hence worth repeating at ceremonies, which also determines what is retained over time. This is comparable to the social constructivist learning process [47]. However the particular theme in which a particular story is used may vary over time as priorities and events change.

Stories are placed in a story-path according to themes. These may be relating to morality, how to uphold the law or suffer penalties, with examples both from the Dreamtime and under the new non-Aboriginal law. Then the story path may be a point in space, a path in time or a story on a theme (see Fig. 4.6).

Alternatively, stories may relate to preservation of the land, and the processes used by ancestors which may relate to a path across the land that people can travel. The story will then describe in sequence the features, seasonal food etc. that can be found and the different aspects of the environment.

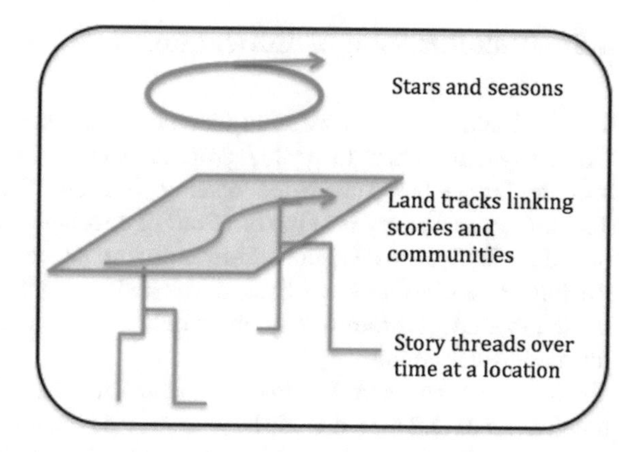

Fig. 4.6 Depiction of story components [6]

4.4.8 Protocols for Tacit Knowledge Sharing

The study of oral knowledge sharing begins with an understanding of the process of tacit knowledge sharing. This storytelling process is used in many Aboriginal Australian communities as a way to carry on knowledge, so it is instructive to understand what works and what protocols are needed to carry out this teaching. In the telling or retelling of a story there are various rules that have relevance to providing stories in a permanent online repository:

Authority to speak: A significant feature of traditional storytelling is only those with authority to speak are able to present a story. Authority comes from 'being there' in person or through a close relation, being part of the group (e.g. kin group) involved in the story or having some personal connection to the story [48].

Community narrative: When a law story is told, many people will contribute the part they know, that they have experience in, so first a theme is established, then many performers add their knowledge.

Deferral to others: When Aboriginal storytellers are speaking, they tend to include or invite other speakers into the story, either as a way of varying the story to keep the listener's attention, as a way of emphasising main points by getting corroboration, or to allow alternative view points to be expressed as a way to help the learner understand.

Knowledge is given not requested: While the teller of the story may start at any point in the narrative, it is their decision where to start. To elicit information a learner must give their knowledge first as a statement of understanding, rather than a question, so the teller knows where to start and how to direct their story.

We will use these criteria to evaluate the interactive tools that were developed. The analysis comes from a series of data collection, workshops, meetings and discussions. There was little opportunity for formalised study of the students or staff working with the system, however we were able to collect feedback from a variety of sources. Basing this on the traditional protocols provides a grounding to the evaluation.

4.5 Enculturation of Education

The final stage of the work is to incorporate the knowledge into all parts of the University curriculum to enable professionals in Australia to be trained in the specific Indigenous knowledge relating to their discipline. The University of Technology in Sydney is at the forefront of developing a University wide policy of including Indigenous Graduate Attributes in all courses. We present here two teaching systems that have been developed to teach Aboriginal experience and culture to students at university, in an online environment and designed to support reflection for learning.

As part of the move to provide modules for this program, we are using community videos and filmed workshops to provide the material to develop games. The first games were developed with Edith Cowan University in Western Australia and focused on experience with the Health system. This followed a story collection project by their Department of Medical Sciences, where community members were asked for their experience of health practice in the state. This project was part of research aimed at reducing the low health and high mortality amongst Aboriginal people. However, while researchers expected stories of mal-practise, the stories were more of lack of comprehension of the needs of Aboriginal patients. Hence while medical support provided for Indigenous patients is often incorrect, this is often due to communication issues, lack of understanding of the patients situation and a complexity of cultural assumptions, rather than deliberate actions.

These stories were also used to develop short scenarios on common issues, which were then filmed with actors. From monitoring teachers' use of the resources, the scenarios describing these general themes were accessed much more often than the individual videos. The teachers we talked to who were using the site stated that this was because there were more detailed teaching resources for these, and they provided a more clear elicitation of single themes. While the stories provided a complete and holistic overview of issues from the contributor, this formed a more complicated story for the (often non-indigenous) learner.

This section looks at the success of these videos and scenarios as a teaching tool, and compares them to interactive games that were developed in the subsequent project. The aim of the videos and the games that were developed was to engage students in cultural dialogue, which can be a challenge for teachers and students.

We grow up with many unconscious biases and when provided with information which is contrary to our beliefs, we resist this new view. Videos can provide the direct story from many different community members, but we do not believe they will necessarily affect people's thinking. If the views presented are not held by the viewer they will simply ignore or discredit the story teller. For instance of one video, many students claimed the speaker was racist as he pointed out that he was an expert on White Australians, whom he had observed for years [49].

Alternatively, games have the potential to break down the barriers, allowing new information in a less threatening way, creating interaction with other gamers or programmed characters and scenarios in the game. However programmed or filmed scenarios will always be representations of real experiences, while multiplayer games where users interact with real people online provide a more authentic representation of the issues. Hence as having storytellers online is not feasible, we tried to incorporate the stories into the game, not just through scenarios, but also as a sequence of videos selected on themes.

In this research we combine the gaming technology with community stories that confront standard views. We develop a tool that teachers can use to modify the material presented, so that they can engage students in interactive games that simultaneously provide a context relevant to their learning and challenge them to think differently.

The project sites are teaching Aboriginal experiences and the effect the culture has on Aboriginal people's values and expectations. There are two sites developed, under the themes of Health and Education. A third is being developed for general cultural communication training across disciplines. Community stories were collected for these themes, to provide an authentic resource for students to learn about other people. For the health site these videos stories are used as the introduction material, with additional comments and questions to guide learning, provided by the project group collecting the stories.

To provide a greater breadth of experience and variety in content for the students we allow community members to continue to upload stories to the repository which can then be selected by teachers as exemplars on the material. For the education and cultural training site we are using existing workshops as the introduction then linking this to community stories to provide more immersive examples.

As we have explained the concept for using stories for knowledge sharing arose from the traditional approach used by Aboriginal people and we continued to adapt these to provide an online learning environment where students can be encouraged to reflect. When considering reflection in learning, we note that stories work on many levels, and how the depth of reflection by the listener will determine the richness of their understanding of the story. There are three basic layers that are present in an Aboriginal traditional story and which can be extracted as people become more experienced in the subject material [11]:

1. Teaching young people about the natural environment
2. Teaching about the relationships between people in a community
3. Teaching about the relationship between a community and the wider world (i.e. the earth and other Aboriginal communities).

As well as this many stories have a spiritual meaning that does not translate well online. With the stories provided in a modern context relating to Aboriginal experience, we find these three same layers of meaning can be discussed in relation to these resources. While we cannot direct the learning, we are relying on the teachers who have read the teaching guides to understand and attempt to steer the students into deeper reflection and consider how games can support this process.

4.5.1 Online Environment Features

We already have some experience of how video story material can be used in learning and the sort of support students might need [50]. We also realised from the initial work that the role play scenarios provided some of the best teaching material, and could be linked more closely to reflective questions.

Using a game engine we can script new scenarios, add questions and multiple choice answer segments with scoring, and repeat scenarios with slight edits. We used reflective questions, as it is believed students offered an extrinsic reward for finding the correct answer were less effective at problem-solving and less confident [51].

The web service developed provides an integrated environment for workshops material, community videos, teacher's playlists, teaching resources, students comments (moderated) and teaching questions to be provided seamlessly to students. The site works similar to blackboard in that a teacher will set up their course and direct their students within that course.

For a course teachers can select from available material: questions; video playlist; comments and teaching resources; as well as help documents; or add their own. In particular they can edit their own playlist or copy the list from a similar course. The focus is on having a shared repository of resources so the teachers can support each other. Also students can guide each other by recommending videos.

We use the Unity game engine, which provides the ability to have the videos to be embedded in a scene, and the questions to be asked of players as they move through the scene, to be accessed from an external website with an api. As the mobile language app can access the python site for a weekly list and examples, the Unity games online can access the python storytellers site for the teacher's list of videos and questions to be shown in the game for their course.

On top of this we wanted to provide support for the reflective steps and the students journal writing, so that their thoughts could be recorded and used later for reflection. The aim was to encourage students to engage with each other, at least within a cohort, to provide support and advise.

The Unity game engine does include an option for networked multiplayer games, where students could share descriptions and reflective comments in a chat session as they navigate a scenario. However there was concern from teachers that such a group could become quite toxic online so we are wary of putting comments online without moderation in what is a volatile area. Thus the process of learning from stories has raised issues relating to cultural safety, of those in the class and those who might read other class notes.

4.5.2 Cultural Safety

The cultural safety aspects of how subjects are raised in class, and how this will affect other students are discussed in resources provided to teachers so they can understand and handle this through the website. We wish to encourage reflection without creating a hostile or critical environment and emphasize the reflective process rather than directly challenge specific content the students bring to class [52]. Most importantly we encourage students learning from each other [53].

To set up such a learning environment online requires setting up clear guidelines to the students of how they may behave towards their peers. The current web system has only been used in conjunction with face-to-face classes, so the existing resources on cultural safety, plus class discussions, have been sufficient. However we may need to provide an environment with some auto-monitoring through language processing resources for teaching fully online.

4.5.3 Features

The system is available open source [54]. For the online environment we are providing support for:

- Aboriginal authors to provide questions with suggested answers for reflection on their stories.
- Teachers to select an ordered playlist of stories for their students and edit questions, answers, resources and help material for their course.
- Students to select stories they think most relevant to their peers and comment on these.
- Multimedia staff to script further scenarios in the Unity 3D gaming environment.

The online stories and scripted scenarios can be used as a complete learning system where the students are directed to playlist and game which they then use to develop their reflective journal, or it can be part of a classroom where the teacher proposes questions and topics for group discussion.

4.5.4 3D Gaming Environment

To enhance the video playlists that teachers can use, we are using a gaming environment to provide some of the features of a simulated environment or actor-role play. It has been shown [51] that games support learning in that, if a learning task is set in the context of a story, independent of the student's preference for that story, the student will learn more. Also students preferred story-context problems to unembellished ones. Hence we are using the Aboriginal stories as part of the game, but also trying to link these into summative scenarios.

For reflective assessment we are not utilising game scoring tools but user records can be kept, such as the videos already viewed and how much was watched. Also students can record comments and forward them to their own email or a tutor through the website.

While scripted scenarios lack the interactive features of gaming, they do provide material similar to cutscenes that can be incorporated in a games environment as we enhance the resources. The scenarios we are using were those scripted and recorded as films in a prior project, so we can use the audio and by using game characters they can be 're-run' with various features altered, such as the number of onlookers. We then ask the students how they would respond to scenarios with more witnesses.

The scenarios are not intended as a depiction of reality, they are a chance for students to consider issues relating to the learning goals, and they are encouraged to submit their reflection to share with other users. The teacher moderates the public viewing of these submissions.

4.5.5 Interactive Components of Gaming

We chose a gaming environment rather than a simulation environment as we wanted to encourage students to immerse themselves in an environment where we could set up complex interactions. While the online version cannot include autonomous agent plugins, there is an option to extend the system to games for download to the computer. In these versions we are developing agent rules to enact Kinship roles and various protocols and greetings used during interactions with Aboriginal people in the professional situation. The agents system is FAtiMA developed at GAIPS, INESC-ID in Lisbon [55].

The agent modelling incorporates different modules, including implementations of: the OCC model for relative emotional states of characters [56], Hofstede's model of cultural attributes [33], and social importance [57].

The first interactive game developed and tested was a short game for health students and showed Aboriginal people using a modern version of kinship based on modern societal roles. In the scene the agents had the role of school or health clinic staff, and their inter-relation and status depended on this role. The player is required to negotiate the people at the clinic, including their workmate from the school, for the information they need about the clinic. This scene uses the idea of kinship and relatedness for what knowledge one can receive, but also the modern aspect of respect of cultural norms, to allow the player to gain social status and so be granted the information they seek.

This is the first application of the agent system to present encultured agents in the Indigenous context. To provide a more complex representation of the culture we will have to redesign certain aspects of the agents system. However the social status system has enabled us to provide examples to community members at workshops and obtain positive feedback that this process can provide some useful learning interactions.

4.6 Teaching About Culture

From the experience of the two programs we worked with for teaching Aboriginal culture we have found that the role play process and reflection based on video stories were both successful in encouraging reflection in the face-to-face context.

However it is hard for any teacher to take the topics discussed in the videos and provide an interesting and stimulating course, particularly as many teachers lack any background knowledge in this area. With the present emphasis on Indigenising the curriculum there are a growing number of such courses. Hence the aim is to develop resources that can be shared between teachers on sites and use the success of these sites to collect more material from community as teaching resources.

This is particularly important to Aboriginal people who request that they be the ones to teach their culture, rather than bringing in 'experts' from the European culture to teach. Certainly few teachers would feel confident to teach a culture that was not their own. However there are often few local Aboriginal people with the confidence to provide a course to students, and they may be reticent to speak others stories. Having the story spoken and presented by the person with authority will reduce this problem.

The sharing of stories online we hope will provide a resource that can be shared at least within the state or region where the stories are collected, although not across the country due to culturally and historically different experiences. This provides the material from which to build a corroboree online. The community contributed stories are then reused by many courses, as the teachers can select the ones most relevant to their domain and can order them in the way they think their students will most benefit from the content. Also teachers can add questions and responses to the list provided by the story author. Finally the scenario editing system provides games that can be shared between courses, as the issues that are handled in such games are usually fairly generic to all disciplines.

However it is hard for teachers to directly edit a scenario game, and it is hard for Aboriginal contributors to evaluate the game, such as interactive scenarios relating to communication protocols, unless they are experienced gamers. We are developing visual tools to reduce the need to programming skills for the creation of games, but also provides a format where the experienced editor can train others in how the interactions work [50].

4.6.1 The Learning Process

In teaching the students about reflective writing or self-authorship at the start of the health course, we use the definition by Sandars [58].

> Reflection is a metacognitive process that occurs before, during, and after situations with the purpose of developing greater understanding of both the self and the situation.

This is a difficult process to teach, and much of the teaching occurs at present in the class with group discussion so it will be hard to support this learning online, without some structured scaffolding for students, such as many questions as guidance early on [59], or using a blog to share ideas and comments to support peer-to-peer learning.

Also at present in the face to face course students have the opportunity to submit drafts of their journal for early comment, so this can continue, with the use of a facility to email tutor with their work from the learning site.

However the online system does have the added advantage, mentioned about, of allowing teachers to provide slight variations in scenarios and repeat them to ask the students to consider how this would affect their response to the situation. Also the scenario can be stopped at some points and the student asked to consider their next step, before the scenario script moves on to provide a possible response.

Also the online system allows us and teachers to add online resources within the learning environment. From previous experience teachers tend to use the scenarios more in teaching than the video stories, and this may be due to the teaching notes that have been developed for these, or because they provide a more succinct version of the story concepts.

When supporting reflection we need to be clear what the students will gain from this process. In the present course reflection the following learning process is followed [49], which we aim to extend to reflective narrative learning [60]:

- Learn from and integrate previous experiences into broader narrative of their life
- Understand their strengths and weaknesses, mistakes and successes
- Challenge underlying assumptions, values and beliefs
- Recognise areas of bias or discrimination
- Acknowledge and face their fears.

In providing online learning environment we have less access to the individual's personal beliefs and experiences, but we do have access to a greater range of stories to which they may be able to relate. Also we can use the peer environment to provide comments and recommendations to others in their professional field as to which stories to watch.

4.6.2 Stages of Reflection

Another way to support students is to guide them through the steps in a reflective process so that this can become their automated response to new situations [61]. The Edith Cowan health course discussed here is using the Gibbs [62] Reflective Cycle (see Fig. 4.7). It is the emotive aspects of reflective learning that will translate least well into a gaming environment.

Fig. 4.7 Gibbs reflective cycle 1998

However the other steps can be provided through the website workflow. As well as providing a series of steps for the student to follow, each exercise can be phrased to be either:

Confronting to force the student to reflect: following the pedagogy of discomfort Michalinos Zembylas [63]; or

Constructivist: Leading them from their own experience, to see from another.

This process will encourage them to consider the views of the storytellers, and so consider their own lack of knowledge. For the final assessment they are required to write an essay on a chosen topic. Racism manifests as non-reflective practise, and hence it is easy to provide feedback to students that highlights the problems with their learning. However when the group discussion is online and highly limited, we do need to monitor whether being isolated from the class during the online learning reduces the options for reflection in students who start with fixed views.

There is an added possible advantage for online learning in that the class group will not be exposed to the immediate reactions to the stories and scenarios of those who are intolerant of other people. This is an aspect of cultural safety which is difficult for a teacher to handle, whether to react and correct the student for their views, or allow them to discuss and develop.

4.6.3 Evaluating the Learning

The face-to-face classes are run at many locations using the online video resources, and the wholly online system is also to be used across many campuses. However we did run some sessions at one campus to evaluate both systems with students and teachers. The issues that arose in discussion and analysis of reflective learning were:

Changing culture: The stories developed for Western Australia were not suitable for teaching culture in NSW as the experience was very different and the approach to dealing with the long term issues is quite different.

Interaction: The online discussion was greatly reduced and we need more tools to allow comments and teachers to easily moderate to increase the flow of discussion. This can be compared to class discussion where the lack of understanding expressed by some students may offend other students.

Privacy of comments: At present comments are private, and the level of discussion with the tutor by individual students online can be more open and more reflective from the start. Being written rather than spoken allows more thought in some cases, and the learning material was open on the screen as they wrote responds to questions.

Loss of feedback on student's individual views: In going from the class to online there was a reduction in the amount of feedback from students. The reduced chance for discussion and expressing ideas reduced the peer-to-peer learning that supports reflection.

Different comments when altering scenario slightly: Perhaps the most successful in terms of reflective response was the use of similar scenarios and then asking the students to reconsider their actions or feelings in the new situation.

Limitations of game versus films: The use of games as a representation of the culture was compared with disfavour to films by some teachers as they were not as 'real'. Students did not raise this, maybe as they are more used to this medium.

4.6.4 Evaluating Tacit Protocols

The second part of the evaluation is to verify adherence to the design principles developed in this work. We list here the features developed for the language (audio) and storytelling (video) sites as there is much overlap in the process.

Authority of speaker: In the language project we need to get permission to use any archival tapes from the eldest living relative/descendant. For the recent recordings of sound and stories we are hoping to run more workshops to have community upload and authorise their own story.

However, we are also working with a team to develop a suitable transcription tool to allow segmentation to extract words and phrases for the dictionary interface, and the present video interface allow segmentation to interleave questions into the video. We will need to retain information on the speaker when each segment of an audio recordings are used. For video each time a speaker first presents a story to a learner, their introduction is included, but subsequent appearances of this speaker are not introduced.

Community narrative: The site can collect a continually updated series of stories that will be tagged with topics relevant to that learning domain. This will provide a student with a variety of formats for information, and a collection of stories from a variety of view points, time periods and experiences. These will more likely provide material that stimulates thought for different learners.

Deferral to others: For the language sites, where there are many versions of a word, including audio versions, we allow various forms to be shown or linked as audio, and the community of language learners and teachers will verify or moderate these.

For the video stories, we are collecting group stories as well as individual stories so that this process of deferral to others knowledge can be included. Also by linking stories by tags we hope to provide some connection between stories for the students.

Knowledge is given not requested: We are encouraging teachers to develop their own thematic lessons and utilise the material as they wish. The themes or language they know and understand is the material they will teach best with. While teachers can share language sheets or playlists of videos, we expect they will use their own where possible.

We would prefer not to edit stories, but realise that longer stories may not be viewed by students. Hence we encourage chopping videos into segments and pausing playlists to ask questions of the students. Also for these questions we are providing answers suggested by the Aboriginal teams on the projects, which are shown after the student considers their own ideas.

4.7 Conclusion

This chapter encompasses a wide range of resources focused on sharing Aboriginal culture online, all developed in line with the analysis of the knowledge sharing process discussed at the start of the chapter. The work is quite broad, traversing cultural teaching through workshops, language and community stories. However the focus is on computer support for oral learning as a way to provide teaching resources that can adapt to the learners and the teacher's focus.

We have also attempted to implement the traditional culture of Australian Aboriginal people into the teaching process wherever possible, not just through material, but also method. The work is also designed to be adaptable by teachers to suit the variety of Aboriginal cultures and histories in Australia.

However the important aspect of learning in games is challenge-based or experiential learning where the student devises strategies to get to their goal. For this we need to be clear about their final goal in the game and provide support for them to learn how to achieve this. In the process we can allow them to indirectly encounter views that challenge their own view.

Through this immersion in culture, thorough language sites, wikis on different peoples, and direct training systems, we hope to improve cultural awareness in Australia of the Indigenous people and the effect of the history of invasion and genocide in this country.

References

1. Clarke, Robin L.: Pollen and charcoal evidence for the effects of Aboriginal burning on vegetation in Australia. Archaeol. Oceania, **18** (1983)
2. ABC: Q&A on ABC 1 June 2015. http://iview.abc.net.au/programs/qanda/FA1407H018S00 (2015). Retrieved 02 June 2015
3. Rosenzwig, R., Thelen, D.: The Presence of the Past Popular Uses of History in America. Columbia University Press, New York (1998)
4. Mellor, D.: Artefacts of memory: oral histories in archival institutions. Humanit. Res. **III**, 1 (2001)
5. Bell, D.: Ngarrindjeri Wurruwarrin: A World That is, Was, and Will be. Spinifex Press, VIC (1998)
6. Kutay, C., Ho, P.: Australian Aboriginal grammar used in knowledge sharing. In: Proceedings of IADIS International Conference on Cognition and Exploratory Learning in Digital Age (CELDA 2009), Rome, Italy, November (2009)
7. Verran, H., Christie, M.: Using/Designing digital technologies of representation in Aboriginal Australian knowledge practice. Hum. Tech. **3**(2), 214–227, May (2007)
8. Reece, G., Nesbitt, K., Gillard, P., Donovan, M.: Identifying cultural design requirements for an Australian Indigenous website. In: Proceedings of the 11th Australian User Interface Conference, Brisbane (2010)
9. Soro A., Brereton M., Taylor J.E., Lee Hong A.G., Roe P.: Cross-cultural dialogical probes. In: Proceedings of the First African Conference on Human Computer Interaction, pp.114–125 (2016)
10. Kutay, C.: Tacit knowledge sharing, UTS YouTube. Retreived from https://www.youtube.com/watch?v=pb8LZyFcbCc (2017)
11. Sveiby K.-E., Skuthorpe, T.: Treading Lightly, Allen and Unwin (2006)
12. Kutay, C.: Trust Online for Information Sharing, KMIS, Barcelona, October 4–7 (2012)
13. Kutay, C.: One Person's Culture is Another One's Entertainment. In: Blackmore K., Nesbitt K., Smith S.P. (eds.) Proceedings of the 2014 Conference on Interactive Entertainment (IE2014). ACM, New York, NY (2014)
14. Nakata, M. et al.: Australian Indigenous Digital Collections: First generation issues. http://hdl.handle.net/2100/631 (2008). Retrieved 10 Oct 2008
15. Kutay, C.: Issues for Australian Indigenous culture online. In: Blanchard, E., Allard, D. (Eds.) Handbook of Research on Culturally-Aware Information Technology: Perspectives and Models. IGI Global (2009)
16. Watson, L.: Recognition of Indigenous Terms of Reference: A contribution to change. Keynote address presented at Cooperation out of Conflict Conference; Creating Difference; Embracing Equality, Hobart 21, 4 September (2004)
17. Finalyson, J.D.: Sustaining memories: the status of oral and written evidence in native title claims, In Finlayson, J.D., Rigsby, B., Bek H.J. (Eds.) Connections in Native Title: Genealogies, Kinship and Group, Centre for Aboriginal Economic Policy Research, Canberra. Research Monograph, No. 13. pp. 85–98 (1999)
18. Attwood, B.: Understandings of the Aboriginal past: history or myth. Aust. J. Polit. Hist. **34** (2), 271 (1988)
19. Gray, P.R.A.: Saying it like it is: oral traditions, legal systems and records. Arch. Manuscripts **26**(2), 248–269 (1998)
20. Minoru, H.: Reading oral histories from the pastoral frontier: a critical revision. J. Aust. Stud. **26**(72), 21–28 (2002)
21. Yunkaporta, T.: Aboriginal pedagogies at the cultural interface. Ph.D. thesis, James Cook University. http://eprints.jcu.edu.au/10974/ (2009). Retrieved 12 Feb 2011
22. Zaman, T., Yeo, A.W., Kulathuramaiyer, N. Harnessing Community's Creative Expression and Indigenous Wisdom to Create Value: Tacit-Implicit-Explicit (TIE) Knowledge Creation

Model. IKTC2011: Embracing Indigenous Knowledge Systems in a New Technology Design Paradigm (2011)
23. Zaman, T., Yeo, A.W., Kulathuramaiyer, N.. Indigenous Knowledge Governance Framework (IKGF): A holistic model for indigenous knowledge management. In: Paper presented at the Second International Conference on User Science and Engineering (i-USEr2011) Doctoral Consortium, Kuala Lumpur (2011)
24. Christie, M.: Grounded and ex-centric knowledges: exploring Aboriginal alternatives to western thinking. In: Edwards, J. (Ed.) Thinking: International Interdisciplinary Perspectives, Hawker Brownlow Education, Victoria (1994)
25. Pascoe, B.: Dark Emu. Black Seeds; Agriculture or accident? Magabala Books, Broome (2014)
26. Winschiers-Theophilus, H., Zaman, T., Yeo, A. Reducing "white elephant" ICT4D projects: a community-researcher engagement. In: Proceedings of the 7th International Conference on Communities and Technologies (C&T '15). ACM, New York, NY, USA, pp. 99–107 (2015). doi:10.1145/2768545.2768554
27. Attwood, B.: Learning about the truth: the stolen generations narrative. In: Attwood, B., Magowan, F. (eds.) Telling Stories, pp. 183–212, Allen & Unwin, Crows Nest, NSW (2001)
28. Zaman, T., Winschiers-Theophilus, H.: Penan's Oroo' Short Message Signs (PO-SMS): Co-design of a Digital Jungle Sign Language Application. Presented at INTERACT (2015)
29. Cox, E. and Kutay, C.: What works in indigenous programs and policy. Developed at Jumbunna Indigenous House of Learning UTS. Available at http://www.whatworks.org.au
30. Hunter, J.: The role of information technologies in Indigenous knowledge management. In: Nakata, M., Langton, M. (Eds.) Indigenous Knowledge and Libraries. Australian Academic & Researchers Libraries: Canberra, July (2005)
31. Zaman, T., Yeo, A.W., Kulathuramaiyer, N.: Introducing Indigenous Knowledge Governance into ICT-based Indigenous Knowledge Management System. In: IPID 8th International Annual Symposium 2013, 9–10th December, University of Cape Town, Cape Town, South Africa (2013)
32. George, R., Nesbitt, K., Donovan, M., Maynard, J.: Evaluating indigenous design features using cultural dimensions. In: Proceedings of the Thirteenth Australasian User Interface Conference (AUIC2012), Melbourne, Australia (2012)
33. Hofstede, G.: Cultures and Organisations. McGraw-Hill, London (1991)
34. Pumpa, M., Wyeld, T.G.: Database and narratological representation of Australian Aboriginal knowledge as information visualisation using a game engine. In: Tenth International Conference on Information Visualization (IV'06), IEEE Computer Society, London, United Kingdom, 5–7 July (2006)
35. Green, J., Woods, G., Foley, B.: Looking at language: appropriate design for sign language resources in remote Australian Indigenous communities. In: Thieberger, N., Barwick, L., Billington, R., Vaughan, J. (Eds.) Sustainable Data from Digital Research. Humanities Perspective on Digital Research. Custom Book Centre, University of Melbourne, (2011)
36. Kutay, C.: HCI study for culturally useful knowledge sharing. In: Proceedings of the 1st International Symposium on Knowledge Management & e-Learning (KMEL) Hong Kong, (2011)
37. Rogers, Y., Scaife, M.: How can interactive multimedia facilitate learning? In: Lee, J. (ed.) Intelligence and Multi Modality in Multimedia Interfaces: Research and Applications, pp. 68–89. AAAl Press, Menlo Park, CA (1998)
38. Pirolli, P., Card, S.: The evolutionary ecology of information foraging. Technical Report, UIR-R97-0 l. Palo Alto, CA: Xerox PARCo (1997)
39. Bishop, J.: Increasing participation in online communities. In a framework for human–computer interaction. Comput. Hum. Behav. **23**, 1881–1893 (2007)
40. Dalang.: http://www.dalang.com.au (2015). Retrieved 10 Aug 2015
41. Cassidy, S.: A RESTful interface to annotations on the web. In: 2nd Linguistic Annotation Workshop, Morroco, May 2008

42. Magowan, F.: Crying to Remember. In: Attwood, Bain, Magowan, Fiona (eds.) Telling Stories. Crows Nest, Allen and Unwin (2001)
43. Bird, S.: Natural language toolkit. http://www.nltk.org/ . Retrieved 10 July 2015
44. Kutay, C.: Language software for providing language dictionary or wordlist, wiki pages and archival material with tools to support teachers developing learning resources. Available on github https://github.com/ckutay/Language (2015). Retrieved 10 Mar 2015
45. Langton, M.: Grandmothers' Law, Company Business and Succession in Changing Aboriginal Land Tenure Systems. In: Yunipingu, Galarrwuy (ed.) Our Land is Our Life, pp. 84–117. University of Queensland Press, St Lucia, Queensland (1997)
46. Schwab, R.R: The calculus of reciprocity: principles and implications of Aboriginal sharing. Centre for Aboriginal Economic Policy Research Discussion Paper No 100 (1995)
47. Berger, P.L., Luckmann, T.: The Social Construction of Reality: A Treatise in the Sociology of Knowledge. Anchor Books, Garden City, NY (1966)
48. Povinelli, E.: Labour's Lot: The power, History and Culture of Aboriginal Action, University of Chicago Press (1993)
49. Alexander, R: per. comm. Edith Cowan University (2013)
50. Kutay, C., Sim, M., Wain, T.: Support for non IT savvy teachers to incorporate games. In: Wang, Jhing-Fa, Lau, Rynson (eds.) Advances in Web-Based Learning ICWL, pp. 141–151. Springer, Kenting Taiwan (2013)
51. Lepper, M.R., Cordova, D.I.: A desire to be taught: Instructional consequences of intrinsic motivation. Motiv. Emot. **16**(3), 187–208 (1992)
52. Kutay, C., Leigh, E.: Intercultural Competence as a Personal Security and Social Issue. To be presented at SIMTecT August, (2015)
53. King, J.: IDEA Item no 10. Asked students to help each other understand ideas or concepts http://ideaedu.org/research-and-papers/pod-idea-notes-on-instruction/idea-item-no-18/ (2015). Retrieved 03 Aug 2015
54. Intertac.: Pylons software for sharing stories and games online, available on SourceForge https://sourceforge.net/projects/intertac/?source=directory (2015). Retrieved 01 Apr 2015
55. Dias, J., Mascarenhas, S., Paiva, A.: FAtiMA modular: towards an agent architecture with a generic appraisal framework. In: Workshop in Standards in Emotion Modeling, Leiden: Netherlands (2011)
56. Ortony, A., Clore, G., Collins, A.: The Cognitive Structure of Emotions. Cambridge University Press, New York (1988)
57. Kutay, C., Mascarenhas, S., Paiva, A., Prada, R.: Intercultural-role plays for e-learning using emotive agents. In: Joaquim Filipe, Ana L. N. Fred (Eds.) ICAART 2013—Proceedings of the 5th International Conference on Agents and Artificial Intelligence, vol 2, Barcelona, Spain, 15–18 February, 2013, pp. 395–400, SciTePress, (2013)
58. Sandars, J.: The Use of Reflection in Medical Education: AMEE Guide No. 44. Med. Teach. **31**(8), 685–695 (2009)
59. Dyment, J.E, O'Connell, T.S.: Assessing the quality of reflection in student journals. Teaching in Higher Education **16**(1), 81–97 (2011)
60. Eynon, B., Gambino, L.M., Török, J.:. Reflection, Integration, and ePortfolio Pedagogy. Retrieved from http://c2l.mcnrc.org/pedagogy/ped-analysis/
61. Boud, D., Keogh, R., Walker, D.: Reflection: Turning Experience into Learning. London, Kogan Page (1985)
62. Gibbs, G.: Learning by Doing: A Guide to Teaching and Learning Methods. Further Educational Unit, Oxford Polytechnic, Oxford (1988)
63. Zembylas, Michalinos: Witnessing in the classroom: the ethics and politics of affect. Educ. Theor. **56**(3), 305–324 (2006)

Chapter 5
Culturally-Aware Healthcare Systems

Langxuan Yin and Timothy Bickmore

Abstract Cultural congruity between patients and healthcare providers has been demonstrated to be an important factor in maximizing the efficacy of healthcare. Similarly, the targeting and tailoring of health messages for a particular culture has been shown to increase their impact on patients, compared to generic messages. These findings indicate the importance of culture in designing messages, interventions, and care protocols intended to increase population health, especially for minority cultures for which generic messages and interventions represent a cultural mismatch. As our healthcare processes become automated—to increase access, decrease cost, and improve care—attention to cultural cues and their effects becomes more and more important. While cultural cues can be encoded in very subtle ways in any computer interface, embodied conversational agents provide a health communication medium in which culture can be explicitly encoded in order to achieve the same benefits of cultural congruity and tailoring seen in human-human interactions. In this chapter we review research on cultural congruity and tailoring in traditional medicine, and how these effects can be achieved with conversational agents. We present the results of an empirical study of the effects of cultural and linguistic tailoring of an animated exercise coach on user ratings of the coach's trustworthiness and persuasiveness. We then review two large-scale systems for longitudinal health behavior change intervention which feature conversational agent-based health coaches tailored for specific minority cultures: Latinos in Northern California and those in the Boston area. We also review results from a pilot study on the effectiveness of one of these systems in promoting physical activity over a four-month period of time. We close the chapter with a research agenda and challenges for future work in culturally-tailored conversational healthcare agents.

L. Yin (✉)
50 Christopher Columbus Dr. Apt 2705, Jersey, NJ 07302, USA
e-mail: omnific9@gmail.com

T. Bickmore
College of Computer and Information Science, Northeastern University,
360 Huntington Avenue, 910-177, Boston, MA 02115, USA

C. Faucher (ed.), *Advances in Culturally-Aware Intelligent Systems and in Cross-Cultural Psychological Studies*, Intelligent Systems Reference Library 134, https://doi.org/10.1007/978-3-319-67024-9_5

Keywords Intelligent virtual agents · Tailored health messages
Behavioral change

5.1 Introduction

Culture is key in medicine. Numerous studies have demonstrated the importance of culture in face-to-face doctor-patient interactions, showing that cultural differences between doctors and their patients can lead to both being less communicative and effective in their conversations [1]. This can be due to a variety of cultural factors, including differences in values, differences in explanatory models of illness, and linguistic barriers [2]. These differences not only affect patient satisfaction: culturally-attuned health communication actually improves patient health outcomes and reduces medical costs [3]. These findings have led to several calls for medical students in the US to be trained in cultural sensitivity.

Culture is also important in written medical information. Studies of "tailored print" interventions in healthcare, in which written healthcare instructions are customized to various patient features, including culture, have shown that they are significantly more effective than generic, non-tailored instructions [4]. Tailoring is hypothesized to work through a number of mechanisms, including increasing attention to the information and the amount of effort spent in understanding it, due to increased perceived relevance and personal involvement with the message [5].

Face-to-face interaction with health providers is a crucially important context for information exchange, given that face-to-face interaction allows providers to dynamically assess patient understanding and tailor their communication accordingly. Face-to-face consultation is especially important for patients with low health literacy, who may have difficulty acting on written healthcare information [6]. However, many individuals may not have access to healthcare professionals when they need it, whether due to cost, schedule, stigma, or logistical barriers. Automated health counseling systems have the potential to provide this access at a relatively low cost. Further, embodied conversational agents, which simulate face-to-face

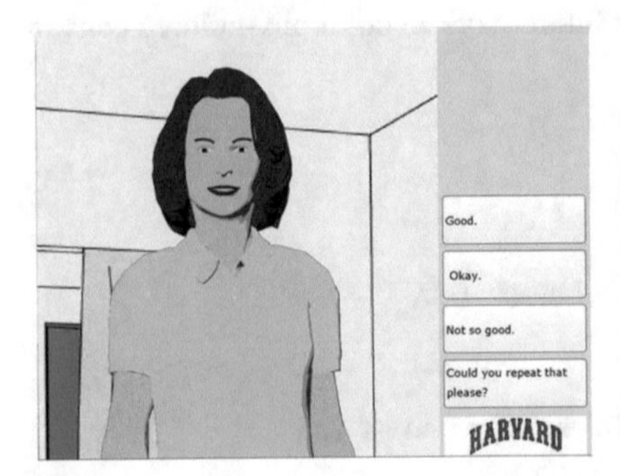

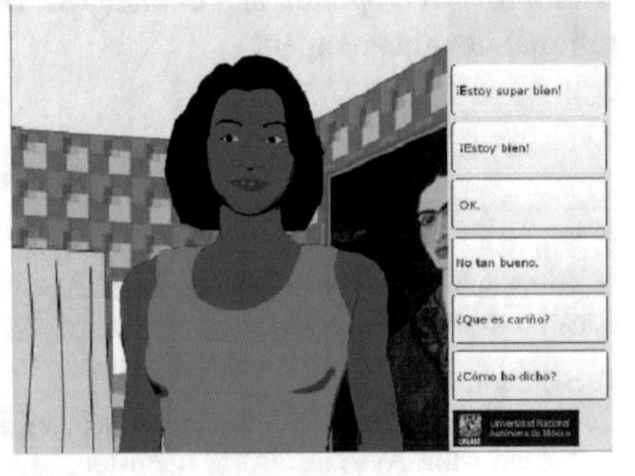

Fig. 5.1 Katherine and Catalina: conversational agents used in experiment

conversation with a provider (Fig. 5.1), may be especially effective. In these systems, the human-computer interface relies only minimally on text comprehension and uses the universally understood format of face-to-face conversation, thus making it less intimidating and more accessible to patients with limited literacy skills. The use of nonverbal conversational behaviors—such as hand gestures that convey specific information through pointing ("deictic" gestures) or through shape or motion ("iconic" and "metaphoric" gestures)—can also provide redundant channels of information for conveying semantic content communicated in speech, and the use of multiple communication channels enhances the likelihood of message comprehension. In addition, all cultures have nonverbal means for marking emphasis (for example, eyebrow raises and "beat" or "baton" hand gestures in American English), and these can be used to highlight the most salient parts of a message, a mechanism hypothesized to assist learners with low literacy skills. Finally, conversational agents provide a much more flexible and effective communication medium than a video-taped lecture or even combined video segments. The use of synthetic speech makes it possible to tailor each utterance to the patient (e.g., using their name, information from their medical record, and other personal information) and to the context of the conversation (e.g., what was just said, the fact that the patient asked the same question 10 min earlier, whether it is morning or evening, etc.).

Given the promise of conversational agents in automating aspects of health communication, and the importance of culture in health communication, the cultural adaptation of health counseling agents and the creation of culture-aware agents are important areas of investigation. In the rest of this chapter, we describe a series of project in which we have developed culturally-tailored health counseling agents and evaluated them in empirical studies.

5.2 An Empirical Study on Culturally Tailored Agents

By 2009, the Intelligent Virtual Agents research community (IVA) has begun paying attention to culture in virtual agent design. Rossen et al. conducted a study in which medical students were asked to interact with Edna, a virtual patient featuring a 55-year old woman, and by varying Edna's skin color, they found that Caucasian subjects tend to show more empathy to the agent with a lighter skin tone [7]. Endrass et al. integrated Japanese and German speech patterns in a virtual agent who spoke a meaningless gibberish language, and discovered that Germans clearly preferred virtual characters speaking in the German manner to those speaking in the Japanese manner [8]. A later study would find consistent results on Americans' preferences of virtual agents speaking in American vs. Arabian manners. Jan et al. attempted to create a model for designing culturally tailored virtual agents by simulating the proxemics, gaze, and turn taking of members of a cultural group, and collected initial evidence that virtual agents' behaviors created by using some of the dimensions specified in their model can be recognized by viewers [9].

Although the above studies only integrated a very small set of features that distinguish cultures from each other, i.e. skin color and certain speech patterns, the results of these studies all pointed in one direction: an individual is likely to recognize a virtual agent that performs the specific behaviors that are common in the individual's culture, and may prefer interacting with such an agent versus agents that don't perform these culturally-specific behaviors. But is this difference meaningful in achieving the instrumental, task-oriented goals of a conversational agent system? Is an agent resembling an individual's cultural member more likely to make a sale, successfully convey knowledge, change the individual's opinion about a subject matter, or convince the individual to change a certain behavior?

To answer these questions, we conducted an empirical study to test the persuasiveness of two conversational agents, Katherine, who resembled an American Caucasian woman, and Catalina, who resembled a Latina. The two agents differed not only in their appearance, including hairstyle, skin color and facial structure, but also in the virtual environment they were in (which would appear to the participant as the agent's house, see Fig. 5.1, and in the way they talked. Specifically, the agents were designed to conduct a conversation with users to try to convince them that the benefits of physical activity outweigh the drawbacks. The agent argued for or against the same statements as shown in Fig. 5.2, but argued in different ways: Katherine focused on participants' well-being, whereas Catalina expressed more interest in participants' family and friends. Beyond the actual conversation, while the agents were walking on to the screen, the participant would also hear a short music clip that is from the agent's culture. During the conversation, the computer program sometimes moved the camera closer to the agent so as to make the participant feel being at a closer proximity with the agent, but the program did this more often for Catalina than for Katherine. The conversations were tailored in these ways because individuals in the Latino cultures are considered to value family and

+ I would have more energy if I exercised regularly.
Exercise is important even if I don't want people to see me exercising.
+ I would feel less stressed if I exercised regularly.
- Exercise is important even if it is time-consuming.
+ Exercising puts me in a better mood for the rest of the day.
- Exercise is important even if exercise clothes make me uncomfortable or embarrassed.
I would feel more comfortable with my body if I exercised regularly.
- Exercise is important even if it is complicated.
+ Regular exercise would help me have a more positive outlook on life.
My doctor tells me that exercise is important.

Fig. 5.2 Ten statements about regular exercise. Statements starting with a "+" are those that the agent argued for, and those starting with a "–" are those argued against. Statements starting with neither were not mentioned

friends more than themselves as compared to their American counterparts, and Latinos also tend to prefer speaking in closer proximity than Americans [10].

Unlike many other agents exhibiting cultural features, Katherine and Catalina were designed to conduct meaningful conversations about health-related behaviors. Therefore, an important question is what language they should speak. Intuitively, to further culturally adapt the agents, Katherine should speak English, and Catalina should speak one of the Latino languages, and given that the most prevalent population in Massachusetts was Puerto Rican, Catalina would probably speak Spanish. However, there is a growing native-born Latino population in the United States, who may consider English as their native tongue, and many Anglo-Americans who can speak Spanish well, thus conducting a controlled trial on the language spoken by the virtual agent appeared to yield significant practical implications as well.

Therefore, our experiment consisted of four conditions, divided on the dimensions of cultural congruity (Katherine vs. Catalina) and language congruity (English vs. Spanish), as shown in Table 5.1. We used the word "congruity" to describe these conditions because we wanted to keep the study participants in mind: for an Anglo-American, Katherine speaking English may be the most congruent, but for a Latino participant, chances are Catalina speaking Spanish would appear more congruent.

When the study began, a participant would be asked to perform a ranking task designed to measure attitudes towards exercise. The statements to be ranked were exactly the same as shown in Fig. 5.2. The participant would then hold a conversation with an agent (Katherine or Catalina) in either English or Spanish, as specified by the study condition they were randomized to. After the agent attempted to convince the participant that the pros outweigh cons of exercising, the participant would be asked to do the ranking test again. The primary outcome was the difference between the two rankings. This is similar to the Desert Survival Problem [11], a method commonly used to test the outcomes of persuasion. Besides persuasion, we also used four 7-point scale Likert questions as a manipulation check for cultural congruity: "How much do you feel that the agent is a member of the American culture", "How much do you feel that the agent is a member of the Latino/Hispanic culture", "How easy was it to understand the agent's language" and "How much do you feel you and the agent are from the same culture".

Due to the bi-lingual nature of this experiment, and the fact that we wanted to randomize any participant into any of the four conditions, we had to place a special restriction on the participants; that they had to speak both English and Spanish. Even if an individual speaks two languages, there is no guarantee that the individual has enough exposure to the culture of his or her secondary language, especially with Spanish, a language being spoken in many different countries over the world, each with a unique culture. Therefore, we tightened the restrictions to recruit only

Table 5.1 The four conditions of the study

Katherine, English	Katherine, Spanish
Catalina, English	Catalina, Spanish

(a) Anglo-Americans born in the United States who had stayed in Spanish-speaking Latin American countries for at least two months with English being their first language, and (b) Latinos/Latinas born in a Spanish-speaking Latin American country who had lived in the United States for at least two months, with Spanish being their first language.

We ended up recruiting 43 participants for the study. Among these participants, 65.0% were Latinos, and the participants' ages ranged from 18 to 65. Katherine was also considered slightly more American than Catalina, although Catalina was not perceived as more Hispanic. Contrary to our expectations, the agent's appearance, and even the agent's simulated home environment, does not matter as much as their command of a language. In other words, as long as an agent speaks a culture's language properly (since the agents spoke through text-to-speech synthesizers and could be considered fluent in whichever language they spoke), she is considered a member of that culture.

We also discovered a significant interaction effect between cultural congruity (i.e. whether the agent's appearance matched a participant's cultural group) and need for cognition (one of the parameters that influence which route the person is likely to follow when processing a persuasive message) on both the effect of persuasion and the agent's perceived trustworthiness. A high need for cognition score indicates the participant is predisposed to being influenced by central route to persuasion, i.e. arguments related to the topic, while a low score indicates the participant is predisposed to being influenced by peripheral cues, i.e. features of the individual delivering the argument that the participant likes or dislikes [12]. As Fig. 5.3 shows, participants with low need for cognition tend to be persuaded by the agent resembling a member of their culture, and participants with high need for cognition tend to be persuaded by the agent resembling someone who is not a

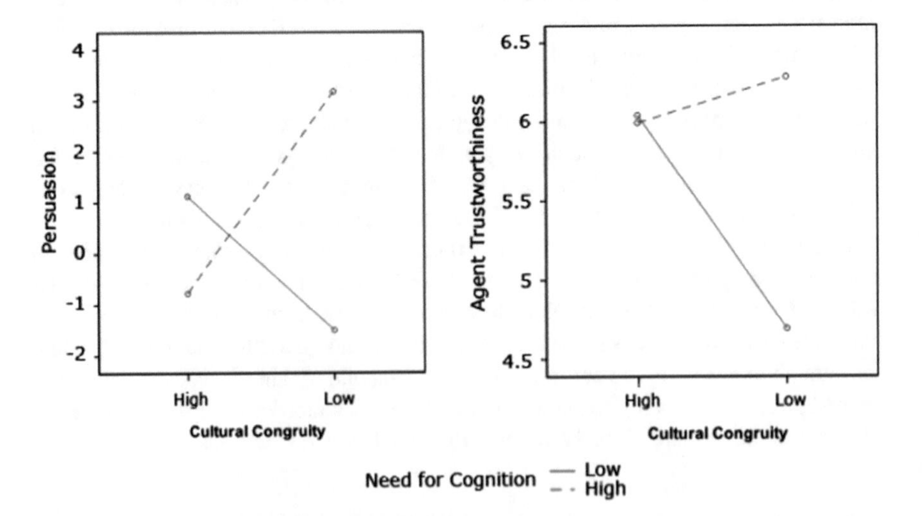

Fig. 5.3 Effect of need for cognition and cultural congruity on persuasion and agent trustworthiness

member of the participants' culture. More details of the experiment and the complete results can be found in our original paper [13].

A couple of important conclusions can be drawn from this study. First, the persuasiveness and perceived trustworthiness of an agent not only relies on the agent's culture awareness and cultural congruity, but also depends on the route to persuasion of the individual interacting with the agent: some individuals may be heavily influenced by a virtual agent resembling a member of their culture, while others may trust an agent representing a member of some other culture. According to the Elaboration Likelihood Model, depending on the individual and the topic in question, an individual may take one of two routes to process a message: the central route, which is used when the individual has the motivation as well as the ability to think about the topic under discussion; and the peripheral route, which is used when the individual has little or no interest in the message [14]. Individuals high in need for cognition tend to prefer the central route to persuasion, while those low in need for cognition tend to prefer the peripheral route. Based on our study, individuals that tend to prefer the central route tend to be influenced by an agent representing a race different from them more than an agent representing their culture. It is not clear why this is the case, and future research is required to fully understand this phenomenon. However, we propose that designers should not assume that an agent representing a user's culture will end up being more trustworthy or more persuasive compared to an agent that appears otherwise. Second, through this study we showed that a virtual agent speaking the user's native language is generally preferred over an agent speaking the user's secondary language, even if the agent does not appear to be a member of the user's culture. This is perhaps because, even if the agent does not appear to "belong to" the user's cultural group, speaking the user's native language well shows a good understanding of the user's culture, and this perceived understanding may have made the agent likable.

These conclusions have influenced some of our later projects using culture-aware agents. In these projects, we always made sure the agent speaks the language of the user's choice, and we designed the agent to appear knowledgeable not only about the user's national culture, but also about the culture of the local community that the user is involved in.

5.3 Exercise Promotion for Underserved Populations: COMPASS

Health behavior change technologies, including smart phone apps and state-of-the-art websites, hold the promise of increasing access to behavioral medicine interventions in a cost-effective and efficient manner. However, these technologies are often designed for individuals with high levels of computer, health, and reading literacy. Thus, they can serve to actually further increase disparities in health for underserved communities by presenting additional barriers to disadvantaged individuals to get the help and information they need.

These barriers can be addressed by using conversational agents as interfaces to automated health services. Conversational agents can replace hard-to-use, conventional human-computer interfaces with the familiar and intuitive format of face-to-face conversation. Conversational agents are especially effective for low literacy populations [15] since they can use hand gesture and other nonverbal behavior to provide redundant channels of communication [16], can use prosody and nonverbal behavior to mark the most salient information, and can use verbal and nonverbal "grounding" strategies to assess user understanding [17].

Conversational agents that are culturally adapted to the user's cultural background amplify the effectiveness of all of these features. The overall "look and feel" of the interface is even more familiar, appealing, and intuitive, increasing acceptance even by individuals with no technical literacy. The agent's use of culturally-appropriate nonverbal behavior (hand gesture, facial display, etc.) make communication even more effective. And, of course, as we have demonstrated in our preliminary study on culturally tailored agents, if the agent is speaking in the user's preferred language, it increases overall comprehension and the efficacy of any health intervention delivered by the agent.

To test these hypotheses in a real-world health behavior change intervention with measurable health outcomes, we developed the COMPASS system. COMPASS is an automated exercise promotion intervention that features a conversational agent in the role of an exercise advisor.

5.4 Implementations of COMPASS

The COMPASS project is divided into two phases: the first phase, COMPASS-1, was a pilot project on 40 Latino adults over 4 months, where a virtual agent was used to encourage the participants to perform regular physical activity in a relatable way; the second phase, COMPASS-2, extended the first phase to one-year long, including a larger amount of physical-activity-related conversations, as well as social dialogues that helped participants form a lasting relationship with the agent. Because the purpose of this project is to promote physical-activity-related behavior change, many of the conclusions from our persuasion study were useful in the design of COMPASS. Some of the design decisions were also influenced by research in the cultural virtual agent community.

5.4.1 Appearance

According to our previous study, a virtual agent's appearance, from skin color to accessories, and even the virtual environment that the agent is in, can influence a user's decision-making, but this influence can be positive or negative, depending on the user's preferred route to persuasion. For the purpose of the COMPASS project, we had no control over our participant population in terms of what type of persuasive message they were prone to. Therefore, our choice of the virtual agent ended up resembling a Latina woman, with a somewhat dark skin tone, in a professional outfit and a virtual environment that resembles an American health counselor's office.

5.4.2 *Language*

From our previous study, it was clear that users liked an agent who spoke their own language. Although this liking was not shown to lead to more trust or persuasiveness, it could contribute to improving rapport between a participant and the agent and forging a relationship between the two parties. In a longitudinal study, such a relationship may lead to prolonged usage of the system, thus leading to high retention. However, although our target users were adults in a Latino community, many of these participants have immigrated to the United States years ago, and some were second generation immigrants. For these participants, Spanish is not always preferred over English. Therefore, we designed the agent Carmen to speak both English and Spanish. Study participants were given the opportunity to choose which language they would like to communicate in with the agent at the beginning of the study, and they could change this setting with the help of a research coordinator at any time during the study. As it turned out, approximately half of the study participants chose English as their preferred language. A few participants, although they preferred to speak Spanish in daily life, chose to speak English with the agent as a means to practice their English, perhaps because the agent could repeat a sentence over and over without getting frustrated or judgmental.

5.4.3 *Social Dialogue*

Because the virtual agent is installed on a kiosk computer, stationed in a community center where participants of the study were recruited from, the only time that the participants were able to interact with the agent was when they were at the community center. Therefore, the participants were only encouraged to interact with the agent once per day, and all the conversations were designed around such a schedule. During the four months of the study, the agent could have up to 120 conversations with each participant. Over so many conversations, the agent must engage participants in some way, or else they may feel bored and stop interacting with the agent altogether. In COMPASS, the agent's primary means of keeping participants engaged is the use of social dialogue. Specifically, the agent would engage in casual conversation at the beginning of most sessions. We designed the dialogues in such a way as to demonstrate the agent's understanding of the Latino culture, by referencing TV shows and celebrities popular in the community. In particular, because our target audience was mostly of Mexican origin, we included content specific to the Mexican culture, as well as things popular in the particular locale the agent was in, such as topics that refer to casinos near San Jose, California, where the community center was located.

5.4.4 *Local References*

Apart from references to the local culture in social dialogues, the agent also showed a great interest and wide connections in the local community. We came to know that a member of the community center was a quite popular local singer. Our research coordinators in San Jose recorded him singing "La Bamba" and we played the recording at the beginning of each interaction session with participants, while the agent was walking on to the screen of the kiosk computer. Our research coordinators also designed several walking routes in the East Side of San Jose, near the community center, to help participants plan their physical activity.

Our efforts in designing a virtual counselor that appeals to and engages a small Latino community ended up creating a unified identity that our participants recognized and related to very well.

5.5 Results of the COMPASS Study

In collaboration with researchers at Stanford University, we conducted a randomized, two-arm, between-subjects experiment to evaluate the COMPASS system, relative to a wait-list control group [18]. Physical activity was assessed in both groups using the Community Healthy Activities Model Program for Seniors (CHAMPS) questionnaire [19].

Forty participants (92.5% Latino) aged 55 and over were enrolled, half in each arm of the study. Those in the intervention group were asked to wear pedometers daily and check in three times a week for four months with Carmen on a touch screen computer in the computer room of the community center.

Self-reported acceptance and satisfaction with the system was very high. In particular, scores on the bond subscale of the Working Alliance Inventory—measure trust in working with Carmen to achieve desired outcomes [20]—was 6.0 ± 0.84 out of 7.

All participants enrolled into the intervention group completed the four month study (100% retention). Four-month increases in reported minutes of walking per week were greater in the intervention arm (mean increase = 253.5 + 248.7 min/week) relative to the control group (mean increase = 26.8 + 67.0 min/week; difference $p = 0.0008$). Walking increases in the intervention arm were substantiated via objectively measured daily steps from the pedometers (slope analysis, $p = 0.002$).

Access to the COMPASS system in the community center was provided for an additional 20 weeks past the end of the 4-month study period. During this extended time, intervention participants accessed Carmen a mean of 14 ± 20.5 additional times (range = 0–4.5 sessions per week). Fully 95% (19/20) interacted with Carmen at some point during this extended period.

5.6 The Heart-Healthy Action Program for Puerto Rican Adults (HHAP)

Heart disease, including coronary heart disease, heart attacks, congestive heart failure, and congenital heart disease, is the leading cause of death in the United States. Hispanics, including Puerto Ricans, are at a greater risk of premature death due to heart disease compared to non-Hispanic Whites (23.5 vs. 16.5%). Changes in lifestyle health behaviors, including physical activity, diet, and stress management, can result in significant reductions in deaths due to heart disease. To address these health disparities, we adapted the Carmen conversational agent to promote lifestyle health behavior change, tailored for Puerto Rican adults aged 45 and older, living in the continental United States. The automated intervention is used as part of a participatory, community-based, multi-level, trans-disciplinary intervention to intervene on the social, cultural and environmental factors that influence cardiovascular health behaviors among Puerto Rican adults. The other elements of the intervention include weekly group meetings with a facilitator to discuss exercise, diet, and stress management, and weekly group Zumba classes for exercise.

5.6.1 Implementations of HHAP

Compared to COMPASS, HHAP covers a much larger range of health-related content, including four interrelated topics: physical activity, nutrition, stress management and health literacy. Therefore, we focused our efforts on developing conversational content that incorporates common issues usually encountered by Latinos in the community. For the design of the virtual agent herself, we used Carmen, the same agent presented in the COMPASS project, but her dialogue patterns were tuned to fit better with a Puerto Rican audience.

5.6.2 Culturally-Tailored Content

The intervention was designed to last one year, which means if a participant interacts with the agent once per day, there could be as many as 365 conversations that must be supported by the automated system. Health counseling on the four topics in the HHAP project does not require so many sessions, and so we designed the agent to give only one contentful counseling session per week. To make sure a participant receives the most informative session every week, each contentful counseling session was hand-crafted by researchers with expertise in each of the four topics. Not only were these conversations well crafted, they also incorporated common themes that came up in the local Latino communities. For example, when the agent discusses nutrition, she sometimes refers to cultural food in Latino countries, and explains the historical reasons why, although some of the foods can be bad for health, they became part of the Latino cuisine. As another example, we developed 12 dialogue sessions of stress management based on the Stress Inoculation Training techniques [21], which included several role-playing sessions directed by the virtual agent. We deliberately eliminated several common training cases that are related to sensitive topics in the community, such as receiving a speeding ticket and the interaction with a police officer that would follow, and instead used health-related examples, such as being diagnosed with diabetes, which participants considered more comfortable and concerned with.

5.6.3 Social Dialogue

Because the counseling only took place once per week, if a participant decided to log in more than once a week, the agent would check the participant's progress toward their health goals and engage in casual conversations. Therefore, filling in the social dialogues became all the more critical in keeping the participants' engagement level high. For this purpose, we acquired the help of a professional fiction writer to craft a series of stories that the agent could share with the participants. Once the stories had been created, we consulted experts of the Latino culture to tailor the content for the Latino community, and especially for Latinos in Boston. The virtual agent told these stories as her own life experience. This type of storytelling has been shown to create trust and rapport between the agent and the participant [22].

5.6.4 Local References

Unlike the COMPASS project, HHAP is currently deployed in four different community centers around Boston, and therefore it is difficult to create an experience that can be perceived as custom tailored to all the participants in the study. However, we did sample four pieces of Latino music that were considered to be popular among the communities that Carmen was to serve, and played these music pieces as Carmen walked on the screen.

A two-arm, randomized evaluation study is currently underway, comparing the overall HHAP intervention to a non-intervention control. The study is expected to complete in 2016. Preliminary results indicate the system is well-liked and used by the Latino study participants. The final results of this study will be discussed in a separate publication.

5.7 Conclusions

Culturally appropriate interaction between human patients and their health advisors is important, and as our experiments and intervention programs suggest, this is important in interactions between people and automated health advisors as well. There are many ways in which conversational agents can be tailored for specific cultures, and many reasons why this tailoring is important: not only to improve accessibility, usability, and retention, but to actually impact outcomes, especially for low literacy and other underserved populations.

Cultural awareness in computer-based health systems is a relatively young research field, and research into culturally-tailored virtual conversational agents has become a focus of attention only in the past decade. It will be many long years before findings from cultural studies are applied in designing culturally-tailored agents, but the impact of these applications on the health benefits of their users has shown promise. An important goal, if not the ultimate goal, of this area of research is to create virtual agents that are perceived as members of a culture by exhibiting culturally-appropriate behaviors in most social situations. To achieve this goal, we believe implementing an underlying framework to guide the agent's behavior intelligently is necessary. Such a behavior framework may inevitably lay its foundations on a somewhat comprehensive understanding of cultures, based on theories such as the Cultural Dimensions [23], or specific ways of describing cultures by means of symbols, rituals, values, metaphors and paradoxes, etc. [24].

References

1. Carrillo, J., Green, A., Betancourt, J.: Cross-cultural primary care: a patient-based approach. Ann. Int. Med. **130**, 829–834 (1999)
2. Schouten, B., Meeuwesen, L.: Cultural differences in medical communication: a review of the literature. Patient Educ. Couns. **64**, 21–34 (2006)
3. Brach, C., Fraser, I.: Can cultural competency reduce racial and ethnic health disparities? A review and conceptual model. Med. Care Res. Rev. **57**(Suppl 1), 181–217 (2000)
4. Noar, S.M., Benac, C., Harris, M.: Does tailoring matter? Meta-analytic review of tailored print health behavior change interventions. Psychol. Bull. **133**, 673–693 (2007)
5. Hawkins, R.P., Kreuter, M., Resnicow, K., Fishbein, M., Dijkstra, A.: Understanding tailoring in communicating about health. Health Educ. Res. **23**(3), 454–466 (2008)
6. Nielsen-Bohlman, L., Panzer, A., Kindig, D.: Health literacy: a prescription to end confusion. The National Academies of Sciences (2004)
7. Rossen, B., Johnsen, K., Deladisma, A., Lind, S., Lok, B.: Virtual humans elicit skin-tone bias consistent with real-world skin-tone biases. In: Proc. IVA 2008. Springer (2008)
8. Endrass, B., Rehm, M., André, E.: Culture-specific communication management for virtual agents. In: Proc. of 8th Int. Conf. on Autonomous Agents and Multiagent Systems (AAMAS 2009), International Foundation for Autonomous Agents and Multiagent Systems, pp. 281–287 (2009)
9. Jan, D., Herrera, D., Martinovski, B., Novick, D., Traum, D.: A computational model of culture-specific conversational behavior. In: Proc. IVA 2007, pp. 45–56. Springer (2007)

10. Albert, R., Ha, A.: Latino/Anglo-American differences in attributions to situations involving touch and silence. Int. J. Intercultural Relat. **28**(3–4), 253–280 (2004)
11. Lafferty, J., Eady, P.: The desert survival problem. Experimental Learning Methods (1974)
12. Cacioppo, J.T., Petty, R.E., Feinstein, J.A., Jarvis, W.B.G.: Dispositional differences in cognitive motivation: The life and times of individuals varying in need for cognition. Psychol. Bull. **119**(2), 197 (1996)
13. Yin, L., Bickmore, T., Cortés, D.E.: The impact of linguistic and cultural congruity on persuasion by conversational agents. In: Proc. Intelligent Virtual Agents, pp. 343–349. Springer (2010)
14. Petty, R., Cacioppo, J.T.: Communication and persuasion: central and peripheral routes to attitude change. Springer Science & Business Media (2012)
15. Bickmore, T., Pfeifer, L., Byron, D., Forsythe, S., Henault, L., Jack, B., Silliman, R., Paasche-Orlow, M.: Usability of conversational agents by patients with inadequate health literacy: evidence from two clinical trials. J. Health Comm. **15**(Suppl 2), 197–210 (2010)
16. McNeill, D.: Hand and mind: what gestures reveal about thought. Cambridge University Press, Cambridge (1992)
17. Clark, H.H., Brennan, S.E.: Grounding in Communication. In: Resnick, L.B., Levine, J.M., Teasley, S.D. (eds.) Perspectives on Socially Shared Cognition, pp. 127–149 (1991)
18. King, A., Bickmore, T., Campero, M., Pruitt, L., Yin, L.: Employing 'virtual advisors' in preventive care for underserved communities: results from the COMPASS study. J. Health Comm. **18**(12), 1449–1464 (2013)
19. Stewart, A., Verboncoeur, C., McLellan, B., Gillis, D., Rush, S., Mills, K., King, A., Ritter, P., Brown, B., Bortz, W.: Physical activity outcomes of champs II: a physical activity promotion program for older adults. J. Gerontol. **56A**(8), M465–M470 (2001)
20. Horvath, A., Greenberg, L.: Development and validation of the working alliance inventory. J. Couns. Psychol. **36**(2), 223–233 (1989)
21. Meichenbaum, D., Cameron, R.: Stress inoculation training. Springer (1989)
22. Bickmore, T., Schulman, D., Yin, L.: Engagement vs. deceit: virtual humans with human autobiographies. In: Proc. IVA 2009, pp. 6–19. Springer (2009)
23. Hofstede, G.: Culture's consequences: comparing values, behaviors, institutions, and organizations across nations. Sage Pubns (2001)
24. Gannon, M.J.: Paradoxes of culture and globalization. Sage Publications (2007)

Chapter 6
Combining a Data-Driven and a Theory-Based Approach to Generate Culture-Dependent Behaviours for Virtual Characters

Birgit Lugrin, Julian Frommel and Elisabeth André

Abstract To incorporate culture into intelligent systems, there are two approaches that are commonly proposed. Theory-based approaches that build computational models based on cultural theories to predict culture-dependent behaviours, and data-driven approaches that rely on multimodal recordings of existing cultures. Based on our former work, we present a hybrid approach of integrating culture into a Bayesian Network that aims at predicting culture-dependent non-verbal behaviours for a given conversation. While the model is structured based on cultural theories and theoretical knowledge on their influence on prototypical behaviour, the parameters of the model are learned from a multimodal corpus recorded in the German and Japanese cultures. The model is validated in two ways: With a cross-fold validation we estimate the power of the network by predicting behaviours for parts of the recorded data that were not used to train the network. Secondly we performed a perception study with virtual characters whose behaviour is driven by the calculations of the network and are rated by members of the German and Japanese cultures. With this chapter, we aim at giving guidance for other culture-specific generation approaches by providing a hybrid methodology to build culture-specific computational models as well as potential approaches for their evaluation.

B. Lugrin (✉)
Human-Computer Interaction, University of Wuerzburg, Am Hubland (M1), 97074 Würzburg, Germany
e-mail: birgit.lugrin@uni-wuerzburg.de

J. Frommel
Institute of Media Informatics, Ulm University, James-Franck-Ring, 89081 Ulm, Germany
e-mail: julian.frommel@uni-ulm.de

E. André
Human Centered Multimedia, Augsburg University, Universitätsstr. 6a, 86159 Augsburg, Germany
e-mail: andre@hcm-lab.de

C. Faucher (ed.), *Advances in Culturally-Aware Intelligent Systems and in Cross-Cultural Psychological Studies*, Intelligent Systems Reference Library 134, https://doi.org/10.1007/978-3-319-67024-9_6

Keywords Virtual Agents · Culture · Non-verbal behaviour
Bayesian Network · Hybrid approach · Evaluation

6.1 Motivation

Human behaviour is influenced by several personal and social factors such as culture and can be referred to as a mental program that drives peoples' behaviour [20]. Non-verbal behaviour as a part of human communication is, amongst others, also influenced by cultural background [45]. Cultural differences on the one hand manifest themselves on an outward level and are, on the other hand, also judged differently by observers of different cultural backgrounds [46].

Examples include the presentation and evaluation of facial expressions, gestures or body postures. In South Korea and Japan, for example, restrained facial expressions and postures indicate an influential person, while in the United States, relaxed expressions and postures give the impression of credibility [45].

As anthropomorphic user interfaces, such as virtual characters or humanoid robots, aim at realistically simulating human behaviour it seems likely that their behaviour conveys different impressions on observers from different cultural backgrounds as well. Taking potential cultural differences into account while designing an agent's non-verbal behaviour can thus help improve their acceptance by users of the targeted cultures.

Building models that determine culture-related differences in behaviour is challenging as the dependencies between culture and corresponding behaviour need to be simulated in a convincing and consistent manner. In the last decade numerous attempts have been made to face these challenges, mainly relying on either theoretical knowledge (theory-based) or empirical data (data-driven). While theory-based approaches model culture-specific behaviours based on findings from the social sciences, data-driven approaches aim to extract culture-specific behaviour patterns from human data to inform computational models. Data-driven approaches bear the advantage that they are based on empirically grounded models. However, a large amount of data is required to derive regularities from concrete instantiations of human behaviour. Theories from social sciences include information that may help us encode culture-specific behaviour profiles. But there are usually missing details that are required for a convincing realization of culture-specific virtual characters or humanoid robots. The objective of the present chapter is to combine the advantages of theory- and data-driven approaches. To this end, we illustrate and evaluate an integrated approach that coherently connects a theory-based and a data-driven approach.

The chapter is structured as follows: In the next section, we briefly introduce models of culture from the social sciences that inspired our work. Then we introduce related work on the computational integration of culture into intelligent systems. In Sect. 6.4, we outline our approach by referring to our own prior work that was used as a basis for the endeavor. Subsequently, we describe the design and implementation of a computational model that is built upon the combined approach.

Two potential ways of evaluating the resulting network are then presented in Sects. 6.6 and 6.7. Finally, we conclude our chapter by reflecting on potential contributions and tribulations of our approach.

6.2 Models of Culture

There is a wide variety of theories and models that explain the concept of culture and what drives people to feel that they are belonging to a certain culture. Many definitions of culture provided by the social sciences conceptually describe cultural differences but stay rather abstract. Some theories describe different levels of culture that address, among other things, that culture does not only determine differences on the surface but also works on the cognitive level. Trompenaars and Hampden-Turner [46], for example, distinguish implicit and explicit levels of culture that range from very concrete and observable differences to a subconscious level that is not necessarily visible to an observer.

Other definitions of culture use dimensions or categories to explain differences between certain groups. These approaches describe culture in a way that facilitates building computational models. Therefore, in this section, we selected cultural theories from the large pool provided by the social sciences which explain culture along dimensional models or dichotomies that help understand culture in a more descriptive manner.

A very well known theory that uses dichotomies was introduced by Hall [18], who classifies cultures using different categories such as their members' perception of space, time or context. Regarding haptics, for example, Hall [18] states that people from high-contact cultures tend to have higher tactile needs than members of low-contact cultures who, vice versa, have more visual needs. These needs can also show on the behavioural level. In some Arab countries, as an example for high-contact cultures, it is a common habit between two males to embrace for greeting or to link arms in a friendly way which can be very unusual behaviour in low-contact cultures.

An approach that describes cultures along dimensions was introduced by Kluckhohn and Strodtbeck [29], who formulate different value orientations in order to explain cultures. One dimension, for example, constitutes the relationship to other people, and describes how people prefer relationships and social organizations to be. Although in this theory a classification of values is provided, the impact on behaviour is described rather vaguely and is therefore hard to measure. Building a computational model with it is thus a demanding task and has not been attempted yet.

Another example of defining culture using dimensions is given by Hofstede [20], who categorized different cultures into a five dimensional model. For the model more than 70 national cultures were categorized in an empirical survey.[1] The *Power Distance* dimension (PDI), describes the extent to which a different distribution of power is accepted by the less powerful members of a culture. The *Individualism* dimension (IDV) describes the degree to which individuals are integrated into a group. The *Masculinity* dimension (MAS) describes the distribution of roles between the genders. The *Uncertainty Avoidance* dimension (UAI) defines the tolerance for uncertainty and ambiguity. The *Long-Term Orientation* dimension (LTO) explains differences by the orientation towards sustainable values for the future. For each of these dimensions, clear mappings are available from national cultures to the cultural dimensions on normalized scales [21]. In [22], Hofstede and Pedersen introduce so-called synthetic cultures that are based on Hofstede's dimensional model. Each synthetic culture observes one of the extreme ends of each dimension in isolation, and conceptually describes stereotypical behaviour of its members.

6.3 Related Work

As pointed out earlier, basically two approaches have been proposed to simulate culture-specific behaviours for synthetic agents: *Theory-based approaches* and *Data-driven approaches*.

Theory-based approaches typically start from a theory of culture to predict how behaviours are expressed in a particular cultural context. A common approach to characterize a culture is to use dichotomies, which are particularly suitable for integration into a computer model [21].

Alternatively, cultures have been characterized by the prioritization of values within a society [42]. Approaches that aim to modulate behaviours based on culture-specific norms and values, typically start from existing agent mind architectures and extend them to allow for the culture-specific modulation of goals, beliefs, and plans.

One of the earliest and most well-known systems that models culture-specific behaviours within an agent mind architecture is the Tactical Language System (http://www.tacticallanguage.com/), which has formed the basis of a variety of products for language and culture training by Alelo Inc. Tactical Language is based on an architecture for social behaviour called Thespian that implements a version of theory of mind [44]. Thespian supports the creation of virtual characters that understand and follow culture-specific social norms when interacting with each other or with human users. While the user converses with the characters of a training scenario, Thespian tracks the affinity between the single characters and the

[1]Originally, Hofstede used a four-dimensional model. The fifth dimension, long term orientation, was added later in order to better model Asian cultures. Meanwhile a sixth dimension, indulgence, was added that described the subjective well-being that members of a culture experience.

human user, which depends on the appropriateness of the user's behaviour. For example, a violation of social norms would result in a decreased affinity value.

To simulate how an agent appraises events and actions and manages its emotions depending on its alleged culture, attempts have been made to enrich models of culture by models of appraisal, see [1] for a survey. An example includes the work by Mascarenhas et al. [31] who aim at the modelling of synthetic cultures that may be obtained by systematically varying particular behaviour determinants. To this end, they extend an agent mind architecture called FAtiMA that implements a cognitive model of appraisal [35] by representations of the Hofstede cultural scales. Based on the extended architecture, agents with distinct cultural background were modelled. In their model, an agent's alleged culture determines its decision processes (i.e., the selection of goals) and its appraisal processes (i.e., how an action is evaluated). For example, an action that is of benefit to others is the more praiseworthy, the more collectivistic the culture is. Using the extended FAtiMA architecture as a basis, the ORIENT [3], MIXER [2] and Traveller [26] applications simulate synthetic cultures with the overall aim to generate a greater amount of cultural awareness on the user's side.

Data-driven approaches rely on annotated multimodal recordings of existing cultures as a basis for computational models of culture. The recordings can be used directly for imitating the human behaviour or statistical patterns can be derived from the data which govern the behaviour planning process.

Such a cross-cultural corpus has, for example, been recorded for multi-party multi-modal dialogues in the Arab, American English and Mexican Spanish cultures [19]. The corpus has been coded with information on proxemics, gaze and turn taking behaviours to enable the extraction of culture-related differences in multi-party conversations. A statistical analysis of the corpus reveals that findings are not always in line with predictions from the literature and demonstrate the need to enhance theory-driven by data-driven approaches.

More recent work by Nouri and Traum [34] makes use of a data-driven approach to map statistical data onto culture-specific computational models for decision making. In particular, they simulate culture-specific decision making behaviour in the Ultimatum Game based on values, such as selfishness, held by Indian and US players collected through a survey. Their work aims to adapt decision making of virtual agents depending on culture-specific values, but does not consider culture-specific verbal and non-verbal behaviours.

While the theory-driven approach ensures a higher level of consistency than the data-driven approach, it is not grounded in empirical data and thus may not faithfully reflect the non-verbal behaviour of existing cultures. Another limitation is that it is difficult to decide which non-verbal behaviours to choose for externalizing the goals and needs generated in the agent minds. The advantage of data-driven computational models of culture lies in their empirical foundation. However, they are hard to adapt to settings different from the ones recorded, as the data cannot be generalized due to a lack of a causal model.

6.4 Approach

To implement a hybrid approach that combines a theory-based with a data-driven approach, we rely on our earlier work that includes implementations of both approaches, but did not yet integrate them in a synergistic manner (see Fig. 6.1).

In scope of the Cube-G project, we recorded a cross-cultural corpus for the German and Japanese cultures [37]. More than 20 participants were video-taped in each culture, each running through three scenarios. So far, only the first scenario, a first time meeting, was considered. For this scenario a student and a professional actor (acting as another student) were told to get acquainted with one another to be able to solve a task together later. Recording one participant and an actor at a time ensured a higher control over the recordings. This way, participants did not know each other in advance and the actor was able to control that the conversation lasted for approximately 5 min. Actors were told to be as passive as possible to allow the participant to lead the conversation and be active in cases where the conversation was going to stagnate.

For the corpora, statistical analyses were performed to identify differences between German and Japanese speakers in the use of gestures and postures, communication management, choice of topics, and the like (e.g. [14, 15, 40]).

For simulation we developed a social simulation environment [10] with virtual characters that portray typical Japanese and German behaviours by gestures and dialogue behaviours. The modelling of the behaviours was based on the observations made in the multimodal corpora.

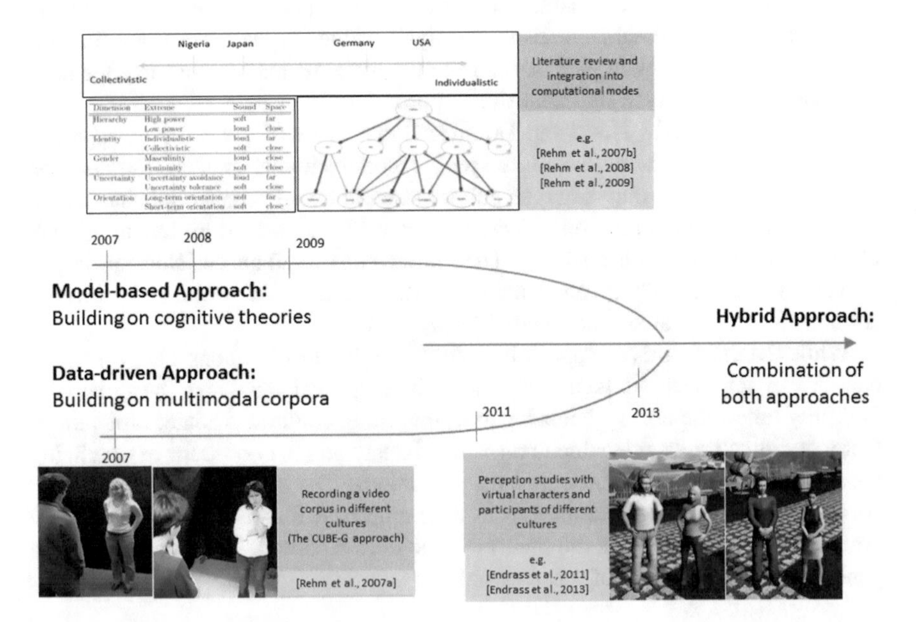

Fig. 6.1 Model-based and data-driven approaches as conducted in our former work

At a later stage, we made use of a data-driven approach and conducted perception studies with virtual characters that simulated the findings of the corpus analysis. Results suggest that users prefer virtual character behaviour that was designed to resemble their own cultural background, e.g. [13, 14]. At this point in time, the characters' behaviour was completely scripted to follow the statistical distribution of the corpus findings, and no computational model was built yet, while each of the studies looked at one behavioural aspect in isolation.

In parallel to the corpus recordings, we started a theory-based approach that developed a parameterized model of cultural variation based on Hofstede's dimensions [39]. To code dominant culture-specific patterns of behaviour, we made use of Bayesian Networks. Culture-specific behaviours were then generated following probability distributions inspired by theories found in the literature. The model was used to adapt the dynamics of gestures, proxemics behaviours as well as the intensity of speech of a group of virtual characters to the assumed cultural background of the user [38].

In this contribution, we combine the two approaches in a synergistic manner. In particular, we will make use of cultural theories to model dependencies between culture-related influential factors and behaviours and employ statistical methods to set the parameters of the resulting model; i.e. probabilities will be learned from recordings of human behaviour. A similar approach was presented by Bergmann and Kopp [4] to model co-verbal iconic gestures, however, without considering culture-specific aspects. We extend their work by adapting parameter settings of a Bayesian Network to a particular culture based on a data-driven approach.

The approach combines advantages of the commonly used theory-based and data-driven approaches, as it explains the causal relations of cultural background and resulting behaviour, and augments them by findings from empirical data. The resulting hybrid model, combines all previously considered behavioural aspects (verbal and non-verbal) in a complete model. Having such a model at hand, we are able to generate culture-specific dialogue behaviour automatically following the statistical distribution of the recorded data.

We are thus, extending our previous approaches in the following ways: (1) by integrating aspects of verbal and non-verbal behaviour into a complete model (2) by augmenting the model with empirical data (3) by validating the resulting model by testing its predictive qualities and performing a perception study with human observers.

6.5 Modeling Culture-Specific Behaviours with Bayesian Networks

We have chosen to model culture-specific behaviours by means of a Bayesian Network. The structure of a Bayesian Network is a directed, acyclic graph (DAG) in which the nodes represent random variables while the links or arrows connecting nodes describe the relationship between the corresponding variables in terms of conditional probabilities [41]. The use of Bayesian Networks bears a number of advantages. In particular, we are able to make predictions based on conditional probabilities that model how likely a child variable is given the value of the parent variables. For example, we may model how likely it is that a person makes use of very tight gestures if the person belongs to a culture that is characterized as highly collectivistic. By using a probabilistic approach for behaviour generation, we mitigate the risk of overstereotyping cultures. For example, a character that is supposed to portray a particular culture would show culture-specific behaviour patterns without continuously repeating one and the same prototypical behavioural sequence. Furthermore, Bayesian Networks enable us to model the relationship between culture-related influencing factors and behaviour patterns in a rather intuitive manner. For example, it is rather straightforward how to model within a Bayesian Network that a member of a collectivistic culture tends to use less powerful gestures. Finally, Bayesian Networks support the realization of a hybrid approach. While the structure of the Bayesian Network—in particular the dependencies between cultural influencing factors and behaviour patterns—can be determined by relying on theories of culture, the exact probabilities of the Bayesian Network can be learnt from recordings of culture-specific human behaviours.

The aim of the Bayesian Network model described in this section, is to automatically generate culture-dependent non-verbal behaviour for a given agent dialogue in the domain of first time meetings. Therefore, the network is divided in two parts:

- influencing factors: those factors that are given for the specific conversational situation such as the cultural background of the interlocutors, and
- resulting behaviours: the specific settings of non-verbal behavioural aspects that are calculated by the network as a result of the given influencing factors.

The Bayesian Network was modeled using the GeNIe modeling environment [12]. Figure 6.2 shows the structure of the network with its influencing factors and resulting behaviours that will be further explained in the following subsections.

6.5.1 Influencing Factors

To be able to construct different agent dialogues that vary with cultural background, influencing factors in our model are further divided into cultural background and conversational verbal behaviour.

In our network model, we rely on Hofstede's dimensional model [21], that captures national cultures as a set of scores along dimensions, providing a quite complete and validated model, especially considering the fuzzy concept of culture (see Sect. 6.2). The model has widely been used as a basis to integrate culture for

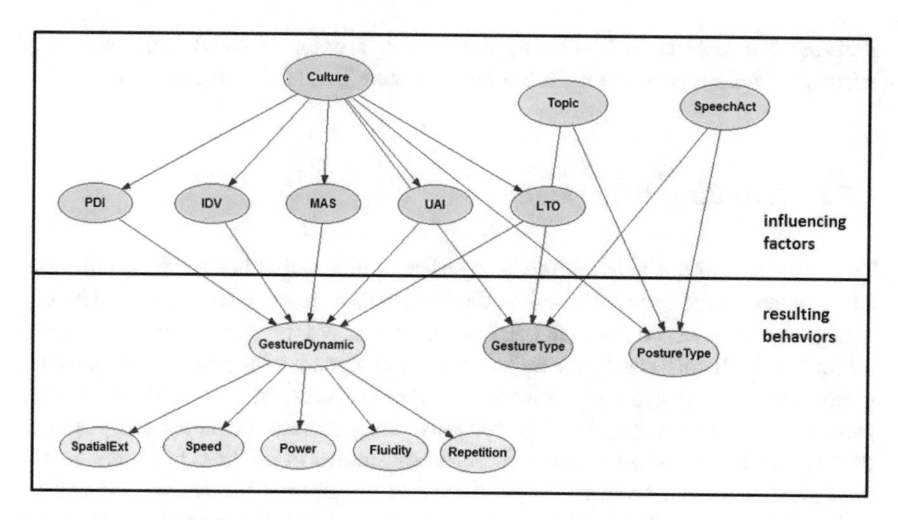

Fig. 6.2 Network model including influencing factors and resulting behaviours for culture-related behaviour generation

anthropomorphic interfaces, e.g. [2, 3, 32]. For our implementation, we categorized the scores on the cultural dimensions, provided by [21], into three discrete values (low, medium, high).

Regarding the content of dialogues, cultural differences can play a crucial role as well. According to Isbister and colleagues [24], for example, the categorization into safe and unsafe topics varies with cultural background. Therefore we expect variations in the semantics of first-time meeting conversations dependent on cultural background. Schneider [43] classifies topics that prototypically occur in first-time meetings as follows: The *immediate situation* holds topics that are elements of the so-called frame of the situation, such as the surrounding or the atmosphere of the conversation. The *external situation* describes topics that hold the larger context of the conversation, such as the news, politics, or recent movies. In the *communication situation* topics are focusing on the conversation partners, e.g., their hobbies, family or career. Potential topics for dialogues in our network model are based on this categorization.

Besides the semantics of speech, the function of each utterance is considered in our network. For computational models, verbal behaviour is often described by speech acts. Core and Allen [9], for example, provide a coding schema that categorizes speech acts along several layers. One layer of their schema, labels the communicative meaning of a speech act. As the whole schema is too complex for our purpose, we use the following subset of communicative functions that meet the requirements of first time meeting scenarios: *statement*, *answer*, *info request*, *agreement/disagreement* (indicating the speaker's point of view), *understanding/misunderstanding* (without stating a point of view), *hold*, *laugh* and *other*.

6.5.2 Resulting Behaviours

The lower part of our network model consists of non-verbal behavioural traits, that are calculated by the model based on the influencing factors (see Fig. 6.2). There is a large number of non-verbal behaviours that could be dependent on cultural background. At this stage, we focus on the upper part of the torso and consider gesture types, their dynamic variation and arm postures, and do not look at other parts of the body yet, e.g. head movements. The generation of adequate postures, gestures, and their dynamics have been widely studied in the field of virtual agents (e.g. [5, 8, 36]) and seem to be good aspects to improve the characters' believability. Regarding cultural differences, for body postures, for example, it has been shown that different cultures perceive different emotions depending on body postures [28].

In [6], Bull provides a coding schema to distinguish different body postures, including prototypical positions of, for example, head, feet or arm. To distinguish arm postures in our model, we employ Bull's categorization for arm postures. In total, 32 different arm positions are described in the schema and included to the network. Table 6.1 shows an extraction of the schema containing postures that were frequently observed during our corpus study and are thus most relevant for our approach.

Gestures can be considered on two levels: which gesture is performed and how it is performed. The most well known categorization of gesture types has been provided by McNeill [33]. Although the categorization is not meant to be mutually exclusive, the types described are a helpful tool to distinguish gestures: *Deictic* gestures are pointing gestures; *Beat* gestures are rhythmic gestures that follow the prosody of speech; *Emblems* have a conventionalized meaning and do not need to be accompanied by speech; *Iconic* gestures explain the semantic content of speech, while *metaphoric* gestures accompany the semantic content of speech in an abstract manner by the use of metaphors; *Adaptors* are hand movements towards other parts

Table 6.1 Extraction of arm postures considered in our model

Arm posture	Description
PHIPt	Put hands to pocket
PHFE	Put hand to face
PHEW	Put hand to elbow
PHWr	Put hand to wrist
FAs	Fold arms
JHs	Join hands
PHB	Put hands back

of the body to satisfy bodily needs, such as scratching one's nose. In our network model, we classify gesture types as a subset of McNeill's categorization [33]: deictic, beat, iconic, and metaphoric. Adaptors were excluded from the network as we are focusing on gestures that accompany a dialogue. Emblems were excluded as they are not generalizable and might convey different meanings in different locations. An example includes the American OK-gesture (bringing the thumb and the index finger together to form a circle). While it means OK in the Northern American culture, it is considered an insult in Latin America, and can be interpreted as meaning homosexual in Turkey. Thus, even if our model could predict that a emblematic gesture type would be appropriate in a given situation, different concrete gestures needed to be selected based on cultural background.

Besides the choice of a gesture, its dynamics can differ and be dependent on cultural background. According to Isbister [23] "*what might seem like violent gesticulating to someone from Japan would seem quite normal and usual to someone from a Latin culture*". To include this phenomenon in our network, we added a node containing a gesture's dynamics which is divided into three discrete values (low, medium, high). The dynamic variation of a gesture can be further broken down into dimensions [17], each describing a different attribute of the movement. Following [30], who investigated gestural expressivity for virtual characters, we employ the parameters spatial extent (the arm's extent relative to the torso), power (acceleration), speed, fluidity (flow of movements) and repetition. Initial values in our network were set in a manner that a high dynamics is more likely to result in a higher value for each of the parameters.

6.5.3 Dependencies

So far we introduced the nodes and their parameters of the influencing factors and resulting behaviours of the network and stated why they were included. In this subsection we explain their dependencies that were modeled for the network.

Although there is a strong evidence that the types of gestures and arm postures are dependent on cultural background, e.g. [45], there are no clear statements in the literature on how exactly McNeill's gesture types or Bull's arm posture types would correlate with Hofstede's dimensions of culture. We thus connected the nodes holding gesture types and arm postures directly to the culture node instead of linking them via Hofstede's dimensions.

Regarding the dynamics of a gesture and its correlation to Hofstede's cultural dimensions, we rely on the concept of synthetic cultures that builds upon Hofstede's dimensions (see Sect. 6.2). For these abstract cultures prototypical behavioural traits are described. For example, the *extreme masculine culture* is described as being loud and verbal, liking physical contact, direct eye contact, and animated gestures. Members of a *extreme feminine culture*, on the other hand, are described as not raising their voices, liking agreement, not taking much room and being warm and friendly. Furthermore, the position on Hofstede's dimensions determines the stereotypical movements of a synthetic culture. Thus, in our model, culture is connected to the gesture dynamic node via the cultural dimensions.

Please note that although it is known from the literature that aspects of verbal behaviour, e.g. the choice of conversational topic, can be dependent on cultural background, in our network model they are considered as influencing factors. This design choice was made, as we aim at generating culture-dependent non-verbal behaviour for a given dialogue and do not want to generate the dialogue itself. Therefore, nodes containing information about the selected verbal behaviour (speech act and conversational topic) are not connected to the culture node but linked directly to the nodes describing the resulting non-verbal behaviour types. The dialogue that is used as an input, can of course contain culture-specific content and thus influence the selected non-verbal behaviour.

6.5.4 Parameters of the Model

Using an automated learning process the network's model described in the previous subsection was augmented with empirical data. For that purpose the findings of the corpus (cf. Sect. 6.4) had to be processed before applying an automated learning process. The Anvil tool [25] allows to align self-defined attributes and parameters to moments in videos and was therefore used to annotate the videos for our former statistical analysis. Different behavioural attributes, as described in the previous subsection, were annotated for the videos. Conversational topics were annotated for the verbal behaviour, as well as speech acts. Further, non-verbal behaviour was annotated, i.e. arm postures, gesture types and dynamics of gestures. Afterwards the cultural background was added to the meta data of the annotations.

For further processing, the different modalities had to be aligned. Therefore, the annotated conversations were divided into conversational blocks (*dataset* in the following). These datasets are defined to refer to a specific speech utterance, specified by speech act and conversational topic. Following the categorization into speech acts, datasets are thus determined by a clause or sub-clause.

As non-verbal behaviour was defined to be determined by the verbal behaviour it accompanies, gestures and arm postures were added to datasets based on timely overlap. Thus, if a arm posture and/or gesture was annotated for the same time as a speech act, it was added to the corresponding dataset. If there was no non-verbal behaviour in the same time frame, an empty token was added to the dataset. In order to reflect that gestures and postures can be maintained for a longer time span, a gesture (or posture) was added to all datasets it overlapped with. Inversely, when multiple gestures or postures overlapped with a dataset's time period, the token with the longest overlap was chosen.

Due to data loss during the years of our endeavor that was caused by multiple annotation files and separate statistics, it was not possible to temporally align the gestures' dynamics with the corresponding speech acts any more, but to use them only quantitatively. Therefore, two different datasets were used for learning the parameters of the network. The aligned dataset was used to learn the joint probability distributions of arm postures and gesture types subject to verbal behaviour and culture. The probabilities of the gestures' dynamics were learned from the unaligned dataset and thus based on culture only.

After the extraction, we had two datasets: the aligned dataset containing 2155 values and the non-aligned dataset containing 457 values. Parameters were learned with the EM-algorithm [11] that is provided by the implementation of the SMILE-Framework [12] underlying the modeling environment used to model our network. This algorithm is able to deal well with missing data and is thus suitable for our purpose as there were some aspects not annotated for every person recorded. In the Japanese part of the corpus, e.g., there were some annotations of speech act and topic missing due to absent translation. The learning process itself was performed in two steps. First the probabilities of the parameters for arm posture and gesture type were learned from the aligned dataset. Afterwards the parameters for gesture dynamics were determined by applying the EM-algorithm with the non-aligned dataset.

6.5.5 Resulting Network

Figure 6.3 exemplifies the calculations of the resulting network with the evidence of cultural background either being set to Japanese (upper) or German (lower). In case no other evidence than cultural background is set within the network, distributions reflect the findings of our former statistical analysis, where we have been looking at behavioural aspects depending on cultural background in isolation (see Sect. 6.4). With this setting, cultural variations in non-verbal behaviour can be reflected in a general manner based on culture only.

Having a model at hand that combines several behavioural aspects instead of looking at them in isolation, further observations can be done in an intuitive manner. By setting additional evidences in the network, e.g. for verbal behaviour, the trained Bayesian Network allows to explore further interdependencies in the data. For example, a correlation of chosen topic and non-verbal behaviour frequency stayed unnoticed in our earlier work. From previous analysis of verbal behaviour [14], we know that the topic distribution is different for the two cultures in the data. While in Japan significantly more topics covering the immediate situation occurred compared to Germany, in Germany significantly more topics covering the communication situation occurred compared to Japan. Setting evidences of the topic nodes and the cultural background, the network reveals that people in both cultures are more likely to perform gestures when talking about less common topics. In particular, the communication situation in the Japanese culture and the immediate situation in the German culture. This effect could be explained by the tendency that talking about a more uncommon topic might lead to a feeling of insecurity that results in an increased usage of gestures. Thus, the network also reveals how culture-related non-verbal is mediated by culture-specific variations in verbal behaviour.

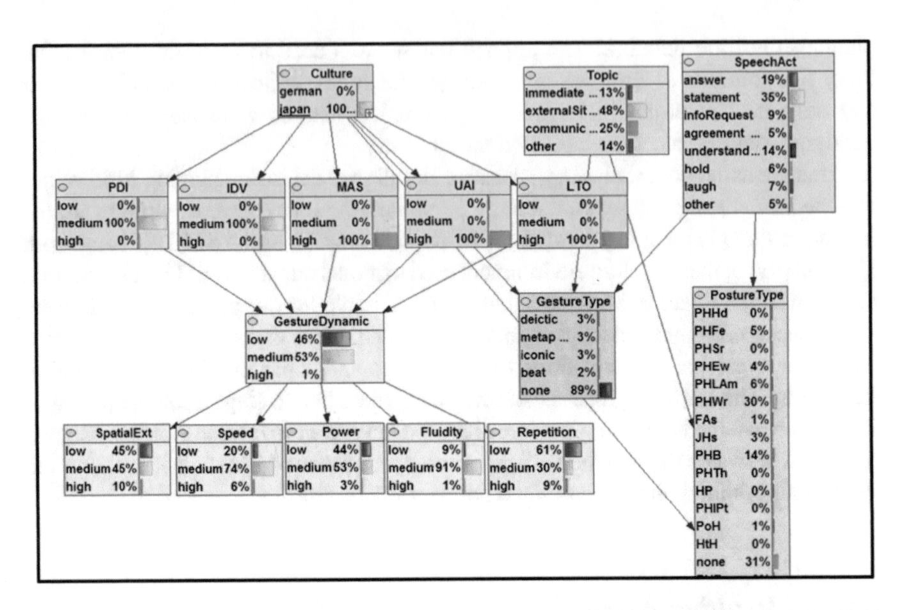

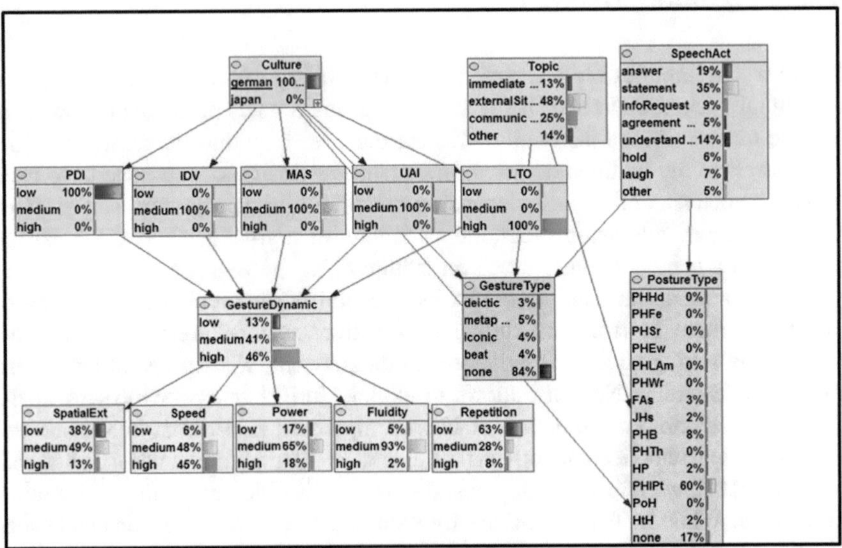

Fig. 6.3 Resulting Bayesian Network with parameters learned from empirical data, with cultural background set to Japanese (*upper*) and German (*lower*) respectively

6.6 Validation of the Network

An important question concerns the validation of the network. Basically, there are two possibilities. In this section, we will evaluate to what extent the network is able to predict characteristic behaviours of a person portraying a particular culture; i.e.

the generated behaviours are compared against the collected corpus. In Sect. 6.7, we will present a perception study in order to find out how human observers respond to the culture-specific behaviours of virtual characters generated with the network. That is the effect of the generated behaviours is evaluated from a user's perspective. We decided to rely on videos for the perception study to be able to evaluate our hypotheses in a controlled setting. As an alternative, we might have users interact with characters showing culture-specific behaviors. For example, in [27] we represented users by avatars which imitate their body movements and postures. System controlled characters respond to the users by dynamically adapting their behavior depending on their own assumed cultural background.

6.6.1 Measuring the Predictive Qualities of the Network

To validate the model, we investigate whether the network is able to predict appropriate culture-specific behaviours for new situations that are not included in the training corpus. To this end, a tenfold-cross-validation was performed. A dataset of 2155 values was used in which non-verbal behaviour (gesture types and arm postures) was aligned with the speech of the participants. The model was trained using 90% of the data while the remaining 10% were used as validation data. This validation process was performed ten times each time leaving out another part of the original dataset. For each dataset the cultural background (German or Japanese) and performed verbal behaviour (speech act type and topic) is given while gesture and arm posture are predicted by the network. Please note that the non-verbal dynamics cannot be validated using this approach due to the missing alignment. Predictions of the network were compared to the behaviour observed in the corpus data. Generally it appears quite unlikely for humans to behave the exact same way in a given situation several times. As a similar variety of behaviour is desirable for virtual characters we consider the network as suitable in case the observed gesture or posture is finding itself in the best three guesses of the network.

Figure 6.4 shows the prediction rates for gesture and arm posture types. Although results look quite promising, with an overall accuracy of 88% for gesture types and 56% for posture types, these results should not be overrated as for many of the observed speech acts no non-verbal behaviour was conducted (resulting in the gesture and posture type *none*). In particular, a gesture was performed only in 11% of our dataset, while a posture was performed for 71% of the values in the dataset.

As a result, another 10-fold-cross-validation was carried out on an adjusted dataset excluding data where no gesture or posture was observed. With it, prediction rates of the network are calculated assuming a gesture or posture should be performed by an agent. In total, 233 speech acts were accompanied by a gesture, 1551 by a posture. Figure 6.5 shows the results. Although a weak trend can be observed, only 34% of the performed gestures were predicted correctly by the

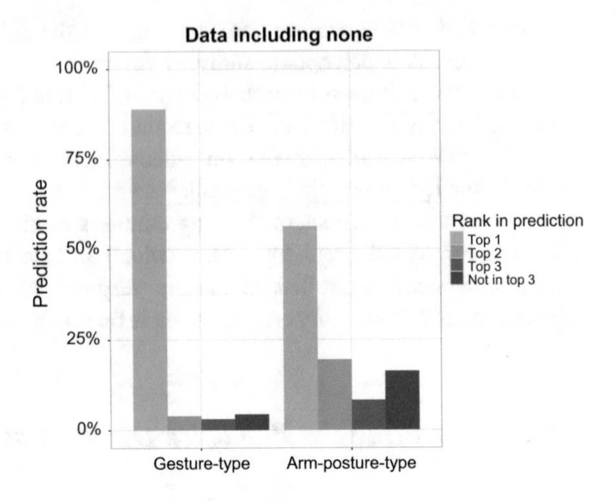

Fig. 6.4 Prediction rates of observed gesture types and posture types being in the first, second or third most likely gesture and posture type

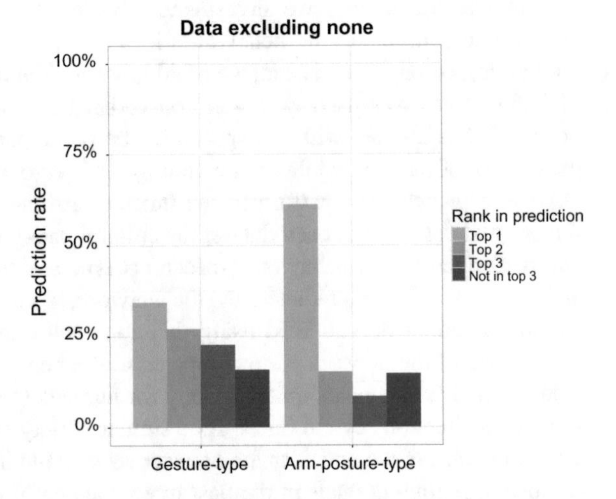

Fig. 6.5 Prediction rates of observed gesture type and posture type being in the first, second or third most likely gesture and posture type, excluding none-elements

network which seems not better than random. For posture types, however, the overall accuracy looks much more promising with 61% correct predictions.

In a further validation step cultural background was reversed to find out whether predicted gestures and arm postures reflect a prototypical cultural background. Thus, a 10-fold-cross-validation was performed with the cultural background set to German in the network while using the Japanese validation data, and vice versa. Including *none*-elements the accuracy of gesture types was still 80% as no gesture was the most likely option for both cultures. Therefore, *none*-elements were excluded again resulting in a drop of accuracy for gesture types to 24% (see Fig. 6.6), resulting in worse predictions of the network compared to the original data set (cf. Fig. 6.5).

For posture types, with reversed cultural background including *none*-elements accuracy drops to 5% while 75% of the observed postures did not even fall in the top 3 predictions. Excluding none-elements, accuracy is less than 2% with reversed cultural background with 92% of the observed posture types not being in the top three guesses (see Fig. 6.6). Thus, for arm postures changing the cultural background leads to a very low predictive power of the network suggesting that postures can be predicted culture-dependently by the network.

For the parameters of the gesture dynamics a 10-fold-cross-validation was performed based on cultural background alone. Thus, probabilities for the levels of dynamics are calculated given that a gesture is performed. More dataset values could be used in this case, as missing translations of verbal behaviour could be ignored, resulting in a dataset containing 457 gestures. Results are shown in Fig. 6.7. Please note that in this case the levels did comprise only three categories. Still, results look promising, suggesting that trends can be predicted for culture-dependent gesture dynamics.

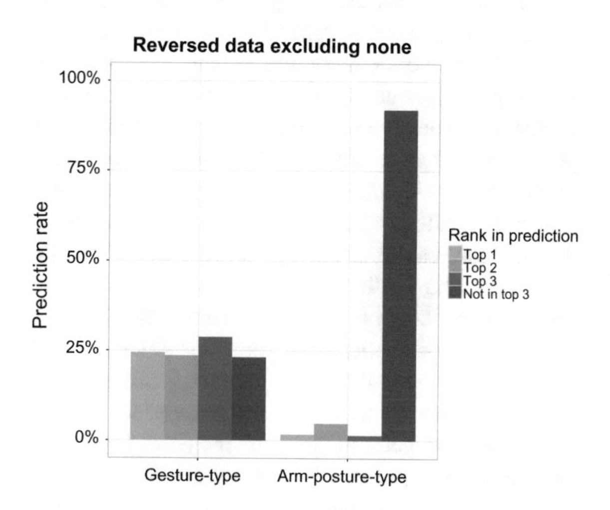

Fig. 6.6 Prediction rates of observed gesture types and posture types being in the first, second or third most likely gesture or posture type with the cultural background set to the reversed culture, excluding none-elements

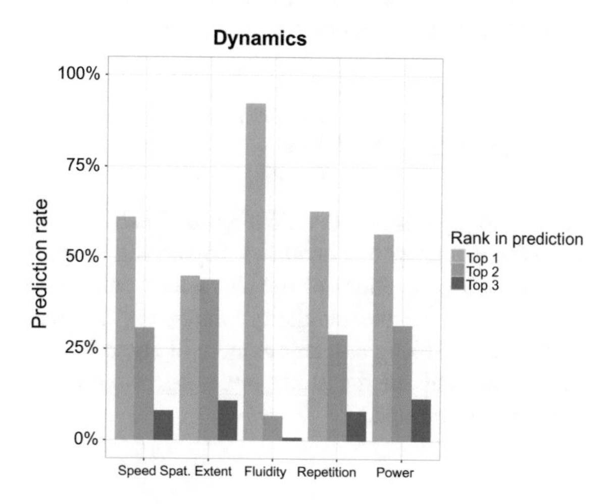

Fig. 6.7 Prediction rates of observed gestural dynamics being in the first, second or third most likely category

6.6.2 Discussion

With the cross-fold validation, we evaluated to what extent the network is able to predict culture-related behaviours of the underlying corpus. Regarding posture types and gestural expressivity, the presented network performed well. Our previous statistical analysis (e.g. [40]), also showed a strong correlation of cultural background and body posture with statistically significant differences between the cultures. In general, postures that regularly occurred in one culture barely occurred in the other culture. Similarly, the data revealed strong differences between the cultures regarding gestural expressivity suggesting that German participants gestured more expressively than Japanese participants.

Regarding gesture types, no reliable predictions could be made by our network. This is also reflected in the underlying data. The former statistical analysis showed that the overall number of gestures is similar in both cultures and no significant differences were found regarding the frequencies of McNeill's gesture types. As a result, the network cannot add to believably simulating culture-specific behaviours focusing on gesture types. This might have been caused by the abstraction of gestures to categories. Even if, for example, a deictic gestures is performed by people of different cultural backgrounds, the concrete execution can be very different. While a deictic gesture, for example, is typically performed using the index finger in Western cultures, this is considered rude in some Asian cultures, where deictic gestures are usually performed using the whole hand.

In sum, the resulting model can only be only as meaningful as the data being used to learn the probabilities. We thus believe that learning a Bayesian Network to enculturate non-verbal behaviours for simulated dialogues is a good approach in case the underlying data contains strong cultural differences. In our case, we believe the network can be used to add posture-types and levels of expressive behaviour to simulated dialogues in order to increase the culture-relatedness of the simulated non-verbal behaviours. For gesture types further research is needed such as going into more depth regarding the concrete performance of specific gestures or their correlation to the semantics of speech rather than speech acts.

6.7 Perception Study

A second approach of validating our network, is to investigate the resulting behaviours with virtual characters, by asking people of the targeted cultural groups to rate their perceptions of the characters' behaviours.

According to the similarity principle [7] interaction partners who perceive themselves as being similar are more likely to like each other. We therefore expect that participants of our study prefer agent conversations that resemble their own cultural background.

In our former work we performed perception studies with virtual characters that followed scripted behaviour based on the statistical analysis of our video corpus. Each behavioural aspect was tested in isolation to find out which of the implemented aspects of behaviour cause the desired effects. Results suggested that observers tended to prefer virtual agent behaviour that is in line with their own cultural background for some of the behavioural aspects (such as postures or prototypical topics), while we did not find significant differences for other aspects (such as gesture types) [13].

In comparison to the scripted perception studies, the present study uses the Bayesian Network described in Sect. 6.5.5 which is able to present culture as a non-deterministic concept and preserve a certain variety in the characters' behaviours. In addition, all behavioural aspects implemented in the network are generated in combination based on cultural background and the underlying dialogue.

6.7.1 Design

The perception study was conducted in a mixed-design with the participant's culture as independent between-subjects variable with two levels (German and Japanese), the agent-culture as within-subjects measure (German and Japanese) and the participants' subjective impression of the characters behaviour and the conversation in general as dependent measures. The participants consecutively watched four videos of conversations of virtual characters. After watching a video they rated their agreement to several statements regarding their impression of the conversation they saw in the video. Hence, the subjective impression per culture was designed as a within-subjects factor with two levels, each calculated as the average of two videos for every variable. The videos were presented to the participants based on a 4×4 Latin square to counterbalance order effects.

6.7.2 Apparatus

For the realization of the perception study we need a demonstrator containing a virtual environment with virtual characters, a conversational setting with verbal behaviour, as well as a simulation of the generated non-verbal behaviours.

6.7.2.1 Virtual Environment

We use our virtual character engine [10] that contains a virtual Beergarden scenario, in which virtual characters can be placed. Regarding the verbal behaviour of the characters, a text-to-speech engine with different voices and languages such as German, English or Japanese can be used.

To simulate a first time conversation similar to the ones recorded in our corpus, two characters were placed into the scenario facing each other. To be as culturally neutral as possible, characters were chosen that do not contain typical ethnically appearances such as blond hair. To avoid side effects evoked by gender, we chose a mixed gender combination for the agent conversations. Thus a female and a male character interact with each other. Figure 6.8 shows a screenshot of the setup.

To use the virtual agent engine with our network model, a dialogue component was added that allows to script dialogues in an XML structure where speech acts and topic categories can be tagged for verbal behaviour, and a cultural background can be set for each character. For dialogues that have been prepared in that manner, the Bayesian Network is able to generate non-verbal behaviours for a given cultural background.

Regarding non-verbal behaviour, over 40 different animations can be performed by the characters, including gestures and body postures. Body postures were modeled to match the arm-posture types included in our network. Figure 6.9 illustrates two typical arm postures that were observed regularly in our video corpus.

To select gestures, existing animations had to be labeled according to the gesture types used by our network. In addition, gestures can be customized by the animation engine to match different levels of expressivity [10]. The speed parameter is adapted by using a different frame rate. Animation blending is used for differences in the spatial extent (blending with a neutral body posture) and fluidity (blending over a shorter or longer period of time). Differences in the repetivity are archived by playing the stroke of a gesture several times.

Fig. 6.8 Two virtual characters facing each other during a conversational setting

Fig. 6.9 Prototypical arm postures displayed by our male virtual character (*left*: The prototypical Japanese posture "*Put hands to wrist*"; *right*: the prototypical German posture "*Put hands into pocket*")

6.7.2.2 First-Time Meeting Dialogue

As the focus of the present chapter lies on non-verbal behaviour that accompanies first-time meeting dialogues, a dialogue needs to be written that contains the casual small talk of such a situation, whilst not holding culture-specific content. Creating such a dialogue is not trivial, since, as we pointed out earlier, dialogue behaviour can heavily depend on cultural background. In one of our previous studies [14], we addressed that issue, and analysed the first-time meetings of our video corpus regarding differences in topic selection. Following Schneider [43], topics have been classified into (1) Topics covering the *immediate situation* which describe elements of the so-called "frame" of the situation. The frame of a small talk conversation at a party, for example, holds topics such as the drinks, music, location or guests. (2) The second category, the *external situation* or "supersituation" includes the larger context of the immediate situation such as the latest news, politics, sports, movies or celebrities. (3) The *communication situation* contains topics that concentrate on the conversation partners. Thus, personal things such as hobbies, family or career are part of this category. The corpus analysis revealed that topics covering the *external situation* were the most common topics in both cultures. In the German conversations the *immediate situation* occurred significantly less compared to the *external situation*, while in the Japanese conversations the *communication situation* occurred significantly less compared to the *external situation*. In a perception study with virtual characters we found out that conversations with a typical German topic distribution were preferred by German participants, while conversations with a typical Japanese topic distribution were preferred by Japanese participants [14]. It is therefore crucial to avoid topic categories in the present study that are not common in one of the targeted cultures. Another issue might be that a dialogue that is written by us might be influenced by our own cultural background and might thus not be general enough to be considered a normal casual small talk conversation in the other culture. To tackle this issue, in [14], we agreed on six English dialogues with our Japanese cooperation partners, that would in general be feasible in both cultures. Please note that for our former aim three of the dialogues contained a prototypical German topic distribution and three of the dialogues contained a prototypical Japanese topic distribution. Therefore, for the present study, we needed to cut down the dialogues in a sense that we only keep parts in which the *external situation* is discussed, while the *immediate situation* and the *communication situation* were avoided. This results in a dialogue that lasts for approximately 60 s, that could occur in both cultures, and that only contains topics that are common in both cultures. As pointed out in Sect. 6.7.2.1, dialogues need to be tagged with their speech acts, to be used by our network. Please see Table 6.2 (left part) for the resulting dialogue and the annotated speech acts. The table additionally contains an example of non-verbal behaviour that was generated by our network for the German and Japanese culture respectively. Please note, that a posture was maintained by the characters until a different animation was selected.

6.7.2.3 Video Generation

The probabilities for non-verbal behaviours are generated by the network depending on the current speech act, topic, and cultural background of the agents. For display with the virtual characters, our demonstrator allows two options. Either the most likely non-verbal behaviour is displayed, or the probability distribution of the network is reflected by displaying a non-verbal behaviour that is chosen based on an algorithm that follows the probabilities. While always choosing the most probable behaviour is very well suited for illustration, it lacks a certain variety in the characters' behaviours. Thus, to present culture as a non-deterministic concept and to reflect the generative power of our network, we use the second option for the videos that shall be shown in our perception study.

This approach can be quite risky, as it might result in a conversation that contains behaviours that are very unlikely for a given cultural background, or produce conversions that do not contain a reasonable amount of animations at all. While these effects would cancel out over a long time period of agent conversations, for an evaluation study with a dialogue of 60 s only the generated behaviour might not be representative enough. Therefore, we were running the network ten times for each culture and recorded videos of the resulting conversations. To prepare our perception study, we manually selected two videos for each culture, that contained a comparable amount of non-verbal behaviours and no animations that were very unlikely for the given cultural background. By selecting two videos per culture, we wanted to assure that we did not accidentally choose a video that would in general be rated better or worse due to the specific animation selection.

Table 6.2 Dialogue and generated prototypical non-verbal behaviour

Interlocutor	Utterance	Speech act	Jap. non-verb.	Ger. non-verb.
Agent A	Do you know Mary for a long time now?	InfoRequest	PHB	PHIPt
Agent B	No.	Answer	None	PHIPt
	Not too long.	Hold	None	PHIPt
	I met her last year at university.	Statement	PHWr	PHIPt
Agent A	Mary looks busy for her part-time job.	Statement	PHB	PHIPt
Agent B	Yes.	Agreement	PHWr	PHIPt
	I heard that she goes to the part-time job 3 times a week.	Statement	Beat	PHIPt
	But one of our friends is missing.	Statement	PHWr	Beat
	She is on a trip to Brazil.	Statement	PHWr	PHIPt
Agent A	Brazil?	Hold	PHB	PHIPt
	It is supposed to be very beautiful there.	Statement	PHWr	Metaphoric
Agent B	Yes.	Understanding	PHB	PHIPt
	And people there are very friendly.	Statement	PHB	PHIPt
Agent A	This is especially good for hiking as far as I know.	Statement	PHWr	PHIPt
Agent B	Right.	Agreement	PHB	PHIPt
	There are many good hiking trips.	Hold	PHB	PHIPt
	There are also the Olympic Games in Brazil this summer, aren't they?	InfoRequest	Deictic	PHIPt
Agent A	I think you are right.	Agreement	PHWr	PHIPt
	They are taking place in Brazil this year.	Statement	PHWr	Deictic

Please note, that the original dialogue had slightly to be modified, as the current location of the proceeding Olympic games had changed over the years

Please see Table 6.2 (right part) for the non-verbal behaviours that were generated by our network for two of the chosen videos of our perception study.

6.7.2.4 Questionnaires

For evaluation, a two-parted questionnaire was developed: Part A included questions focused on the observed conversation, part B requested demographical data.

In part A of the questionnaire, participants were asked to rate the characters in the video as well as their perception of the conversation on 7-point-Likert scales,

Table 6.3 Participants had to state their agreement regarding their perception of the characters and the conversation in general (Part A of the questionnaire)

Statements regarding the perception of the characters
The characters' behaviour was natural
The characters' behaviour was appropriate
The characters were getting along with each other well
Statements regarding the perception of the conversation
The characters' movements matched the conversation *I liked watching the conversation*
I would like to join the conversation
The conversation or a similar conversation would be realistic in my own life /my friends life

ranging from "strongly disagree" to "strongly agree" (see Table 6.3). At the end, participants were provided a comment box for further opinions.

In part B of the questionnaire, participants were asked to provide demographical data on:

- age
- gender
- the country they currently live in
- the country they have lived in mostly in the last 5 years
- their ability to understand spoken English well and
- their ability to understand the spoken English of the videos well (the latter two were rated on a 7-point-Likert skale)

6.7.3 Procedure

The study was embedded in an online survey.[2] At first it was explained that participation was totally voluntary, that they could withdraw at any time, and that they should turn on their speakers to be able to listen to the proceeding conversations. Then they were introduced to the scenario they were going to see. The imaginary scenario involved two virtual characters, that have just met each other for the first time in a social setting. They were introduced to each other by a common friend that has just left to pick up something to drink for all of them. To get participants acquainted with the virtual environment, the visual appearance of the characters, as well as their ability to conduct non-verbal behaviours they were shown a "neutral" video first. This video showed the two characters greeting each other by introducing themselves stating their names. In order to assure participants had their speakers

[2]The study was created with SoSci Survey (Leiner 2014) and made available to the participants on www.soscisurvey.com.

enabled and understood the language, they had to enter the characters' names as free-response items to be able to further proceed in the study.

After the introduction, participants were told that they were going to watch four videos of the same conversation. Videos could be watched several times if the participants wanted to do so. After each video, participants had to answer part A of our questionnaire.

After the last conversation, participants were asked to provide the demographic data requested in part B of our questionnaire. Finally, participants were thanked for their participation. Overall, completion of participation took participants 10–15 min.

6.7.4 *Participants*

There were 56 respondents taking part in the online-survey. After excluding four respondents that did not finish the questionnaire, there was a final sample of 52 participants (29 German and 23 Japanese). The culture of participants was assessed by asking where they lived mostly the last 5 years and where they currently live. Two participants did currently live in another country than the country they lived in mostly in the last five years. For these participants latter values were used as they fit the scope of the study better. 27 of the participants were male (11 Japanese-male, 16 German-male) and 25 female (12 Japanese-female, 13 German-female), with an age range between 19 and 41 ($M = 24.98$, $SD = 5.29$). All participants were volunteers, recruited via social networks or e-mail lists. On average the self-reported English skills for Japanese participants ($M = 4.22$, $SD = 1.20$) was significantly lower than for German participants ($M = 6.31$, $SD = 0.76$), $t(36) = 7.44$, $p < 0.001$, $r = 0.78$. As Levene's test indicated unequal variances ($F = 4.613$, $p = 0.037$) degrees of freedom were adjusted from 50 to 36. Regarding their understanding of the spoken English of the videos Japanese participants' ratings ($M = 4.87$, $SD = 1.25$) were also significantly lower than the ratings of the German participants ($M = 6.76$, $SD = 0.51$), $t(28) = 6.790$, $p < 0.001$, $r = 0.79$. Degrees of freedom were adjusted from 50 to 28 as Levene's test showed unequal variances ($F = 20.023$, $p < 0.001$). Since all self-reported English skills were high enough, no participant had to be excluded due to language skills.

6.7.5 *Results*

In order to draw conclusions based on the generated behaviour and not on a specific video, the mean of both videos was calculated for each agent-culture and every item of the subjective impression. For example, the subjective naturalness of the prototypical German behaviour was calculated as the mean of the subjective naturalness of both videos showing prototypical German behaviour. Figures 6.10 and 6.11

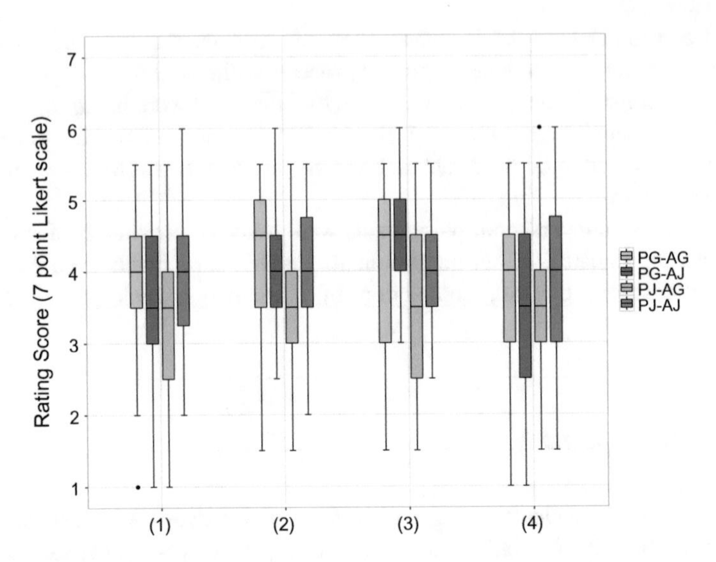

Fig. 6.10 Results of the subjective impression of the German agents (AG) and Japanese agents (AJ) for German participants (PG) and Japanese participants (PJ) with regard to the statements from Table 6.3: (*1*) "*The characters' behaviour was natural.*", (2) "*The characters' behaviour was appropriate.*", (*3*) "*The characters were getting along with each other well.*", (*4*) "*The characters' movements matched the conversation.*"

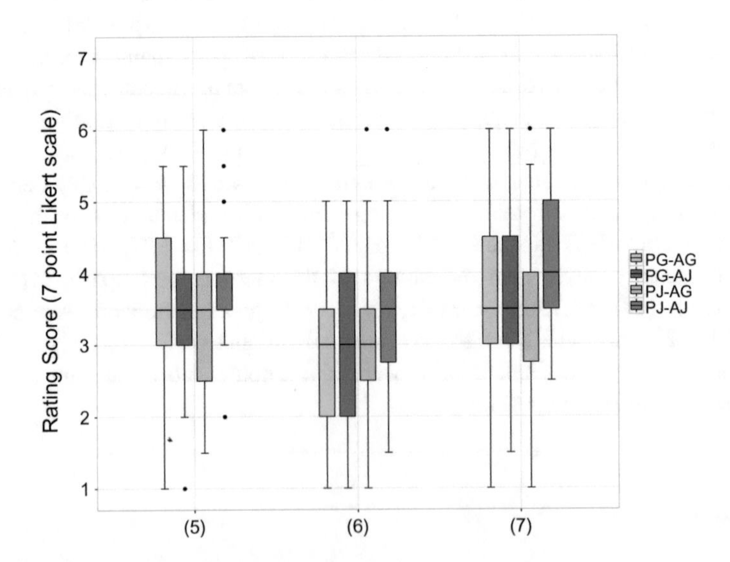

Fig. 6.11 Results of the subjective impression of the German agents (AG) and Japanese agents (AJ) for German participants (PG) and Japanese participants (PJ) with regard to the statements from Table 6.3: (*5*) "*I liked watching the conversation.*", (*6*) "*I would like to join the conversation.*", (*7*) "*The conversation or a similar conversation would be realistic in my own life/ my friends life.*"

show the ratings for German and Japanese agents. Mixed ANOVAs were conducted for every item, each with the agent-culture as independent within-subjects measure and the participants' culture as independent between-subjects variable.

The subjective naturalness of the characters was not significantly different between the agent-cultures, $F(1,50) = .727, p = .398$, as well as for the participants culture, $F(1,50) = .221, p = .640$. However, there was a significant interaction of the agent-culture and participant-culture, $F(1, 50) = 4.173, p = .039$. Participants did rate the agent-culture based on their own cultural background higher than the conversations of the other agent-culture. That means that Japanese participants did rate the Japanese conversations more natural than the German conversations while Germans' ratings did show the reverse effect.

There was no significant effect of agent-culture on subjective appropriateness, $F(1, 50) = 2.110, p = .153$. Participants' culture had no effect, $F(1, 50) = 1.723, p = .195$, and neither did the interaction, $F(1, 50) = 2.110, p = .153$.

Regarding the statement "*The characters were getting along with each other well*" there was a significant main effect of the agent-culture, $F(1, 50) = 5.744, p = .020$. The conversations of the Japanese agents were overall rated better. The participants' culture, however, did not show significant effects, $F(1, 50) = .3.431, p = .070$, and neither did the interaction, $F(1, 50) = .001, p = .981$.

The participants' ratings whether the characters' movements matched the conversation did not differ significantly depending on the participants' culture, $F(1, 50) = .303, p = .584$, on the agents' culture, $F(1, 50) = .051, p = .823$, or the interaction of participants' and agents' culture, $F(1, 50) = .787, p = .379$.

Participants did not like watching conversations of either agent-culture significantly more, $F(1, 50) = 1.425, p = .238$. There was also no effect of the participants' culture, $F(1, 50) = 1.002, p = .322$, and no effect of the interaction, $F(1, 50) = 1.958, p = .168$.

Concerning the statement "*I would like to join the conversation*" there was no significant effect on the participants' rating depending on the agents' culture, $F(1, 50) = 2.772, p = .102$. However, there was a significant effect of the participants' culture, $F(1, 50) = 4.196, p = .038$. Japanese participants did agree significantly more liking to join the conversation than Germans did. There was no interaction effect of agent-culture and participant-culture, $F(1, 50) = .518, p = .475$.

Participants' ratings whether such or a similar conversation could happen in their own life did significantly depend on agent-culture, $F(1, 50) = 6.327, p = .015$, with the Japanese conversations being rated more realistic. The effect of participants' culture, $F(1, 50) = .194, p = .661$, and the interaction effect, however, were non-significant, $F(1, 50) = 2.992, p = .090$.

6.7.6 Discussion

The survey was conducted to investigate whether the network can generate culture-specific non-verbal behaviour that is perceived differently by human observers of different cultures. This question comprises different hypotheses, i.e. if

(1) agent-culture, (2) participant-culture, and (3) the interaction thereof has an effect on the participants' subjective impression of the characters' conversations. The results show that the hypotheses can be confirmed in part.

In general, the ratings of the subjective impression of the Japanese agents were higher than for the German agents for all statements. However, significant effects were only found for "*The characters were getting along with each other well*" and "*The conversation or a similar conversation would be realistic in my own life /my friends' life*". This suggests that the network did indeed generate different behaviours for the characters as the videos with Japanese agents were rated better. However, this might also imply that the generated behaviour for the Japanese agents fits the conversation better than for the German agents. For example gestures might have been more suitable to the semantics of speech or their timing was better. This is in fact a limitation of the automatic generation of non-verbal behaviour. Therefore, it appears necessary to incorporate further techniques into an application using an automatic generation approach in order to validate that the selected non-verbal behaviour is suitable.

Participant culture did also have a significant effect on the subjective impression for one statement. Japanese participants were significantly more interested in joining the conversation than Germans did. Further research is necessary to explain whether this is a cultural difference or whether some design aspect of the conversations influenced the ratings.

The expectation that the participants of our study prefer agent conversations that resemble their own cultural background led to the hypothesis that there should be an interaction effect of participant-culture and agent-culture on the subjective impression. Figures 6.10 and 6.11 show that the participants' culture did mostly lead to higher ratings of agents of their own culture, albeit these effects were not significant except for the subjective naturalness confirming the hypothesis partially. As suggested, this might be due to the similarity principle [7] that states that interaction partners who perceive themselves as being similar are more likely to like each other. This principle could also apply in case participants perceive agents as being similar to themselves.

Although participants watched a neutral video first to get acquainted with the scenario, ratings still might have been influenced by details of our demonstrator. This impression is strengthened by some open-ended comments of the participants mentioning details such as graphics and sound of the videos, facial expressions or interpersonal distance of the characters—although these aspects were the same in all videos, including the neutral video.

6.8 Conclusion and Future Work

In this chapter, we presented an approach to generate culture-dependent non-verbal behaviour for virtual characters that is theory-based as well as data-driven. The approach combines advantages of procedures commonly used, as it explains the causal relations of cultural background and resulting behaviour, and augments them by findings from empirical data. Therefore, we built on our former work where we have explored both approaches (theory-based as well as data-driven) separately.

To realize this endeavor, we relied on a Bayesian Network. While the structure of the network along with categorizations of behavioural aspects have been constructed based on existing theories and models, the parameters were learned from annotated data. The resulting network generates non-verbal behaviours based on observations for the German and Japanese cultures for given conversations. With the network we are able to keep a certain variability in behaviour by making predictions based on conditional probabilities.

Later in the chapter, we showed and performed two ways of validating the resulting model: regarding its predictive power and the perception of human observers of the generated behaviours.

The more technical validation of the network shows promising results for some of the behavioural aspects (postures and gestural dynamics), while it fails in predicting culture-related choices of gestures-types. Comparing these outcomes with our previous statistical analysis of the video data, it reflects that the resulting model can only be as meaningful as the underlying data. In those cases, where strong culture-related differences can be observed in the data, training a Bayesian Network seems a good approach to generate culture-related differences for simulated dialogues. In cases where no significant differences were found, e.g. gesture types, the network fails in predicting culture-specific behaviour. We therefore have to research the usage of gestures in more detail and analyse, for example, their concrete performance (e.g. handedness or hand-shape), their correlation to the semantics of speech, or their timely synchronization.

In the second evaluation we examined the perception of participants of the targeted cultures for different versions of behaviour generated by our network. We expected participants to prefer virtual characters with simulated culture-specific behaviour that is in line with prototypical culture-specific behaviour of the participant's culture. The survey partly confirmed our expectations. We therefore hope that the localization of the characters' behaviours can help improve their acceptance by users of the targeted cultures.

With the present chapter we want to provide guidance for other research aiming at integrating culture-specific behaviour into virtual character systems by describing a complete approach. The resulting model may be expanded by further aspects of culture-specific behaviours. In our future work, we aim at adding additional non-verbal behavioural traits that are known to be dependent on cultural background, such as head nods. In a similar way, other diversifying factors, such as gender or age, can be added.

References

1. André, E.: Preparing emotional agents for intercultural communication. In: Calvo, R., D'Mello, S., Gratch, J., Kappas, A. (eds.) The Oxford Handbook of Affective Computing. Oxford University Press (2014)
2. Aylett, R., Hall, L., Tazzymann, S., Endrass, B., André, E., Ritter, C., Nazir, A., Paiva, A., Hofstede, G.J., Kappas, A.: Werewolves, Cheats, and Cultural Sensitivity. In: Proc. of 13th Int. Conf. on Autonomous Agents and Multiagent Systems (AAMAS 2014) (2014)
3. Aylett, R., Paiva, A., Vannini, N., Enz, S., André, E., Hall, L.: But that was in another country: agents and intercultural empathy. In: Proc. of 8th Int. Conf. on Autonomous Agents and Multiagent Systems (AAMAS 2009) (2009)
4. Bergmann, K., Kopp, S.: Bayesian decision networks for iconic gesture generation. In: Ruttkay, Z., Kipp, M., Nijholt, A., Vilhjálmsson, H.H. (eds.) Proc. of 9th Int. Conf. on Intelligent Virtual Agents (IVA 2009), pp. 76–89. Springer (2009)
5. Buisine, S., Courgeon, M., Charles, A., Clavel, C., Martin, J.C., Tan, N., Grynszpan, O.: The role of body postures in the recognition of emotions in contextually rich scenarios. Int. J. Hum. Comput. Interac. **30**(1) (2014)
6. Bull, P.: Posture and Gesture. Pergamon Press, Oxford (1987)
7. Byrne, D.: The attraction paradigm. Academic Press, New York (1971)
8. Cassell, J., Vilhálmsson, H., Bickmore, T.: BEAT: The behaviour expression animation toolkit. In: Proc. of 28th Annual Conf. on Computer Graphics (SIGGRAPH 2001), pp. 477–486. ACM (2001)
9. Core, M., Allen, J.: Coding Dialogs with the DAMSL Annotation Scheme. In: Working Notes of AAAI Fall Symposium on Communicative Action in Humans and Machines, pp. 28–35. Boston, MA (1997)
10. Damian, I., Endrass, B., Huber, P., Bee, N., André, E.: Individualized Agent Interactions. In: Proc. of 4th Int. Conf. on Motion in Games (MIG 2011) (2011)
11. Dempster, A.P., Laird, N.M., Rubin, D.B.: Maximum likelihood from incomplete data via the EM algorithm. J. Roy. Stat. Soc. B (Methodological) pp. 1–38 (1977)
12. Druzdzel, M.J.: SMILE: Structural Modeling, Inference, and Learning Engine and GeNIe: A development environment for graphical decision-theoretic models (Intelligent Systems Demonstration). In: Proc. of the 16th National Conf. on Artificial Intelligence (AAAI-99), pp. 902–903. AAAI Press (1999)
13. Endrass, B., André, E., Rehm, M., Nakano, Y.: Investigating culture-related aspects of behavior for virtual characters. Auton. Agent. Multi-Agent Syst. **27**(2), 277–304 (2013). doi:10.1007/S10458-012-9218-5
14. Endrass, B., Nakano, Y., Lipi, A., Rehm, M., André, E.: Culture-related topic selection in SmallTalk conversations across Germany and Japan. In: Vilhjálmsson, H.H., Kopp, S., Marsella, S., Thórisson, K.R. (eds.) Proc. of 11th Int. Conf. on Intelligent Virtual Agents (IVA 2011), pp. 1–13. Springer (2011)
15. Endrass, B., Rehm, M., Lipi, A.A., Nakano, Y., André, E.: Culture-related Differences in Aspects of Behavior for Virtual Characters across Germany and Japan. In: Proc. of 10th Int. Conf. on Autonomous Agents and Multiagent Systems (AAMAS 2011), pp. 441–448 (2011)
16. Field, A.: How to Design and Report Experiments. Sage Publications, UK (2003)
17. Gallaher, P.E.: Individual differences in nonverbal behavior: Dimensions of style. J. Pers. Soc. Psychol. **63**(1), 133–145 (1992)
18. Hall, E.T.: The Hidden Dimension. Doubleday (1966)
19. Herrera, D., Novick, D.G., Jan, D., Traum, D.R.: Dialog behaviors across culture and group size. In: Stephanidis, C. (ed.) Universal Access in Human-Computer Interaction. Users Diversity—6th International Conference, UAHCI 2011, Held as Part of HCI International 2011, Orlando, FL, USA, July 9–14, 2011, Proceedings, Part II, *Lecture Notes in Computer Science*, vol. 6766, pp. 450–459. Springer (2011)

20. Hofstede, G.: Cultures Consequences—Comparing Values, Behaviours, Institutions, and Organizations Across Nations. Sage Publications, UK (2001)
21. Hofstede, G., Hofstede, G.J., Minkov, M.: Cultures and Organisations. SOFTWARE OF THE MIND. Intercultural Cooperation and its Importance for Survival. McGraw Hill (2010)
22. Hofstede, G.J., Pedersen, P.B., Hofstede, G.: Exploring Culture – Exercises, Stories and Synthetic Cultures. Intercultural Press, United States (2002)
23. Isbister, K.: Building bridges through the unspoken: Embodied agents to facilitate intercultural communication. In: Payr, S., Trappl, R. (eds.) Agent Culture: Human-Agent Interaction in a Multikultural World, pp. 233–244. Lawrence Erlbaum Associates (2004)
24. Isbister, K., Nakanishi, H., Ishida, T., Nass, C.: Helper agent: Designing an assistant for human-human interaction in a virtual meeting space. In: Turner, T., Szwillus, G. (eds.) Proc. of Int. Conf. on Human Factors in Computing Systems (CHI 2000), pp. 57–64. ACM (2000)
25. Kipp, M.: Anvil - A Generic Annotation Tool for Multimodal Dialogue. Eurospeech **2001**, 1367–1370 (2001)
26. Kistler, F., André, E., Mascarenhas, S., Silva, A., Paiva, A., Degens, N., Hofstede, G.J., Krumhuber, E., Kappas, A., Aylett, R.: Traveller: An interactive cultural training system controlled by user-defined body gestures. In: Kotzé, P., Marsden, G., Lindgaard, G., Wesson, J., Winckler, M. (eds.) Human-Computer Interaction—INTERACT 2013—14th IFIP TC 13 International Conference, Cape Town, South Africa, September 2–6, 2013, Proceedings, Part IV, *Lecture Notes in Computer Science*, vol. 8120, pp. 697–704. Springer (2013)
27. Kistler, F., Endrass, B., Damian, I., Dang, C.T., André, E.: Natural interaction with culturally adaptive virtual characters. J. Multimodal User Interfaces **6**(1–2), 39–47 (2012)
28. Kleinsmith, A., Silva, P.D., Bianchi-Berthouze, N.: Recognizing emotion from postures: Cross-cultural differences in user modeling. In: User Modeling 2005, no. 3538 in LNCS, pp. 50–59 (2005)
29. Kluckhohn, K., Strodtbeck, F.: Variations in value orientations. Row, Peterson, New York, United States (1961)
30. Martin, J.C., Abrilian, S., Devillers, L., Lamolle, M., Mancini, M., Pelachaud, C.: Levels of representation in the annotation of emotion for the specification of expressivity in ECAs. In: Proc. of 5th Int. Conf. on Intelligent Virtual Agents (IVA 2005), pp. 405–417. Springer (2005)
31. Mascarenhas, S., Dias, J., Prada, R., Paiva, A.: One for all or one for one? the influence of cultural dimensions in virtual agents' behaviour. In: Z. Ruttkay, M. Kipp, A. Nijholt, H.H. Vilhjálmsson (eds.) Intelligent Virtual Agents, 9th International Conference, IVA 2009, Amsterdam, The Netherlands, September 14–16, 2009, Proceedings, *Lecture Notes in Computer Science*, vol. 5773, pp. 272–286. Springer (2009)
32. Mascarenhas, S., Silva, A., Paiva, A., Aylett, R., Kistler, F., André, E., Deggens, N., Hofstede, G.J., Kappas, A.: Traveller: an intercultural training system with intelligent agents. In: Proc. of 12th Int. Conf. on Autonomous Agents and Multiagent Systems (AAMAS 2013) (2013)
33. McNeill, D.: Hand and Mind—What Gestures Reveal about Thought. University of Chicago Press, Chicago, London (1992)
34. Nouri, E., Traum, D.R.: Generative models of cultural decision making for virtual agents based on user's reported values. In: Bickmore, T.W., Marsella, S., Sidner, C.L. (eds.) Intelligent Virtual Agents—14th International Conference, IVA 2014, Boston, MA, USA, August 27–29, 2014. Proceedings, *Lecture Notes in Computer Science*, vol. 8637, pp. 310–315. Springer (2014)
35. Ortony, A., Clore, G., Collins, A.: The Cognitive Structure of Emotions. Cambridge University Press, Cambridge, UK (1988)
36. Pelachaud, C.: Multimodal expressive embodied conversational agents. In: Proc. of 13th annual ACM Int. Conf. on Multimedia, pp. 683–689 (2005)
37. Rehm, M., André, E., Nakano, Y., Nishida, T., Bee, N., Endrass, B., Huan, H.H., Wissner, M.: The CUBE-G approach - Coaching culture-specific nonverbal behavior by virtual agents. In: Mayer, I., Mastik, H. (eds.) ISAGA 2007: Organizing and Learning through Gaming and Simulation (2007)

38. Rehm, M., Bee, N., André, E.: Wave like an egyptian: accelerometer based gesture recognition for culture specific interactions. BCS HCI **1**, 13–22 (2008)
39. Rehm, M., Bee, N., Endrass, B., Wissner, M., André, E.: Too close for comfort?: adapting to the user's cultural background. In: Proceedings of the international workshop on Human-centered multimedia, pp. 85–94. ACM (2007)
40. Rehm, M., Nakano, Y., André, E., Nishida, T., Bee, N., Endrass, B., Wissner, M., Lipi, A.A., Huang, H.H.: From observation to simulation: generating culture-specific behavior for interactive systems. AI & Soc. **24**(3), 267–280 (2009)
41. Russell, S.J., Norvig, P.: Artificial Intelligence—A modern approach (3. internat. ed.). Pearson Education (2010)
42. Sagiv, L., Schwartz, S.H.: Cultural values in organisations: Insights for Europe. European Journal of International Management **1**(3), 176–190 (2007)
43. Schneider, K.P.: Small talk: Analysing phatic discourse. Hitzeroth, Marburg (1988)
44. Si, M., Marsella, S., Pynadath, D.V.: Thespian: Modeling socially normative behavior in a decision-theoretic framework. In: Gratch, J., Young, R. Aylett, D. Ballin, P. Olivier (eds.) Intelligent Virtual Agents, 6th International Conference, IVA 2006, Marina Del Rey, CA, USA, August 21–23, 2006, Proceedings, *Lecture Notes in Computer Science*, vol. 4133, pp. 369–382. Springer (2006)
45. Ting-Toomey, S.: Communicating across cultures. The Guilford Press, New York (1999)
46. Trompenaars, F., Hampden-Turner, C.: Riding the waves of culture—Understanding Cultural Diversity in Business. Nicholas Brealey Publishing, London (1997)

Chapter 7
Mental Activity and Culture: The Elusive Real World

Gert Jan Hofstede

Abstract How does culture affect mental activity? That question, applied to the design of social agents, is tackled in this chapter. Mental activity acts on the perceived outside world. It does so in three steps: perceive, interpret, select action. We see that when culture is taken into account, objective reality disappears to a large extent. Instead, perception, interpretation and action selection can differ in many ways between agents from different cultures. This complicates the design of artificially intelligent systems. On the other hand, theory exists that can help us deal with these complications. All people have a shared set of drives and capacities, on which cultures are built. Good knowledge exists on how culture affects perception, interpretation, and action. Empirical research has uncovered major distinctions in social life across cultures. One could say that intelligent agents with different cultures live in the same social world, but in systematically different social landscapes. This social world—in the form of generic sociological theory—and these differences—in the form of cross-cultural theory—can be used for designing these agents. The state of the art is still tentative. The chapter gives examples from recent literature that can serve as points of departure for further work.

Keywords Culture · Social agents · Mental activity · Drives
Social landscape · Generic sociological theory · Cross-cultural theory
Reality · Perception · Interpretation · Action selection

7.1 Introduction

A central premise in artificial intelligence is that "there is one world out there". This world can be observed by intelligent entities, they can reason about their observations, and then act upon their reasoning. In an artificial environment with more than one intelligent entity perception, interpretation and action thus act on a shared world.

G.J. Hofstede (✉)
Department of Social Sciences, Applied Information Technology Group & SiLiCo,
Wageningen University, Wageningen, The Netherlands
e-mail: gertjan.hofstede@wur.nl

C. Faucher (ed.), *Advances in Culturally-Aware Intelligent Systems and in Cross-Cultural Psychological Studies*, Intelligent Systems Reference Library 134, https://doi.org/10.1007/978-3-319-67024-9_7

The central premise in this chapter is that in the real world of human beings, there can be as many observed worlds as there are individuals. In a maxim: "There is no such thing as reality, only perception". Just like perception, interpretation and action selection differ across individuals. Perception is the key factor though, since how can people communicate effectively if they perceive different worlds?

If individuals share the same culture, they live in the same 'social landscape'. Their differences in perception are smaller. They can negotiate them if they want to. They can provide feedback to one another in a language understood and in a way accepted by the other.

The problem of achieving a shared perception is confounded across cultures, because a common language and a common ontology of the observed world may be lacking. In other words, entities have different perception, and lack a way of finding this out. To make things worse, a common set of standards for acceptable communicative behaviour could also be lacking, so that agents misperceive one another's intentions.

Fortunately, there exists a body of cross-cultural research that systematizes differences in perception, interpretation, and action selection across cultures. This research was not done for use in artificial intelligence though. For such use it has to be reworked. This has already been done in pioneering applications, and some of those will be discussed.

This chapter is structured as follows. In Sect. 7.2, 'The Elusive Real World', we provide some examples of differing perception, interpretation and action selection across cultures, to show how deep-seated these issues are. Then in Sect. 7.3, 'Theory', we discuss theory that could be adapted for intelligent systems, with particular emphasis on the theory on national culture by Hofstede and Minkov as one of the usable cross-national frameworks. In Sect. 7.4, 'Recent Work', we go on to show examples of recent work that has operationalized culture in artificially socially intelligent systems. There is a discussion (Sect. 7.5) with some reflection on theory needed versus system ambition. Finally we draw conclusions (Sect. 7.6).

7.2 The Elusive Real World

Interaction in the social world, in particular during formative years, aims at creating shared understanding. Children learn to 'behave'. As a result of shared socialization, people come to believe that there is a single world and that it can be understood in a single way. In reality it is merely shared between members of the same group or culture. This has now been amply shown by social psychologists [1, 2]. In what follows I shall be giving mainly anecdotal examples though, in order to convey the notion that these differences are of an everyday, anywhere nature.

7.2.1 *Perception*

Let us begin by looking at perception. Even basic elements of perception turn out to be cross-culturally quite different. For instance, can you see what rule was used to create Fig. 7.1? Once you have seen it, it is very simple. Some experience helps to see it, other experience hinders.

Perception is also linked to the language in which questions are asked. One could even say that people have different personalities depending on the language in which they think [3].

Here is an anecdotal piece of my own evidence. My first international job was as a receptionist at the European Institute for Advanced Studies in Management in Brussels. A sudden vacancy had occurred just before Christmas 1975 and nobody else could be found at short notice possessing the required language skills. I was at the time a 19 year old lad. My job as a receptionist and telephone person placed me in full view of visitors. Yet quite a few of them addressed me as 'mademoiselle'. Apparently they perceived me in role only, and that role was a female one.

7.2.2 *Interpretation*

Second, we turn to interpretation. An illustrative case concerns a student from Cameroon who had once visited me in my office on the second floor. When leaving, she embarked on a detour. I pointed her to the nearest stairway. She said "I thought it was not for me". I asked her why and she declared "I saw a staff member use it" [4]. The point here is that the very same observation of a person on a stairway leads to different interpretations. It would make me think "there is a stairway nearby and so I can take it as my fastest way out", whereas it made her think "there is a stairway nearby for use of important people, and since I am not important, I should look for another way out". She also used an implicit value saying that the social landscape is organised along lines of importance, and an implicit norm specifying that depending on their importance, people should not use public space in the same way. In fact perception and interpretation are hard to separate; our interpretations appear to us like perceptions. I would perceive the fact that a person walked that stairway, while she would perceive the fact that a more important person than herself walked it.

```
A   EF HI KLMN      T VWXYZ
 BCD  G  J    OPQRS U
```

Fig. 7.1 The alphabet on two lines. Why this pattern? The rule is revealed at the end of the chapter

Table 7.1 What the designer sees is not what the user gets [5]

User	Designer	
	Intended	Unintended
Perceived	1 Known to both	3 Unintended user content
Unperceived	2 Lost upon user	4 Unknown to both

A serious empirical study of these issues of perception and interpretation for designers and users of agent systems was carried out [5]. They found that cross-cultural differences between designers and users can cause trouble. They can cause a system to be perceived and interpreted by the users in ways about which the designers have no clue. They depicted this mismatch in perception using an adapted version of the Johari windows [6], see Table 7.1. They also found that such differences varied systematically across countries.

In the article from which we took Table 7.1, the designers had created a script with a teacher and a student. The students come to ask the teacher for an extension of a deadline for handing in a result. The designers had created scenarios that were supposed to distinguish between participants from culturally masculine countries and those from culturally feminine countries. It turned out that the actual participants (from Germany, Japan, Netherlands and Thailand) perceived different aspects, in particular having to do with respectfulness of both parties, and which created a gap between the European countries and the Asian ones—correlated with the culture dimensions of individualism in particular.

Let us take a closer look at Table 7.1. Ideally, one would like all content to be in quadrant 1. If cultures differ then the relative size of quadrant 1 will diminish. Quadrant 3 in particular could cause problems, since a system could convey messages of which the designers are not aware. This is precisely what happened in the case of the student's extension. The moral is that when designing this kind of system, one should user-test it across all cultures in which it could be used.

7.2.3 Action Selection

Third, we consider action selection. Differences are basic here too. Typically, an action does not show the reasons for which it was chosen. Also typically, the wider symbolical context ('who', 'when') has greater prominence to the observers than the detailed content ('what') of the action. This can lead to misinterpretations of actions by those who perceive them.

Let us take a very simple instance. The average tendency to answer 'yes' rather than 'no' to questions ('acquiescence bias') varies systematically across cultures. Depending on culture, a 'yes' could either mean 'yes, I agree', or 'yes, I respect you'. A younger person in a hierarchical culture would usually take care not to answer 'no' to a senior person, just because it is improper to do so. The extent to which saying 'no' is interpreted as impolite as opposed to honest, depends on culture. It follows that agents interpreting a 'yes' or a 'no', or implicit variants of these messages, would need culture-aware rules to decode signals for affirmative, noncommittal, or contradicting intention.

Or let us look at business: Organisational behaviour studies have shown that the aims of business leaders vary systematically across cultures [7]. For instance, according to MBA students from many countries, the top aim of business leaders from their country was as follows. Brazil: game and gambling spirit; China: respecting ethical norms; Germany: responsibility towards society; India: continuity of the business; USA: growth of the business (ibid, p. 324). It stands to reason that these differences in aims are associated with differences in perception, interpretation and action selection. Suppose, for instance, that acquiring a lucrative firm with a somewhat dubious reputation is being considered: depending on the top aim, the choice would likely be different. Other groups have not been studied so systematically, but the fact that such differences also hold for politics is demonstrated daily through international affairs.

The list above could be easily expanded to hundreds of pages. There is no end to the number of examples one could give of cross-cultural differences in perception, interpretation and action selection. It is important to realise that cultures are actually created and maintained, or changed, through myriad acts of perception, interpretation and action selection by individuals in a population. We can conclude that differences in perception, interpretation and action selection are a phenomenon to take seriously in modelling mental activity across cultures.

7.3 Theory

In order to assist us in modelling, we do not only need theory on perception, interpretation and action selection per se. We need theory on how these three differ across cultures. By definition it follows that our simulations need to involve the group as level of analysis.

Yet we need to implement the theory in individual intelligent agents. This seems like a paradox, but is not. What we need to do is distinguish between shared attributes for all agents, shared attributes by group, and individual attributes (Fig. 7.2). This figure is an elaboration of the statement by Kluckhohn and Murray "Every man is in certain respects (a) like all other men, (b) like some other man, (c) like no other man" [8]. It shows which phenomena pertain to which of the levels.

Figure 7.2 thus shows how each level of aggregation has its own relevant concepts. Note that culture is a universal phenomenon. All people have the capacity for culture and spend their childhood, and their adulthood to a much lesser degree, acquiring and imparting culture. They do that in the groups in which they operate.

Which of the three levels are needed if we want to model culture? Since we compare the group-level phenomenon of culture, we certainly need theory at the level of the group. This theory cannot stand on its own feet though; it does not touch the basis of the pyramid of Fig. 7.2. The agents can only have culture if they

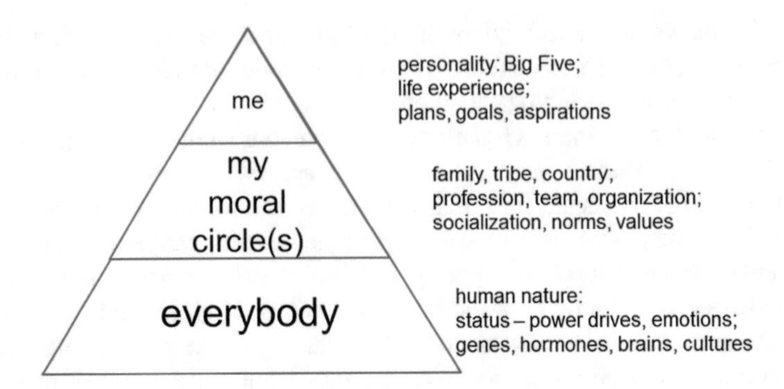

Fig. 7.2 Three levels of uniqueness for humans. 'Big Five' refers here to the personality model by McCrae and Costa [9]

have a common foundation upon which culture can build. It follows that for understanding cross-cultural differences we will need some generic human behaviour upon which the cultures can act. If we are not interested in specific individuals or personalities, we do not need theories at the level of the individual.

Cross-cultural differences are sometimes modelled as differences between individuals for want of group-level constructs, and we believe this conflation between the level of the group and the individual is undesirable, since it creates a conceptual lock-in. It prevents us from comparing across groups.

If possible, we need theory that has proven its capacity to explain real-world phenomena. In methodological parlance this is called a large nomological network. In this book about mental activity and culture, the focus lies on theory that allows comparing across groups. This focus implies that two kinds of theory are needed:

- Some generic theory at the level of all humans is needed; this theory provides the elements to compare.
- Some theory is needed that differentiates groups from one another.

Depending on the actual application and its specifics, other elements may be needed such as social norms, ritual, or physical context, reference groups and social identity [10–12].

7.3.1 Level of All Humans

In this chapter, we avoid proximate mechanisms mentioned in Fig. 7.2 such as genes, hormones, and brain circuitry. Their link with mental activity will no doubt be further elucidated in years to come. For drawing our big picture we can cut the corner. A sociological theory that addresses commonality between all human beings is status-power theory by Theodore D. Kemper [13]. According to Kemper, we live in a status world. Being deserving of status is a deep human need. This is enacted through conferrals and claims. Giving and receiving proper status is essential to human social life. 'Status' in this phrase is not just a scalar attribute but also a social commodity that is exchanged. It is given freely and involuntarily to the deserving. Traits that are deserving of status could be group membership, attractiveness, age, gender, clothing or many others. Conferring status voluntarily would normally be referred to as love, respect, politeness, attention, or by a thousand other words. Status can also be claimed by behaviour, such as joking or boasting to get attention, or by appearance, such as dressing up. And of course, humans are masters of disguise; we can be hypocritical about status conferrals, for instance by giving compliments with nasty undertones. Anyway: from the first act of education that a baby receives—let it cry, or pick it up—we all live in a status world.

At this point some readers may be a bit lost. I sympathize. When I first read Kemper it took me some time to get used to the non-standard use of the word 'status'. As a Dutchman I do not like to be suspected of seeking status. But I do seek love, I do love my loved ones, I do wish to be loved by them—and then I realized all this falls under Kemper's theory. What won me over to the theory's worth is his account of friendship versus love. When in love, we want nothing better than to give all the status in the world to the object of our love—whether that object reciprocates or not. Being in love is an ecstasy of status-conferral to the loved one. But with friends, we know they will give us status and we them, and that is a precondition for a continued friendship. Friendship is about free mutual conferral of status in commensurate amounts, unconstrained by social desirability.

Actually, Kemper's model has a second side: power. Whenever the status game breaks down, so that nobody freely gives us the status we crave and deserve, we are driven to obtain that status in other ways. This could be through force, guile, tantrums, or other means, depending on the situation; and all this is summarised by Kemper as 'power moves'. Power comes into play when, to speak with a Rare Bird song from 1969, 'there's not enough love to go round'. Power moves always come at a cost though; they create resentment. Therefore, power moves are often disguised as status conferrals. The nicer I am to you, the more likely you are to freely do what I want you to. Politeness and good manners are about staying away from power.

7.3.2 *Social Importance Dynamics*

For now we shall limit ourselves to the status side. A model for intelligent agents was developed to capture the status side of Kemper's model: Social importance dynamics [14, 15]. 'Social importance' is the operationalization of all the elements that could contribute to the status that one agent attributes to another, or that one agent believes it receives from another. For instance, taking a case from the song "Le parapluie" by French singer Georges Brassens, if it was raining I might not offer a place under my umbrella to a stranger, unless that stranger were very attractive; this would be modelled in the social importance model by the fact that a stranger has a social importance below the threshold for offering a place under my umbrella, whereas an attractive stranger has a social importance above that threshold. Similar rules would hold for strangers versus friends or family members. As a result, in a Kemperian status world, a balance between status claims and status conferrals is always sought (Fig. 7.3). An inspection of Fig. 7.3 will show first, that it is a simplification, and second, that status worthiness depends a lot on group membership, on culture, and on perspective. Each agent could have their own Fig. 7.3, updated based on experience. Some of the content of Fig. 7.3 is scripted by culture, e.g., the relative status due to the sexes. An extensive account of the sexes in the USA is presented by Ridgeway [16]. Ridgeway also shows that whereas gender is always a background variable connected to deservingness of status, some situations are scripted more by the actual task. In Fig. 7.3 this would be donor, or beggar. Finally, some attributions depend on whose side you are on. One person's terrorist is another person's freedom fighter.

With these caveats, the figure is a powerful way of charting social perceptions, and could be used in specific cases to be modelled. It can serve as a basis for investigating the status landscape in a particular situation. It can also serve to model the movement of a character through this landscape: through its actions, an agent could travel through the figure. For instance:

- A polite or courteous agent will gain deservingness (and move up in Fig. 7.3)
- An agent who violates local norms will lose deservingness (and move down)
- An agent who asserts himself will be considered as claiming more status (and move to the right)

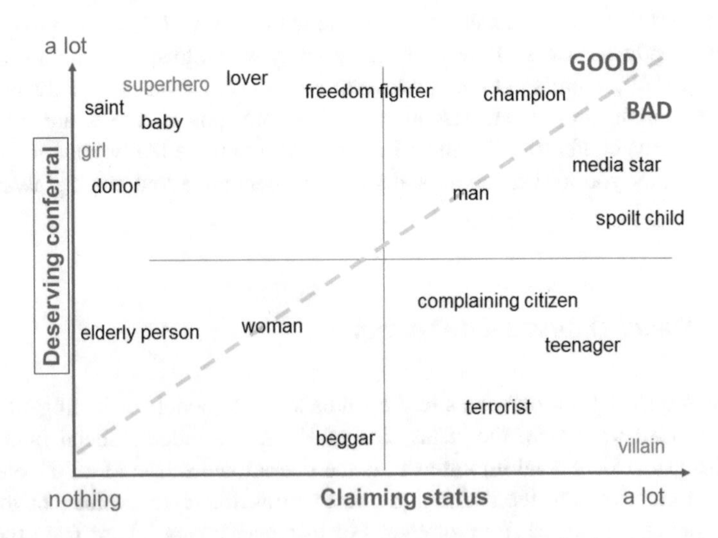

Fig. 7.3 Status claims versus status deservingness, and some social roles. *Upper left* of the diagonal are good roles, *bottom right* are bad roles

- An agent who claims more status than their current deservingness will create ill will in others, and a tendency to refuse their status claim (they are below the Good/Bad diagonal)

This last point is important: it points to the fact that when a person is far removed from the Good /Bad diagonal in the eye of others, those others will try to remedy that situation though their status conferrals. To the extent they are able to, they will praise the good and punish the bad.

This modelling approach is closely linked to emotions, since emotions are the mechanism that tell people whether their claims for status are being met (if not they might feel offended), or exceeded (if so they might feel grateful). Such social emotions are taken on board in the OCC model that has also been used in intelligent agents [17].

7.3.3 Level of the Group

Whatever the group, it cannot live without creating and maintaining culture. Culture allows a group's members to understand one another's' status deservingness, status claims, and status conferral intentions. It gives them a shared status-power world in which to live. Groups in which members enter at birth such as countries, tribes, or families, are by far the most influential in shaping a person's deep culture. The cultures of professions, teams, and organizations are not without importance, but more concerned with common understanding of practices than with deep-seated values [7].

We shall limit ourselves to the Hofstede framework of national cultures, although alternatives exist that could certainly also be used. See e.g., Minkovs overview of all the main comparative cross-cultural frameworks (Minkov 2013). To date, the Hofstede dimensional model, now counting six dimensions, is still the one with the largest nomological network [18, 19]. It replicates well, it explains phenomena in all realms of social life. The first four dimensions, the ones from the original 1980 book, have been most used in other studies. A recent meta-analysis of 598 studies representing over 200,000 individuals found continued predictive effect of these dimensions on many organizational outcomes, as well as on emotions and attitudes [20]. Incidentally, the strong link with emotions implies that these dimension operate at a basic level of our psyche. If people have different emotions in different cultures, and given that emotions affect mental processes, it follows that thought processes differ in different cultures. We shall return to this point.

Let us first introduce Hofstede's dimensions of national culture. In brief, the model consists of six 'dimensions of value', that is, big issues that a society has to resolve and develop shared meanings about in order to survive. They are:

First introduced in Geert Hofstede's original 1980 book [21]:

- Independence (Individualism versus Collectivism)
- Hierarchy (Power Distance)
- Aggression and gender roles (Masculinity versus Femininity)
- Anxiety (Uncertainty Avoidance)

First introduced in the Chinese Value Study [22]:

- Long-term Orientation

First introduced by Minkov based on World Values Study [23]:

- Indulgence versus Restraint

The first five of these dimensions have all been replicated in several to many studies, which is further testimony to their validity. The Hofstede model is conveniently accessible to modellers due to its simplicity. It was found by a bottom-up ecological analysis, not relying on any preconceived ontology of values, but only by factor analysing group averages on a great variety of questions. In other words, the data have spoken for themselves, not been forced into preconceived dimensions. The result is a six-dimensional space. Each dimension is projected onto a continuum of values roughly between 0 and 100. The figures are publicly available for research purposes at www.geerthofstede.eu.

7.3.4 Using the Theory

The tricky part is for users of the model to interpret the meaning of each dimension. The name of a dimension cannot capture all its ramifications, and so a close contact with empirical data remains important.

It is helpful to think of society as presenting itself to human minds as a social landscape [10, 11]. This landscape is a landscape of status-power relations. These relations hold both within and across groups. For every person, it is essential to know how much status is due to them, as well as to every other person in the groups to which they affiliate. Actually, most of social life is about this. Even during the most technical, instrumental of activities, people are also indicating liking, respect, disgust: in one word, status. When large changes in status deservingness take place, rituals attended by many people underline them and make sure they are socially shared. This holds for ritual battles in sports, for marriages and funerals, for promotions and shifts of power. Rituals also exist in the small: Gossip, for instance, is mostly about status deservingness. It often denounces those who claim more status than they deserve. Modelling rituals is discussed in more detail in [24].

Culture, in this metaphor, determines the details of the social landscape. How steep are the hierarchies, how tight the connections, how free the relationships? Such questions about the layout of the social landscape pertain to culture.

How much to model?

Most applications require only a small subset of the social landscape to be built. Developers might be tempted to only stick to the explicit elements of a system. For instance, in a simulation of a multi-storey office, agents might observe a stairway and use it to move around a space. When introducing culture, however, we have to introduce a social landscape. An agent will need to ask social questions:

- Is that stairway accessible to me, do I 'perceive' it as a way to move around?
- Under what social conditions can I use that stairway?

In status terms these questions would become

- What is the amount of status for which it is appropriate to use that stairway?
- Am I below the required status, above it, or in the proper window?

This last question occurred to me when recently on a visit to Zimbabwe I wanted to visit the toilet in a supermarket. I, a well-dressed elderly white male, was told by a local staff member to take the door that said 'staff only'. I did so hesitantly and found a clean toilet facility. In my part of the world, I would have expected a door saying 'men'—but then I live in a much more horizontal status world.

If real subjects are available for mock-ups of a situation, doing simulation gaming with them can be a great way to discover elements of their social landscape that would probably have escaped the designer [25].

7.4 Recent Work

We shall now move from normative statements to a description of some recent work with which the author is familiar. This may give readers a bit more of a foothold. Of the three examples presented, 'Social importance dynamics' has the broadest perspective and the richest model of the social world; 'negotiation across cultures' has the richest implementation of culture; 'consumer behaviour depending on personality and culture' has the strongest validation in empirical data.

7.4.1 Social Importance Dynamics: TRAVELLER

Social importance dynamics were implemented in a virtual world application for cross-cultural learning in which the user plays the role of a young person travelling the world in search of grandfather's treasure [15]. The game uses a Kinect x-box, and is operated through gestures. It can also be played with a keyboard. The keyboard version can be accessed at http://ecute.eu/traveller/.

The idea of TRAVELLER is to have generically social agents in a virtual world, whose culture can be modified by sliders that depict the Hofstede dimensions of national culture. The social importance model mentioned above [14] lends itself well to modelling cross-cultural variation in status rules. In its generic form the model decides:

- how one agent knows the social importance of another depending on ascribed or observable characteristics
- how one agent updates its beliefs about that social importance depending on another one's actions
- how much social importance an agent needs in order to be granted certain favours.

What TRAVELLER provides is a system that adapts the threshold of these rules to culture. For instance, which ascribed characteristics are worth a place under my umbrella in the rain, which observable characteristics, and which actions? Any reader would have an idea about these matters. There would be systematic cross-cultural differences though. Would you offer such a place to a subordinate, an attractive person, somebody who just told you the way? In order to make such decisions, it is necessary to know the symbolic value of the action to the agents. Based on a good understanding of culture one can make educated guesses. For instance, in a hierarchical culture, one would not normally think of offering either one's subordinate or one's boss a place under the umbrella, since that might connote equality. Whenever empirical testing is feasible it should be undertaken, as argued based on Fig. 7.3 It is easy for designers to overlook symbolic connotations that come to the mind of people from different societies [5].

The social importance model is still a pioneering effort. Most artificially intelligent systems today model a narrowly defined task rather than a generic social setting. In such a system, mapping the system's actions to human motivation and to culture is less tricky.

7.4.2 Social Importance and Power Distance

The social importance model includes rules for the salience of attributes of status-worthiness. Here we highlight the dimension of Power Distance; the model also includes Individualism and Masculinity. In the current model, these culture effects are specified numerically. In a more general format they could appear as in Fig. 7.4.

Figure 7.4 makes the point that the status landscape has a self-fulfilling quality. In societies of large power distance, agents look for status differences, and reinforce them through their conferrals and claims of status. They might use honorary titles for the elderly and their parents and teachers, while expecting their youngers and children to do it for them, and this behaviour will reinforce norms of asymmetry in status relations.

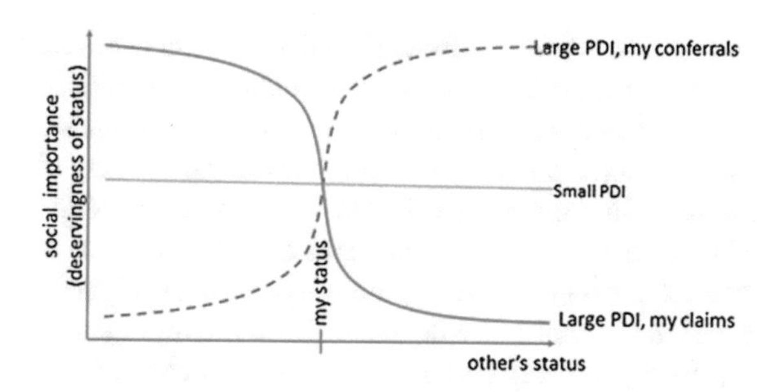

Fig. 7.4 Perceived social importance due to other agents (y-axis) from an agent's ('my') point of view, depending on relative social status. The x-axis shows my status in the middle, and other agents' possible status as lower, equal, or higher. In cultures of small PDI (Power Distance Index), social importance hardly depends on status-relevant attributes (*middle line*), so that status claims and conferrals are always about the same. At large PDI, I strongly adapt my status claims and conferrals to perceived social importance of others. The result is that my status claims fall with high-status others, and rise with low-status others. For my conferrals it is the reverse

Another point implicit from Fig. 7.4 is that agents in societies of large power distance will try to avoid to be involved in situations that imply equal status when they are with agents whose social status is obviously different from their own. For instance, professors and students, or bosses and subordinates, are unlikely to mingle in public places such as bars and restaurants. By the same token, agents from egalitarian cultures will try to avoid situations of obvious difference in status. They might e.g., dress down, or avoid elevating the teacher's chair, or drive in a big chauffeured car. Or take the case of the Cameroonian student who avoided the 'staff' stairway. She was trying to avoid claiming too much status. She acted in accordance with the line 'large PDI, my claims' in Fig. 7.4.

Concluding we can say that the culture dimension of power distance, in the mind of an agent, blows up or shrinks vertical differences in social status between people. It creates a more or less vertical social landscape.

Similar reasoning can apply to the other dimensions of culture and be implemented in intelligent systems. In a realistic context, all dimensions of culture together affect status deservingness. Kemper and the author discussed this and came to the following hypothesis [26]:

- Individualism-collectivism is a society's implicit specification for the unit that has the right to claim and receive status: the individual, or the group.
- Large versus small power distance is the idea that people are intrinsically unequal, and the willingness by people to accept status and/or power domination by those who are placed above one, or to dominate or disregard those placed below one.
- Masculinity versus femininity is a preference for either power-oriented or status-oriented social relations. In a masculine culture, wielding forceful power gives status, while in a feminine culture, soft power gives status. In all societies, men have more status than women, but the width of the gap varies with masculinity.

- Uncertainty tolerance is the rigidity with which status-power rules are mandated to be followed. In an uncertainty avoiding culture, deviations from unwritten rules ('strange' behaviours) can cause loss of status worthiness.
- Long versus short term orientation is a matter of how change in status-power rules is accepted. In long-term oriented cultures, everything is expected to change, including these rules. In short-term oriented cultures, all things moral, and this includes status-power rules, are assumed to be eternal.
- Indulgence versus restraint is the degree of control over organismic satisfaction. For Kemper, the organism is one of the entities that can claim status, and in an indulgent society the idea is that one should listen to its voice.

These relationships can be used for designing agents with the social importance model that can alter their behaviour across cultures in rich ways. Figures analogous to Fig. 7.4 can be created for each dimension of culture. In TRAVELLER, the first three dimensions were operationalized, but only on a case-by-case basis, not using generic primitives.

7.4.3 Negotiation Across Cultures

Another example model of mental activity across cultures concerns the context of negotiating about products with a hidden quality attribute [27, 28]. This model is based on the TRUST AND TRACING GAME [29]. Two intelligent, non-embodied agents from two countries meet and negotiate. If a good is sold that the seller claims is of high quality, the buyer may request a trace to reveal the true quality.

In this application there is no ambiguity of perception; it's all in the interpretation and action selection. Figure 7.5 gives a condensed overview of the process.

Figure 7.5 can be read from left to right. First, agents determine a short-term trade goal (buy or sell? High quality or basic quality?). Next, they search for acceptable trade partners with complementary needs that are willing to negotiate.

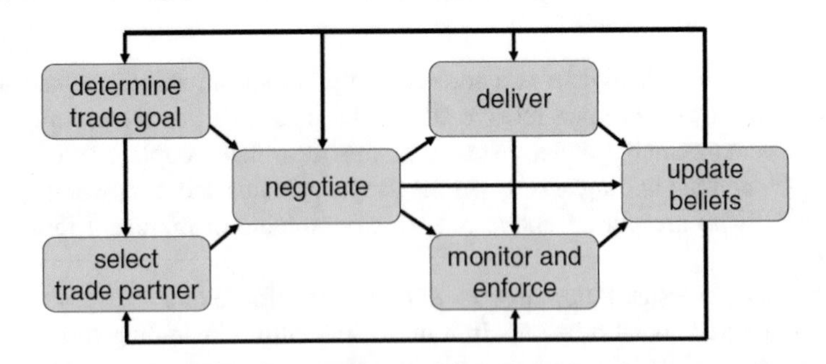

Fig. 7.5 Process model of negotiation model from TRUST AND TRACING GAME [30]. Note that delivery could either be truthful or deceptive on the part of the seller, and the buyer could decide to monitor (trace)

After negotiating a contract, it comes to delivery, where a seller may deliver according to contract or defect when selling high quality. Buyers have the choice to trust the seller to comply with the contract, to opt for low quality purchase, or to spend resources on monitoring and enforcing deliveries of purported high quality. Finally, after these trade interactions, the trading agents reflect upon the effectiveness of their trade, on the trustworthiness of their trade partners, and on their own decisions. On the basis of that reflection they update their beliefs.

Efforts were taken to create a plausible simulation. For the task of negotiation, a validated model existed: ABMP [31]. The negotiations were set in the context of a production chain that had been modelled in the same way for a simulation game with real participants, THE TRUST AND TRACING GAME [32].

The main design task for operationalizing culture was carefully checking all the steps of the negotiation process and considering how culture could be expected to play a role. This was done carefully, one dimension at a time, resulting in one article per dimension, except Indulgence which had not been included in the model at the time when the study began. Specifically: Individualism [33], Masculinity [34], Uncertainty Avoidance [35], Long-Term orientation [36], and Power Distance [37]. Combining these sub-models resulted in a matrix that gave the influence of each dimension of culture on each parameter. This matrix is reproduced in Table 7.2, not for spelling out the details but in order to give the reader an idea about the level of detail.

Table 7.2 has the following columns:

- 'Dimension index', specifying the first five dimensions of culture from Hofstede [38].
- 'Culture and related characteristics', specifying one extreme of a dimension per row, in some cases also specifying the trait of the partner most salient to this dimension of culture
- 'Cultural factor to be taken into account', the computational form of the previous column, to be used in one of four calculations
- 'Effect on', specifying the four calculations that use the cultural factors.
  - Deceit threshold: how deceit-prone is the agent when selling
  - Inclination to trace: how distrustful is the agent when buying
  - Negative update factor: trust decline in case of cheating
  - Positive update factor: trust increase in case of truthful delivery

Results with fictitious partners from a varied set of countries, using Hofstede dimension scores as a proxy for culture [39], were plausible. It was difficult to obtain participants for testing against the real world, since each game session required 25 people. The simulation game had been played sufficiently often only with participants from the Netherlands and the USA. The real-world results from those sessions were replicated in the model and gave matching results. In fact this amount to an empirical test of the dimension of Masculinity versus Femininity, the only one on which these two societies differ markedly. The USA, a masculine society, saw a tendency to go for high quality rather than low, and a strong tendency

Table 7.2 Effects of Hofstede's dimensions of culture and relational characteristics on deceit and trust parameters (+ indicates increasing effect; – indicates decreasing effect) [30]

Dimension index	Culture and relational characteristics	Cultureal factor to be taken into account	Effect on			
			Deceit thresh old	Inclination to trace	Negative update factor	Positive update factor
PDI	Large power distance	PDI^*				
	-with higer ranked partn.	$\max\{0, \text{PDI}^*(s_b - s_a)\}$	+	–		
	-with lower ranked partn.	$\max\{0, \text{PDI}^*(s_a - s_b)\}$		–		
	Small power distance	$1 - \text{PDI}^*$				
UAI	Uncertainty avoiding	UAI^*			+	–
	-with stranger	$\text{UAI}^* \cdot D_{ab}$	–	+		
	Uncertainty tolerant	$1 - \text{UAI}^*$				
IDV	Individualistic	IDV^*				
	Collectivistic	$(1 - \text{IDV}^*)$			+	
	-with in-group partner	$(1 - \text{IDV}^*)(1 - D_{ab})$		–		
	-with out-group partner	$(1 - \text{IDV}^*)D_{ab}$	–			
MAS	Masculine (competitive)	MAS^*	–	+	–	
	Feminine (cooperative)	$1 - \text{MAS}^*$		–		
LTO	Long-term oriented	LTO^*	+	–	+	
	Short-term oriented	$(1 - \text{LTO}^*)$				
	-with well-respected part	$(1 - \text{LTO}^*)_{sb}$	+	–		
	-with other partners	$(1 - \text{LTO}^*)(1 - s_b)$	–			

to trace the goods. Efforts to validate these cross-cultural patterns using results from a wide range of participants to a simpler real-world game [40] are still on-going.

We can conclude that this simulation scores well on empirical grounding, but less well on validation. This state of affairs is hard to improve upon. Validating this kind of system against real actors, including the effect of all six dimensions of culture, would require sample sizes in the tens of thousands, well matched across countries, taken from a wide range of cultures.

7.4.4 Consumer Behaviour Depending on Personality and Culture

A third example to be discussed here is consumer behaviour [41]. An agent-based model was created of non-embodied consumers purchasing cars. Its internal architecture was based on the framework MASQ for social agents [42]; see Fig. 7.6. The MASQ framework allows distinguishing between mental attributes of individuals (upper left, mind quadrant) and of groups (cultural attributes, lower left, culture quadrant). These latter were shared between all individuals in the simulation, depicting a single culture.

On the left are the two mental quadrants. The individual's mind quadrant reads in a straightforward way. In the top box, four processes influence the mind's state largely cyclically. They are affected by three boxes of parameters: personality, wealth (since this is salient for the example of car purchasing), and the individual's culture, which might differ from the culture of the group.

The culture quadrant has two variable boxes: culture dimensions of the group that form the unwritten rules of trade, historical status quo that gives norms. These feed into the process 'establish beliefs' that influences the mind.

On the right are the 'Body' model elements that pertain to the outside world. The ConsumersSellers Space at the top is where individual buyers and seller of cars may meet and trade. The Consumers Space gives the existing population of consumers and their cars. This can be perceived by the agent minds.

The cars were categorised according to a number of attributes. These included safety, environment-friendliness and social status connotation. Hypotheses were generated on the attractiveness of these attributes depending on culture and personality. Each agent had random values for three personality attributes: extroversion, openness and agreeableness. Agents could meet and influence one another's behaviours. For culture, only the dimension of power distance was modelled. The empirical base consisted of eleven European cultures that show great variation in Power Distance scores. The model reproduced car purchasing behaviours well (Fig. 7.7). We can conclude that this model has good face validity.

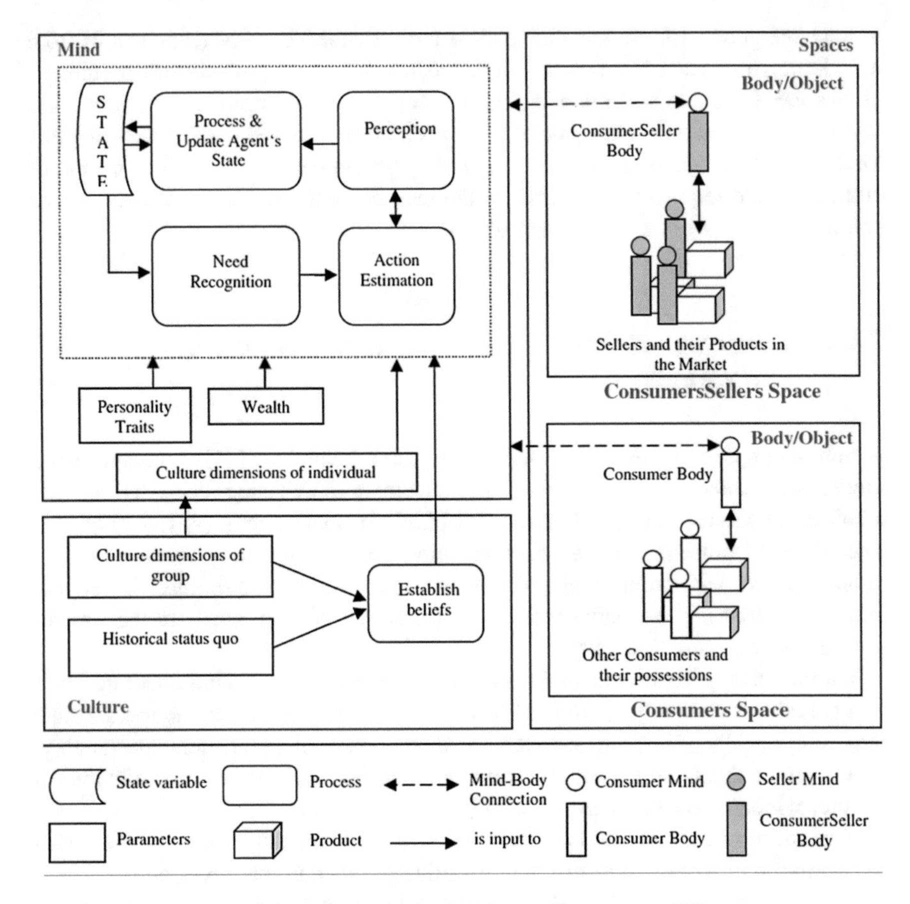

Fig. 7.6 Conceptual model for the consumer decision making process [41]

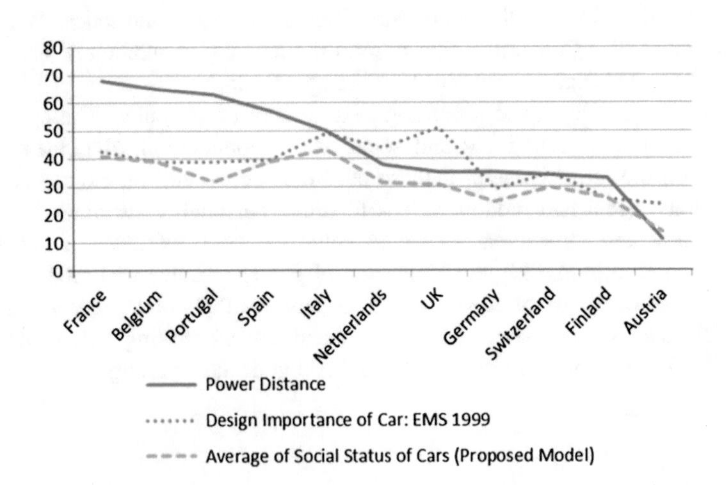

Fig. 7.7 Design importance according to EMS (European Media and Marketing survey) 1999, and average social status of purchased cars as a function of cultural Power Distance [41]

7.5 Discussion

We have in this chapter made some general points, and presented three example applications. Obviously there is a vast world of unexplored possibilities waiting to be discovered using not only the theories presented here but also other ones, and using thousands of creative ideas. This chapter does not intend to lay the law. It answers to the tendency in artificial intelligence to stay away from the complexity of the social world when modelling mental activity. It intends to seduce the reader into realising that mental activity is deeply influenced by the social world. This social influence on mental activity does not need to remain inaccessible to formalisation in intelligent systems. We need to take it on board for continued relevance of the intelligent systems that we are building in education, health care and other sectors. There are a number of generic theories available for such purposes, waiting to be used. So far, the systems shown here have shown good validity, both in an empirical and in an intuitive sense. This kind of work has promise, and I hope to convince others of this.

A crucial distinction to keep in mind when modelling social influences on mental activity is level of analysis. Do influences pertain to only one individual? Very often that which seems an individual decision or thought is actually embedded in group judgement or values. Typically the level of these groups is not operationalized in models. This is precisely where a lot of progress can still be made.

There are numerous topics that remain un-dealt with in this chapter, and this does not imply they are not important. We mention non-verbal behaviour and its recognition, speech generation or recognition, to name two obvious and culture-relevant topics.

Finally, we would like to stress that if and when intelligent systems are going to be used in everyday contexts such as health care, then a general-purpose kind of status-power intelligence is needed. Such systems for general-purpose interaction need to 'behave', just like human individuals and groups.

7.6 Concluding Remarks

This chapter has argued that mental activity across cultures cannot take the existence of a single observed world for granted, nor the existence of universal rules of inference from observation to interpretation to action selection.

The chapter has introduced theory at two levels of analysis that can be used in modelling mental activity: the group, and all of humanity. It also showcases some recent agent-based modelling work that shows promise.

There is still a lot of conceptual work to be done, and it is likely that many conceptual alleys will be explored and will yield bits of insight. On the implementation side, each application has to be carefully validated as well as is feasible. This is a thorny issue since simulations of necessity leave out a lot of detail, and the real world comes with all this detail—unless one has large datasets that are so well matched that many differences average out. Lack of validation should not stop progress though. The author is of the opinion that in these cases, in order to advance, we should be careful not to let rigour kill relevance.

Solution to Fig. 7.1

The rule for Fig. 7.1: Letters with curves go to the bottom line, letters without them go to the top line.

As you may imagine, this is usually easy to see for people who are not familiar with the Latin alphabet. Users of the alphabet tend to disregard the shape of letters.

It would also be easier to see for those who tried carving letters in wood: curves are harder to make.

References

1. Smith, P.B., Fischer, R., Vignoles, V.L., Bond, M.H.: Understanding social psychology across cultures: Engaging with others in a changing world. Sage (2013)
2. Smith, P.B., Bond, M.H., Kagitcibasi, C.: Understanding social psychology across cultures sage (2006)
3. Ramírez-Esparza, N., Gosling, S.D., Benet-Martínez, V., Potter, J.P., Pennebaker, J.W.: Do bilinguals have two personalities? A special case of cultural frame switching. J. Res. Pers. **40**, 99–120 (2006)
4. Hofstede, G.J., Pedersen, P.B., Hofstede, G.: Exploring Culture: Exercise. Stories and Synthetic Cultures. Intercultural Press, Yarmouth, Maine (2002)
5. Degens, D.M., Endrass, B., Hofstede, G.J., Beulens, A.J.M., André, E.: What I see is not what you get: why culture-specific behaviours for virtual characters should be user-tested across cultures. AI & society: J. Hum. Mach. Intell. 1–13 (2014)
6. Luft, J., Ingram, H.: The Johari Window: A graphic model of awareness in interpersonal relations. Hum. Relat. Training. News. **5**, 6–7 (1961)
7. Hofstede, G., Hofstede, G.J., Minkov, M.: Cultures and organizations. Software of the Mind. McGraw Hill, New York (2010)
8. Kluckhohn, C., Murray, H.A.: Personality in nature. Culture and Society. Knopf, New york (1948)
9. McCrae, R.R., Costa, P.T.j.: Personality in adulthood: A five-factor theory perspective. Guildford, New York (2003)
10. Degens, D.M., Hofstede, G.J., McBreen, J., Beulens, A.J.M., Mascarenhas, S., Ferreira, N., Paiva, A., Dignum, F.: Creating a World for Socio-Cultural Agents. In: Bosse, T., Broekens, J., Dias, J., van der Zwaan, J. (eds.) Emotion Modeling: Towards Pragmatic Computational Models of Affective Processes, vol. LNAI 8750, pp. 27–43. Springer, Zürich (2014)
11. Dignum, F., Hofstede, G.J., Prada, R.: Let's get social! From autistic to social agents. In: 13th International conference on autonomous agents and multi-agent systems (AAMAS), pp. 1161–1164. IFAAMAS (2014)
12. Hofstede, G.J.: GRASP agents: social first, intelligent later. AI & Society (accepted 2016)
13. Kemper, T.D.: Status, Power and Ritual Interaction; A Relational Reading of Durkheim. Goffman and Collins. Ashgate, Farnham, UK (2011)
14. Mascarenhas, S., Prada, R., Paiva, A., Hofstede, G.J.: Social importance dynamics: A model for culturally-adaptive agents. In: Aylett, R., Krenn, B., Pelachaud, C., Shimodaira, H. (eds.) Intelligent Virtual Agents, vol. 8108, pp. 325–338. Springer (2013)

15. Degens, D.M., Hofstede, G.J., Mascarenhas, S., Silva, A., Paiva, A., Kistler, F., André, E., Swiderska, A., Krumhuber, E., Kappas, A.: Traveller–intercultural training with intelligent agents for young adults. In: Proceedings of the 8th international conference on foundations of digital games (2013)
16. Ridgeway, C.L.: Framed by gender: How gender inequality persists in the modern world. Oxford University Press (2011)
17. Ortony, A., Clore, G., Collins, A.: The Cognitive Structure of Emotions. Cambridge University Press, UK (1998)
18. Minkov, M., Hofstede, G.: The evolution of Hofstede's doctrine. Cross Cultural Management: An International Journal **18**, 10–20 (2011)
19. Minkov, M.: Cross-cultural analysis: The science and art of comparing the world's modern societies and their cultures. Sage, Los Angeles (2013)
20. Taras, V., Kirkman, B.L., Steel, P.: Examining the impact of Culture's consequences: A three-decade, multilevel, meta-analytic review of Hofstede's cultural value dimensions. J. Appl. Psychol. **95**, 405–439 (2010)
21. Hofstede, G.: Culture's Consequences. International Differences in Work-Related Values. Sage, Beverly Hills (1980)
22. Hofstede, G., Bond, M.H.: The Confucius connection: From cultural roots to economic growth. Org. Dyn. **16**, 5–21 (1994)
23. Minkov, M.: What makes us different and similar, A new interpretation of the World Values Survey and other cross-cultural data. Klasika i Stil, Sofia (2007)
24. Hofstede, G.J., Mascarenhas, S., Paiva, A.: Modelling rituals for Homo biologicus. ESSA 2011 (7th Conference of the European Social Simulation Association), Montpellier (2011)
25. Hofstede, G.J., Caluwé, L.d., Peters, V.: Why simulation games work-In search of the active substance: A synthesis. Simulation & Gaming 41, 824–843 (2010)
26. Hofstede, G.J.: Theory in social simulation: Status-Power theory, national culture and emergence of the glass ceiling. Social Coordination: Principles, Artefacts and Theories, pp. 21–28. AISB, Exeter (2013)
27. Hofstede, G.J., Jonker, C.M., Verwaart, T.: Cultural Differentiation of Negotiating Agents. Group Decis. Negot. **21**, 79–98 (2012)
28. Hofstede, G.J., Jonker, C.M., Verwaart, T.: Computational Modeling of Culture's Consequences. In: Bosse, T., Geller, A. (eds.) Multi-Agent-Based Simulation XI, vol. 6532, pp. 136–151. Springer, Berlin / Heidelberg (2011)
29. Meijer, S.A., Hofstede, G.J., Omta, S.W.F., Beers, G.: The organization of transactions: research with the Trust and Tracing game. Journal on Chain and Network Science **8**, 1–20 (2008)
30. Hofstede, G.J., Jonker, C.M., Verwaart, T.: A Cross-Cultural Multi-agent Model of Opportunism in Trade. In: Nguyen, N.T., Kowalczyk, R. (eds.) Transactions on Computational Collective Intelligence II, pp. 24–45. Springer, Berlin Heidelberg, Berlin, Heidelberg (2010)
31. Jonker, C.M., Treur, J.: An agent architecture for multi-attribute negotiation. In: International joint conference on artificial intelligence, pp. 1195–1201. Citeseer (2001)
32. Meijer, S., Hofstede, G.J., Beers, G., Omta, S.: Trust and Tracing game: learning about transactions and embeddedness in a trade network. Production Planning & Control **17**, 569–583 (2006)
33. Hofstede, G.J., Jonker, C.M., Verwaart, T.: Individualism and Collectivism in Trade Agents. Lectures Notes in Artificial Intelligence, vol. 5027, pp. 492–501. Springer, Berlin (2008)
34. Hofstede, G.J., Jonker, C.M., Meijer, S.A., Verwaart, T.: Modelling Trade and Trust across Cultures. In: Trust Management: 4th International Conference, iTrust 2006: Proceedings, Pisa, Italy, May 16–19, 2006, pp. 120–134. Spinger Verlag (2006)
35. Hofstede, G.J., Jonker, C.M., Verwaart, T.: Uncertainty Avoidance in Trade. In: Proceedings of 2008 Agent-Directed Simulation Symposium (ADS'08), Ottawa, Ontario, Canada, 14–16 April 2008, pp. 143–152 (2008)

36. Hofstede, G.J., Jonker, C.M., Verwaart, T.: Long-Term Orientation in Trade. Complexity and Artificial Markets, pp. 107–119. Springer-Verlag, Berlin, Heidelberg (2008)
37. Hofstede, G.J., Jonker, C.M., Verwaart, T.: Modelling Power Distance in Trade. In: David, N., Sichman, J.S. (eds.) Multi-Agent-based Simulation IX, International Workshop, MABS 2008, Revised Selected Papers, pp. 1–16. Springer, Berlin (2009)
38. Hofstede, G.: Culture's Consequences, Comparing Values, Behaviors, Institutions, and Organizations Across Nations. Sage, Thousand Oaks (2001)
39. www.geerthofstede.com
40. Hofstede, G.J., Verwaart, D., Jonker, C.M.: Lemon car game. In: Proceedings of the 30th Conference ISAGA 2008 International simulation and gaming association. Kaunas University
41. Roozmand, O., Ghasem-Aghaee, N., Hofstede, G.J., Nematbakhsh, M.A., Baraani, A., Verwaart, T.: Agent-based modeling of consumer decision making process based on power distance and personality. Knowl.-Based Syst. **24**, 1075–1095 (2011)
42. Ferber, J., Stratulat, T., Tranier, J.: Towards an Integral Approach of Organizations in Multi-Agent Systems. In: Dignum, V. (ed.) Handbook of Research on Multi-Agent Systems, pp. 51–75. IGI Global, Hershey (2009)

Chapter 8
Affective Body Movements (for Robots) Across Cultures

Matthias Rehm

Abstract Humans are very good in expressing and interpreting emotions from a variety of different sources like voice, facial expression, or body movements. In this chapter, we concentrate on body movements and show that those are not only a source of affective information but might also have a different interpretation in different cultures. To cope with these multiple viewpoints in generating and interpreting body movements in robots, we suggest a methodological approach that takes the cultural background of the developer and the user into account during the development process. We exemplify this approach with a study on creating an affective knocking movement for a humanoid robot and give details about a co-creation experiment for collecting a cross-cultural database on affective body movements and about the probabilistic model derived from this data.

8.1 Introduction

Humans are very good in expressing and interpreting emotions from a variety of different sources like voice, facial expression, or body movements. In this chapter, we concentrate on body movements and show that those are not only a source of affective information but might also have a different interpretation in different cultures. Work on modeling the behavior of multicultural agents primarily relies on the analysis of video recordings of multimodal face to face interactions between humans, where the videos have been collected in different cultures. This poses some questions concerning the cultural biases of the analysis due to the cultural background of the annotators. Although this is most of the time a successful approach, there have always been some discomforts about it due to the following reasons:

- Subjectivity of the annotation

M. Rehm (✉)
Aalborg U Robotics, Technical Faculty of IT and Design,
Aalborg University, Aalborg, Denmark
e-mail: matthias@create.aau.dk

C. Faucher (ed.), *Advances in Culturally-Aware Intelligent Systems and in Cross-Cultural Psychological Studies*, Intelligent Systems Reference Library 134, https://doi.org/10.1007/978-3-319-67024-9_8

- Cultural bias of the annotation and the implementation/design of the agents
- Applicability of results to agents (they might represent a culture of their own)

These challenges are enhanced when we move from virtual to physical agents, i.e. robots. Here the difference between humans and robots is more apparent, due e.g. to limited expressive channels or reduced degrees of freedom. For instance the Nao platform which we are using has fewer joints in arms and legs than a human, making movements look different, independent on how careful movements have been designed. To cope with these multiple viewpoints in generating and interpreting body movements in robots, we suggest a methodological approach that takes the cultural background of the developer and the user into account during the development process. We exemplify this approach with a study on creating an affective knocking movement for a humanoid robot. This includes details about a co-creation experiment for collecting a cross-cultural database on affective body movements, about the machine learning approach employed on this database and on an evaluation study to verify the applicability of this approach.

8.2 Related Work

8.2.1 Culturally-Aware Technology

The emergence of the term Culturally-Aware Technology is relatively recent (see e.g. [3]) and encompasses technology where culture has been taken into account either during the design and development process but most often as a factor influencing the interaction with the system. Treating culture as a parameter for an interactive system is not a trivial task, because depending on the research discipline one is looking at for help in defining the concept (e.g. anthropology, psychology, business studies, …), numerous and diverse definitions and delimitations are to be found. Research on Culturally-Aware technology is very diverse and addresses issues such as cultural data management [37], enculturated design [6] and interaction [35], culture-based decision making for interaction [36] and intercultural education ([4]; [33]).

Most of the research on Culturally-Aware technology is grounded on existing cultural frameworks such as Hall's work on verbal and non-verbal communication [14] or Hofstede's system of values [18]. Hofstede defines culture as a five-dimensional concept, where national cultures are attributed a specific value on each of the dimensions, thus simplifying cross-cultural comparisons. For instance, members of collectivistic cultures (low value on the individualism dimension) tend to stand closer together in face to face interactions then members of individualistic cultures (high values on the dimension) [19]. Taking this idea for granted, it becomes possible to predict behavioral tendencies based on the position of a culture in this five-dimensional space. There are many shortcomings of this theory regarding its application in ICT, esp. related to the sample used for the empirical analysis and the fact that it is tailored to business and management. Nonetheless, Hofstede's work has been successfully adapted in the area of cultural usability (e.g. [24]) and virtual agents (e.g. [15]; [34]).

Other promising cultural frameworks exist, but have not been exploited in depth yet for developing Culturally-Aware technology. Evolutionally approaches for instance can be viewed as "cognitive friendly" and would be good candidates for intelligent systems. Examples include the dual inheritance theory [26] or the epidemiology of representations [39]. Another promising set of frameworks are those that focus on intercultural training, which is also one of the main applications areas of Culturally-Aware technology so far. While not defining culture as a parameter useful for interactive systems, they aim at identifying the aspects of cultural behavior that are relevant for achieving communicative goals in intercultural encounters (e.g. [2]; [38]).

For the current study we do not analyze the concept of culture further, but remain at the very abstract level of national cultures, following the assumptions that members of a given national culture will behave similar to a certain degree and will follow similar heuristics for generating and interpreting nonverbal behavior. We are not speculating how such a set of heuristics has been established (for different ideas on this process see e.g. [26, 39, 44]). To be on the safe side, we are restricting ourselves to a specific subgroup in each culture, namely university students, thus creating comparable samples for the studies. We are aware of the shortcomings of such an approach (e.g. [17]) but are convinced that it is sufficient as a starting point.

Kleinsmith and colleagues [20] highlight the importance of cross-cultural design with an experiment that investigates the differences in perceiving affective body movements or, to be more precise, different expressive postures. They could show that participants from different cultures were able to identify the emotional content of the postures to a certain degree but with a low accuracy between 44.9 and 65% over four emotions (anger, fear, happy, sad) and three cultures (Japan, Sri Lanka, US). Additionally, they showed that participants from different cultures differed significantly in the intensity ratings for the emotions. As a consequence, they use a system of low-level numerical features to describe the different postures and by running a discriminant analysis were able to identify culture-specific feature patterns for the affective postures that allow for a highly accurate culture-specific classification (Japan: 90%; Sri Lanka: 88%; US: 78%).

In our own previous work, we have shown the need for an enculturated interaction for virtual agent systems and have especially highlighted the importance of non-verbal behavior like gestures or body movements [35]. We found evidence that the way gestures are performed differs significantly between cultures. The proposed virtual agent architecture takes this variance into account and adapts the agents' non-verbal behavior to the cultural background of the users, who in several experiments confirmed preferring agents that exhibit non-verbal behavior that is in line with their own expectations (e.g. [34]; [23]). We are strongly convinced that this effect will be seen with every type of embodied technology and will thus also be apparent in human robot interactions.

In a series of studies, it has been shown by Evers and colleagues that culture does indeed play an important role in human robot interaction (e.g. [11]; [42]; [43]). Despite this fact, there is a noticeable lack of work on cultural aspects in human robot interaction. In this paper, we present our methodology of capturing, generating and testing culture-specific interactions with robots, which can be used as a guideline for work on cross-cultural human robot interaction.

8.2.2 Affective Body Movements

Although much is known about affective facial expressions, the information on how affective information is conveyed and interpreted through body movements is diverse and inconclusive. According to Gross and colleagues [13], the main reason for this fact is the use of actors to present stimuli to observers. They argue that the actors' rationale for why a specific movement conveys an emotion is not registered leaving the parameters that govern the decision for performing a movement obscure. Moverover, they could show that the actors' intuition about what makes a movement emotional does not necessarily coincide with the observers' interpretation of the movement.

What is needed is thus a set of objective parameters for describing the movement. Several examples of such parameter sets exist, developed for quite different phenomena like emotion, culture, or personality [30]. Efron [8] for instance distinguishes between spatio-temporal aspects (e.g. speed, spatial extent), interlocutional aspects (e.g. distance), and co-verbal aspects to analyze cultural influences on body movements. Gallaher [12] describes a similar set of parameters with categories expressiveness (e.g. speed, frequency), expansiveness (e.g. spatial extent, distance), coordination (fluidity), and animation (e.g. standing erect vs. with slumped shoulders). Based on this set of parameters, Gallaher derives information about personality from body movements. Regarding emotional connotations of body movements, De Meijer [7] e.g. focuses on parameters like speed and power of movements while Walbott [41] concentrates on activity, spatial extension, and power.

Based on this analysis it becomes difficult to map specific parameters or parameter combinations to specific social signals. Instead, every movement will (at least partly) convey several aspects of a person's background (culture, personality) and current state (emotion) at the same time. The same will be true for the interpretation of such movements by an observer. Thus, it is safe to assume that there will be differences in what is regarded as an affective body movement across cultures based on the expressivity of this movement.

Gross and colleagues present Laban movement analysis as a means to objectively describe the quality of a movement. Laban (e.g. [22]) developed a notation system for body movements to describe the movement of dancers but due to its generality it can be used for any type of body movement. Movement qualities are described by the dimensions of effort and shape. They capture how a movement is performed (effort) and how the body is shaped during the movement (shape).

Effort denotes movement qualities with four parameters: time, weight, space, and flow. Time describes the speed of the movement, weight the energy or power that is put into the movement, space denotes if the movement is direct or indirect, and flow describes if the movement is bound or free. The shape dimension is concerned with how the whole body is affected by the movement and comprises three parameters describing the form of the body (towards vs. away from body center), how the body is related to the environment (gathered vs. scattered), and the movement path in space (spoke vs. arc).

Several attempts have already shown that Laban movement analysis is a good starting point for modeling movement characteristics of humanoid robots on a thoroughly objective basis, especially if movements are considered as a communicative tool, e.g. expressing affective information. Nakata and colleagues [27] present a first approach of utilizing Laban's movement parameters for expressing emotional content of a dance performed by a small robot that is able to raise and lower its arms, to raise and lower its head, and has two wheels to move around. Even with this very restricted setup in terms of available degrees of freedom they could show that the Laban parameters correlate with the perception of the emotional content. Takahashi and colleagues [40] report on the design of emotional body movements for a robot that is primarily used for communication purposes and resembles a teddy bear. Similar to the previous example, the robot has very limited movement abilities. The arms can be raised or lowered as well as moved to the front of the body. The head can tilt and shake. Despite this low expressivity, recognition rates for three of the six emotions (joy, fear, sadness) are acceptable, highlighting the potential of Laban's parameters for movement control. Another recent approach of utilizing Laban's movement parameters is described by Masuda and Kato [25]. Concentrating on four emotions (pleasure, anger, sadness, relaxation) they show that the modification of a basic movement relying on Laban's parameters allows adding an affective connotation that is perceivable by observers.

Other approaches do not rely on Laban movement analysis to define affective body movements. Beck and colleagues [1] use motion capture for creating highly realistic movements for a Nao robot. Actors are used for expressing different emotional content. In the evaluation phase, it is tested if the key poses exhibited by a Nao robot, i.e. static postures of the robot, are correctly classified by observers. This approach suffers from the same problems that are generally described for the area of affective body movements, i.e. it is unknown why actors perform the movements as they do and thus the features for creating affective movements remain unknown. Nomura and Nakao [28] investigate how movements that express emotions are perceived by different user groups, younger (18–23) and older users (64–79). Their study suffers from the adhoc nature in which the affective body movements have been created. There is apparently no theoretical or empirical basis for the design of the movements, which makes it difficult to interpret their results. Häring and colleagues [16] describe an evaluation study that looks into the recognizability of different body movements in regard to their emotional content and in relation to other multimodal output like sound and light. Movements seem to have been created relying on deMeijer's analysis but it remains unclear how the features described in [7] relate to the stimuli used in the evaluation. Moreover, movements seem to have been tested not in isolation but always combined with eye colors and sound, making it difficult to separate the effects of movement characteristics from the effects of the eye color and sound.

Summing up, we can say that the work on affective body movements for robots relies on two sources, either the intuition of the developer or the utilization of Laban's effort and shape model. None of the approaches so far has addressed the cross-cultural challenges. From work in virtual agents we know that non-verbal behavior has a strong influence on the acceptance and willingness to interact. It has to match the cultural heuristics that structure the user's perception and sense-making processes. Thus, to be able to deploy our robots in diverse contexts and cultures it is necessary to investigate cultural influences in the design and interaction process. In the remainder of this paper we suggest a methodological approach for addressing this challenging task.

8.3 Developing Embodied Interactions

The experimental design follows the development cycle that has been suggested for embodied interactions by Cassell [5]. Figure 8.1 visualizes the main steps in the development while highlighting cultural influences coming from different sources (see also [32]). To disentangle the influences of the various cultural backgrounds of developers, designers, participants in experiments, and last but not least, users, we propose a three step process that allows capturing the culture-specific aspects of the creation and perception of affective body movements. This three step design combines two main sources of inspiration coming from empirical psychological research and embodied agent research.

Elfenbein and Ambady [10] argue that the cultural background has to be taken into account as a separate variable in an experiment, both on the side of the stimulus provider as well as the observer. We have taken this idea up in earlier work [21], which also serves as inspiration for integrating cultural influences into the whole development cycle. In this article, we focus on the three steps that are visualized in Fig. 8.1: (i) describing the collection of a cross-cultural database of affective body movements in a co-creation experiment with users from the target cultures (study); (ii) using this database to create a probabilistic model for generating affective body movements (model); (iii) evaluating the model with users from the target cultures (test).

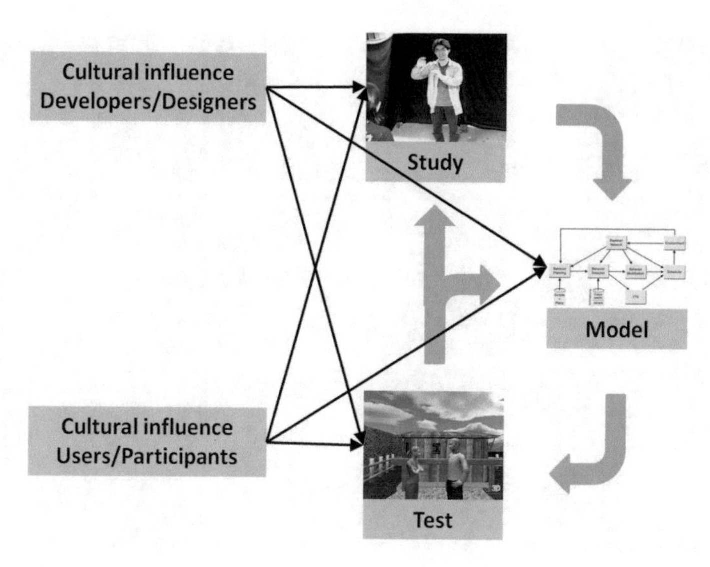

Fig. 8.1 Different sources of cultural influences on the development cycle for embodied interactions here exemplified for interactions between virtual agents

8.4 Data Collection

For the study described in this paper, we concentrate on the effort dimension, because the robot under investigation (Nao) does not allow control of the shape parameters due to the rigid nature of the body. We assume that this will be the case with most robotic systems.

Following [13], the experiment concentrates on one specific movement, a knocking. Figure 8.2 visualizes the basic movement for the Nao robot. By manipulating the effort parameters of the Laban system, this basic movement can now be changed to create an angry knocking, a sad knocking, etc. In the co-creation experiment, participants are asked to create these affective body movements by manipulating the Laban parameters of time, weight, and space.

One problem with the co-creation method is the strain on participants, e.g. with three parameters and a five point scale, participants have to choose already between 120 different parameter combinations for each emotion. It would be difficult to keep the participants' interest long enough to get enough data for all emotions. Thus, we decided to break the co-creation experiment up into different parts starting with a high level of abstraction, where each parameter is only manipulated on a three point scale.

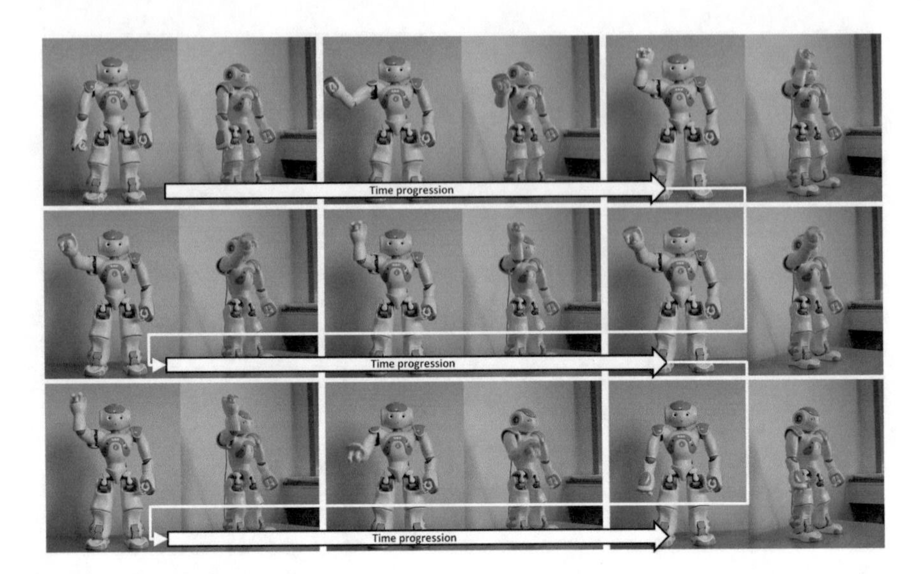

Fig. 8.2 Knocking movement

8.4.1 Participants

Data collection started in Denmark, Germany, Japan, and Greece. It is envisioned as a continuing endeavor in order to collect a large cross-cultural database. So far, data has been collected from 69 participants (17 Danish, 27 German, 14 Japanese, 11 Greek), 24 female and 45 male, with an age range from 20 to 45 (mean: 28.5, SD: 6.44).

8.4.2 Apparatus and Material

A Nao H25 v3.3 robot from Aldebaran Robotics was used to show the affective body movements.

8.4.3 Procedure

Before participants can start the experiment, they are required to answer some demographic questions. Aiming at a cross-cultural database, we are in need of some specific demographic information like the participant's nationality and the country, where s/he has lived for the longest time during the last five years (see Fig. 8.3 (left) for the corresponding web interface). This is interesting information because it can be assumed that after a longer period of time, assimilation of and blending with cultural heuristics of the host culture has taken place, invalidating this data set for the purpose at hand. Other information that is collected include age, gender and hands-on experience with robots.

During the co-creation part of the data collection, participants are presented with six tasks in random order. Each task requires them to create a different affective meaning of the knocking movement, each corresponding to one of Ekman's basic emotions [9] (see Fig. 8.3 (right) for the corresponding web interface). Because the parameter names time, weight and space have been shown in a pilot test to be not readily understandable by participants, they were changed to speed, power and path for the experiment. Each parameter could be changed on a three point scale: speed (slow to fast), power (weak to strong), path (direct to indirect). For each movement, participants were then asked to indicate their satisfaction with the movement on a five point Likert scale.

Each parameter modification is directly visible in the video window, which shows the robot's movement according to the current parameter values. The video shows the robot from two perspectives—a front view and a side view—to allow for a better insight into how parameter modifications change the movement.

Result of the co-creation experiment is a cross-cultural database with effort parameters related to affective body movements. It takes cultural influences into account by using the study participants as co-designers of the body movements, which then will reflect relevant cultural patterns in the parameter values.

8.4.4 Results and Discussion

The main goal is the collection of a cross-cultural database, which can be used to train the model for generating affective body movements. The collected data is analyzed to extract information about cross-cultural differences in the design of affective movements. The underlying hypothesis for the analysis is: *There will be differences in the parameter values between different cultural groups.*

This hypothesis is based on observations from a previous project [34] where we analyzed human face to face interactions in the German and Japanese cultures and found significant differences in the way gestures were performed. The analysis was based on expressivity parameters that have been shown to relate to cultural background, emotions as well as personality (see [31] for an analysis) and that are loosely related to the Laban parameters used in the current experiment.

Figure 8.4 gives an overview of the average parameter values across the participating cultures for the six basic emotions. Three different significance levels are visualized for pairwise comparisons. As expected, we find significant differences regarding all parameters with an emphasis on the weight and space parameters. Differences are found across all emotions. Thus, the hypothesis could not be rejected. The results show that culture presents a relevant influence factor in the design of affective body movements.

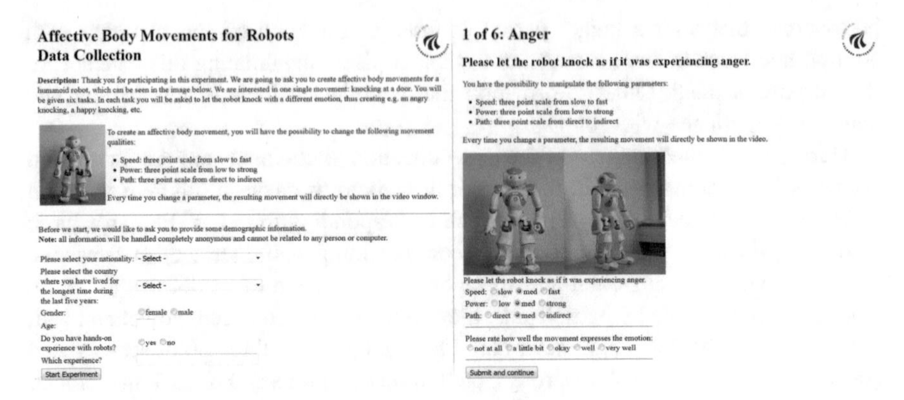

Fig. 8.3 Web interfaces for collecting demographic information (*left*) and for the co-creation experiment (*right*)

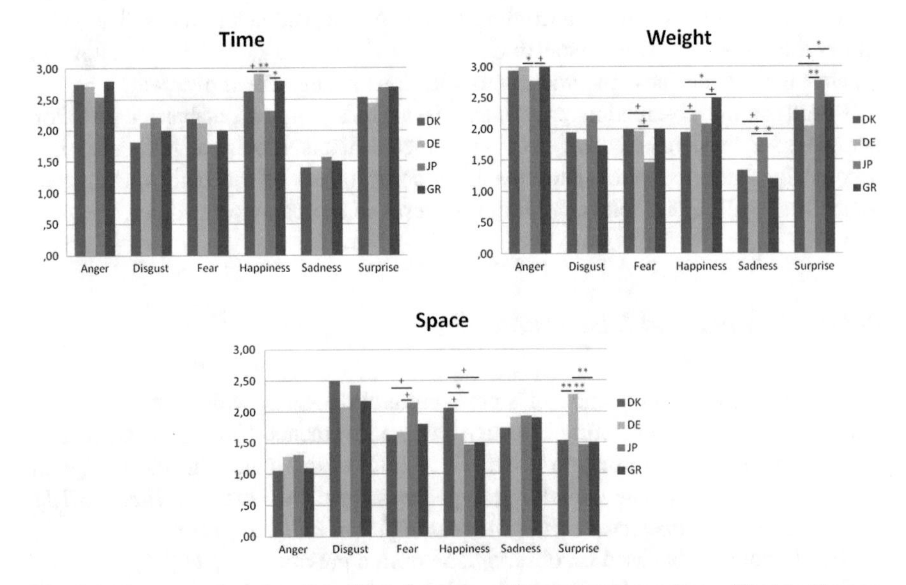

Fig. 8.4 Average values for effort dimensions across four exemplary cultures (Danish, German, Japanese, Greek). Significance levels: +: $p < 0.1$; *: $p < 0.05$; **: $p < 0.01$

For motivational reasons, participants could only manipulate the parameters on a three point scale, which is a simplification of the actual precision of the parameters. Based on the results, it becomes possible now to re-design the experiment on a less abstract level, taking the information about cultural preferences into account and offering participants more precision around the value ranges that have been identified for their cultural groups.

8.5 Model

Based on the cross-cultural database collected in the previous step, it becomes now feasible to derive a probabilistic model for generating affective body movements for a humanoid robot, taking into account the cross-culural differences revealed by the co-creation experiment.

We decided for a two step process, first creating a decision tree for each culture allowing for capturing the relative relevance of each Laban parameter for the interpretation of the body movements. Then, we build a Bayesian network with the data, allowing for capturing the inherent uncertainty and ambiguity in culture-related decisions when applied to individuals.

The rationale for this approach is that cultural influences can only be described as abstractions on a group level, in our case taking national culture into account. Thus, specific individuals may and will deviate from these general heuristics of behavior depending on their individual social interaction history and experience with other (sub-)cultures. Using Baysian networks allows for reflecting the unreliability and ambiguity that is a trademark of phenomena like culture and emotions.

8.5.1 *Decision Tree*

Following ideas by Ochs and colleagues [29], J48-decision trees were calculated on the cross-cultural data base. The previous analysis revealed significant differences between the cultures, thus a separate tree was calculated for each culture. Figure 8.5 depicts the four trees, highlighting the relative importance of the Laban parameters for the emotional interpretation of the robot's body movements in each of the four cultures.

The root node of all trees is weight, making it the most discriminating feature for deciding on the emotional interpretation of a movement. For the European cultures, low weight results in an interpretation of sadness whereas for the Japanese culture, low weight of the movement is a sign of fear. On the other hand, high weight interpretations always include anger as a relevant emotion for all cultures. But Fig. 8.5 also reveals differences in the relative importance and interpretation of features. For instance, high weight determines the emotion for the German sample to be anger, whereas for the Danish culture space is another parameter that has to be taken into account before a decision can be made. For the Greek culture instead, time is the discriminating factor. The next section gives details on how this information about the relative importance of features is used to build up a probabilistic model for generating affective body movements.

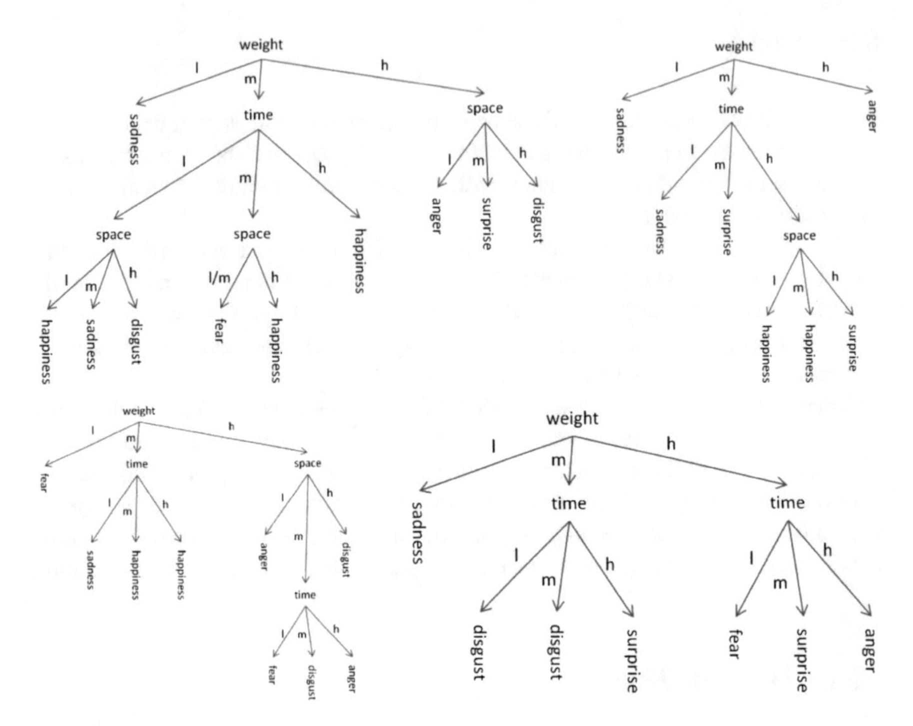

Fig. 8.5 Decisions trees (J48 pruned) for the four cultures (*above* Danish, German; *below* Japanese, Greek)

8.5.2 Bayesian Network

Using Bayesian networks allows for reflecting the unreliability and ambiguity that is a trademark of phenomena like culture and emotions. Moreover, Bayesian networks allow for drawing causal as well as diagnostic inferences, i.e. they can be used to generate culture-specific body movements but at the same time can serve for recognizing the emotional content of body movements or making an educated guess about the cultural background of the user.

Using the data from the co-creation experiment, the network's probabilities can be derived. Additionally, the knowledge from the decision trees is utilized by increasing the probabilities for feature combinations that correspond to leaf nodes in the decision trees. For instance, according to the information from the decision trees, the emotion associated with body movements of low weight is always sadness for the exemplary cultures Danish, German, and Greek. Thus, the probability is set to 0.99 for feature combinations involving low weight. There is also some probability for the emotion sadness with other feature combinations, which do not show up in the decision tree. In these cases, the probabilities are calculated from the data collected during the co-creation experiment. For instance for the feature combination medium weight, high time, low space, the probability for sadness is 0.15 for the Japanese culture.

An overview of the network is given in Fig. 8.6. It models the interrelation between three different phenomena, the emotional interpretation of body movements (1) based on cultural background (2) and Laban parameters (3). Thus, setting evidence for two of those phenomena allows deriving a probability distribution for the remaining phenomenon, making the Bayesian network useful for generating affective body movements as well as for interpreting those movements, either in terms of emotional content or cultural background.

Figure 8.7 presents examples for these three cases. If we are e.g. interested in generating a happy movement for a German user, we would set evidence for the emotion and the cultural background and can derive the values for the Laban parameters that will control the movement of the Nao robot. The result can be seen in Fig. 8.7 at the top. The most likely combination of Laban parameters would be medium weight (probability of 76%), high time (probability 91%) and either low or medium space (50 or 44% respectively).

Given an emotional interpretation (e.g. fearful movement) and evidence for the Laban parameters [e.g. medium weight and fast movement (value high for time parameter)], the network gives us a probability distribution over the cultures which were included in the data collection. The result is shown in Fig. 8.7 in the middle, where the Danish and German culture are the most likely candidates (40 or 30% respectively).

Or we might want to know the most likely emotional interpretation of a body movement defined by e.g. medium weight and low spatial extent for a Japanese user. The lower part of Fig. 8.7 shows the resulting probabilities in the network if the evidences are set accordingly. The most likely interpretation would be the emotion happiness with a probability of 67% followed by sadness with 32%.

8.6 Evaluation

In order to evaluate the database as well as the model derived from it, we conducted another web experiment. Participants were presented with emotional displays generated from the model and then had to rate the emotional content of the given movement. The experiment thus investigates how well the model is able to generate Culturally-Aware emotional movements, working with the following hypothesis:

> Participants will identify emotional movements generated from the model with a higher probability than chance.

For each emotional display, participants will have to decide between six different emotional labels, i.e. the result should be significantly higher than 16.7%. Independent variable will be the emotional movement, dependent variable the participant's subjective ratings.

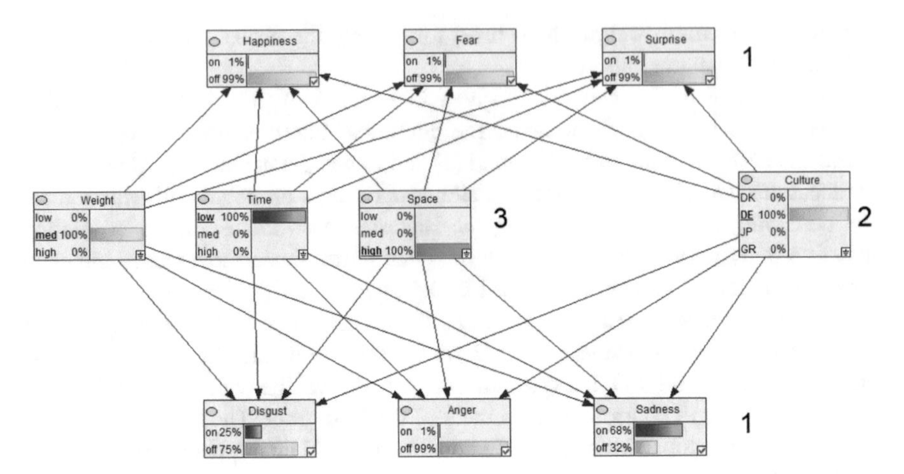

Fig. 8.6 Bayesian network modeling the relation between emotions (*1*), cultural background (*2*), and Laban parameters (*3*)

8.6.1 Participants

55 students were recruited for a web-based experiment in three different countries (Denmark: 8; Japan: 37; Germany: 10).

8.6.2 Apparatus and Materials

A Nao H25 v3.3 robot from Aldebaran Robotics was used to show the affective body movements. The movements have been generated by the model and then videos had been recorded for use in the web-based experiment. Participants had to fill out the same demographic information as during the data collection (see above). For each movement a questionnaire was administered, where participants were asked if one of the following emotions has been expressed by the robot in the video clip: anger, hate, happiness, sorrow, surprise, fear, other. Each emotion question had a five point Likert scale associated ranging from “I strongly think so” to “I do not think so”.

8.6.3 Procedure

First, participants were introduced to the study aim and were shown a still image from the video along with the questionnaire that they were supposed to answer. Then they had to give some demographic information. When they were not from

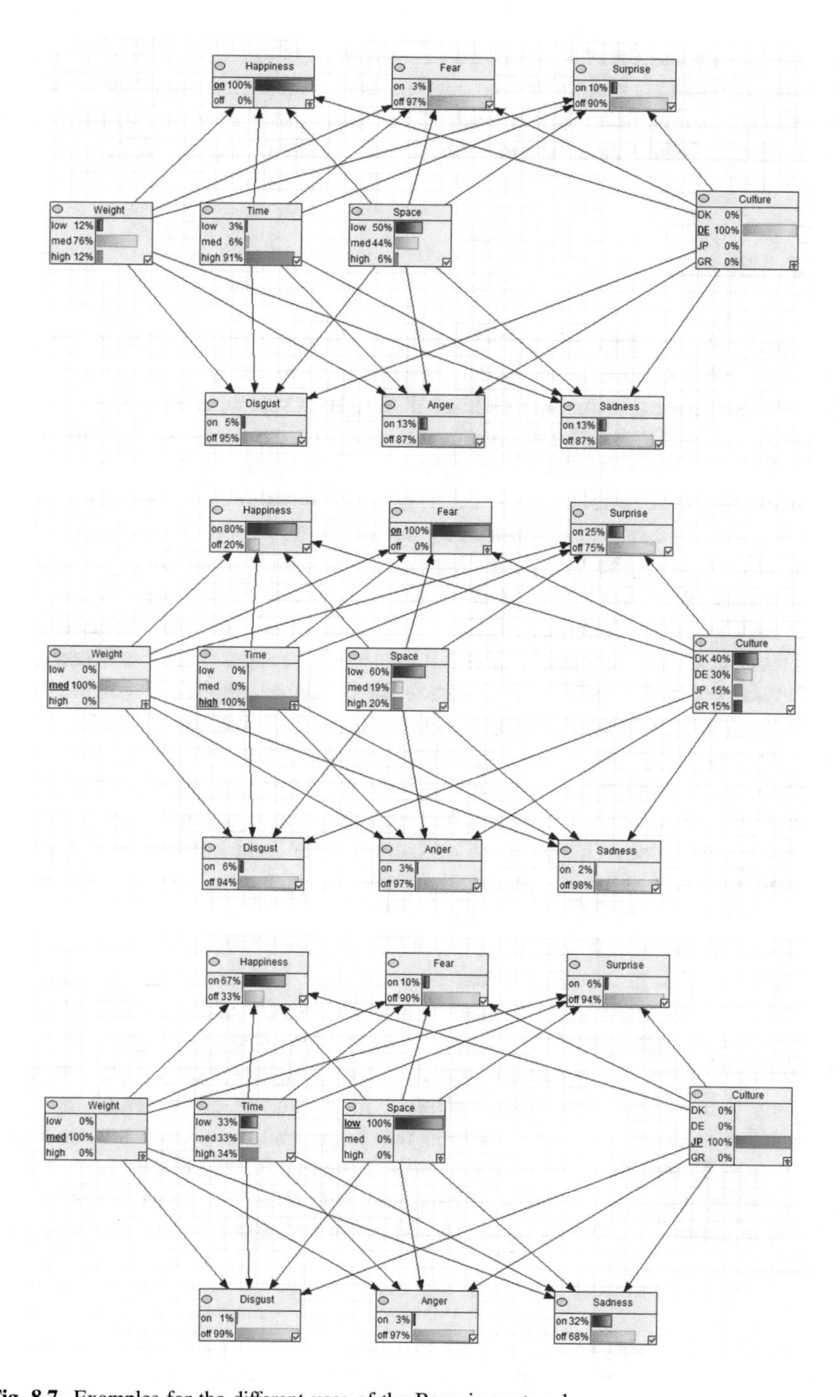

Fig. 8.7 Examples for the different uses of the Bayesian network

one of the cultures that the model so far supports (DK, DE, JP, GR), they were redirected to the data collection web page instead. Otherwise, a series of six videos was shown to them one after the other. For each video they had to fill out the questionnaire, before the next video was made available (see Fig. 8.8).

8.6.4 Results

297 valid data sets have been used for the analysis, 42 from Denmark, 50 from Germany, and 205 from Japan. 33 data sets had to be excluded because participants did not finish the experiment, which resulted in data sets with no ratings.

Tables 8.1, 8.2, 8.3 and 8.4 present the results of the model evaluation, where Table 8.1 summarizes results for all cultures and the Tables 8.2, 8.3 and 8.4 give the culture-specific results. Each table presents measurements for recognition, accuracy, and precision of the user interpretation in relation to the target emotion.[1] Recognition denotes the recognition rate for the target emotion, i.e. the percentage of correct classifications for the given emotion. Accuracy also takes correct negative classifications into account, e.g. if the target emotion is anger and the user has to interpret a happy movement and does not interpret it as anger, then this is taken into account as a correct negative classification for the target emotion. Precision at last takes into account false positive classifications, e.g. if the user in the previous example does interpret the movement as anger, this a false positive classification. For instance, the recognition rate for happiness for the Japanese sample (Table 8.4) for condition one is 50%. This means that for half of the happy movements, the user also interpreted those as happy. Precision on the other hand is 43.6%, which means that more than half of the movements that were interpreted as happy movements by the user were actually depicting other emotions.

For each measure two conditions are presented. Condition one presents results if only the user's first choice, i.e. the emotion with highest rating, is taken into account. Because interpreting the emotional content of a decontextualized movement is not an easy task [13], the second condition also counts it as a correct classification if the target emotion has the second highest rating. This correction was only done, when there was a difference in the ratings of the remaining emotions. If for instance, the target emotion was anger and the participant rated happiness with 5 and all other emotion with 2, then this was treated as a wrong classification. If the participant rated happiness with 5 and e.g. anger with 4 and all other emotions with 2, then this was treated as correct classification (for the second condition).

[1]The measures (recognition, accuracy, and precision) have been calculated from the confusion matrices, which are given in the appendix as Tables 8.5, 8.6, 8.7, 8.8, 8.9, 8.10, 8.11 and 8.12. In the appendix the reader can also find the formulas for recognition (usually called recall), accuracy, and precision.

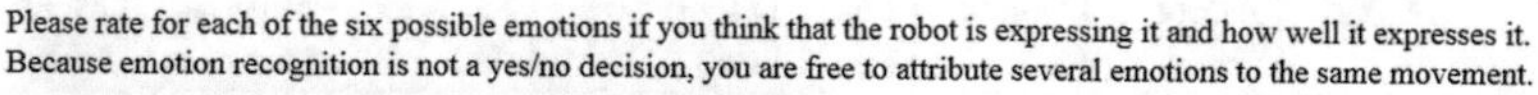

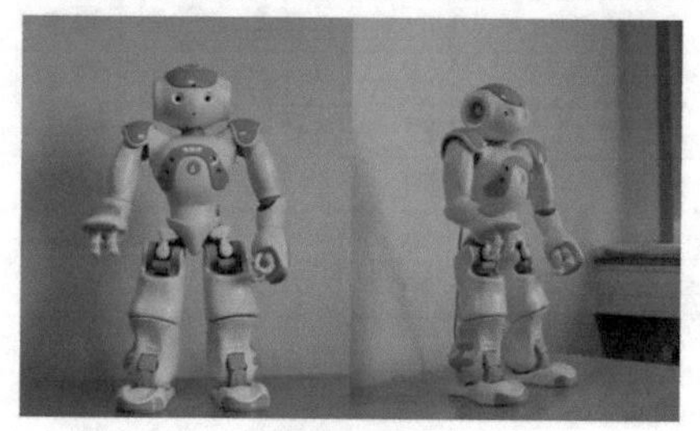

Fig. 8.8 Web experiment for model evaluation

Table 8.1 Classification results for all cultures

Measure	Cond.	Anger (%)	Disgust (%)	Fear (%)	Happiness (%)	Sadness (%)	Surprise (%)	Mean (%)
Recognition	1	52.1	20.8	50	49	58.8	17.7	41.4
	2	60.4	47.9	68	53.1	78.4	31.4	56.5
Accuracy	1	77.1	82.2	76.8	82.8	85.5	81.8	81.1
	2	77.8	87.5	84.5	85.5	91.9	85.5	85.5
Precision	1	31.3	40	36.2	48	57.7	42.9	42.5
	2	38.2	65.7	54	56.5	75.5	66.7	59.4

Table 8.2 Classification results for Danish culture

Measure	Cond.	Anger (%)	Disgust (%)	Fear (%)	Happiness (%)	Sadness (%)	Surprise (%)	Mean (%)
Recognition	1	50	14.3	38	16.7	88	14.3	36.7
	2	50	57.1	63	16.7	100	28.6	52.5
Accuracy	1	69.1	76.2	81	85.7	81	83.3	79.4
	2	73.8	85.7	85.7	85.7	92.9	85.7	84.9
Precision	1	23.1	20	50	50	50	50	40.5
	2	27.3	57.1	62.5	50	72.7	66.7	56.1

Table 8.3 Classification results for German culture

Measure	Cond.	Anger (%)	Disgust (%)	Fear (%)	Happiness (%)	Sadness (%)	Surprise (%)	Mean (%)
Recognition	1	75	12.5	50	66.7	55.6	0	43.3
	2	87.5	25	62.5	77.8	77.8	12.5	57.2
Accuracy	1	90	80	68	88	84	78	81.3
	2	92	82	80	90	90	82	86
Precision	1	66.7	25	25	66.7	55.6	0	39.8
	2	70	40	41.7	70	70	33.3	54.2

Table 8.4 Classification results for Japanese culture

Measure	Cond. (%)	Anger (%)	Disgust (%)	Fear (%)	Happiness (%)	Sadness (%)	Surprise (%)	Mean (%)
Recognition	1	47.1	24.2	52.9	50	52.9	22.2	41.6
	2	55.9	51.5	70.6	52.9	73.5	36.1	56.8
Accuracy	1	70.7	83.9	78.1	81	86.8	82.4	80.5
	2	75.1	89.3	85.9	84.4	92.2	86.3	85.5
Precision	1	27.6	50	38.3	43.6	62.1	50	45.3
	2	34.6	73.9	55.8	52.9	78.1	72.2	61.3

The results show that for nearly all cultures and emotions, participants were better than chance in identifying the movements that have been generated by the model for the different emotions. Where this was not the case, results usually got better when including the second choice of the user in the analysis (condition two). Exceptions are happiness for the Danish culture and surprise for the German culture. Surprise was the most difficult target emotion for all cultures and was also the one with the noisiest data in the data collection. It seems to be difficult to attribute surprise to the movement that is used in the experiment. Why happiness is problematic in the Danish sample is unclear.

Accuracy is generally good (over 70%) for all emotions and cultures. Thus, emotional interpretations are not often used for the wrong target emotions. Precision is lower on average (around 50% on average), which means that correct and incorrect use of the target emotion label occur equally often. There are though large variations between cultures and emotions. For instance, precision for anger in the German sample is 70% (condition two) whereas it is only 33.3% for surprise. On the other hand precision for anger is 34.6% (condition two) for the Japanese sample, where the precision for surprise is 72.2% (condition two).

8.6.5 Discussion

The results show that the hypothesis can be retained, i.e. participants were able to identify the target emotions with a higher probability than chance. The results for precision also show that some emotions are easier to identify than others (see anger and surprise results above), and that cultures vary in the specific emotions that are easier to identify.

Although the results are very promising, there are some caveats that have to be addressed next. First of all, the number of culture-specific data sets is highly variable with relatively few samples for the Danish and German cultures and sufficiently large sample for the Japanese culture. The results show the same trends for all cultures, but it is still necessary to adjust the sample numbers for the non-Japanese cultures.

Secondly, so far the experiment just included evaluation of culture-specific stimuli. Although this is in line with ideas proposed earlier, e.g. by Elfenbein and Ambady [10] or Koda et al. [21], the current study does not shed any light on the question if there is an in-group advantage for stimuli that are congruent with the model. To investigate this question, another experiment is needed, where participants also will have to rate stimuli that are not created for their own culture.

8.7 Conclusion

In this chapter we argue that culture is a relevant parameter for creating interactive and especially embodied systems like humanoid robots. There is evidence that the generation and interpretation of non-verbal signals like body movements is unconsciously governed by cultural heuristics. What is unclear is, if such cultural heuristics also play a role in human robot interaction and to what degree. The challenge lies in entangling the mesh of possible cultural influences that play a role in developing, designing, as well as interacting with a humanoid robot. In this paper, we have presented a methodological approach that has matured over the last years in various cross-cultural projects and that allows for embracing the challenge of designing systems for cross-cultural interaction. We have exemplified this approach with a study on affective body movements. What is apparent from this study is that doing it right takes time and effort because at each step of the development, possible cultural influences have to be analyzed and kept at bay by integrating them tightly in the development process. Based on the data collected in a co-creation experiment, a computational model was derived that is used to drive the nonverbal behavior of a humanoid robot. The model was evaluated with a perception experiment across three different cultures and results showed that the model indeed generated recognizable affective movements.

Acknowledgements Special thanks go to Prof. Yukiko Nakano (Seikei University, Japan), Prof. Tomoko Koda (Osaka Institute of Technology, Japan), Prof. Katharina Rohlfing (Univeristy of Paderborn, Germany), Amaryllis Raouzaiou (National Technical University of Athens, Greece), Prof. Birgit Lugrin (University of Wuerzburg, Germany), and Markus Häring (Augsburg University, Germany) for their recruiting efforts for the studies described in this paper.

Appendix Recognition (Recall), Accuracy, and Precision

For calculating the three measures, the following notions are needed: true positive (tp), false positive (fp), true negative (tn), false negative (fn). The color codes in Table 8.5 exemplify to which part of the confusion matrix they refer to if the target emotion is anger. The following codes are used: tp (green), fp (yellow), tn (grey), fn (red).

Table 8.5 Confusion matrix summarizing results for all cultures (condition 1)

Target	Anger	Disgust	Fear	Happiness	Sadness	Surprise	Total
Anger	25	1	8	6	4	4	48
Disgust	11	10	14	8	2	3	48
Fear	4	6	25	5	9	1	50
Happiness	12	3	4	24	3	3	49
Sadness	2	2	11	5	30	1	51
Surprise	26	3	7	2	4	9	51
Total	80	25	69	50	52	21	297

Table 8.6 Confusion matrix summarizing results for all cultures (condition 2)

Target	Anger	Disgust	Fear	Happiness	Sadness	Surprise	Total
Anger	29	1	6	5	4	3	48
Disgust	6	23	9	8	0	2	48
Fear	4	5	34	3	3	1	50
Happiness	12	3	4	26	3	1	49
Sadness	1	1	4	4	40	1	51
Surprise	24	2	6	0	3	16	51
Total	76	35	63	46	53	24	297

$$\text{recognition (recall)} = \frac{tp}{tp+fn}$$

$$\text{accuracy} = \frac{tp+tn}{all}$$

$$\text{precision} = \frac{tp}{tp+fp}$$

Table 8.7 Confusion matrix for Danish culture (condition 1)

Target	Anger	Disgust	Fear	Happiness	Sadness	Surprise	Total
Anger	3	0	2	0	1	0	6
Disgust	4	1	0	0	1	1	7
Fear	0	1	3	1	3	0	8
Happiness	3	1	0	1	1	0	6
Sadness	0	1	0	0	7	0	8
Surprise	3	1	1	0	1	1	7
Total	13	5	6	2	14	2	42

Table 8.8 Confusion matrix for Danish culture (condition 2)

Target	Anger	Disgust	Fear	Happiness	Sadness	Surprise	Total
Anger	3	0	2	0	1	0	6
Disgust	2	4	0	0	0	1	7
Fear	0	1	5	1	1	0	8
Happiness	3	1	0	1	1	0	6
Sadness	0	0	0	0	8	0	8
Surprise	3	1	1	0	0	2	7
Total	11	7	8	2	11	3	42

Table 8.9 Confusion matrix for German culture (condition 1)

Target	Anger	Disgust	Fear	Happiness	Sadness	Surprise	Total
Anger	6	0	1	0	1	0	8
Disgust	0	1	4	2	0	1	8
Fear	0	2	4	1	1	0	8
Happiness	0	0	0	6	1	2	9
Sadness	1	0	3	0	5	0	9
Surprise	2	1	4	0	1	0	8
Total	9	4	16	9	9	3	50

Table 8.10 Confusion matrix for German culture (condition 2)

Target	Anger	Disgust	Fear	Happiness	Sadness	Surprise	Total
Anger	7	0	0	0	1	0	8
Disgust	0	2	3	2	0	1	8
Fear	0	2	5	1	0	0	8
Happiness	0	0	0	7	1	1	9
Sadness	1	0	1	0	7	0	9
Surprise	2	1	3	0	1	1	8
Total	10	5	12	10	10	3	50

Table 8.11 Confusion matrix for Japanese culture (condition 1)

Target	Anger	Disgust	Fear	Happiness	Sadness	Surprise	Total
Anger	16	1	5	6	2	4	34
Disgust	7	8	10	6	1	1	33
Fear	4	3	18	3	5	1	34
Happiness	9	2	4	17	1	1	34
Sadness	1	1	8	5	18	1	34
Surprise	21	1	2	2	2	8	36
Total	58	16	47	39	29	16	205

Table 8.12 Confusion matrix for Japanese culture (condition 2)

Target	Anger	Disgust	Fear	Happiness	Sadness	Surprise	Total
Anger	19	1	4	5	2	3	34
Disgust	4	17	6	6	0	0	33
Fear	4	2	24	1	2	1	34
Happiness	9	2	4	18	1	0	34
Sadness	0	1	3	4	25	1	34
Surprise	19	0	2	0	2	13	36
Total	55	23	43	34	32	18	205

References

1. Beck, A., Stevens, B., Bard, K.A., Canamero, L.: Emotional body language displayed by artificial agents. ACM Trans. Interact. Intel. Syst. **2**(1), 1–29 (2012)
2. Bennett, M.J.: A developmental approach to training for intercultural sensitivity. Int. J. Intercult. Relat. **10**(2), 179–195 (1986)
3. Blanchard, E.G., Mizoguchi, R., Lajoie, S.P.: Structuring the cultural domain with an upper ontology of culture. In Blanchard, E.G., Allard, D. (eds.) Handbook of Research on Culturally-Aware Information Technology: Perspectives and Models, pp. 179–212. IGI Global, Hershey PA (2010)

4. Blanchard, E.G., Ogan, A.: Infusing cultural awareness into intelligent tutoring systems for a globalized world. In: Nkambou, R., Mizoguchi, R., Bourdeaur, J. (eds.) Advances in Intelligent Tutoring Systems, pp. 485–505. Springer, Berlin (2010)
5. Cassell, J.: Body language: lessons from the near-human. In Riskin, J. (ed.) Genesis Redux: Essays in the History and Philosophy of Artificial Intelligence, pp. 346–374. University of Chicago Press (2007)
6. Clemmensen, T.: A framework for thinking about the maturity of cultural usability. In: Blanchard, E.G., Allard, D. (eds.) Handbook of Research on Culturally-Aware Information Technology: Perspectives and Models, pp. 295–315. IGI Global, Hershey PA (2010)
7. de Meijer, M.: The contribution of general features of body movement to the attribution of emotions. J. Nonverbal Behav. **13**(4), 247–268 (1989)
8. Efron, D.: Gesture, Race and Culture. Mouton and Co (1972)
9. Ekman, P.: Basic emotions. In Dalgleish, T., Power, M. (eds) Handbook of Cognition and Emotion, Chap. 3, pp. 45–60. Wiley, Chichester (1999)
10. Elfenbein, H.A., Ambady, N.A.: When familiarity breeds accuracy: cultural exposure and facial emotion recognition. J. Pers. Soc. Psychol. **85**(2), 276–290 (2003)
11. Evers, V., Maldonado, H., Brodecki, T., Hinds, P.: Relational vs. group self-construal: untangling the role of national culture in HRI. In: Proceedings of Human Robot Interaction (HRI), pp. 255–262 (2008)
12. Gallaher, P.E.: Individual differences in nonverbal behavior: dimensions of style. J. Pers. Soc. Psychol. **63**(1), 133–145 (1992)
13. Gross, M.M., Crane, E.A., Fredrickson, B.L.: Methodology for assessing bodily expression of emotion. J. Nonverbal Behav. **34**, 223–248 (2010)
14. Hall, E.T.: Beyond Culture. Doubleday (1976)
15. Hall, L., Lutfi, S., Nazir, A., Hodgson, J., Hall, M., Ritter, C., Jones, S., Mascarenhas, S., Cooper, B., Paiva, A., Aylett, R.: Game based learning for exploring cultural conflict. In: AISB (2011)
16. Häring, M., Bee, N., André, E.: Creation and evaluation of emotion expression with body movement, sound and eye color for humanoid robots. In: Proceedings of the 20th IEEE International Symposium on Robot and Human Interactive Communication, pp. 204–209 (2011)
17. Henrich, J., Heine, S.J., Norenzayan, A.: The weirdest people in the world? Behav. Brain Sci. **33**, 61–135 (2010)
18. Hofstede, G.: Cultures Consequences: Comparing Values, Behaviors, Institutions, and Organizations Across Nations. Sage Publications, Thousand Oaks, London (2001)
19. Hofstede, G.J., Pedersen, P.B., Hofstede, G.: Exploring Culture: Exercises, Stories, and Synthetic Cultures. Intercultural Press, Yarmouth (2002)
20. Kleinsmith, A., De Silva, P.R., Bianchi-Berthouze, N.: Recognizing Emotion from Postures: Cross-Cultural Differences in User Modeling. In Ardissono, L., Brna, P., Mitrovic, A. (eds.) User Modeling, pp. 50–59. Springer, Berlin (2005)
21. Koda, Tomoko, Ishida, Toru, Rehm, Matthias, André, Elisabeth: Avatar culture: cross-cultural evaluations of avatar facial expressions avatar culture: cross-cultural evaluations of avatar facial expressions. AI & Society, Spec. Issue Enculturat. HCI **24**(3), 237–250 (2009)
22. Laban, R.: The Mastery of Movement. Dance Books (2011)
23. Lipi,A.A., Nakano, Y., Rehm, M.: Culture and social relationship as factors of affecting communicative non-verbal behaviors. Trans. Jpn. Soc. Artifi. Intell. **25**(712–722), 6 (2012)
24. Marcus, A., Hamoodi, S.: The impact of culture on the design of arabic websites. In: IDGD'09 Proceedings of the 3rd International Conference on Internationalization, Design and Global Development, pp. 386–394. Springer, Berlin, Heidelberg (2009)
25. Masuda, M., Kato, S.: Motion rendering system for emotion expression of human form robots based on Laban movement analysis. In: Proceedings of the 19th IEEE International Symposium on Robot and Human Interactive Communication, pp. 324–329 (2010)

26. McElreath, R., Henrich, J.: Dual inheritance theory: the evolution of human cultural capacities and cultural evolution. In: Dunbar, R., Barrett, L. (eds.) Oxford Handbook of Evolutionary Psychology. Oxford University Press (2007)
27. Nakata, T., Mori, T., Sato, T.: Analysis of impression of robot bodily expression. J. Robot. Mechatron. **4**(1), 27–36 (2002)
28. Nomura, T., Nakao, A.: Human evaluation of affective body motions expressed by a small-sized humanoid robot: comparison between elder people and university students. In: Proceedings of the 18th IEEE International Symposium on Robot and Human Interactive Communication, pp. 363–368 (2009)
29. Ochs, M., Niewiadomski, R., Pelachaud, C.: How a virtual agent should smile? morphological and dynamic characteristics of virtual agent's smiles. In Allbeck, J., Badler, N., Bickmore, T., Pelachaud, C., Safonova, A. (eds.) Intelligent Virtual Agents, pp. 427–440. Springer, Berlin (2010)
30. Rehm, M.: Developing enculturated agents—pitfalls and strategies. In: Blanchard, E.G., Allard, D. (eds.) Handbook of Research on Culturally-Aware Information Technology. IGI Global (2010)
31. Rehm, M.: Non-symbolic gesture usage for ambient intelligence. In: Human-Centric Interfaces for Ambient Intelligence. Elsevier (2010)
32. Rehm, M., André, E., Nakano, Y.: Some pitfalls for developing enculturated conversational agents. In: Jacko, J.A. (ed.) Human-Computer Interaction, Part III, HCII 2009, pp. 340–348. Springer, Berlin (2009)
33. Rehm, M., Leichtenstern, K.: Gesture-based mobile training of intercultural behavior. Multimedia Syst. **18**(1), 33–51 (2012)
34. Rehm, M., Nakano, Y., André, E., Nishida, T., Bee, N., Endrass, B., Wissner, M.: Afia Akhter Lipi, and Hung-Hsuan Huang. From observation to simulation—generating culture specific behavior for interactive systems. AI & Soc. **24**, 267–280 (2009)
35. Rehm, M., Nakano, Y., Koda, T.,Winschiers-Theophilus, H.: Culturally aware agent communication. In: Zacarias, M., de Oliveira, J.V. (eds.) Human-Computer Interaction: The Agency Perspective, pp. 411–436. Springer, Berlin (2012)
36. Reinecke, K., Bernstein, A.: Improving performance, perceived usability, and aesthetics with culturally adaptive user interfaces. Trans. Comput.-Hum. Interact. **18**(2), 1–29 (2011)
37. Ruotsalo, T., Aroyo, L., Schreiber, G.: Knowledge-based linguistic annotation of digital cultural heritage collections. IEEE Intell. Syst. **24**(2), 64–75 (2009)
38. Salmoni, B.A., Holmes-Eber, P.: Operational Culture for the Warfighter: Principles and Applications. Marine Corps University Press (2008)
39. Sperber, D.: Explaining Culture: A Naturalistic Approach. Blackwell Publishers Ltd. (1996)
40. Takahashi, K., Hosokawa, M., Hashimoto, M.: Remarks on designing of emotional movement for simple communication robot. In: Proceedings of the IEEE International Conference on Industrial Technology (ICIT), pp. 585–590. (2010)
41. Wallbott, H.G.: Bodily expression of emotion. Eur. J. Soc. Psychol. **28**, 879–896 (1998)
42. Wang, L., Rau, P.-L.P., Evers, V., Robinson, B.K., Hinds, P.: When in Rome: The role of culture and context in adherence to robot recommendations. In: Proceedings of the 5th ACM/IEEE international conference on Human-robot interaction, pp. 359–366 (2010)
43. Weiss, A., van Dijk, B., Evers, V.: Knowing me knowing you: exploring effects of culture and context on perception of robot personality. In: Proceedings of the 4th international conference on Intercultural Collaboration (ICIC), pp. 133–136. (2012)
44. Wenger, E.: Communities of Practice: Learning, Meaning, and Identity. Cambridge University Press (1998)

Chapter 9
Modeling Cultural and Personality Biases in Decision-Making

Eva Hudlicka

Abstract Cultural, personality and affective biases in decision-making are well documented. This chapter describes a method for modeling multiple decision biases resulting from cultural effects, personality traits and affective states, within the context of a symbolic cognitive-affective agent architecture: the MAMID methodology and architecture. The approach emphasizes the role of affect in decision-biases, as the primary mediating factor of a wide range of biasing effects, and lends itself to exploring alternative mechanisms mediating a wide range of decision biases. The approach provides a uniform framework for modeling both *content* and *processing biases,* in terms of parameter vectors that control processing within the architecture modules. The effects of these biases are encoded in specific values of architecture parameters, which then influence the processing of the distinct architecture modules, including the architecture topology itself. The associated simulation environment enables the modeling of a wide variety of decision-makers, in terms of distinct personality and cultural profiles, and consequent affective profile and affect-induced decision-biases. The key contribution, and distinguishing feature, of the MAMID modeling approach is the parameter space it provides for representing the interacting effects of multiple types and sources of biases, and the potential of this approach for modeling the fundamental mechanisms that mediate decision-biases.

Keywords Modeling decision-making biases · Cultural, personality and affective biases · Cognitive-affective architecture · Mechanisms of affective biases · Parameter-based modeling of affective biases

E. Hudlicka (✉)
Psychometrix Associates & College of Information and Computer Sciences,
University of Massachusetts-Amherst, Amherst, MA, USA
e-mail: Hudlicka@cs.umass.edu

C. Faucher (ed.), *Advances in Culturally-Aware Intelligent Systems and in Cross-Cultural Psychological Studies*, Intelligent Systems Reference Library 134, https://doi.org/10.1007/978-3-319-67024-9_9

9.1 Introduction

Cultural and personality effects and biases in decision-making are well documented (see reviews [1, 2]). Particularly dramatic examples of cultural effects have been identified in aviation, where a number of accidents have been attributed cultural factors, such as adherence to strict hierarchical communication patterns which prevented co-pilots from informing the pilot of critical problems, that eventually resulted in fatal accidents (e.g., Korean Air Flight 801).

Increasing reliance on human-machine systems across a broad range of environments, including decision-support systems in healthcare, behavioral technologies, serious gaming, aviation, space missions, process control and the military, necessitates that these systems be aware of cultural and personality biases. Decision-support systems in particular would benefit from an ability of the system to recognize and react to cultural and personality decision-biases. This is the case for systems that support long-term collaborative efforts (e.g., Groupsystems' ThinkTank), but is especially critical for systems designed to support decision-making in real-time, high-tempo and high-stress environments (e.g., space missions, aviation, healthcare, military).

An ability to model decision-biases in human behavior models and user models embedded in decision-support systems would enhance their abilities to provide customized decision-support, and user-adapted interaction, for a range of decision-maker profiles. In addition, such models would enhance our understanding of these biases, both their mechanisms [3], and their triggers and effects. Such improved understanding would in turn enable the development of training systems that would help decision-makers counteract the deleterious effects of biases, across a range of operational settings.

This chapter discusses an integrated approach to modeling a broad range of decision biases, within the context of a human behavior model, embedded in a cognitive-affective agent architecture: the MAMID methodology and architecture. The key contribution, and distinguishing feature, of the MAMID modeling approach is a uniform parameter space it provides for representing the interacting effects of multiple types and sources of biases. The effects of these biases are encoded in specific values of these architecture parameters, which then influence the processing of the distinct architecture modules, including the architecture topology itself.

The primary focus is on modeling cultural and personality biases, and biases due to transient affective states (e.g., anxiety and stress, boredom, frustration, anger). Affective biases are included because emotions often act as the intervening variables for a number of decision biases. The modeling approach also draws a distinction between *content-based* and *process-based* biases, in an effort to provide a categorization of biases that is based more on their deeper mediating mechanisms, rather than their observable surface features.

The chapter is organized as follows. Section 9.2 provides background information about specific cultural factors that have been identified in cultural anthropology and decision-making research. Section 9.3 discusses different types and sources of biases, emphasizing the role of emotion as the intervening variable, and providing examples of specific trait and state linked biases. Section 9.4 introduces a categorization of biases into content-based and process-based biases, and discusses the benefits of this organization for modeling a broad range of biases, and for elucidating their mechanisms. Section 9.5 introduces the MAMID framework for modeling a broad range of interacting decision-biases, and provides an illustrative example demonstrating its capabilities to model multiple types of biases. Section 9.6 discusses the potential. applications of the MAMID modeling framework. Section 9.7 concludes the chapter with a summary of MAMID capabilities and applications.

9.2 Cultural Influences on Cognition, Emotion and Behavior

Cultural factors represent important behavior determinants and exert influence on cognitive processing, affective processing (including the generation and expression of emotions), and, ultimately, behavior. The influence of culture has received much attention lately, and is of interest to modelers of cognition, emotion and behavior for a variety of reasons, ranging from basic research questions regarding the underlying mechanisms to applied objectives across a broad range of areas, including diplomacy, public policy, education and training, healthcare, entertainment, business and the military. The applications of interest include improved understanding of business negotiations in multi-cultural business settings; diplomacy and peace negotiations in government; and multi-national task forces, peacekeeping operations, as well as a range of military operations (e.g. asymmetric warfare). Extensive literature exists in psychology and cultural anthropology that addresses the characteristics of cultures, methods used in cultural anthropology research, and the effects of culture on decision-making and social interactions (e.g., [4–9]).

In spite of the vast literature addressing cultural issues, there is relative paucity of attributes defined at a sufficient level of specificity to enable computational modeling and inferencing; that is, cultural characteristics which could be operationalized to enable computational modeling of the effects of culture on individual decision-making and, more importantly, computational models with predictive capabilities.

Perhaps the most prominent set of "cultural" factors is the set proposed by Hofstede [6], which consists of the following: Power Distribution, Individualism-Collectivism, Femininity-Masculinity, Uncertainty Avoidance, and Short versus Long-Term Orientation. This list was recently augmented by Klein et al. [10], and termed the "cultural lens" model. Klein's set of factors augments the

Table 9.1 Examples of existing 'cultural differences' factors

Researcher	Cultural differences factors
Hofstede	Power distance Individualism-collectivism Feminity—masculinity Uncertainty avoidance Short versus long-term orientation
Klein	Power distance Independence—dependence (individualism—collectivism) Uncertainty avoidance Time orientation (short vs. long-term orientation) Activity orientation (process/people vs. task/outcome oriented) Counterfactual thinking versus hypothetical reasoning (?) Dialectical reasoning

Hofstede factors to include counterfactual thinking versus hypothetical reasoning, and dialectical reasoning. Table 9.1 summarizes these sets of cultural factors.

Power distance is the degree to which both the powerful and the powerless accept the existing power structure; and reflects the degree to which each group willingly participates in the established power structure, that is, the powerful accept their power and the powerless accept their powerlessness. *Individualism—Collectivism* captures the "every man for himself" versus "we will protect you but will expect loyalty" attitudes; that is, it reflects the individual's orientation towards self vs the group. *Femininity—Masculinity* captures the emphasis on "achievement and material success, versus harmony and caring" [41, p. 247]. We can also think of this in terms of an emphasis on task and outcome versus people and process. *Uncertainty avoidance* captures the desire for absolutes and truth versus tolerance of, and comfort with, uncertainty and ambiguity. *Short versus Long Term Orientation* reflects the planning horizon, and the ability to wait for 'rewards', vs the need for immediate gratification.

Hofstede's dimensions were obtained from a factor-analytic study of the results of a questionnaire administered to 120,000 IBM employees in 40 countries. Power distance is high for Latin America, Asian, and African countries, smaller for Western Europe and the US. Individualism is high in western countries; low in less industrialized and eastern countries. Masculinity high in Japan, Germanic European countries, moderate in English-speaking countries, low in Nordic and Netherlands, and moderately low in Latin and Asian countries.

There are several problems with the existing cultural factors that limit their utility and applicability to computational modeling. Specifically:

- The *factors are identified at very abstract levels*, which have little or no predictive value for specific behaviors selected by particular individuals or groups, operating within particular situational and environmental constraints. For example, how will knowing that a particular individual comes from a culture with low power distribution help us predict whether s/he is likely to participate in a violent demonstration to be held tomorrow?

- The *factors are insufficiently operationalized* to serve as a basis for computational modeling and behavior prediction. In other words, it is difficult to translate factors such as power distribution, femininity/masculinity, or dialectical thinking into instances of specific rules, or some other artificial intelligence inferencing and representational formalisms, capable of modeling decision-makers from specific cultures.
- The *factors lack adequate breadth* to capture all of the relevant and necessary cultural variables. In spite of the existing literature, the actual number of specific cultural factors identified is surprisingly small. In addition to the two problems above, the identified factors do not begin to capture the complexity of cultural influences, nor do they adequately address the problems of multiple and possibly conflicting cultural influences, or specific individual factors that may amplify (or inhibit) a particular cultural influence.

An interesting recent attempt to address some of these issues, and to develop more predictive approaches to capturing cultural effects, has explored a notion introduced by Karabaich [11] that proposes to consider each group to which an individual belongs as representing a distinct culture; that is, the assumption that *every group creates its own culture* [12]. In this context, the term culture is not limited to nations or ethnic groups, as has generally been the case in the literature [6, 7], but is instead broadened to include any group that influences the individual's behavior. This approach is motivated by the observation that national and ethnic groups are in fact not as diagnostic with respect to behavior prediction as are smaller groups to which an individual belongs (e.g., student group, social group, political group, family, etc.).

Regardless of the number and types of cultural factors that influence cognitive and affective processing, and behavior, it must be emphasized that any cultural influence ultimately functions at the individual level, and must therefore be translated to one or more individual differences factors, which then influence behavior via specific effects on cognitive and affective processing. A thorough understanding of these individual differences, and a determination of the mappings of the cultural factors onto these individual differences, are therefore critical for modeling the effects of cultural factors on decision-making. Section 9.3 outlines a categorization of these individual differences that is suitable for modeling decision-making and culturally-induced decision-biases.

9.3 Decision Biases

Research in judgment and decision-making has identified a number of *biases* in decision-making. Kahneman, Tversky and colleagues [13] identified a number of judgment and decision-making biases in individual decision-making, including the following: *availability bias* (the tendency to base decisions on the most readily available evidence, rather than the most appropriate evidence, hence the *primacy* and *recency* bias resulting from biased memory recall), *confirmation bias* (the tendency to prefer, or actively seek out, evidence that supports one's hypotheses, expectations and goals), and *framing effects* (the tendency to be influenced by, and respond differently to, the wording (as opposed to the substance) of a question or a decision).

Biases have been identified at various levels of processing complexity, ranging from biases on fundamental perceptual and cognitive processes (e.g., attention, memory), to biases on high-level cognitive processing (e.g., situation assessment, problem solving, decision-making). A number of individual differences contribute to decision-biases, including cultural and personality factors, and of course distinct affective states. As mentioned above, emotions in fact represent the primary mediating variables for a wide range of decision-biases. These factors are discussed below.

9.3.1 Cultural and Personality Decision Biases

A number of personality trait sets have been proposed that aim to capture correlations between specific trait values and particular biases. These range from relatively low-level traits such as the Five Factor model (Openness, Conscientiousness, Extraversion, Agreeableness, Neuroticism) [14], to more aggregated traits such as task-based versus process-based styles of leadership. Several cultural trait sets have also been proposed, as outlined above, most notably Hofstede's Uncertainty Avoidance, Individualism-Collectivism, Power Distribution, Short versus Long-term Orientation, and Femininity-Masculinity [6].

Recent attempts to correlate the cultural factors with individual personality traits suggest that the observed behavioral regularities associated with high-level cultural factors such as Uncertainty Avoidance may eventually be explained in terms of individual personality traits, and even specific mechanisms; e.g., correlations have been identified between Uncertainty Avoidance and the Five Factor Model factor of neuroticism [15].

Examples of personality-linked biases include: preference for self and affective-state stimuli associated with high-neuroticism and high-introversion, bias toward threat-cues associated with high-neuroticism, and sensitivity to reward and higher risk tolerance associated with high extraversion [16]. There are of course endless examples of cultural variability in beliefs, values and behavioral norms. More recently, differences in reasoning have also been identified, and summarized in a comprehensive review by Peng et al. [2]. Peng and colleagues propose a categorization of inferencing biases due to cultural effects, as follows:

- Inductive Reasoning
  - Covariation judgment (ability to detect and evaluate associations among cues)
  - Causal attribution

    - Social domain
    - Physical domain
  - Person perception
    - Impression formation
    - Inference of mental states
  - Categorization
    - Category coherence
    - Category learning
    - Category of self
- Deductive and Formal Reasoning
  - Syllogistic reasoning
  - Dialectical reasoning

Tables 9.2 and 9.3 summarize some of the cultural effects on inferencing. However, these results must be interpreted with caution, since the research on cultural differences in human inferencing is in the early stages, and the data summarized in these tables are based on a limited number of studies. The findings must therefore be considered preliminary. As the study of cultural differences continues, additional factors that play a role in the observed differences will be identified. For example, effects that may once have been thought to result from cultural differences may in fact be due to exposure to traditional Western-style education.

Cultural decision-biases are usually attributed to differences in knowledge, beliefs, values, attitudes, goals and behavioral norms. *Personality-linked biases* are also reflected in these types of differences, but are also evident in differences in processing, and differences in affective dynamics. For example, a decision-maker with high degree of extraversion and low degree of neuroticism may be biased toward the external environment, preferentially focusing on task cues and schemas. In contrast, a low-extraversion, high-neuroticism decision-maker may be biased toward self-cues and schemas, resulting in a focus on his own internal states. In high-stress situations, these differences will likely result in differences in decision outcomes. Whereas the extraverted/low-neuroticism decision-maker may be more likely to remain focused on the task, the introverted/high-neuroticism decision-maker may have a tendency to focus on his/her internal state, resulting in task neglect. In addition, the associated negative emotions experienced by this decision-maker will exert further biasing effects on his/her decision-making, by influencing speed and capacity of attention and working memory, and further exacerbating the existing self-bias.

Personality and cultural traits represent long-lasting, permanent influences. In contrast to these *trait-based* effects on decision-making, psychologists and decision-scientists also study *state-based* effects. These are transient effects due to some temporary state of the decision-maker, most often states with a strong affective component (e.g., stress, anxiety, boredom, surprise, frustration, anger,

Table 9.2 Findings regarding cultural differences in human inference: inductive reasoning (ability to generalize from limited data)

Category of inference	Findings
Covariation judgment (Identifying correlations between cues)	Ji et al. (1999) Chinese versus Americans Simple stimuli presented on computer screen Chinese more confident about judgments Chinese more correct in judgments Chinese showed no primacy effect Americans showing strong primacy effect
Causal attribution (Identifying causal relations between cues) social	Miller [28] Americans versus Hindu Indians Fundamental attribution error evident in Americans Hindu Indians attributes behavior to social roles, obligations, physical environment Attributed to different beliefs regarding causality (content difference)
	Morris et al. [29] Americans versus Chinese Fundamental attribution (mass murderers, computer animations of fish) Americans attributed behavior to individual dispositions Chinese attributed behavior to environment
	Lee et al. [30] Americans and Hong Kong Chinese Sports writers' descriptions of events American writers focus on individuals Hong Kong writers on situational factors
	Nisbett [31]; Jones and Harris [32] Americans and Koreans Judgment of another person's attitude Americans assume due to disposition Koreans assume due to contextual influences
Causal attribution physical	"Asian folk physics is 'relational', emphasizing fields and force over distance" "Western folk physics focuses on nature of object itself, rather than its relation to the environment" Peng et al. [2, p. 252]
	Peng and Knowles (2000) Chinese versus Americans Force-over-distance explanations (aerodynamic, hydrodynamic, magnetic) Americans referred more to nature of object Chinese referred more to the field
Person perception	Chiu et al. [33] Hong Kong Chinese versus Americans Judgment of self as fixed versus changing Americans assume fixed, enduring traits Chinese assume changing self
	Ames and Peng [34] Chinese versus Americans Type of information used in person perception judgments Americans focused on evidence provided by target Chinese focused on evidence provided *about* the target by others

Table 9.3 Findings regarding cultural differences in human inference: deductive reasoning (ability to draw logical conclusions from a given set of data)

Category of inference	Findings
Syllogisms	Luria [35] (Russia); Cole [36] Africa Subjects did not 'engage' syllogistic problems at the theoretical level (i.e., if asked to deduce something based on a presented syllogism, they would frequently 'think out of the box' and suggest that the experimenter go find out for himself; why would x be true, etc.) – Real-world (culturally-relevant) grounding of topic makes a large difference in success on task
Dialectical reasoning	Asians: "changing nature of reality and enduring presence of contradictions" versus Western: "linear epistemology built on notions of truth, identity, and noncontradiction" Resolving contradiction: Chinese—seek compromise; Americans: seek exclusionary (either-or) 'truth' and resolution Peng and Nisbett [37] Assumption in Eastern dialectical epistemology: – principle of change—everything is always in flux (thus x may not be identical with itself because it may change over time) – principle of contradiction—opposing qualities co-exist – principle of holism—everything is linked to everything else and isolating phenomena may lead to misleading conclusions – Folk wisdom: greater frequency and preference for dialectical (apparently contradictory proverbs) among Chinese than Americans
Social contradictions/conflicts	– Americans tended to 'blame one side' versus Chinese tended to 'see fault in both'

sadness), but also physical states (e.g., fatigue). Affective biases represent an important category of decision-biases and are discussed below.

9.3.2 *Affective Decision Biases*

Influence of emotions on decision-making is well-documented and emotions play an important role as the intervening variables of a wide range of decision biases. Both short-term *emotions* and longer-term *moods* influence decision-making. Some of the more robust findings include the effects of anxiety and fear, anger and frustration, and positive and negative mood. These effects are evident in both the fundamental cognitive processes (e.g., attention, memory), but also in the higher-level processes mediating decision-making, planning, learning and problem-solving. Examples of biases include: changes in attention capacity, speed and bias; changes in the speed and capacity of working memory; differential activation of specific perceptual and cognitive schemas that mediate the perception and processing of particular stimuli. Examples of specific biasing effects include: association between positive affect and lack of anchoring ('anchoring' refers to the biasing effect of a single piece of information (often the first item encountered), referred to as the 'anchor', during decision-making) overestimation of positive events, underestimation of negative events [17]; fear- and anxiety-linked attentional and perceptual bias toward the detection and processing of threatening stimuli [18]; anxiety-linked focus on possible failure, and subsequent choice of protective behavior; anxiety-linked preference for analytical processing; mood-congruent bias in memory recall [19], and predictions of future positive or negative events [16]; and anger-linked increase in risk-tolerance, and attributions of hostility. Table 9.4 provides a summary of some of these basing effects.

Recent research has explored the relationship between cultural traits and emotions, focusing in particular on emotion regulation. Matsumoto and colleagues [20], have found correlations between the cultural trait of Uncertainty Avoidance and the individual's ability to control the generation and manifestations of emotions, primarily negative emotions, via a variety of regulatory coping strategies, such as re-appraisal or physiological methods (e.g., breathing, relaxation). Their findings indicate that low Uncertainty Avoidance (UA) correlates with more re-appraisal schemas for anxiety reduction, and more diverse coping strategies for emotion regulation, whereas high UA correlates with more rigid adherence to emotion display rules.

9.4 Content and Process Biases

The list of biases outlined above, along with the existing categorizations, provide a foundation for the development of computational models decision biases. However, the existing categories may not be the most useful ones for the development of computational models, or for understanding the mechanisms that mediate the broad range of identified biases.

For modeling purposes, it is more useful to divide decision biases into *content-based* and *process-based*. During the course of decision-making, both categories of biases may of course operate simultaneously. *Content-based* biases relate to differences in the specific content and organization of the knowledge schemas that mediate perception, and the cognitive processes involved in decision-making, and other higher-level cognitive processes. These schemas represent long-term beliefs, values, goals, preferences, and attitudes of the decision-maker. The long-term influences of culture and personality are most easily represented in the specific contents and organization of these schemas. For example, a decision-maker coming from a high-collectivism culture (e.g., China) would be expected to have more memory schemas related to the well-being of the larger group (e.g., perceptual structures for assessing group state, and goals for group well-being). In contrast, a decision-maker from a high-individualism culture (e.g., US) would have more memory schemas devoted to perceptions and goals regarding his/her own well-being and individual goal priorities. Regarding the effects of personality traits: a high-neuroticism individual would have a larger proportion of schemas related to

Table 9.4 Effect of emotions on attention, perception and decision-making: examples of empirical findings

Anxiety and attention and WM	**Anger and attention, perception, decision-making and behavior**
Williams et al. [18]; Mineka and Sutton [38]. Narrowing of attentional focus Reduced responsiveness to peripheral cues Predisposing towards detection of threatening stimuli Reduced capacity of working memory	Increases feelings of certainty Increases feelings of control and ability to cope Induces shallow, heuristic thinking Induces hostile attributions to others' motives and behavior Induces an urge to act
Arousal and attention Edland [39]	**Affective state and memory**
Faster detection of threatening cues Slower detection of non-threatening cues	Bower [19]; Blaney [40] Mood-congruent memory phenomenon (positive or negative affective state induces recall of similarly valenced material)
Positive affect and problem solving	**Negative affect and perception, problem-solving, decision-making**
Isen [41]; Clore [42]; Kahn and Isen [43]; Mellers et al. [17]; Gasper and Clore [44] Promotes heuristic processing Clore [42] Increased likelihood of stereotypical thinking, unless held accountable for judgments [17] Increased estimates of degree of control Overestimation of likelihood of positive events/underestimation of likelihood of negative events Increased problem solving Facilitation of information integration Promotes variety seeking Promotes less anchoring, more creating problem-solving [17] Longer deliberation, use of more information, more reexamination of information Promotes focus on 'big picture'	Williams et al. [18]; Streufert and Streufert [45]; Gasper and Clore [44] Depression lowers estimates of degree of control Anxiety predisposes towards interpretation of ambiguous stimuli as threatening Use of simpler decision strategies Use of heuristics and reliance on standard and well-practiced procedures Decreased search behavior for alternatives Faster but less discriminate use of information—increased choice accuracy on easy tasks but decreased on more difficult tasks Simpler decisions and more polarized judgments Increased self-monitoring Promote focus on details

self- and threat-monitoring, as well as more differentiated set of schemas regarding the representation of self-relevant and threat-related states, in contrast to a low-neuroticism individual, who would likely have more environment and task-oriented schemas.

In contrast, *process-based biases* relate to the way stimuli are perceived and processed during decision-making, and the manner in which existing knowledge is used during the inferencing mediating decision-making. The primary sources of process-based biases are the current states of the decision-maker, most notably, the affective states. Effects of traits are also evident here however, since personality traits exert an influence on both the nature, but especially the dynamics, of affective states, the latter influencing their intensity and onset and decay rates. Process-based biases are most readily implemented in terms of transient changes to the fundamental processes that mediate decision-making, altering the speed and capacity of attention and working memory, as well as the likelihood that particular content will be processed. For example, anxiety-linked threat bias is associated with faster processing of high-threat stimuli, subsequent bias toward selecting a high-threat interpretation, and a subsequent preference for self-preservation goals.

Categorizing decision-biases into content versus process based implies distinct mediating mechanisms, and each category has distinct implications for modeling. Content-based biases are most appropriately modeled via differences in the structure, content and overall organization of the memory schemas mediating decision-making, and inferencing in general; that is, the long-term knowledge, within which stable and permanent differences in cultural and personality traits are reflected. In contrast, process-based biases are best represented by temporary alterations in the processes that mediate decision-making, and inferencing in general, and are most appropriately modeled via transient, parameter-controlled variations in those processes.

The MAMID framework and architecture for modeling decision-biases provides a means of modeling both categories of biases, as described below.

9.5 Framework, Methodology and Architecture for Modeling Process and Content Biases

The MAMID framework and modeling methodology provides an approach for modeling a broad range of decision-biases, arising from both state and trait effects. The distinguishing feature of the MAMID approach is its *parameter space*, which provides a uniform means for representing the combined influences of a variety of state and trait effects. This in turn enables MAMID to model a broad range of *process-based decision-biases,* including cultural trait-linked biases (e.g., Uncertainty Avoidance-linked to preference for higher-confidence cues), personality trait-linked biases (e.g., Neuroticism-linked self-focus and predisposition to negative emotions), and state-linked biases (e.g., anxiety-linked threat bias in attention and perception, anger-linked increased risk-tolerance). MAMID also supports the modeling of *content-based decision biases*, by enabling the representation of alternative knowledge schemas in the architecture's long-term memory, which represents the domain knowledge. Differences in the content and organization of these schemas, represented in terms of belief nets, are associated with differences in traits, both cultural and personality. Below we describe the MAMID cognitive-affective architecture, and the associated MAMID modeling methodology, that enables the modeling of multiple, interacting types of decision biases, arising from multiple sources: cultural characteristics, personality traits, and affective (and other) states.

9.5.1 MAMID Cognitive-Affective Architecture

The MAMID architecture provides a computational modeling and simulation environment within which the effects of multiple, interacting traits and states on decision-making can be represented [3, 21–24]. Since emotion acts as an intervening variable in many decision biases, MAMID includes an explicit representation of emotions, their generation, and their effects on decision-making. MAMID has been instantiated within two domains, a peacekeeping scenario, where MAMID models individual commanders [22], and a search-and-rescue game task, loosely based on the DDD task developed by Aptima, where MAMID models individual team members, as they search for a 'lost party', while encountering a series of obstacles and adverse events [25].

The MAMID architecture models recognition-primed decision-making, in terms of several modules which progressively map the incoming stimuli (cues) onto the outgoing behavior (actions), via a series of intermediate internal representational structures, termed *mental constructs* (e.g., situations, expectations, and goals) (Fig. 9.1). The architecture consists of the following modules: *Sensory Pre-processing*, translating the incoming raw data into high-level task-relevant perceptual cues; *Attention*, selecting a subset of cues for further processing; *Situation Assessment*, integrating individual cues into an integrated situation assessment; *Expectation Generation*, projecting the current situation onto possible future states; *Emotion Generation*, dynamically deriving the affective state from a combination of external and internal stimuli; *Goal Manager,* selecting the most relevant goal for achievement; and *Action Selection*, selecting the most suitable action for achieving the highest-priority goal within the current context.

The MAMID Cognitive-Affective architecture is a symbolic architecture modeling recognition-primed decision-making. It implements a see-think/feel-do process via a series of modules that maps incoming cues onto outgoing actions, via a series of intermediate mental constructs (situations, expectations, goals). It models emotion generation via cognitive appraisal and focuses on modeling the effects of states (primarily emotions) and traits onto cognitive processing via a series of parameters that influence processing within the individual architecture modules.

Each module has an associated long-term memory (LTM), consisting of either belief nets or rules, which represents the knowledge necessary to transform the incoming mental construct (e.g., cues for the "Situation Assessment" module) into the outgoing construct (e.g., situations for the "Situation Assessment" module). The currently-activated set of constructs within each module constitutes that module's working memory. Each module has parameters that determine the speed of processing and the capacity of the module's working memory. Parameters associated with the mental constructs determine the ranking of each construct, thereby influencing the likelihood of the construct's processing within a given simulation cycle. This enables the modeling of specific content biases, such as the bias towards processing threatening cues.

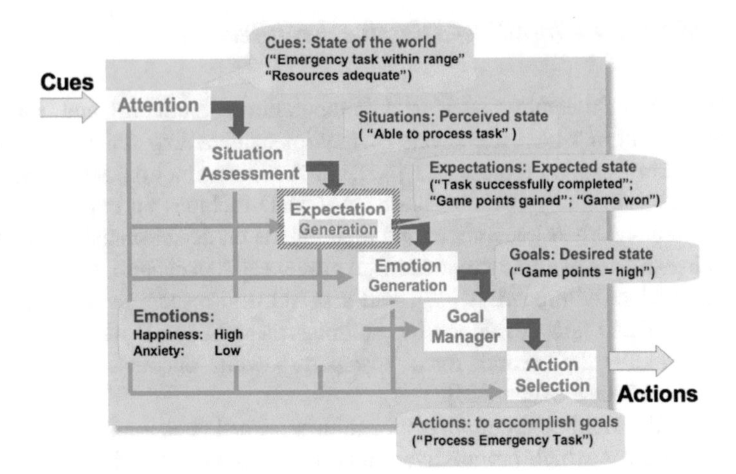

Fig. 9.1 Schematic illustration of the MAMID cognitive-affective architecture

9.5.2 Framework for Modeling Multiple Types of Decision Biases

The underlying assumption of the MAMID modeling approach is that the combined effects of a broad range of decision biases can be modeled *by modifying the fundamental properties of the processes and structures* mediating decision-making [23, 24]. Examples of the properties necessary for *process-based biases* are the *speed* of the individual modules (e.g., fast or slow attention), and the capacities of the working memories associated with each module, which determine the number of mental constructs processed within each execution cycle. Examples of these properties for the *content-based biases* are the structures available for storing long-term memory (LTM) schemas, their contents, and their overall organization.

The speed and capacities of the MAMID architecture modules are controlled by a series of parameters, whose values are derived from the decision-maker's state and trait profile, expressed as a vector of specific values for a distinct configuration of cultural and personality traits and emotional states. The functions calculating the parameter values are currently defined via weighted linear combinations of the specific factors influencing the associated process; e.g., value of the *rank attribute* of a mental construct (e.g., cue, situation) is a function of the following construct attributes: salience, confidence, threat level, decision-maker's anxiety level, individual history, neuroticism and uncertainty avoidance. The traits and states used to calculate a particular parameter, and their associated weights, are based on empirical data. Modeling different types of decision-makers is thus accomplished by changing the associated trait and state profiles, and mapping these onto the parameter vectors. Different values of the parameters cause different 'micro' variations in processing (e.g., number and types of cues processed by the Attention Module), which eventually result in 'macro' variations in observable behavior. For example, a highly anxious decision-maker misses a critical cue due to attentional narrowing and selects the wrong action. Figure 9.2 illustrates how MAMID models threat bias effects associated with neuroticism and anxiety. Figure 9.3 illustrates the changing affective profiles of low- and high-anxious decision-makers, within the search-and-rescue task, as a function of the situations encountered, and the resulting values of the working memory capacity parameter of one of the processing modules.

The MAMID parameter space thus provides a uniform framework within which the combined influences of cultural traits, personality traits, and affective states (as well as other states such as fatigue) can be represented. By providing the capabilities to represent a broad range of traits and states, and integrate their biasing effects on perception, cognitive processing, and action selection, the MAMID modeling framework enables the integration of cultural, personality and affective influences on decision-making.

The decision-maker's high neuroticism trait, both independently and combined with the high state of anxiety, modifies the architecture parameters to cause preferential processing of threatening cues (by ranking them more highly in the attention module) and preferential processing of threatening situations and expectations (by ranking them more highly in the Situation Assessment and Expectation Generation modules, respectively). This then biases overall processing towards threat content, and influences goal and action selection accordingly.

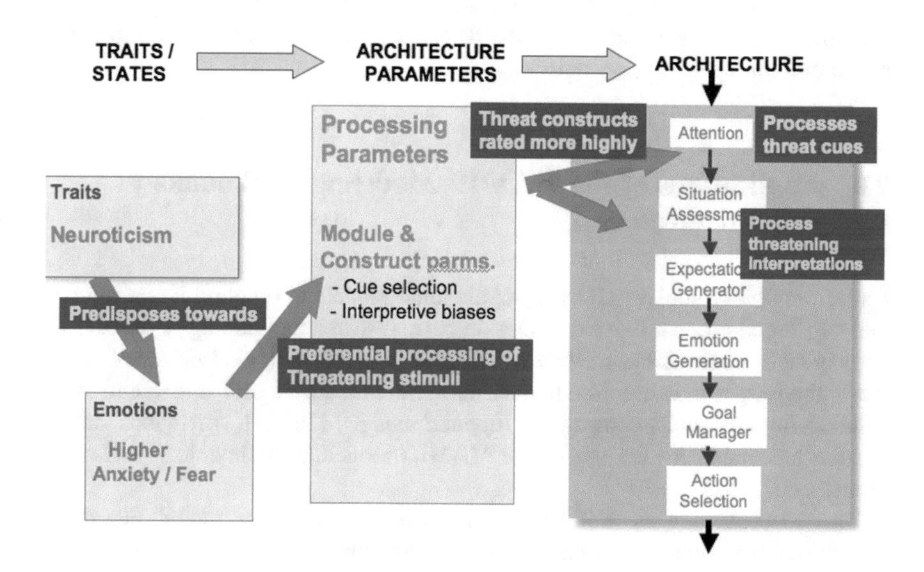

Fig. 9.2 Modeling threat bias in terms of MAMID's parameter-based approach

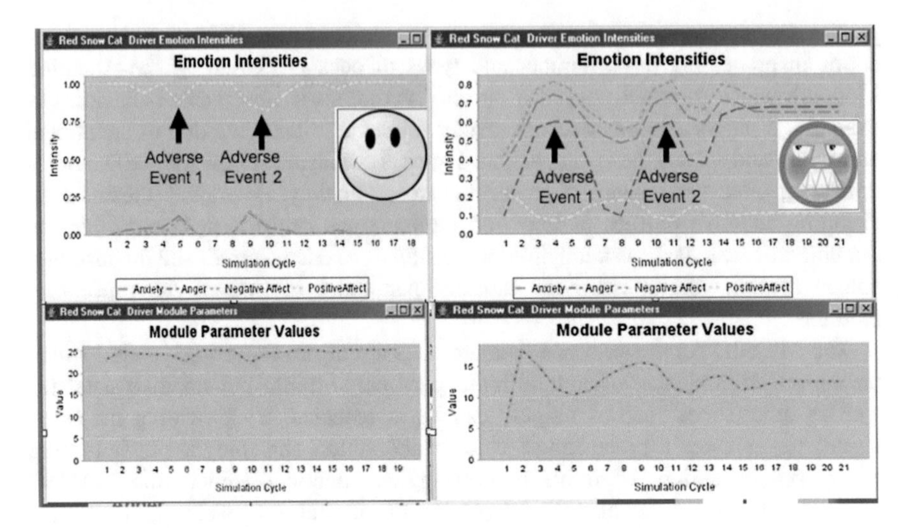

Fig. 9.3 Variations in two distinct decision-makers' affective profiles (*top*), and corresponding module parameter values (*bottom*), during a search and rescue task

The figure illustrates the different patterns of emotion intensities derived for two decision makers: "positive" trait/state profile decision-maker on the left, and "negative" trait/state profile decision-maker on the right. The top part of the figure shows the timeline of the distinct affective states derived in MAMID, as the decision-maker encounters adverse events in the simulation. The bottom part of the figure illustrates some of the module parameter values, which are influenced by the decision-makers trait and state profile, and in turn influence processing within the MAMID modules and, ultimately, the decision-maker's behavior within the simulated environment.

9.6 Applications of the MAMID Modeling Methodology and Architecture

I have previously suggested that computational models of emotion be categorized into two broad groups: those focusing on basic research and aiming to identify the fundamental mechanisms mediating affective processing (*research models*) and those focusing on enhancing human-machine interaction and agent and robot believability and social competence (*applied models*) [26, 27]. Analogous categorization is suitable for the use of the MAMID modeling methodology and architecture outlined above.

The methodology is well-suited for *basic research*, to model alternative hypotheses regarding the mechanisms whereby cultural and personality traits, as well as affective states, exert biases on perception, cognition and decision-making.

The methodology and architecture is equally well suited to develop *more realistic virtual agents and robots*, to enhance their affective and social believability and competence, and thus improve their effectiveness in human-agent and human-robot interactions. In addition, the MAMID methodology and architecture are also well-suited for *developing affective user models*, to facilitate affect-adapted user interaction across a broad range of human-machine systems, including gaming, education and training, and decision-support. Below I focus on the use of the MAMID approach for enhancing agent and robot believability, and for developing affective user models to support affect-adaptive user interaction, and briefly describe examples of specific applications.

9.6.1 Modeling More Believable Agents and Robots

The ability of the MAMID approach to model multiple, interacting factors that influence a range of cognitive processes makes it well-suited to model both trait (cultural and personality) and state (emotions and other states) effects in virtual agents and social robots, and thus to enable the development of agents and robots exhibiting distinct types of personalities and cultural profiles, as well as the ability to display a variety of emotions. This enables the agents and robots to exhibit distinct personalities and behavior as required for a particular cultural and operational context. For example, in behavioral technologies (e.g., mobile mental health apps for coaching or behavior change) or training environments, it is important that the user be able to relate to the agent, to enhance the effectiveness of the helping or training interactions. The ability to rapidly construct agents and robots with customized cultural and personality traits facilitates the creation of multiple agent types, so that the user can select the trait/state profile that best matches their individual preferences; e.g., an individual from a particular culture (e.g., European vs. Asian vs. US) and personality type (e.g., extraverted vs. introverted) may prefer to interact with an agent that displays similar characteristics. These capabilities also enable the agents/robots to display more complex reactions to both humans, and to other agents, in simulated environments, and to adapt their behavior to the human's/agent's own cultural, personality trait and state profile. Together, these capabilities enhance the agents' and robots' believability, by improving their social and affective realism and competence.

A context within which these capabilities are particularly relevant is gaming, including serious gaming, where games are used for educational, training and skill acquisition purposes. The MAMID methodology is well suited for developing non-playing characters with a variety of personality and affective profiles, and thereby enhancing their complexity and behavior variability. This in turn creates gaming environments that are more realistic and engaging. In games used for entertainment only, this contributes to an improved user experience during the gameplay. In serious games, this capability contributes to more engagement and more effective learning and training.

Some training environments explicitly require the rapid creation of agents with distinct cultural, personalities and affective profiles. For example, training environments are being developed to support the training of communication skills in different cultures, for business, diplomatic and military applications, and for the training of social skills, to help individuals overcome social anxiety or to learn basic social interaction and conversation skills (e.g., for individuals with Asperger's syndrome). The MAMID methodology facilitates the rapid creation of a variety of agent state and trait profiles, and associated distinct behavior patterns, via simple manipulation of the architecture parameters. This capability, coupled with the ability to create increasingly realistic depictions of virtual agents within virtual reality environments, then facilitates the creation of agents with both visual and social realism, and thereby contributes to their believability and effectiveness in human-agent interaction.

9.6.2 *Affective User Modeling for Affect-Adaptive User Interaction*

The MAMID methodology and architecture can also be applied to affective user modeling, to support affect-adaptive interaction with the user, in a variety of contexts, including behavioral technologies, gaming, training and decision-support. To date, the majority of affective user modeling relies on statistical and machine learning approaches, to collect and analyze user data and construct user profiles based on identified patterns of behavior. A less frequent, and much more challenging approach, is to actually attempt to simulate or emulate the user behavior, so that a wider range of user behaviors can be modeled and more finely tuned adaptations can be implemented. Clearly, it is currently not within the state-of-the-art to model the behavior of a specific human in general. However, within a narrow context, such as a serious game-based training environment, and with sufficient baseline data, it may be possible to model a subset of the user's behavior, and to thereby support more effective affect-adaptive agent behavior. By facilitating the rapid modeling of multiple traits and states, the MAMID methodology and architecture lends itself well to this application.

9.7 Conclusions

This chapter discussed an approach for modeling decision biases in a symbolic agent architecture. The approach is based on the assumption that a broad range of interacting decision biases can be modeled in terms of parameters that manipulate properties of the fundamental processes mediating decision-making; e.g., speed, capacity, content bias. The approach is suitable for modeling traits (cultural and personality) and states (emotions), for both content and process biases. Its distinguishing feature is a parameter space that provides a uniform means of representing effects of multiple, interacting trait- and state-based biases that influence decision-making by exerting influence on the constituent cognitive processes: attention, situation assessment, expectation generation, goal selection and action selection.

This approach is relevant to both basic research and applied contexts. By explicitly modeling individual cognitive processes mediating decision-making, and emotion generation and effects, the approach lends itself to modeling the mechanisms mediating decision-biases, thereby supporting basic research in decision-making and the mechanisms of decision-biases. In-depth understanding of these mechanisms would enable identification of attributes that contribute to decision errors, in individuals and teams. This would then enable the design of more effective human-machine systems in operational contexts, and more effective training systems. By facilitating the specification of a broad range of agent types, defined by distinct cultural, personality and affective profiles, the MAMID methodology is well-suited for rapidly creating a wide variety of agent and robot types, in simulation- and game-based environments, thereby contributing to the creation of more believable and socially and affectively realistic agents. Such agents, embedded within virtual reality and gaming environments, can then be used to enhance the engagement and effectiveness across a wide range of training, learning and skill acquisition contexts, as well as to enhance engagement in games and VR environments for entertainment purposes.

References

1. Hudlicka, E.: Summary of Factors Influencing Decision-Making and Behavior. Psychometrix Report TR0605. Psychometrix Associates, Blacksburg, VA (2005)
2. Peng, K., Ames, D.R., Knowles, E.D.: Culture and human inference: perspectives from three traditions. In: Matsumoto, D. (ed.) Handbook of Culture and Psychology. Oxford University Press, NY, NY (2001)
3. Hudlicka, E.: Modeling the mechanisms of emotion effects on cognition. In: Proceedings of AAAI Fall Symposium on BICA, TR FS-08-04, 82–86. AAAI Press, Menlo Park, CA (2008b)
4. Codevilla, A.M.: The Character of Nations. Basic Books, NY (1997)
5. Cooper, C.R., Denner, J.: Theories linking culture and psychology: Universal and community-specific processes. Annu. Rev. Psychol. **49**, 559–584 (1998)
6. Hofstede, G.: Cultures and Organizations: Software of the Mind. McGraw Hill, NY (1991)
7. Matsumoto, D.: The Handbook of Culture and Psychology. Oxford, NY (2001)
8. Renshon, S.A., Duckitt, J.: Political Psychology: Cultural and Crosscultural Foundations. NY University Press, Washington Square, NY (2000)
9. Triandis, H.C., Suh, E.M.: Cultural influences on personality. Annu. Rev. Psychol. **53**, 133–160 (2002)
10. Klein, H.A: National differences and naturalistic decision making. In Proceedings of the 46th Meeting of the HFES. Baltimore, MD (2002)

11. Karabaich, B.: Target Analysis Concept. Technical Memo. Karabaich Strategic Information Systems, Leavenworth, KS (1996)
12. Hudlicka, E.: Personality and Cultural Factors in Gaming Environments. Psychometrix Associates Technical Memo 0306. Psychometrix Associates, Blacksburg, VA (See also DMSO Workshop on Personality and Culture Factors in Military Gaming, Washington, DC, July 2003) (2003b)
13. Kahneman, D., Slovic, P., Tversky, A.: Judgment under uncertainty. Cambridge, UK (1982)
14. Costa, P.T., McCrae, R.R.: Four ways five factors are basic. Personality Individ. Differ. **13**, 653–665 (1992)
15. Hofstede, G., McCrae, R.R.: Personality and culture revisited: Linking traits and dimensions of culture. Cross-Cultural Research **38**, 52–88 (2004)
16. Matthews, G., Derryberry, D., Siegle, G.J.: Personality and emotion. In: Hampson, S.E. (ed.) Advances in Personality Psychology. Routledge, London (2000)
17. Mellers, B.A., Schwartz, A., Cooke, A.D.J.: Judgment and decision making. Annu. Rev. Psychol. **49**, 447–477 (1998)
18. Williams, J.M.G., Watts, F.N., MacLeod, C., Mathews, A.: Cognitive Psychology and Emotional Disorders. John Wiley, NY (1997)
19. Bower, G.H.: Mood and memory. Am. Psychol. **36**, 129–148 (1981)
20. Matsumoto, D., Yoo, S.H., LeRoux, J.A.: Emotional & intercultural communication. In: Kotthoff, H., Spencer-Oatley, H. (eds.) Handbook of Applied Linguistics, Mouton (2005)
21. Hudlicka, E.: Reasons for emotions. In: Gray, W. (ed.) Advances in Cognitive Models and Cognitive Architectures, Oxford University Press, NY (2007a)
22. Hudlicka, E.: Modeling effects of behavior moderators on performance. In Proceedings of 12th Conference on Behavior Representation in Modeling & Simulation, Phoenix, AZ (2003a)
23. Hudlicka, E.: This time with feeling: Integrated model of trait and state effects on cognition and behavior. Appl. Artif. Intell. **16**(1–31), 2002 (2002)
24. Hudlicka, E.: Modeling emotion in symbolic cognitive architectures. In: AAAI Fall Symposium Series, TR FS-98-03. AAAI Press, Menlo Park, CA (1998)
25. Hudlicka, E.: Application of Human Behavior Models to Risk-Analysis and Risk-Reduction in Human-System Design. TR 0507. Psychometrix, Blacksburg, VA (2007b)
26. Hudlicka, E.: Guidelines for developing computational models of emotions. Int. J. Synth. Emotions **2**(1) (2011)
27. Hudlicka, E.: What are we modeling when we model emotions? In: Proceedings of AAAI Spring Symposium on "Emotion, Personality and Social Behavior", TR SS-08-04, pp. 52–59. AAAI Press, Menlo Park, CA (2008a)
28. Miller, J.: Culture and the development of everyday social expla- nation. J. Pers. Soc. Psychol. **46**, 961–978 (1984)
29. Morris, M., Nisbett, R., Peng, K.: Causal attribution across domains and cultures. In: Sperber, D., Premack, D. (eds.) Causal Cognition: A Multidisciplinary Debate, pp. 577–614. Clarendon, New York (1995)
30. Lee, F., Hallahan, M., Herzog, T.: Explaining real life events: How culture and domain shape attributions. Pers. Soc. Psychol. Bull. **22**, 732–741 (1996)
31. Nisbett, R.E.: Essence and accident. In: Cooper, J., Darley, J. (eds.) Attribution Processes, Person Perception, and Social Interaction: The Legacy of Ned Jones, pp. 169–200. American Psychological Association, Washington, DC (1998)
32. Jones, E.E., Harris, V.A.: The attribution of attitudes. J. Exp. Soc. Psychol. **3**, 1–24 (1967)
33. Chiu, C.Y., Hong, Y.I., Dweck, C.S.: Lay dispositionism and implicit theories of personality. J. Pers. Soc. Psychol. **73**, 19–30 (1996)
34. Ames, D., Peng, K.: Culture and Person Perception: Impression Cues that Count in US and China. UC Berkeley, Berkeley, CA (1999)
35. Luria, A.R.: Psychological expedition to central Asia. Science **74**, 383–384 (1931)
36. Cole, M.: Cultural Psychology: A once and Future Discipline. Belknap Press of Harvard University Press, Cambridge (1996)

37. Peng, K., Nisbett, R.E.: Culture, dialectics, and reasoning about contradiction. Am. Psychol. **54**, 741–754 (1999)
38. Mineka, S., Sutton, S.K.: Cognitive biases and the emotional disorders. Psychol. Sci. **3**(1), 65–69 (1992)
39. Edland, A.: On cognitive processes under time stress: A selective review of the literature on time stress and related stress. Reports from the Department of Psychology. University of Stockholm, 68, 1–31 (1989)
40. Blaney, P.H.: Affect and memory. Psychol. Bull. **99**(2), 229–246 (1986)
41. Isen, A.M.: Positive Affect and Decision Making. In: Haviland, J.M., Lewis, M. (eds.) Handbook of Emotions. The Guilford Press, New York, NY (1993)
42. Clore, G.L.: Why emotions require cognition. In: Ekman, P., Davidson, R.J. (eds.) The Nature of Emotion. Oxford University Press, Oxford (1994)
43. Kahn, B.E., Isen, A.M.: The influence of positive affect on variety-seeking among safe, enjoyable products. J. Consum. Res. **20**, 257–270 (1993)
44. Gasper, K., Clore, G.L.: Attending to the big picture: Mood and global versus local processing of visual information. Psychol. Sci. **13**(1), 34–40 (2002)
45. Streufert, S. and Streufert, S.C. (1981). Stress and information search in complex decision making: Effects of load and time urgency. ONR Technical Report, Penn State College of Medicine
46. Ji, L., Peng, K., Nisbett, R.E.: Culture, control, and perception of relationships in the environment. J. Pers. Soc. Psychol. **78**(5), 943–955 (2000)

Chapter 10
Considering the Needs and Culture of the Local Population in Contemporary Military Intervention Simulations: An Agent-Based Modeling Approach

Jean-Yves Bergier and Colette Faucher

Abstract In the context of increasingly complex human environments faced by the armed forces tasked with implementing peacebuilding missions in various locations around the world, operations of influence, which are non-violent undertakings aimed at conquering the support of the locals through use of trust, persuasion and material assistance, are more than ever a vital approach for successful military action. As such, they require specific and complete training and designing a computational tool to this end offers opportunity for innovative modeling of cultural complexity. This chapter presents and describes how a local culture and its specifics are delineated and represented in the SICOMORES multi-agent system and covers the issue of human needs as a socio-cultural phenomenon. It is proposed that culture has to be approached at several levels that are entwined and has to be represented accordingly in the agent population, while the culturally dependent aspect of needs can be bypassed in order to produce a universal set of needs for the context of conflict affecting civilian populations. It is then possible to computationally use these theoretically well-grounded elements to generate with a high level of detail the effects of actions of influence directed at the agents as cognitive processes.

Keywords PsyOps · CIMIC · COIN · Intelligent agents
Social-cognitive simulation · Culture · Needs

J.-Y. Bergier (✉)
45, Rue Brancion, 75015 Paris, France
e-mail: jeanyves.bergier@gmail.com

C. Faucher
LIP6 (Laboratoire d'Informatique de Paris 6), UPMC (Université Pierre et Marie Curie), Bureau 26:00, 503, 4, Place Jussieu, 75005 Paris, France
e-mail: Colette.Faucher@lip6.fr

C. Faucher (ed.), *Advances in Culturally-Aware Intelligent Systems and in Cross-Cultural Psychological Studies*, Intelligent Systems Reference Library 134, https://doi.org/10.1007/978-3-319-67024-9_10

10.1 Introduction: A System for Simulating Actions of Influence in the Context of Asymmetric Warfare

International interventions are quick to spawn controversy in an increasing interdependent world. Budget cuts and public opinion are constraining factors of the operations timescale, and globalization often brings and ensures intensive scrutiny. For each military endeavor, the context asks not only for efficacy but for almost immediate efficiency. At the same time results are increasingly difficult to obtain and assess, especially in the short run, in conflicts involving increasing numbers of diverse local actors extensively disregarding the limit between civilian and fighting spheres, up to a point that these categories and the very concept of victory are ultimately questioned [1].

One of the answers provided by the militaries is an emphasis on the cross-cultural skills and cultural awareness of the soldiers, as indicated by the unmistakable growth in number of culture oriented training and decision support computational tools. These aim at giving the soldiers a better capacity at understanding and fitting into the complex human environment they are about to face, for supposedly improved and quicker performance. Lots of these programs are limited to learning of culture specifics and development of conversational abilities [2, 3], while others implement a certain model of culture though representation of composite societies [4, 5].

The militaries also have other, more direct means to reach some sort of efficiency in dealing with a culturally determined setting. Once deployed, a particularly relevant instrument specifically designed to delineate a socio-cultural context and pinpoint how and where to act for gaining support of the locals, while avoiding cultural mistakes harming the peacemaking process and security of the Force, are the operations of influence. Particularly related and sensitive to local culture issues, they are a potentially powerful set of non-kinetic actions and functions oriented on dealing with complexity and pacifying though persuasion, and as such deserve a special attention provided the nature and stakes of today asymmetric conflicts. Hence, the need for a training system intended for influence specialists. The SICOMORES (SImulation COnstructive et MOdélisation des effets des opérations d'influence dans les REseaux sociaux) project addresses this need by modeling the civilian parts of a theater of war and generating in it the effects of actions of influence in a manner characterizing and stressing how they are affected by local culture. Three kinds of actions are modeled: psychological operations (PsyOps), civil-military cooperation operations (CIMIC) and key leader engagements.

But how is culture to be defined? Even once the normative and evolutionist concept of culture as opposed to barbarism is set aside, and culture approached as what characterizes and organizes in a specific manner the life of particular social groups, there seems to be no practical answer. A general worldview? A set of underlying and subconscious principles of social life? A package of explicit values and norms? The features of a particular societal structuring and the organization principles underlying it? A set of observable practices and customs? Everything in human life that does not strictly depends on human nature and as such, is not universal? The trivial, right answer is that as a complex research object, culture can be all of these and can be expressed and analyzed as per each of these aspects. While being realistic, this statement is no adequate guideline to the computer scientist trying to develop a theater of war model aiming broadly at integrating the cultural factor to make salient cultural mechanisms impacting military operations effects. It is necessary to point out when and how cultural factors play a part. In this chapter we will discuss how culture is integrated in our agent based model, i.e. how some of these different approaches are implemented, particularly focusing on the issue of cultural specification. Section 10.2 details the creation of a culturally infused virtual population. Section 10.3 surveys the definition of human needs beyond cultural variations, and Sect. 10.4 provides and details some computational uses of these elements for generating the effects of operations on the agents.

10.2 Multi-agent Human Framework: Socio-Cultural Issues

In order to relevantly simulate the effects of the different actions of influence our system concerns itself with, a virtual realistic population is needed. Multi-agent frameworks lend themselves well to the study of effects of influence and propagation of information in intricate contexts [6]. They allow modeling numerous and heterogeneous agents with a distribution of traits across as many dimensions as desired. Furthermore, embedding agents in both social and physical spaces is quite easy in such models. As SICOMORES aims at computationally representing micromechanisms governing cognitive attitudinal reactions, and microbehaviors of information sharing, of socially and culturally characterized persons exposed to various psychological messages and operations, its agent-based model comprises only agents representing individuals. This is obviously of importance for the representation of culture.

10.2.1 A Culturally Tailored and Relevant Population Network

In accordance with empirical input and choice of theoretical bases (see Sects. 10.3 and 10.4), each agent in the population of SICOMORES is an instance i of the INDIVIDUAL frame and is described by a set of attributes:

- Social features: Gender, age, social level ([1, 10]), religion, ethnicity, political opinion, role in the family, leader status.

- Cultural features: Cultural values system {(type of cultural feature, cultural feature, importance ([1, 10]))}
- Reachability features: Language(s), Literacy, reachable by radio, reachable by television, reachable by text message.
- Psychological features: Opinion toward the Force ([1, 10]), intellectual level ([1, 10]), needs {(need, satisfaction degree ([1, 10]))}

Some agents can have a special status such as chief of family and/or leader (political or religious), which may be cumulated. More importantly, all these agents have to be connected so that they can exchange information. In the domain of social interactions modeling it is admitted that the nature and structure of the network govern the process to a great extent, regardless of the chosen type of propagation model (epidemiological or not). In addition, the collective dimension of culture and its influence on information propagation between individuals is by definition inescapable since it determines communication practices. Thus, linkage between agents has to integrate the cultural factor to form a structurally appropriate network, and a method for doing so has to be identified. We suggest the best orientation for culturally infusing a networked virtual population is to aim at grasping as realistically as possible the social life of its members, i.e. their sociality. Sociality is a culture-dependent and complex phenomenon in any context, as the social links can be highly heterogeneous. Sociality is primarily grounded in the belonging to various groups and the form of the relations is crucial. Subsequently, a multilayer network is required to capture the ego-network of sociality of any individual. Also, a typology of the social links at the society level is needed to classify and highlight the links important to the social life of its members in order to define and generate the layers, hence shedding some light on the cultural functioning of this collective entity. It then makes necessary to choose a specific socio-cultural context. For SICOMORES the selected area is sub-Saharan Africa, as an extensive literature exists for this scale and provided that the features of the agents can be specified more precisely, at the nation level. Five dimensions of sociality are retained as layers for the group networks making up the population (see [7]):

- Family layer
- Neighborhood layer
- Friendship layer
- Partisan layer (political)
- Partisan layer (religious)
- War-time layer

All of these types of links are deemed *sine qua non* to a representation of sub-Saharan social life, excepted the war-time layer which is associated to the war situation but is yet particularly relevant to the sub-Saharan area as it captures the salient mechanism of communitarianism. Others, as business links, have been set aside considering they are more than often subsumed in others, more structuring types of relations. Each of these networks is characterized on the scales of three criteria: geographic anchoring (sine qua non to the existence of the links or not), formalization of leadership (explicit or not) and temporality (lasting or not). These specifications provide guidelines to craft the algorithms of generation (see [8]). Such a method for creating the network and the network itself already offers a representation of culture: the different links thus generated indeed encapsulate a specific cultural context. The nature and characteristics of the social links making up an individual's sociality are not only culture dependent, they implicitly and explicitly express certain cultural values and traits. In sub-Saharan Africa the fact that the family is traditionally the basic social and relational structure from which all others relations seem more or less derived tells a lot about the area value system. At the same time, this typology is usable to implement any sub-Saharan national scenario, and such a method for circumscribing sociality by producing a social links classification grounded on social sciences can be rightly replicated for any context.

10.2.2 Considerations on Culture Modeling in the Context of War-Torn Societies

Culture can be considered from various perspectives. A collective phenomenon by essence, its role can be studied at many levels such as individual, group, society, cultural area, and the relevance of each may be discussed in itself or regarding a particular issue. Choices have to be made if this social phenomenon is to be represented in a multi-agent population. As stated before, the favored level for representation of culture in the social network is the cultural area but the generation algorithms for the different dimensions allow more specification by the use of national data. Moreover, the architecture is flexible enough to permit any change in the number or nature of dimensions to meet scenario requirements. But while expressing culture by the modeling of social structures is certainly relevant, it is not enough. A culture can also be defined and expressed by beliefs, values and attitudinal recommendations it promotes.

As shown by the INDIVIDUAL frame, at the individual level all agents of the SICOMORES network are specific but are never isolated from the socio-cultural context in which they live, as they share each and every cultural feature with the other agents, while individually associating a particular numerical value between 1 and 10 to each of these features. These values are randomly generated and the set of numerical values of any agent expresses his own personality and appreciation of the societal value system (i.e. local culture). Cultural and psychological variation between individuals is modeled as the variation between the respective importance they attach to the different cultural traits constituting the societal culture, their common cultural background.

Such a representation of culture is far from being without practical and theoretical justification. It is conceived as reminiscent of some of the theories developed by the culture and personality movement. These studies often emphasize the coherence of cultures, considering they are defined by the consistent manner in which they combine different cultural traits rather to form a *pattern* than being a mere juxtaposition of these traits [9]. More importantly, this movement envisions individuals and their culture as two inseparable and interacting realities which cannot be understood without one another. Each individual has his own way of internalizing and living his culture, developing a personality on the basis of a culturally situated *basic personality structure* [10, 11]. Such basic personality is approached by these researchers as a psychological and behavioral reality, but we state it can be sufficiently described as/in a set of traits.

Also, the model not only has to be realistic regarding its representation of culture, it has to be developed as a framework for simulating persuasive communication processes. Although the nature of the role of culture in the receiving of messages is still discussed, the reality of such a role is undisputable at the current state of research [12]. The issue is to distinguish between universal mechanisms and processes subjected to change from culture to culture. Here we can stick to the ever valid statement of Kluckhohn and Murray: "Every man is in certain respects (a) like all other men, (b) like some other men, and (c) like no other man" [13]. Psychological universals and cognitive processes ("like all other men") are represented in SICOMORES by algorithms of cognitive treatment built on social psychology (see Sect. 10.4). Personality ("like no other man") is as stated modeled as the distinctive set of numerical values associated to cultural features, which is computationally used in cognitive treatments (see Sect. 10.4).

Overall, the level of consideration important at this point for representing culture is the collective unit ("like some other men"). How to define the relevant collective unit(s) is then the important issue as it determines the set of cultural features describing the agents. For SICOMORES the national, or societal, level was chosen. Obviously, our representation of culture does not currently integrate cultural characterization of different groups inside the considered society, i.e. the modeling of subcultures. This is primarily a modeling choice, as quite a few systems modeling contemporary theaters of war seem to not to go beyond the image of a definitely and deeply fragmented society, by structuring their virtual population as an ensemble of distinctive socio-cultural groups without any real shared trait. In the XLand scenario of the SHOUT system simulating dissemination of PsyOps messages, [14, 15] the fact that both ECHO and DELTA communities are Christians is of no real effect on the simulation as the socio-cultural traits of both groups are measured by Hofstede dimensions with values that put them apart (communities, deemed ethnic, can have various level of proximity), while in the POLIAS system simulating the effects of PsyOps on attitude dynamics, individuals represented as agents are characterized by the different social group they belong too, and only inter-group attitudes are modeled [16]. The aim of any military operation being supposedly to restore social peace to the considered society, a peace which usually was a reality before communitarianism spiraled into violence, it seems only a logical option to generate first in the simulation a representation of what is culturally common to all the local civilians. Conflict within a single society, as the State is failing or failed, is the most common situation encountered by soldiers deployed abroad and has become the current archetypal form of armed violence, with great consequences on the form and stakes of operations [17, 18]. Even if a society is clearly in disarray at the time of the military intervention, most of the time a certain cultural unity exists which is the very basis for a possible national reconciliation and which cannot be overlooked in the modeling of reactions to actions of influence.

Moreover, it has to be pointed that analysts more than often insist that in many cases the supposedly unbridgeable cultural divides between feuding groups are simply the result of violent discourses and propaganda by charismatic leaders making a political use of some cultural traits to favor agendas reaching beyond cultural issues. In sub-Saharan Africa the pervasive community principle makes such schemes easier, but community identities are no absolutes and social disaggregation and civil war can come from any kind of social, economic and political frustration [19]. As our system aims at modeling the real mechanisms underlying the aspects of conflicts, thus contributing to their understanding, it is necessary to go beyond appearances and simple reading grids, and a cultural one just does not seem the most relevant in our chosen context. Overemphasizing cultural differences between involved groups is common when trying to understand a conflict, and can be detrimental when considering social mechanisms and psychological mechanisms of influence [20]. Overlooking commonalities would be a bias of analysis.

More importantly, the different groups which are crystallized by the conflict and/or between which conflict can exist are represented in our system, as each node has an ethnicity, a religion and/or a political opinion. These characteristics are a form of (sub)cultural specification, and the way they dictate the activation or de-activation of links in all dimensions of the network as the social cohesion indicator of the considered society varies overtime make antagonist groups a sensible social reality inside the virtual population [7]. In fact, socio-cultural groups are clearly represented in the population through characteristics of agents and social linkage mechanisms for generating the layers of the network, but are not culturally specified in term of value system for all the aforementioned modeling reasons. This is a maybe subtle but yet very necessary modeling approach as it allows a society to be represented not as an juxtaposition of groups but as an ensemble of citizens in which group identities exist that can become more or less salient in the course of events.

Moreover, the architecture of the system SICOMORES is flexible enough to allow any future refinement on this issue if indispensable: definition of rules regarding the generation of nodes (for instance, that any agent characterized by religion A must have a value of 5 or higher in cultural feature B) or addition of extra cultural elements for representing more detailed group value systems.

10.3 Theoretical Background: Toward a Model of Needs

To help the Force's integration in its human environment and more broadly the achievement of the goals defined by the Military Strategy of Influence, civil-military cooperation actions (CIMIC) alleviate the suffering resulting from the conflict by addressing the needs of the civilian population. Possibilities are numerous indeed in this field: construction or rehabilitation of schools, hospitals and other public buildings, distribution of various supplies, creation of access to electricity or clean water, organization of elections to restore healthy governance.... Assessment of needs allows the Force to create shared interests between soldiers and locals [21]. The concept of needs is rightly at the core of CIMIC doctrine, and incidentally the most recent advances in poverty measurement and in specific fields researching development tend to advocate a needs-based approach to assess well-being, a very thorny notion on which no real consensus exists in academic literature. Some CIMIC operations produce their effect on a very short term basis, others are ambitious investment projects involving many actors for a lasting result, but all impact the needs and well-being of local people in ways that have to be accurately measured.

Hence, a relevant model of needs is needed to represent the state of each agent regarding such issues and its evolution, if the effects of these operations are to be correctly generated. A key element describing our agents is subsequently their list of needs and the respective satisfaction degrees of these needs. Such a list should be defined in the most rigorous way. In fact, the very possibility to delineate a list can be rightly discussed. Indeed, what seems more culturally dependent than needs? In an initial approach, needs, as they are felt and expressed by people around the globe, are the results of their continued existence in a specific social, economic, cultural and physical environment. What similarity could be expected between the needs of a Japanese businessman and a Nigerien breeder? In this manner, it could be argued that the various needs describing the state of each agent in the population should be scenario-dependent, to be set by an expert in a fashion similar to the cultural features. However, even a quick overview of the very prolific and diverse literature dealing with the matter of human needs is enough to question this statement. Such a perspective has been chiefly embraced by anthropologists and sociologists, who primarily study human being as part of a group and cultures in their singularity. It is also related to a certain economic angle theorizing needs as the product of a specific production system. While having been undoubtedly fruitful, these approaches do not stand alone: psychologists are more concerned by an immanent human nature beneath the diversity of cultures or individual personalities and preferences [22]. Surely the taste in foods can differ depending on the culture and/or the individual, but every human has to eat. Thus, the long-standing distinction between needs, usually theorized as non-relative, and wants, usually theorized as subjective and depending on the preferences of particular (groups of) people [23]. But while the limit is relatively easy to place when considering only physiological needs, how can other needs be identified and classified? The old question of the distinction between nature and culture is already lurking here, swampy as ever: any attempt to reach human nature beyond culture seems by definition quite unachievable considering that culture is not just superimposed on nature, as Levi-Strauss highlights it: “In a sense, it (culture) replaces life, and in another it uses it and transforms it to produce a synthesis of a different kind” [24].

Despite these reservations, numerous attempts to identify basic human needs have been made. The role of culture in the process of expression of needs and in the designation of means to satisfy them in specific contexts is not questioned here. However, the possibility of an always usable list is too appealing not to consider for the computer scientist trying developing an elegant conceptual model, meaning as simple as possible yet efficient and relevant in its grasping of the possibilities of reality. Such considerations ask for an examination of the legitimacy of some of the most essential models of needs, leading to a clearer definition of a need. Ultimately the question lying here is: is there a possibility for a solidly theoretically and empirically grounded universal set of needs?

10.3.1 Maslow Hierarchy of Needs: An Outdated Approach

Maslow theory of needs [25, 26], is primarily a study of human motivation to action, postulated as dependent on human needs. Undeniably influential, this paradigm opened new possibilities for exploring the concept of well-being, outlining that supposedly universal needs could be specified and were organized in a strictly hierarchical manner by their respective priority to be fulfilled, meaning that a need has to be addressed before the next on the list can be. The identified categories of needs are, in the right order:

- Physiological needs
- Safety needs
- Belongingness and love needs
- Esteem needs
- Self-actualization

Spreading well beyond the social psychology research community and even the academic world, this model enjoys a decidedly eminent position even seventy years after its formulation. It is often used in social and behavioral modeling to this day [27], sometimes appearing as an inescapable paradigm: inspiring a graphical language intermediate between conversational description and software model for agent-based behavioral frameworks [28], or being used to define elegance in system architecture [29].

But, for instance, some research on network learning demonstrates the inadequacy of Maslow's hierarchy to the digital age [30]. In fact, while having been relatively congruent with clinical and experimental data at the time of its publication, Maslow's theory has been early and since largely criticized. Overall the model has been attacked on three main points:

Some of its core elements such as the self-actualization concept are deemed unscientific and built upon an idealized view of human nature, hence being far from universal and operational [31, 32] and needing at least to be quite heavily reconstructed and elaborated [33]. The model itself is not supported by consistent empirical evidence and is overall very difficult to test empirically [34].

More importantly, the directional nature of the model can be questioned. Not relying on empirical research or truly credible data, the projected process from physiological satisfaction to self-actualization is suspected by some to be grounded on Maslow's own life experience, seeming personal and phenomenological [35]. Whaba and Bridwell studies indeed invalidate the hierarchy as presented [32]. Kenrick et al. [36], even though not completely rejecting the possibility of a hierarchy, show that affirming that a need would have to be satisfied before the next is clearly mistaken. Some psychological need can be satisfied while some physiological needs are not or not enough.

More recently, an extensive survey by Diener and Tay [37], while agreeing on the probable existence of universal needs, points out that Maslow's chosen needs are indeed wrong and that he was probably biased to a certain extent by his own culture. This confirms some other empirical experiences questioning the different categories of needs [38, 39]. Hofstede [40] and Mook [41] also argue that Maslow theory is ethnocentric and that a person raised in a less individualistic social environment could not be considered fully developed in such a framework.

It can be concluded here that Maslow's model has been proven invalid to such a degree that further research is needed to identify another, usable set of needs. On the basis of all these criticisms, a programmatic summary can be formulated:

> Because of the paradoxes inherent in human nature, a directional model based on a predetermined hierarchy is not adequate as a theory for motivation. The complex interrelationships between psychological development, personal and situational factors, social networks, the historical context, and the ecological environment must be integrated to create a broad and flexible model of human needs that is responsive to all of the factors that impact motivation of human behavior [35].

This statement would call for a specific research project, but at least settles the fact that a non-hierarchical, non-directional set of needs is required here.

10.3.2 Max-Neef Taxonomy of Human Needs

Working in the field of ecological economics, Max-Need tries to lay the foundation for a possible systematization of needs [42]. While stating that any classification is to be regarded as provisional, his model is an attempt to approach the needs as finite in number and simply derived from membership in a common humanity, thus advocating a virtually universal set of needs. Arbitrariness in the listing is to be avoided by the use of socio-cultural sensitivity and by ensuring the taxonomy is understandable, critical, operational, propositional, and combines scope with specificity. Contrary to "satisfiers" which are particular ways of satisfying these fundamental needs and vary with cultures and circumstances, such needs are constant:

- Subsistence
- Protection
- Affection

- Understanding
- Participation
- Leisure
- Creation
- Identity
- Freedom

These needs are expressed according to axiological categories. According to existential categories, they are classified as being, having, doing and interacting, and combining these two sets allows to produce a matrix of needs and satisfiers. What is noticeable in such a framework is the strong rupture with Maslow's model:

> Fundamental human needs must be understood as a system, the dynamics of which does not obey hierarchical linearities. This means that, on the one hand, no need is per se more important than any other; and, on the other hand, that there is no fixed order of precedence in the actualization of needs (that need B, for instance, can only be met after need A has been satisfied). Simultaneities, complementarities and trade-offs are characteristic of the system's behavior. There are, however, limits to this generalization. A pre-systemic threshold must be recognized, below which the feeling of a certain deprivation may be so severe, that the urge to satisfy the given need may paralyse or overshadow any other impulse or alternative. [42]

Such a stance is far more consistent with recent research and answers to the above mentioned requirements of a realistic model of human needs. But it is not the latest attempt in this regard.

10.3.3 Dimensions of Poverty and Development: Sen, Nussbaum and the Capabilities Approach

At the crossroads between philosophy and economics, this framework first developed by Amartya Sen [43, 44] can be fairly approached as a theory of justice. Before him Rawls promoted quite a similar perspective [45, 46] by defining primary goods stated to be those that the citizens need and desire as rational and free members of the society (basic mental and bodily abilities, basic political liberties and freedom of thought, freedom of movement, prerogatives of office of responsibility, income and wealth, social basis of self-respect). Critical of Rawls, Sen posits that it is not enough to examine what individuals possess, advocating giving attention to action and the capacity to use these goods to lead the life one might want to lead. Well-being depends on what individuals are able to do or not, hence the notion of capability. Two distinct yet close concepts are central to the theory [47]:

Functionings are states of "being and doing" such as having shelter. They should be distinguished from the commodities employed to attain them.

Capability refers to the set of valuable functionings that an individual has effective access to. A person's capability represents the effective freedom of an individual to choose between different functioning combinations that he values, that is between different kinds of life.

This paradigm is rightly a prominent one in development research and has been extensively employed by the United Nations Development Program, which theorizes development as capability extension. But these notions, while being academically fruitful, are not a typology of needs. Such a framework even aims at going beyond basic needs approaches (need versus capability approach, see [23]). Nevertheless and interestingly enough, the capability theory has spawned a debate regarding the timeliness of a list of potentially universalisable capabilities. Nussbaum in particular, has provided one of the most influential versions of the theory [48] and made several attempts at delineating a list of "central capabilities" [49]. The most mature one [50] consists of these core capabilities:

- Life: not dying prematurely
- Bodily Health: being able to have good health, including reproductive health
- Bodily Integrity: being able to move freely from place to place; to be secure against violent assault
- Senses, Imagination, and Thought: being able to use the senses, to imagine, think, and reason in a way informed and cultivated by an adequate education
- Emotions: being able to have attachments to things and people outside ourselves
- Practical Reason: being able to form a conception of the good and to engage in critical reflection about the planning of one's life.
- Affiliation: Being able to live with and toward others, to recognize and show concern for other humans, to engage in various forms of social interaction; Having the social bases of self-respect and non-humiliation
- Other Species: being able to live with concern for and in relation to animals, plants, and the world of nature
- Play: being able to laugh, to play, to enjoy recreational activities
- Control over one's Environment: political (Being able to participate effectively in political choices) and material (being able to hold property and having property rights and to seek employment on an equal basis with others)

This exercise surely clarifies the concept of capability, but Sen and Nussbaum differ on the acceptability of such a project of list. Nussbaum accuses Sen of inconsistency in the fact that he admits some freedom/capabilities are always and everywhere a prerequisite for well-being, without admitting the possibility to list them. She adds that her own list has no claim to be definite and invariable as a metaphysical statement would be, but that a set of central capabilities should be outlined as a guideline for political projects [51]. However, she considers that human need is a relatively stable matter and that there is real hope of delimiting basic human needs that will remain constant over time… [52]. Sen argues that the difficulty is not lying with listing important capabilities but "with insisting on one predetermined canonical list of capabilities, chosen by theorists without any general social discussion or public reasoning" [53]. He insists that human beings can pursue extremely various ends [54] and that the context of use of the capabilities is of outmost importance and should be specified [53, 55]. The capability approach being built upon an evaluation principle, each society should be let with the possibility of weighting capabilities in accordance to its ethical and political reasoning [53, 55]. These points suggest a capability model should be produced for each specific socio-cultural context, a viewpoint contrary to our goal of reaching an unchanging set of needs.

10.3.4 A Model of Needs Suitable for Military Operation Simulations

The argument regarding the legitimacy and viability of lists in the framework of the capabilities approach, while not exactly settled, seems to have directed research towards the possibility of a middle path reaching beyond two irreconcilable perspectives. Recommendation to adopt a flexible approach where the methodology for determining the list, depending on the objectives and circumstances of the research or development project, becomes the main concern and legitimization by way of default, while being an answer to criticisms regarding possibility of a well-defined set of universal needs, illustrates that. Hence, a consensual list would be simply useful for internationally comparing surveys [56]. At the same time the approach of basic human needs is resisting and the capabilities approach is not unchallenged [23]. Therefore, it seems that while a drastic doubt remains about the definition of needs related to the very human nature, the production of a set of “dimensions of well-being”, a concept practically close to a list of fundamental human needs, is acceptable as long as theoretical and practical justifications of some sort are sought and the context is clear: “*I will argue that when we look philosophically at the coexisting components of well-being we come upon an important practical and theoretical tool which is, very simply, a rough set or list of dimensions. As a tool, like a set of crescent wrenches, there are times when nothing else will do the job*” [57].

It is therefore justified, in the absence of specific empirical data, to carefully select some lists of dimensions as theoretical basis among the various possibilities (39 different models listed in [56]), first and foremost examining the methodology used to build the sets of needs and their operational potential. In addition, it appears that even in frameworks claiming to be universalisable, some do not hesitate to modify of make salient specific needs (McIntosh 2007 adapting Max-Neef matrix in [58]).

Besides the work of Max-need and Nussbaum, two other models stand out.

10.3.4.1 Narayan Voices of the Poor Dimensions

The list of dimensions proposed by Narayan [59, 60] benefits from a high degree of confidence as it is the result of an extensive cross-cultural work of survey conducted on many different countries and overall aggregating the views on well-being, poverty and institutions of 60.000 participants, among them people from very poor, geographically remote and/or illiterate populations. Drawing upon participatory surveys data to complement more theoretical approaches is obviously a prerequisite for formulating an universalisable set of needs. The list goes as follows:

- Material well-being: having enough
  - Food
  - Assets
  - Work
- Bodily well-being: being and appearing well
  - Health
  - Appearances
  - Physical environment
- Social well-being
  - Being able to care for, bring up, marry and settle children
  - Self-respect and dignity
  - Peace, harmony, good relations in the family/community
- Security
  - Civil peace
  - A physically safe and secure environment
  - Personal physical security
  - Lawfulness and access to justice
  - Security in old age
  - Confidence in the future
- Psychological well-being
  - Peace of mind
  - Happiness
  - Harmony (including a spiritual life and religious observance)
  - Freedom of choice and action

Some formulations are clearly reminiscent both of the approaches of Max-Neef (having, being) and the capabilities (being able to).

10.3.4.2 The Doyal-Gough Theory of Human Need

The model of Doyal and Gough [23, 61] is built on both philosophical and anthropological justifications and the taking into account of contemporary debates and indicators. Explicitly in favor of a universal basic human needs framework and critical of some points of the capabilities approach, it states and focuses on the existence of "preconditions" of well-being rather than on the concept of well-being itself, adopting a hierarchical perspective in which two needs believed to be universalisable ("*universalisable preconditions for non-impaired participation in any form of life*") are defined: physical health and autonomy of agency, the latter being "*the ability to make informed choices about what should be done and how to go about doing it*" and covering cognitive capacity and opportunities to engage in socially significant activities. The model then proceeds to build a conceptual link between these two basic needs and a subset of universal characteristics of needs satisfiers, also called intermediate needs, which are not subject to cultural variations. The first six correspond to the physical health need, the others to the need for autonomy:

- Nutritional food/water
- Protective housing
- Work
- Physical environment
- Health care
- Security in childhood
- Significant primary relationships
- Physical security
- Economical security
- Safe birth control/childbearing
- Basic education

These needs are not fungible, meaning that the satisfaction of one of them cannot compensate for any dissatisfaction of another, thus refuting a concept of well-being as a sum of the satisfaction values of the different needs.

10.3.4.3 SICOMORES Set of Needs

By cross-analyzing the most credible and convincing models (i.e. Max-Neef, Nussbaum, Narayan and Gough) in order to spot regularities in their sets, it appears that they are often partially overlapping, which builds confidence in their respective results [57] and more importantly decisively provides directions in the building of a list holding a universalization potential [23]. By trying at the same time to adapt them to the particular context of CIMIC operations among civilian populations confronted with war destruction, the following list of dimensions can be generated:

- Water
- Food
- Housing
- Health
- Security
- Social linking (affection, relations, social belonging)
- Social regulation, participation and liberties (justice, political expression)

- Identity (construction of self, individual and social)
- Possibilities for action (access to energy and transport infrastructures)

This list of dimensions of well-being is used to describe needs satisfaction of every agent in the virtual population of SICOMORES. Physiological needs corresponding to imperatives of sustenance are detailed as the situation regarding this dimension is often so degraded in the first stages of a peacekeeping operation or even later that different CIMIC operations have to be independently implemented to restore and/or facilitate access to such basic goods. Also, "Possibilities for action" are not envisioned here as fundamental human needs, but are an indicator aimed at measuring and making salient the effects of operations specifically targeting transport infrastructures and energy assets and networks, allowing locals to broaden their capability for action on everyday life. Retaining such a dimension makes sense in the context of contemporary societies affected by asymmetrical conflicts. We are confident such a model of needs is grounded enough in social and natural science so that it grasps the reality of human needs in a valid and, at the current state of the art and regarding our project, an optimal way.

10.4 Computational Use of Culture and Needs to Generate the Effects of Actions of Influence in SICOMORES

10.4.1 PsyOps Operations: Culture and Physical Well-Being in Message Treatment

The theoretical framework used for modeling the process of message treatment by an agent is the Elaboration Likelihood Model [62, 63], a leading social psychology model in persuasive communication research. The model states that the level of cognitive resources an individual allocates to the assessment of a particular persuasive message depends on his ability and motivation to do so. While not explicitly making use of the concept of culture, the theory highlights the role of interest in the theme and personal relevance in the determination of the motivation to process, as personal relevance generates involvement. The more an individual feels concerned by a message, the more attention he is likely to grant it.

10.4.1.1 Motivation to Process the PsyOps Message

In the intercultural situation of a foreign military force communicating with locals, it can be assumed that any persuasive content emphasizing or on the contrary downgrading or flouting to a certain extend a cultural trait characterizing the receivers will likely proportionally trigger the attention of any local agent, as it makes the message appearing personally involving. Indeed, cultural traits being deeply related to one's identity and mental structures, an explicit or implicit normative reference to such traits should logically produce the feeling of involvement.

To support this claim we can refer to the concept of intrinsic source of personal relevance as defined by Celsi and Olson [64]: "*In contrast, intrinsic sources of personal relevance (ISPR) are relatively stable, enduring structures of personally relevant knowledge, derived from past experience and stored in long-term memory. This knowledge represents perceived associations between objects and/or actions and important self-relevant consequences, such as the attainment of goals and/or maintenance of values*". Although used here in the context of consumer research, this notion of a relatively stable and relevant stock of associations between objects and/or actions related to the maintenance of values can be described as very close to a stock of culturally determined attitudes, derived from continuous immersion in a socio-cultural environment. Interestingly, Celsi and Olson remarked that the ELM theory seemed more concerned with situational sources of personal relevance (immediate environment activating self-relevant consequences, goals and values) and the effects of involvement in general on persuasion, but did not implied these intrinsic sources are excluded from the ELM framework: they are a component of involvement. Johnson and Eagly [65] on the other hand criticized the ELM for considering only what they termed outcome relevant involvement (related to the pursuit of certain goals by the receiver), and not the other kinds of involvement they theorized: value relevant involvement (related to the reference made by the message to values and attitudes important to the receiver) and impression relevant involvement (related to the social image of himself the receiver wants to promote). Not only the second category of involvement seems to validate the idea of a possible issue involvement as cultural involvement, but also Petty and Cacioppo [66] argued that all these categories of involvement induced similar effects on persuasion processes, thus asserting the integrative virtue of the ELM framework in this regard. While here is not the place to settle the debate, what is important is that the role that goals and values, cognitions that can be both heavily dependent on a socio-cultural context, can play in the process is widely accepted: "*Furthermore, the more important the value, goal, sibling, or possession is to the self, the higher the level of involvement with a message on that topic*" [66].

What can be concluded here is that the feeling of personal relevance can be generated in a receiver by different factors. At this point and consistently with our chosen theoretical basis, our model does not integrate specific effects on persuasion of any specific kind of involvement. The point here is that the issue involvement can be modeled has a consequence of cultural features (values in the broadest sense) being concerned by the message, as these can be considered as a valid source of involvement (an intrinsic source of involvement, or the basis for value relevant involvement as termed by Johnson and Eagly). There is therefore a clear theoretical justification for the computational mechanisms of involvement proposed here.

As any message has to capture the attention of a receiver before being processed, it is important to make this constraint appear in the computational treatment. The measure of motivation of an individual agent i to scrutinize a PsyOps message p is thus divided in two steps. The first is the computation of the Degree of Appeal determining if the individual is interested enough (and to which extent) to complete the processing (with a high level of cognitive activity or not) after superficial exposure to the message, and the second being the Level of Personal Relevance. Cultural features are used for computing both.

Various variables enter in the measure of the Degree of Appeal. Here is only detailed how the idea of interest in the theme is first measured, by Involvement by Cultural Features (ICF).

The number and respective importance for i of the concerned cultural features (flouted and promoted) are used. Cultural features associated with a value above 7 are assumed to be particularly meaningful to the considered agent, and a message involving them is more likely to retain his attention even upon a superficial examination. Let q be the number of cultural features of $\{cf_1, \ldots, cf_n\}$ with a value above 7 in Cultural values system(i) and $imp_1, \ldots, imp_n$ their respective importance.

$$ICF\,(i,p) = \frac{1}{10}\cdot\left(\frac{1}{n}\cdot\left(\sum_{i=1}^{n} imp_i\right) + \frac{q}{n}\cdot\left(10 - \left(\frac{1}{n}\cdot\sum_{i=1}^{n} imp_i\right)\right)\right) \tag{10.1}$$

If the Degree of Appeal is above a certain threshold, the Ability to process the message (see below) and The Level of Personal Relevance are computed. For the latter, let n be the number of concerned Cultural Features and $imp_1, \ldots, imp_n$ their respective importance:

$$\text{Level of Personal Relevance}\,(i,p) = \frac{1}{n}\cdot\frac{1}{10}\cdot\left(\sum_{i=1}^{n} imp_i\right) \tag{10.2}$$

This second component of motivation is then aggregated with the Degree of Appeal to obtain the Motivation of the Receiver to Process the Message to be used in the treatment according to the conclusions of the ELM model.

10.4.1.2 Ability to Process the PsyOps Message

The ELM model posits that the ability to engage in extensive message scrutiny is as important in the process as motivational variables. The roles of factors such as mood or distraction on ability have been studied, but relevantly integrating those in the model was neither possible nor desirable at this point. However, cognitive capacity can be impacted by other situational elements.

In the SICOMORES model of cognitive processing, if the medium conveys only written text and the receiver is illiterate, then the process stops as there is no ability to examine the message. If not, the Ability of the receiver to process the message (ARPM) is computed. To this effect the Cognitive Hindrance (CH) is first calculated. This steps aims at simulating the impairing effect exerted on message evaluation by the failure to satisfy certain needs. A war situation usually means disruption of economic activities leading to shortages of supplies, displacement of people and/or destruction of lodging leading to stress and exhaustion, and overall tough living conditions that can prevent receivers from processing a message to the best of their cognitive abilities. Four needs from our list are considered relevant to compute CH in such a context, determining only the physiological ability to operate: water, food, housing (related to amount and quality of rest) and health. They are respectively named n_w, n_f, n_h and n_h. Their respective values range between 1 and 10 and only needs among those four with a value under 5 are considered, as those with a value of 5 or higher are considered fulfilled enough not to generate any cognitive interference.

Let q be the number of needs with a value strictly lower than 5 and N1 the set of these needs $N1 = \{n_1, \ldots, n_q\}$

Let q′ be the number of needs with a satisfaction degree lower or equal to 2.

$0 \leq q \leq 4, 0 \leq q' \leq 4, q' \leq q$

$$\text{Cognitive Hindrance}\,(i,p) = \frac{1}{20} \cdot \left(\sum_{i=1}^{q} (5 - sd(n_i)) + \frac{q'}{q} \cdot \left(20 - \sum_{i=1}^{q} (5 - sd(n_i)) \right) \right) \tag{10.3}$$

As the needs model of our system is deemed universal, i.e. usable in any scenario, so is this computation, an obviously advantageous feature of this message treatment model. Ability is then calculated by aggregating CH with other relevant variables such as the complexity of the message and the intellectual level of the receiver.

10.4.2 CIMIC Operations

For generation of effect of CIMIC operations, we consider three categories of agents in the area of effect of a particular CIMIC: the leaders, who are deemed to have a sort of statutory obligation to form an opinion on any operation taking place in their area of influence, the beneficiaries, whose needs values are affected by the operation, and participants, agents optionally recruited as a workforce for carrying out the operation. For each of these, the extent to which a particular CIMIC contravene and/or highlights certain values part of the local culture is evaluated and computed.

Let us examine the case of an agent I impacted by a CIMIC c:
Let $ecf_1, \ldots, ecf_p$ be the cultural features accentuated by the CIMIC operation.
Let $impf_1, \ldots, impf_p$ be their respective importance.
Let $ece_1, \ldots, ece_q$ be the cultural features flouted by the CIMIC operation.
Let $impe_1, \ldots, impe_q$ be their respective importance.

The Degree of Cultural Conformity (DCC) for an individual i is defined as follows:

$$DCC\,(i,c) = \sum_{i=1}^{p} impf_i - \sum_{i=1}^{q} impe_i \tag{10.4}$$

DCC measures if the reaction of the considered agent to the CIMIC operation is positive or negative. If above or below certain thresholds of satisfaction, its value is updated (computation of Revised Degree of Cultural Conformity RDCC) if some features of high significance to the individual (characterized by a value above 7) are concerned by the CIMIC. Involving such features is deemed to increase the likeliness of a pronounced reaction.

If the agent is a beneficiary, his needs satisfaction (NS) is increased according to the particular type of the CIMIC (for instance, access to a hospital increases the health need of those in the area of effect). We posit the aggregate degree of cultural conformity-needs *ADCCN* (*i*, *c*) of individual i:

$$ADCCN\,(i,c) = [60.\,RDCC\,(i,c) + 40.\,NS\,(i,c)]/100 \tag{10.5}$$

The Opinion toward the Force attribute is updated at the final stage of the treatment, taking into account only cultural conformity for non-beneficiary agents and both cultural conformity and needs satisfaction, though *ADCCN* (*i*, *c*), for beneficiary agents.

10.5 Conclusion

In a world where the disappearance of armed conflicts is but a tremendously illusory perspective, the militaries have to adapt to the evolution of the ever-changing forms of war. In this regard, they unquestionably perceived the increasing complexity of social, economic and cultural situations they are to face in any intervention as the local populations are systematically and in many ways at the core of contemporary conflicts. They subsequently recognized a need for specific training and therefore for training systems able to cope with such complexity. In such an industrial context, it is easy to realize that conceptual issues related to societies and their functioning, more thought-provoking than issues linked to the simulation of a classic military opponent, have to be addressed. The aim of this chapter was thus to present the SICOMORES system and some of its contributions regarding multi-agent approaches, but also to demonstrate that developing such a system generates both an opportunity and an imperative to elaborate on these issues and to make an enlightened use of transdisciplinary material in an original and critical approach. Multi-agent modeling combined with multilayer network creation allows representation of different scales of abstraction and so the use of concepts relevant

and challenging at the same time. Here we detailed theoretical underpinnings of our model regarding the two entwined issues that are culture and needs and how they were represented and computationally used in our framework for measuring cognitive phenomenon, such as personal relevance or cognitive hindrance.

Acknowledgements This work is funded by the French Ministry of Defense (DGA—Direction Générale de l'Armement) in the framework of the DGA RAPID Project SICOMORES.

References

1. Angstrom, J., Duyvesteyn, I. (eds.): Understanding Victory and Defeat in Contemporary War. Routledge, New York (2007)
2. McMillan, J., Waker, A., Clarke, E., Marc, Y.: Cultural awareness training for marine corps operations: the CAMO project. In: Schmorrow, D., Nicholson, D. (eds.), Advances in Cross-Cultural Decision Making, CRC Press Taylor & Francis Group, Part 1, pp. 277–286 (2013)
3. Stuart, S.: Using Video Games to prepare for the Culture Shock of War. (2014). PC.com http://www.pcmag.com/article2/0,2817,2472395,00.asp
4. Bennet, W.: Media Influence Modeling in Support of conflict Modelling, Planning, and Outcome Experimentation (COMPOEX). Presentation to the Military Operations Research Society Irregular Warfare Analysis Workshop, MacDill AFB (2009)
5. Silverman, B., Bharathy, G., Nye, B., Kim, G., Roddy, P., Poe, M.: Simulating state and sub-state actors with statesim: synthesizing theories across the social sciences. In: Sokolowski, J., Banks, C. (eds.) Modeling and Simulation Fundamentals: Theoretical Underpinnings and Practical Domains. New-York, Wiley STM (2010)
6. Kott, A., Citrenbaum, G., (eds.): Estimating Impact, A Handbook of Computational Methods and Models for Anticipating Economic, Social, Political and Security Effects in International Interventions. Springer, New York (2010)
7. Faucher, C., Bergier, J.-Y., Forestier, M.: Capturing sub-Saharan African sociality in social networks to generate a culturally realistic population, IEEE International conference on Culture and Computing, Kyoto University Japan (2015)
8. Forestier, M., Bergier, J.-Y., Bouanan, Y., Ribault, J., Vallespir, B., Faucher, C.: Generating multidimensional social network to simulate the propagation of information, Proceedings of the 2015 IEEE/ACM International Conference on Advances in Social Networks Analysis and Mining, 1324–1331 (2015)
9. Benedict, R.: Patterns of Culture. Houghton Mifflin Harcourt, Boston (1934)
10. Kardiner, A., Linton, R.: The Individual and His Society. Columbia University Press, New York (1939)
11. Linton, R.: The Cultural Background of Personality. Appleton-Century-Crofts, New York (1945)
12. Kolodziej-Smith, R., Friesen, D., Yaprak, A.: Does Culture affect how people receive and resist persuasive messages? research proposals about resistance to persuasion in cultural groups. Glob. Adv. Bus. Commun. **2**(1), Article 5 (2013)
13. Kluckhohn, C., Murray, H.A.: Personality in Nature, Culture and Society. Knopf, New York (1948)
14. Svenmarck, P., Lundin, M., Sjöberg, E., et al.: Message Dissemination in Social Networks for Support of Information Operations Planning. (2010). www.dtic.mil/cgi-bin/GetTRDoc?AD=ADA582148

15. van Vliet, T., et al.: Generic message propagation simulator: The role of cultural, geographic and demographic factors. In: Schmorrow, D., Nicholson, D., (eds.), Advances in Cross-Cultural Decision Making, pp. 416–429. CRC Press Taylor & Francis Group (2011)
16. Brousmiche, K.-L., Kant, J.-D., Sabouret, N., Fournier, S., Prenot-Guinart, F.: Modeling the impact of beliefs and communication on attitude dynamics: a cognitive agent based approach, Social Simulation Conference—10th European Social Simulation Association Conference, Barcelona (2014)
17. Angstrom, J., Duyvesteyn, I. (eds.): Modern War and the Utility of Force: Challenges, Methods and Strategy. Routledge, New York (2010)
18. Norheim-Martinsen, P., Nyhamar, T. (eds.): International Military Operations in the 21st Century, Global Trends and the Future of Intervention. Routledge (2015)
19. Marie, A.: Communauté, Individualisme, Communautarisme: Hypothèse sur quelques Paradoxes Africains. Sociologie et Société **39**(2), 173–198 (2007)
20. Rhoads, K.: The culture variable in the influence equation. The Public Diplomacy Handbook, March (2008)
21. Coopération civilo-militaire.: Doctrine Inter Armées, DIA-3.10.3(A)_CIMIC (2012)
22. Matarasso, M.: Les frontières socio-culturelles des besoins humains. Les cahiers de la publicité **2**(1), 21–36 (1962)
23. Gough, I.: Climate Change and Sustainable Welfare: An Argument for the Centrality of Human Needs. Centre for Analysis of Social Exclusion, London School of Economics. (July 2014). http://sticerd.lse.ac.uk/dps/case/cp/casepaper182.pdf
24. Lévi-Strauss, C.: Les Structures Elémentaires de la Parenté. Presses Universitaires de France (1949)
25. Maslow, A.: A theory of human motivation. Psychol. Rev. **50**, 370–396 (1943)
26. Maslow A.: Motivation and personality. Harper & Row, New York (1954, 1970nd edn.)
27. Bracewell, D.: The Needs of Metaphor. In: Kennedy, W., Agarwal, N., Shanchieh, J.Y. (eds.) Social Computing, Behavioral-Cultural Modeling, and Prediction. 7th International Conference SBP, Springer, 229–236 (2014)
28. Denny, N.: Maslow: A graphical framework for communicating models of human behavior, Proceedings of the 18th Conference on Behavior Representation in Modeling and Simulation, pp. 123–128. Sundance, UT, 31 March–2 April 2009
29. Salado, A., Nilchiani, R.: Using Maslow's hierarchy of needs to define elegance in system architecture. Proceedings of the Conference on Systems Engineering Research **16**, 927–936 (2013)
30. Bishop, J.: An analysis of the implications of Maslow's Hierarchy of Needs for networked learning design and delivery, International Conference on Information and Knowledge Engineering, 49–54 (2016)
31. Berkowitz, L.: Social motivation. In Lindzey, G., Aronson, E. (eds.), Handbook of Social Psychology, 2nd edn., vol 3, Reading, Addison-Wesley, MA (1969)
32. Wahba, M., Bridwell, L.: Maslow reconsidered: A review of research on the need hierarchy theory. Organ. Behav. Human Perform. **15**(2), 212–240 (1976)
33. Heylighen, F.: A cognitive-systemic reconstruction of Maslow's theory of self-actualization. Behav. sci. **37** (1992). http://pespmc1.vub.ac.be/papers/maslow.pdf
34. Cofer, C., Appley, M.: Motivation: Theory and Research. Wiley, New-York (1964)
35. Reid-Cunningham, A.: Maslow's Theory of Motivation and Hierarchy of Human Needs: A Critical Analysis, Ph.D. qualifying examination. (2008). http://www.scribd.com/doc/8703989/Maslow-s-Hierarchy-of-Needs-A-Critical-Analysis#scribd
36. Kenrick, D., Griskevicius, V., Neuberg, S., Schaller, M.: Renovating the Pyramid of Needs: Contemporary Extensions Built Upon Ancient Foundations. Perspect. Psychol. Sci. **5**(3), 292–314 May (2010)
37. Diener, E., Tay, L.: Needs and subjective well-being around the world. J. Pers. Soc. Psychol. **101**(2), 354–365 (2011)
38. Friedlander, F.: Underlying Sources of Job Satisfaction. J. Appl. Psychol. **47**, 246–250 (1963)

39. Schaffer, R.: Job satisfaction as related to need satisfaction in work. Psychol. Monogr. **264** (1953)
40. Hofstede, G.: The cultural relativity of the quality of life concept. Acad. Manag. Rev. **9**(3), 389–398 (1984)
41. Mook, D.G.: Motivation, The Organization of Action, Norton and Company, NewYork (1987)
42. Max-Neef, M.: Human Scale Development. The Apex Press, New York and London (1991) http://www.area-net.org/fileadmin/user_upload/papers/Max-neef_Human_Scale_development.pdf
43. Sen, A.: Commodities and Capabilities. Elsevier Science Publishers, Oxford (1985)
44. Sen, A.: Inequality Reexamined. Oxford University Press, Oxford (1992)
45. Rawls, J.: A theory of justice. Harvard University Press, Cambridge MA (1971)
46. Rawls, J.: Political liberalism. Columbia University Press, New-York (1993)
47. Alkire, S.: Capability and functionings: definition & justification, HDCA Introductory Briefing Note (2005)
48. Nussbaum, N., Sen, A. (eds.): The Quality of Life. Oxford University Press (1990)
49. Nussbaum, M.: Aristotelian social democracy. In: Douglass, R., Mara, G., Richardson, H. (eds.) Liberalism and the Good. Routledge, London (1990)
50. Nussbaum, M.: Women and Human Development: The Capabilities Approach. Cambridge University Press, Cambridge (2000)
51. Nussbaum, M.: Capabilities as fundamental entitlements: Sen and social justice. Feminist Econ. **9**(2–3), 33–59 (2003)
52. Nussbaum, M.: Frontiers of Justice: Disability, Nationality, Species Membership, The Belknap Press, Harvard University Press, Cambridge, Massachusetts (2006)
53. Sen, A.K.: Capabilities, lists, and public reason: continuing the conversation. Feminist Econ. **10**(3), 77–80 (2004)
54. Sen, A.: Rationality and Freedom. Harvard University Press, Cambridge, MA (2002)
55. Sen, A.: Human rights and capabilities. J. Human Dev. **6**(2), 151–166 (2005)
56. Alkire, S.: Choosing dimensions: the capability approach and multidimensional poverty. In: Kakwani, N., Silber, J. (eds), The Many Dimensions of Poverty. pp. 89–119. Palgrave Macmillan, New York (2008). http://www.albacharia.ma/xmlui/bitstream/handle/123456789/31537/1339Choosing%20dimensions%20the%20capability%20approach%20and%20multidimensional%20poverty2006.pdf?sequence=1
57. Alkire, S.: Dimensions of human development. World Dev. **30**(2), 181–205 (2002)
58. McIntosh, A.: The Wheel of Fundamental Human Needs. (2007) http://www.alastairmcintosh.com/general/resources/2007-Manfred-Max-Neef-Fundamental-Human-Needs.pdf
59. Narayan-Parker, D.: Voices of the Poor, Vol. 1: Can Anyone hear Us. World Bank, Washington, DC (2000)
60. Narayan, D.E.A.: Voices of the Poor, Vol. 2: Crying out for Change. Oxford University Press for the World Bank, New York (2000)
61. Doyal, L., Gough, I.: A theory of human need. Macmillan, Basingstoke (1991)
62. Petty, R., Cacioppo, J.: The Elaboration Likelihood Model of Persuasion. In: Berkowitz, L. (ed.) Advances in Experimental Social Psychology, vol. 19, pp. 123–205. Academic Press, New-Yok (1986)
63. Petty, R., Wegener, D.: The Elaboration Likelihood Model: current status and controversies. In: Chaiken, S., Trope, Y. (eds.) Dual-process Theories in Social Psychology, pp. 41–72. The Guilford Press, New-York (1999)
64. Celsi, R.L., Olson, J.C.: The role of involvement in attention and comprehension processes. J. Consum. Res. **15**, 210–224 (1988)
65. Johnson, B., Eagly, A.: Effects of Involvement on Persuasion: A Meta-Analysis. Psychol. Bull. **106**(2), 290–314 (1989)
66. Petty, R., Caccioppo, J.: Involvement and persuasion: Tradition versus Integration. Psychol. Bull. **107**(3), 367–374 (1990)

Chapter 11
Simple Culture-Informed Cognitive Models of the Adversary

Paul K. Davis

Abstract Simple cognitive models of the adversary are useful in a variety of domains, including national security analysis. Having alternative models can temper the tendency to base strategy on the best-estimate understanding of the adversary, and can encourage building a strategy that is better hedged and more adaptive. Best estimates of adversary thinking have *often* been wrong historically. Good cognitive models must avoid mirror-imaging, which implies recognizing ways in which the adversary's reasoning may be affected by history, culture, personalities, and imperfect information, as well as by objective circumstances. This paper describes a series of research efforts over three decades to build such cognitive models, some as complex computer programs and some exceptionally simple. These have been used to represent Cold-War Soviet leaders, Saddam Hussein, Kim Jong Il, and modern-day leaders of al Qaeda. Building such models has been a mixture of art and science, but has yielded useful insights, including insights about the sometimes-subtle influence of leaders' decision-making culture.

11.1 Introduction

11.1.1 Intent of Paper

This paper describes episodic research over several decades to build models representing an adversary so as to better understand him and his possible reasoning (the generic "him," whether an individual or a group, and whether male or female). This can aid in developing strategies to influence his actions—e.g., deterring him from aggression. When building such models it is necessary to consider the

P.K. Davis (✉)
Pardee RAND Graduate School, Santa Monica, CA 90407-2138, USA
e-mail: pdavis@rand.org

C. Faucher (ed.), *Advances in Culturally-Aware Intelligent Systems and in Cross-Cultural Psychological Studies*, Intelligent Systems Reference Library 134, https://doi.org/10.1007/978-3-319-67024-9_11

influence of culture, but doing so has often been difficult and contentious. Leaders may be part of several cultures, each of which is only poorly understood by outsiders. Further, objective circumstances matter, even though viewed through culture influenced lenses. The paper's examples involve leaders in such diverse cultures as Cold-War Soviet Communist leaders, a Muslim Baathist Iraqi despot, members of the North Korean Kim Dynasty, and Middle Eastern Islamist terrorist leaders. The dominant considerations in reasoning have sometimes been objective, sometimes cultural, and sometimes idiosyncratic. More often, all of these have contributed.

Before proceeding it is useful to draw a contrast. A common modern-day approach for research on systems that have thinking adversaries is agent-based modeling. Agent-based simulations can represent large numbers of interacting entities that make decisions about next actions. These can be used, for example, in describing "artificial societies [1];" business interactions in an information economy [2]; complex societies with military, social, and economic turmoil [3]; and complex business challenges [4]. This paper is instead about "simple models" that can be largely explained with a half-dozen viewgraphs or with small computer programs written in high-level visual-programming languages. Such models are intended to inform policy analysis. A historical example of "simple" is game theory's depiction of the prisoner's dilemma. This paper, however, goes well beyond the domain of economic rational actors. Further, it is less about rigorously posing and solving a problem than about understanding and perhaps influencing adversaries. Even more important, because of myriad uncertainties, the models I describe are not reliably predictive. Instead their function is to help us understand and act, but with humility and caution because of uncertainties.

11.1.2 Structure of Paper

The structure of the paper is as follows. Section 11.2 makes the case for why cognitive models are needed. Section 11.3 describes long-ago complex versions based on 1980's artificial-intelligence concepts and technology. Sections 11.4 and 11.5 describe early efforts to "skim the cream" of such work in the study of Saddam Hussein (1990–1991) and Kim Jong Il (1994). Section 11.6 describes an application integrating qualitative social-science information about terrorism. Section 11.7 discusses how such work can be extended to uncertainty-sensitive computational models. Section 11.8 offers principles for work of this general nature, which is a mix of art and science. Consistent with the intent of the larger volume, throughout the paper, I discuss how cultural considerations play a role—sometimes dominant, sometimes only contributory.

11.2 The Need for Cognitive Models

A core assumption in analytic work of the twentieth century was that competitors behave in ways consistent with economic rationality: comparing the costs and benefits of options and choosing the course of action that maximizes subjective expected utility. The related theory is associated with such figures as John von Neuman, Oskar Morgenstern, and Leonard Savage. From the beginning, it was recognized that real people do not necessarily make decisions in that way. Thus, the distinction was drawn between normative (i.e., prescriptive) models and descriptive models: what people *should* decide is…, but what they *actually* decide is often different. Mainstream economics has been driven by the rational-actor model for the better part of a century. An immense body of psychological research exists on the descriptive side, usually associated with Nobel Prize winner Daniel Kahneman and his early collaborator Amos Tversky [5, 6]; The work is often referred to as about heuristics and cognitive biases. As has been decisively demonstrated, people aren't actually so rational. Even some economists have grudgingly accepted this, as reflected by the new sub-field of behavioral economics [7, 8].

The prescriptive versus descriptive distinction, then, has been important. Equally significant but less well recognized, however, is that the prescriptions of rational-actor theory have grievous shortcomings. The practical shortcomings were highlighted in the 1950s by Herbert Simon, who introduced the terminology of "bounded rationality," noting that decisionmakers did not have and could not obtain the information necessary for the idealized rational-actor calculations, that they could not make the complicated calculations anyway, and that in reality they *necessarily* used shortcuts—seeking "satisficing" solutions that were good enough [9]. His observations crossed the boundary between prescription and description. Nonetheless, the rational-actor model continued to be seen as the normative ideal and a description of behaviors, which can be "as if" decided rationally even if the actual process is more complex [10]. Late in the twentieth century, this assumption was fundamentally challenged by Gary Klein and Gird Gigerenzer [11–13]. Their research described humans as often using more intuitive decision processes. They *celebrated* this because, often, quick, intuitive decisionmaking is precisely what is needed. It may also be amazingly perceptive as discussed in a popular book [14]. Although disputes were strong between the heuristics-and-biases school and the intuitive/naturalistic school, synthesis was clearly possible [15]. Kahneman's most recent book partly synthesizes the two schools' themes by noting distinctions between thinking fast and slow, depending on circumstances [16].

An additional consideration of interest to policy makers, managers, and those who advise them is that the effort to do "rational" decision making and to have "rational" decision processes often devolves, in practice, to stultifying processes that lack imagination, creativity, and effectiveness. The options considered may be mundane and the result of log-rolling within the organization; the costs and benefits may be calculated with simplistic metrics; and the calculations may ignore uncertainties. Such processes may purport to be rational, but are not—if "rational" relates to wisdom (see also [17]). Another problem is particularly relevant to this paper: when developing strategy to deal with an adversary, seemingly rational processes may miss the mark entirely because of mirror-imaging: assuming that the adversary reasons in the same way as the analysts, despite the adversary having a different vantage point, being in a different culture, and having his own history and idiosyncrasies. Consider the following:

1. *Cross-corporate negotiations* in which the underlying issues are not just profitability but the separate corporations' cultures and self-concepts (think of the failed mergers of Daimler-Benz and Chrysler, or of AOL and Time Warner). The issues are discussed with fictional companies in a thoughtful paper by Nigel Howard [18].
2. *Competitive military actions* by national rivals relating to arms, territory, navigation rights, and upport of third countries (think of the U.S. and China).
3. *Crisis actions* by military antagonists who are concerned about avoiding catastrophe, but who are also concerned about, e.g., preserving power and saving face (think of the U.S. and Soviet Union during the 1962 Cuban Missile Crisis).
4. *A family-level battle of the spou*ses over something objectively unimportant, but in the context of emotional past events (we can all think of many examples).

Outcomes for such clashes will be better if the sides understand each other in terms that go well beyond assuming narrow economic rationality. As researchers, we may therefore see value in constructing "cognitive models" of adversaries, competitors—or even spouses. What follows draws on my national-security research, but the ideas are more general.

11.3 Background: Large A.I. Models in Analytic War Gaming

11.3.1 Structure of the Approach

In the early 1980s, the U.S. Department of Defense sponsored an activity that generated what was called the RAND Strategy Assessment System (RSAS) [19]. This was a global analytic war game covering conventional war through general nuclear war. It was a large and automated computer simulation, but it was also game-structured with "objects" for what would be Red, Blue, and Green teams in human war games. It permitted substitution of humans for agents, or vice versa. For example, a simulation might have agent-driven decisions by Red and third countries (Green), but decisions of a human team for Blue (Fig. 11.1).

The RSAS allowed independent decisions (whether by humans or agents) by NATO, the Warsaw Pact, and individual nations. Rather than trying to "optimize," the agents used heuristic artificial intelligence methods. We drew a sharp distinction

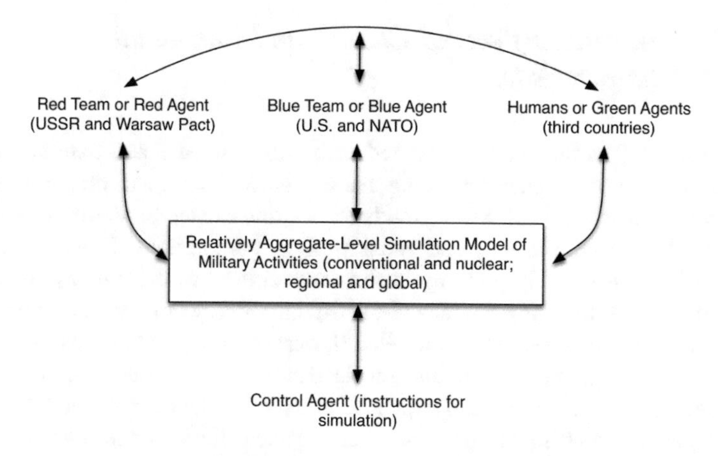

Fig. 11.1 Architecture of the RAND strategy assessment system (1980s)

between strategic-level decisions and operational-level decisions [20, 21]. We saw the strategic decisions as requiring "national command level models" (NCL models) that would ordinarily just monitor developments, but that could choose to escalate the level or scope of conflict, or to otherwise change strategy. Although the models were rule-based, they were very different from the production-rule or expert-system models of the era because they were structured from a top-down strategic perspective. This structuring was in preference to having an inference engine trying to make sense of disorganized rules.[1]

The operational-level models were more like commanders following a war plan, which might have contingent branches and be otherwise complicated because of the need to direct many different force operations. The agent commanders were more or less "following a slotted script" [23, 24]. They would adapt to developments, but the adaptations would be relatively straightforward. For example, during the simulation the commander would each day allocate ground-force reserves to sectors where they were most needed and would apportion air forces across missions such as bombing air bases or attacking ground forces depending on the phase of the plan and event-driven needs.

The NCL models had to have a strategic construct (i.e., a "cognitive model"). Ours were based on escalation-ladder structures and the current and projected status of combat and conflict levels. Thus, a model might characterize its situation as "We are still conducting conventional conflict, but we're losing." This might lead to escalation if it seemed likely that the escalation would improve outcome, taking into account the other side's response.

Given this structure, how could it be "filled in" with content? In particular, what should the Red NCL model, the model of the Soviet Union's leadership, look like? How should it reflect differences in personality, history, and culture?

---

[1]Some of this discussion draws on material in Chap. 3 of a recent National Academy report [22].

11.3.2 Uncertain Effects of Culture and Individual Characteristics

During the 1970s a fierce debate existed within the United States national-security community with one segment insisting that the Soviet Union was preparing to fight and win a nuclear war [25]. Strong evidence for this existed at the military level in the form of doctrine, practice, and general-officer admonitions. It was argued that Soviet political leadership grew up in the same culture as the military leadership, was subject to the planning realities of the Soviet military culture, and remembered having successfully survived World War II, despite thirty million losses. The last point, it was suggested, meant that the leadership was Culturally-Aware of the ability to survive even after catastrophes [26]. Another segment of the community, however, argued that Soviet leadership had no such illusions and understood well the realities of mutual assured destruction [27]. They argued that the horrific memory of World War II made the Soviets more risk-averse, not less.

Given such disagreements among foreign-policy experts, how could the Soviet leadership be modeled? We did so by constructing *alternative* models (alternative “Ivans and Sams”) reflecting the different postulated mindsets [21]. Before expressing these in computer code, we constructed essays, temperament check lists, and illustrative decision trees to strengthen our sense of how the alternatives reasoned. The models' rules would need to reflect implicit assumptions and such devices helped us achieve a degree of coherence. We also avoided simple stereotyping. A “warfighting” Ivan presumably didn't *want* war, much less nuclear war, and a more deterrence-accepting Ivan presumably would fight strongly and use whatever means proved necessary. That is, circumstance and, e.g., perceptions of the adversary (the US and NATO) would interact with predilictions. The simulation reflected this complexity. Depending on details of scenario, the “warfighting” Ivan might end up terminating conflict and the less-warfighting Ivan might end up escalating to general nuclear war. Such model behavior frustrated some observers, especially those steeped in the fight-and-win Soviet military literature, but the context and path dependence of decisions seemed appropriate and consistent with history.

Overall, we concluded that Soviet leaders were far better understood as intelligent human beings with more or less universal characteristics of reasoning (and cognitive biases) than as cultural stereotypes, much less stereotypes drawn by U.S. analysts of Soviet military culture. Khrushchev's memoirs, for example, reveals thinking that reflected the Soviet view of the world (encirclement by enemies, constant pressures from the malevolent western powers, aggressiveness by the U.S., and a willingness to cope with whatever adversity arose, even war) [28]. This was consistent with Soviet history and culture, as well as his own experiences. However, Khrushchev's reasoning was ultimately similar to Kennedy's: it was essential to find a way to avoid war, while allowing everyone to save face adequately. Remarkable discussions between former U.S. and Soviet leaders after the Cold War reinforce this imagery. The primary differences were rooted in history and psychology (e.g., the tendency to impute malevolent motives). Both sides were afraid of the other and could much more readily "see" the threatening behavior of the other side than that of themselves [29].

But what if the Soviet leadership had been different? After all, leaders change over time. In 2003, the Soviet leader, Yuri Andropov—previously, director of the KGB—was seriously concerned about the potential of a U.S. first strike. He worried that the U.S., rather than the Soviet Union, believed that it could fight and win. Since other officials did not share his fears, it is difficult to claim that he was captive to Soviet culture. Nonetheless, the fact that he had a distinctly different perspective reinforces the need to consider alternative models of adversary leadership when developing strategy. During the Andropov period, the U.S. and Soviet Union went through a serious crisis in 2003 without the United States even recognizing that a crisis existed. U.S. and NATO forces were exercising in ways that included nuclear escalation and the Soviet leadership feared that the exercises were cover preparations for an actual attack [30]. Only later did the U.S. come to understand all this. President Reagan in his later memoirs wrote:

> Three years had taught me something…Many people at the top of the Soviet hierarchy were genuinely afraid of America and Americans… many of us in the administration took it for granted that the Russians, like ourselves, considered it unthinkable that the United States would launch a first strike against them. But…I began to realize that many Soviet officials feared us not only as adversaries but as potential aggressors who might hurl nuclear weapons at them in a first strike….(Reagan [31], pp. 588–589)

As an example of work with the RSAS and their Red agents, we conducted experiments with limited nuclear options. In addition to having alternative Red and Blue models, each Blue had alternative models of Red, each of which had a simpler model of Blue, which had an even simpler model of Red. Similarly for Blue. In some cases, Blue would use a limited nuclear option to "re-establish deterrence," as in NATO doctrine. Red, however, would perceive the act as Blue having initiated nuclear war and would initiate all-out general nuclear war. In other runs, depending on details and model, Red would de-escalate or simply proceed. Playing through mainstream scenarios, however, cast doubts on NATO's concepts and plans for nuclear use shortly before collapse of its conventional defenses [20, 32], suggesting that such late use might be ill-advised for reasons discussed below. We now know that similar conclusions emerged from sensitive high-level U.S. human war games conducted in the 1980s, games that ended in general nuclear war [30].

11.3.3 Skimming the Cream with Simpler Adversary Models

Although we saw the RSAS as a technical success and greatly enjoyed building it because of the many substantive and technical challenges, I suspected (correctly, as it turned out) that it was too big and complex for use in government, rather than a Ph.D.-loaded think tank with people willing and eager to consider unconventional views (such as Soviet leadership understanding deterrence rather than reflecting Soviet doctrinal writings), and also too expensive to maintain.

A primary cause of complexity was that the NCL models had to be able to wake up at each time step, assess the situation, and consider options under arbitrary circumstances so that the simulation could proceed uninterrupted. Programming to achieve that would not have been so difficult if we had trivialized the substance, but to be realistic (even at low resolution), an agent had to look at the worldwide situation (developments in one theater might be favorable in one and unfavorable in another), to communicate and "negotiate" with allies, to consider options ranging from changes of military strategy at a given level of conflict to one involving, say, nuclear escalation, and so on. Further, in evaluating options, the model had to consider the other side's response as well as the likely decisions of numerous other countries. Even in a simplified approach, it was necessary to pay attention to alliances, permission rights, and the independent nuclear-use decisions of the UK and France. Further, judgements about the likely outcome of one or another option's military outcome were based on "look-ahead calculations" (simulations within the simulation) or subtle heuristics. On top of this, the RSAS contained a complex multi-theater model of combat, which became the Joint Integrated Combat Model (JIICM) that has now been used for more than twenty years by the U.S. Department of Defense and U.S. Allies.

I suspected that the "big" insights from the decision models could be obtained more simply. Using the earlier example where we learned that NATO's late use of nuclear weapons was problematic, the reason was ultimately simple: if Blue used one or a very few nuclear weapons in an effort to reestablish deterrence by raising stakes, but did so only when it was about to lose badly in the conventional conflict, then Red could see the same reality—that NATO was about to collapse. It might therefore conclude that its way to major victory within a few days was by merely plunging ahead—perhaps responding only minimally if at all to NATO's demonstrative nuclear use. In contrast, a somewhat earlier NATO first use would have created a much bigger dilemma for Red. Indeed, in some of the simulations, Red did indeed terminate conflict: the risks that Red perceived were too great to do otherwise. Couldn't we understand that without all the complex apparatus and simulation? An opportunity soon arose to address the question.

11.4 Saddam Hussein

11.4.1 An Approach to Building Simple Models

By the late 1980s the U.S. was considering different possible adversaries, particularly Iraq's Saddam Hussein. Colleague John Arquilla and I constructed models of Saddam that could be reduced to a few viewgraphs [33]. We structured them around considerations that we believed would be on Saddam's mind as he contemplated alternatives. For example, we constructed Fig. 11.2 to indicate with a cognitive map based on taking seriously Saddam's statements and taking a strategic view of what was in fact going on. The convention in such diagrams is that more of a variable at the start of an arrow tends to cause more of the variable at the end of an arrow, unless a minus sign exists, in which case more leads to less. The map describes a bad situation that is weakening Iraq and undercutting Saddam's ambitions—one seen as due significantly to a U.S.-Gulf-State *conspiracy*.

As in the earlier RSAS work we constructed alternative models because there were fundamental disagreements about how Saddam actually reasoned. Table 11.1 shows a 1990 depiction of the two models we used. This table was to summarize our mental models. Internalizing its content then led to specific decisions in structuring the analytic models. Model 1 was similar to the then-prevailing intelligence-community best estimate. Model 2 was our own construction. Model 2 proved to be more accurate when, in August 1990, Saddam invaded Kuwait.

A core concept of the approach was that we did not claim that one or the other model of Saddam was "right," but rather than strategy development should take seriously that Saddam's reasoning might be like either, or a combination. Appreciating that would encourage building a strategy that laid the basis for adapting to new information. As in a great deal of RAND work, the admonition is to seek a strategy that is flexible, adaptive, and robust (FARness), rather than a strategy tuned to some dubious best estimate. That is the key to planning under uncertainty [34].

Moving forward, we concluded that even a leader attempting to be rational is doing well if merely he considers upside, downside, and best-estimate outcome possibilities for several options.[2] That is, we saw Saddam as perhaps having a de facto cognitive structure such as in Table 11.2. A given model of Saddam would give different weights to the most-likely, best-case, and worst-case outcomes estimated for an option when making his net assessment. Later work generalized this scheme to allow the combining rule to be more complex than mere linear-weighted sums.

We asked in 1990, before and during the crisis that led to Saddam's invasion of Kuwait, how Saddam might assess the likely outcome of different options, such as do nothing, mount a smallish invasion, or mount a full-scale invasion of Kuwait and Saudi Arabia? Although it would be possible to use a combat model to estimate such things, we thought it better to draw on more general knowledge to make the estimates subjectively, thereby including all-important "soft" variables ignored by combat models.

As the crisis unfolded, we structured Saddam's estimate of the Worst Case (calling it Risks) as in Fig. 11.3. In doing so we were calling upon then-current real-world factors visible at the time. During the 1990 buildup before Saddam

[2]This sense was confirmed in a 2004 unpublished review by me, Brian Jenkins, John Arquilla, Michael Egner, and Jonathan Kulick of high-level decisionmaking that drew on the memoirs of top officials, noting frequent shortcomings in the ability to look at both upside and downside potential as well as the alleged best estimate.

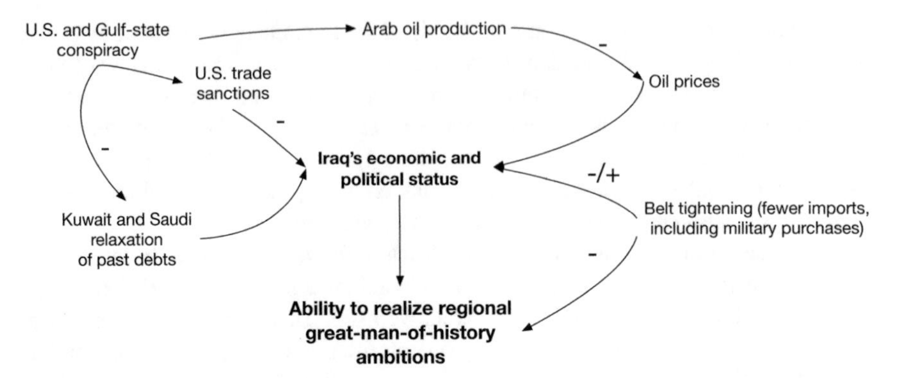

Fig. 11.2 A model of Saddam Hussein assessing the situation in 1990. *Source* Adapted from Figure G.2.1 of National Research Council and Naval Studies Board [33]

Table 11.1 Alternative models of Saddam Hussein (1990)

Attribute	Model 1	Model 2
Ruthless, power focused; emphasizes *Realpolitik*	••	••
Ambitious	••	••
"Responsive;" seeks easy opportunistic gains	••	•
Impatiently goal-seeking; likely to seek initiative	•	••
Strategically aggressive with non incremental attitudes		••
Contemptuous of other Arab leaders	•	••
Contemptuous of U.S. will and staying power		••
Financially strapped and frustrated	••	••
Capable of reversing himself strategically; flexible (not suicidal)	••	••
Clever and calculating (not a hip shooter)	••	•
Pragmatic and once burned, now cautious	••	•
Still risk taking in some situations	•	••
Grandiosely ambitious	•	••
Paranoid tendencies with some basis	•	••
Concerned about reputation and legitimacy in Arab and Islamic worlds	•	
Concerned only about being respected for his power		••
Sensitive to *potential* U.S. power not immediately present	••	•

Note Number of bullets indicated degree to which model reflects the row's attribute. The table is reconstructed from Table G.2.2 of National Research Council and Naval Studies Board [33] and earlier RAND work [35]

actually invaded, we used this structure and concluded with some alarm that deterrence was quite weak [35]. Saddam might reason (as did most U.S. experts at the time) that even if the U.S. was willing to defend Kuwait (unlikely), Saudi Arabia would not cooperate. Further, Saddam would see no warning signs to indicate actual U.S. resolve. The only political warning was minimal and ambiguous and the only military warning was a very weak military exercise, as

Table 11.2 Generic decision table

	Estimated outcome						Net assessment of option	
	Most likely Case		Best case		Worst case		Model 1	Model 2
Option	Model 1	Model 2	Model 1	Model 2	Model 1	Model 2		
1								
2								
3								
4								

Note This is a more general version of a table used in the original work [33]

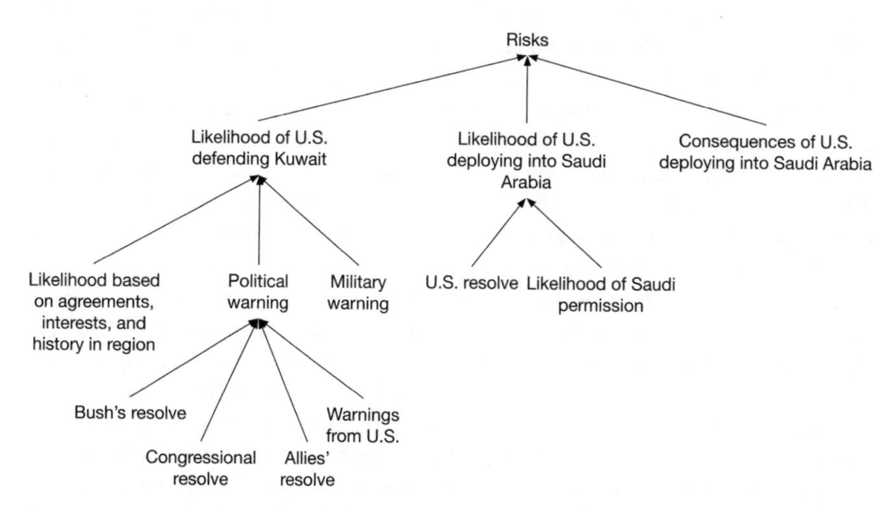

Fig. 11.3 Structure of possible Saddam reasoning about downside risks of invading Kuwait. *Source* Redrawn from Figure G.2.2 of National Research Council and Naval Studies Board [33], taken from Davis and Arquilla [35]

noted later by other authors [36, 37, 38, 39]. Saddam did in fact invade. The U.S. government was surprised because its "mental model" of Saddam was akin to that of Model 1 in Table 11.1, with no hedging against error. A primary value of cognitive modeling of the adversary is to highlight the need to hedge against misunderstanding the adversary.

11.4.2 The Roles of Personality and Culture?

When studying and modeling Saddam Hussein, we had to contend with different views on the issues of personality and culture. Some analysts at the time saw Saddam as reckless and risk-taking; others did not. Some regional experts insisted that Saddam would never do this or that because, in their view, that would be inconsistent with his culture. Hmm. *Which* culture?

Saddam was a Baathist leader in an Arab Muslim country. Which culture would influence him, and how? What seemed to us most relevant was that Saddam had gained power with a combination of violence and wile that would make the worst Mafia leaders proud. Yes, Iraq was a Arab Muslim country, but—on the scale of things—rather secular. Would he never retreat or surrender because of culture? Really? Should we imagine irrationality? Although we attempted to fold in some cultural considerations in our modeling, we concluded—in large part from his speeches and history—that the most dominant considerations were rational, but through the lens of a particular type of personality recognizable among cultures. In this we were supported by the work of Jerrold Post (previously head of a profiling unit at CIA), who developed a detailed psychiatric profile of Saddam and concluded that Saddam was best understood as a malevolent narcissist [40], a diagnosis with numerous implications. Today, we know a great deal about Saddam due to extensive interviewing and document recovery after the Gulf War in 2003. My personal reading of the evidence is that the broad cultural lens (Arab Muslim, etc.) was the wrong way to see Saddam. The rational but malignant narcissist label seems apt, when coupled with misperceptions. The second of our two Saddam models was close in many respects.

Taken together, the wars in Iraq in 1991, 2003, and subsequently, provide evidence for the role of culture. The aftermath of the 2003 invasion by the U.S. and allies saw Iraq break into a Sunni-Shia conflict that persists to this day and will probably lead eventually to a dissolution of Iraq. That history reflects many decades of tension that experts on Iraq warned about before the 2003 invasion. Even so, we should avoid the error of assuming that the dark side of cultural clashes will inevitably dominate events. There is good reason to believe that the Balkan wars of the 1990s were not inevitable due to the much discussed ancient ethnic hatreds. Rather, the vestiges of those hatreds were exploited by Slobodan Milosevic for his personal agenda [41]. Similarly, the tragedy that has befallen Iraq in the last half-dozen years was not inevitable because of Shia-Sunni history, but rather was the consequence of leaders, such as the past President of Iraq, Nouri al-Maliki, failing to rise above that history. That said, anyone who ignores the dark sides of history and culture is likely to make poor bets.

11.5 Modeling North Korean Leaders in the Context of Nonproliferation Negotiations

11.5.1 Cognitive Modeling When One Lacks Personal Detail

Another application of the simple-modeling approach was an attempt to understand what the then-new leader of North Korea, Kim Jong-Il, might do in negotiations about nuclear weapons. At the time, the United States was putting the vast weight of its negotiating capital into an attempt to get North Korea to cease and desist from nuclear development, and to reveal an dismantle prior developments. Could cognitive modeling help? Unfortunately, we were unable to obtain significant information about Kim Jong-Il personally, although much more information came out over the next decade [42, 43]. Thus, our "cognitive model" had to be based more on a combination of political science and a broad strategic understanding of the Korean-peninsula issues, than on something more personal.

We attempted to be dispassionate and to understand how the situation would be viewed by Kim Jong Il. That is, we could not understand his ideosyncracies at that point, but we could—with effort—view the strategic situation from his vantage point. We concluded that to the North Korean leader the issue was not "proliferation" but rather considerations such as suggested by the cognitive map in Fig. 11.4 [44].

After considering alternative versions of such cognitive maps, we constructed more nearly hierarchical cognitive models such as that in Fig. 11.5, which framed the thinking in terms of natural objectives for the despotic leadership of North Korea. This structure had much in common with later work described in an excellent study of proliferation issues by Stanford University's Scott Sagan [45]. The primary observation to make is that "proliferation" or "non-proliferation" is not the point when framed in this cognitive model. Instead, the model identifies objectives such as security, national power and prestige, and the Kim rulers' power and aspirations.

Despite considering alternative models, we concluded (Table 11.3) that North Korea would be very unlikely to truly give up nuclear weapons. Having such weapons would be seen as too fundamental to regime survival and deterrence of the United States. North Korea might agree to something and cheat (Option 5), but not truly give in. On many other issues, the alternative models would reach different decisions, but not on this. In this instance, it seemed to us that strategic considerations were dominant. The facts of the matter are still not clear because no authoritative inside history of North Korea has been released or is likely to be released. Some evidence exists to the effect that, for a period at least, Kim Jong Il was interested in negotiations and potentially willing to go a long way on nuclear weapons [46]. That window, if it existed, closed rather quickly and, as is now well known, North Korea has developed nuclear weapons. It clearly seems them as an important element in deterring South Korea and the United States.

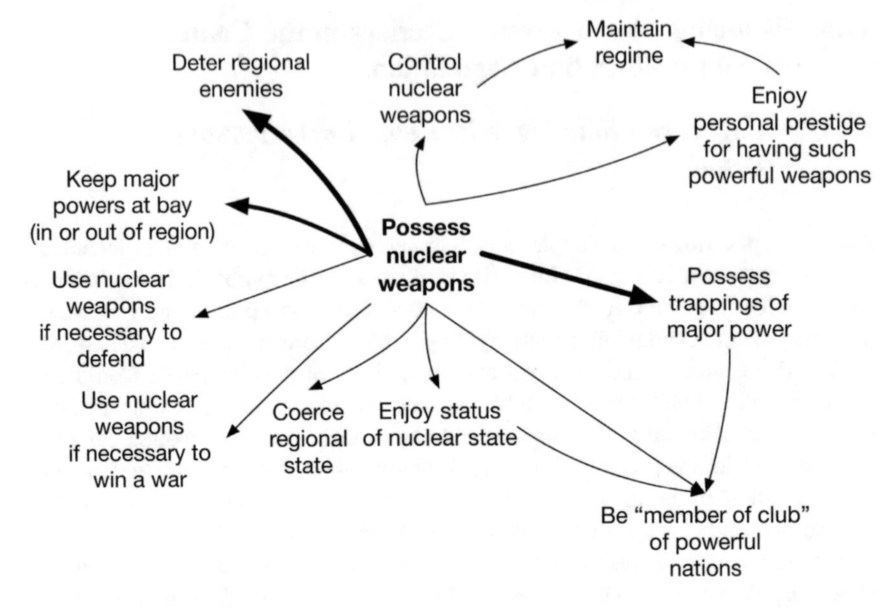

Fig. 11.4 Cognitive map of Kim Jong Il. *Source* Redrawn from Fig. 2 of Arquilla and Davis [44]

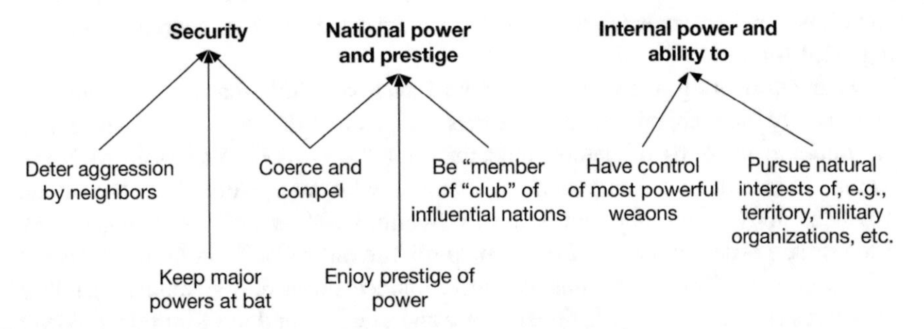

Fig. 11.5 Issues affecting decisions about nuclear proliferation. *Source* Redrawn from Fig. 1 of Arquilla and Davis [44]

11.6 Using Cognitive Models to Understand Terrorism and Public Support for Terrorism

11.6.1 Background

After the September 11, 2001 attack on New York's Trade Center and the Pentagon, RAND and the Institute for Defense Analyses were asked to run a joint study on whether a "deterrent strategy" should be part of the U.S. effort to combat al Qaeda. My colleagues and I sketched a "system view" of terrorism that allowed us to say that, while deterring Osama bin Laden was not in the cards, deterrence and

Table 11.3 Option comparison with a cognitive model of North Korea

Option	Most likely outcome	Worst-case Outcome	Best-case outcome	Net assessment of option
1. Forego nuclear weapons and WMD	Very bad	Very bad	Bad	Very bad
2. Forego nuclear weapons but develop other WMD	Bad	Very bad	Marginal	Bad
3. Sign NPT but continue modest nuclear program	Marginal	Very bad	Good	Marginal
4. Acquire nuclear weapons	Marginal	Very bad	Good	Marginal
5. Agree to forego nuclear weapons if and only if conditions are met, but cheat (covert acquisition)	Good	Very bad	Good	Good
6. Agree to forego nuclear weapons if and only if conditions are met	Bad	Very bad	Good	Marginal

Source Table 3 of [44], for the model that has objectives of long-term survival and reasonable prosperity, place in history, and eventual unification preserving power of North

other influences might be very important with respect to other elements of the terrorist system (e.g., its logisticians, financiers, theologians, and other enablers) [47]. It seemed evident that we should experiment with some cognitive modeling, but other priorities caused that to be deferred for some years.

In 2007 the Department of Defense asked RAND to review the social science that should inform counter-terrorism efforts. This was a time when many people were offering half-baked or flatly incorrect claims that terrorists were crazy, the result of poverty, or the special consequence of Islam. Our review was an opportunity to call on the rich body of literature by social scientists who had actually studied terrorism for years in the field, including my co-editor Kim Cragin [48]. The book began as an edited collection of essays on aspects of the problem (what are the root causes of terrorism, how do the terrorists become radicalized, why does the public support them, etc.). We found, however, that the collected chapters didn't cohere and that the individual chapters had structural problems. We then introduced the factor-tree methodology as described below, akin to building cognitive models, to tighten and organize discussion. The factor trees became the mechanism for internal debate and, then, for briefing our research to both academic social scientists and senior military and civilian leaders. Sometimes, simple graphics can be powerful for communication even if their information content is no greater than a corresponding essay. Beyond communication I had the ambition of moving social science discussion away from statistics (e.g., is terrorism statistically correlated with poverty?) and toward causal modeling. The "model" would be qualitative because that is the nature of the more profound knowledge of terrorism, but it would have degrees of other features that we look for in analytic work: structure, logical flow,

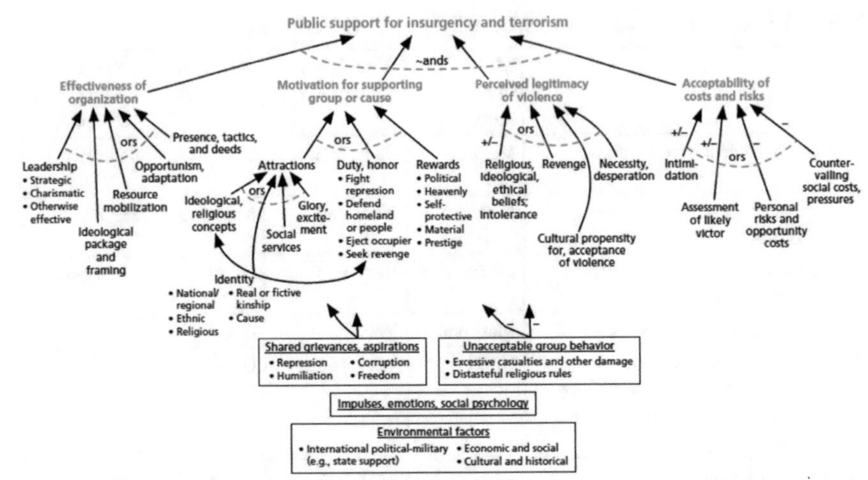

Fig. 11.6 A factor tree for public support of terrorism

defined concepts, falsifiability, and reproducibility. In my view, causal models—even if soft and imperfect—are needed to inform policy decisions.

11.6.2 The Approach of Factor Trees

Figure 11.6 illustrates a factor tree [49] for public support of terrorism. In some ways, this is like a cognitive model of the abstraction that we call "the public." Ot is merely a graphical depiction of factors and their relationships." However, the factors shown are intended to be comprehensive and based on the research base (experts are very good at identifying factors, but often not in making predictions). Also, the factor tree is a multi-resolution qualitative *causal* model. It is for a snapshot in time.

Considerable work is necessary to construct such qualitative models [49]. Fiery arguments arise about whether to include or exclude individual factors, approximating completeness is challenging without a mathematical theory to guide the process, and the same concepts can be described in different ways. Nonetheless, it was possible to generate diagrams and have them extensively peer-reviewed.

We also found ways to accomplish a limited form of validation [49]. We conducted four new case studies to see whether, in those new cases, the same factors arose. Had we left some out? Were some of the factors unique to a particular episode in history? We found that the factor tree held up well, although we learned and refined. As expected, the relative importance of factors varied with case but, as theory predicted, the relative importance of factors also varied over time. Why? As

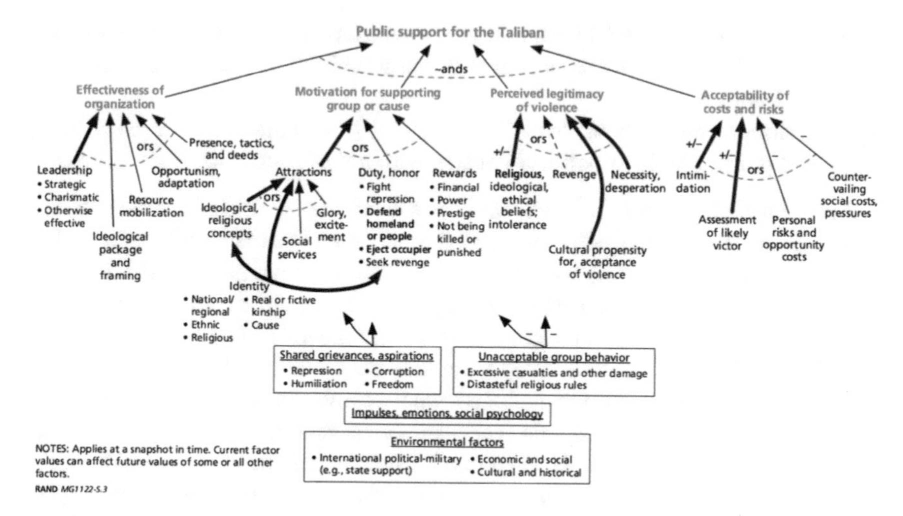

Fig. 11.7 A specialized factor tree for Afghani public support of the Taliban

my colleague Eric Larson emphasized after drawing on social-movement theory, the insurgent organization and the government are in a competition for the minds of the public: they will look for all the levers they can manipulate. Thus, if a lever has not previously been used, it may be in the future.

To illustrate the difference between generic and context-specific factor trees, Fig. 11.7 shows the factors that we found (in work by Zach Haldeman) to be important, circa 2010, for public support of the Taliban in Afghanistan. Identity, leadership, religion, and culture all played an important role (as indicated by arrow thickness), as well as intimidation and judgments about likely victor. Other actors, such as personal gain, were much less important in the period studied.

11.6.3 Reflecting Culture

As with the other examples, representing cultural issues proved both difficult and contentious, but also important. Some particular instances come to mind:

- Many American political scientists, even terrorism experts, strongly resisted the notion that religion was playing a big role in Al Qaeda's terrorism. They had been educated to believe that religion was usually just a cloak for motivations that were more broadly political or selfish (as in power-seeking). They correctly pointed to the long history of terrorism in which other, non-religious, factors were actually dominant even when religion was invoked. This view was supported by research indicating that many of those in al-Qaeda had only the most superficial knowledge of Islam or any other religion [50].

- Some other political scientists, however, characterized what was happening as a Clash of Civilizations deeply rooted in the teachings of a significant strand of Salafi Islam.
- This disagreement about "who is the enemy?" continues today, as is being played out in the 2016 political debates within the United States and Europe.

My own thinking has been affected by a tendency to think in system terms. Although modeling "al Qaeda," we needed to recognize that al Qaeda's top leaders (bin Laden and al Zawahiri) were likely quite different than many others in the al Qaeda organization. From their speeches, from al Zawahiri's writing, and from careful biographies [51], it was clear that they were indeed driven by religious ideas rooted in the Salafist tradition. They were *also* deeply affected by the history and culture of their region, which included what many Muslims have seen as the humiliating decline of the Islamic world in recent centures [52], by the colonial period, and what they saw as the clever and insidious continued colonization of the middle east through the Wests' manipulation of their "masked agents" (the despots of Saudi Arabia, Egypt, and other regional states) [53]. I found al Zawahiri's writing on such matters striking enough to assign it to graduate students. With only a modest amount of imagination, they could imagine it being appealing to impressionably young students of the Middle East, as Sayyid Qatb's writing had influenced a college-age bin Laden.

Spiritual Advisors. My colleague Eric Larson studied the writings of Muslim thinkers spiritually influential within al Qaeda, demonstrating the rigor of the discussions and their deep basis in Islamic writings. That basis is unquestionable, even though the beliefs are held by only a small portion of the Islamic community. All major religions have their dark side.

Foot Soldiers. As for spear-carriers and foot soldiers within al Qaeda, motivations vary drastically (as in the motivations for support of terrorism in Fig. 11.7). Many were reasonably depicted as just a "bunch of guys" [50], but motivations included a desire for glorious action and violence, enjoyment of the organization's camaraderie, revenge, and various others. A sense of Muslim identity was particularly important.

As of 2016, it is the Islamic State that is most prominently discussed in connection with terrorism. I have made no effort to build models of ISIL leadership, but again there would be conflicting considerations. Many within ISIL's leadership are true believers in a particular version of Salafist theology [54]. They are deadly serious about the Caliphate and their desire to adopt the features that they imagine characterized the excellent period of Islamic history in the seventh century. Their behaviors and rhetoric are consistent, even down to the level of their attitudes about beheadings, women, and sex slaves. At the same time, some in ISIL, including military leaders, are left-overs from Saddam's brutal Baathist military, hardly known for its religiosity. These leaders are probably adopting the religious mantle because it suits their purposes as they seek power. To model ISIL leader Abu Bakr al-Baghdadi, then (he even holds a doctorate in Islamic studies), is not the same as modeling other leaders or ISIL as a whole. More broadly, the expressed motivations of various ISIL participants vary widely, from thrill-seeking to religion [55], as expected from the earlier work (see motivations in Fig. 11.6).

11.7 A Computational Implementation of Factor Tree Models

When showing our factor trees, we urged the analytic community to see structuring qualitative factors as more important than doing the quantitative analysts beloved by analysts. We argued that it is possible to *understand* the reasoning of an adversary, and to identify ways to influence it, without being able to predict reliably the adversary's conclusions. Our arguments resonated, particularly with leaders having operational experience in complex campaigns.

11.7.1 Public Support of Terrorism (A Cognitive Model of "the Public")

It was with trepidation, then, that we began building a computational implementation of the factor-tree model [56]. Was this a repudiation of our earlier message? Not really. We chose public support for terrorism as our example. Turning the factor tree into a computational model brought out all the traditional challenges of modeling, and then some. We had to define the variables, indicate how to measure them (using qualitative scales), construct functional forms for their interactions, etc. The latter was especially difficult because no one claims to understand those functional forms reliably. A few "building-block" functional forms, however, went a long way in allowing us to represent the kinds of interactions that we recognized as occurring. For example, if "the public" is disaggregated into disputing factions, is the net result a shift to the view of the stronger factor or is it instead a watered-down mix? These are profound issues in social science, but these bounding cases can be treated with simple functions. It's just that we don't necessarily know which function will be more accurate. Further, we are dealing with complex adaptive systems, so the behavior of which is not predetermined.

We made uncertainty a fundamental, explicit, and difficult-to-avoid feature. We tried to preemptively eradicate the tendency to see models as predictive with some sensitivity analysis as "optional." Figure 11.8 shows an illustrative output. The items at the top are variable parameters. Thus, the user can do exploratory analysis, varying these simultaneously rather than merely doing sensitivity analysis. For the particular settings shown, and looking at the rightmost bar, public support is 9 (very high) if motivation for the cause is very high (horizontal axis), intimidation by the insurgent group is very high (the "key variable" distinguished by color of bars),

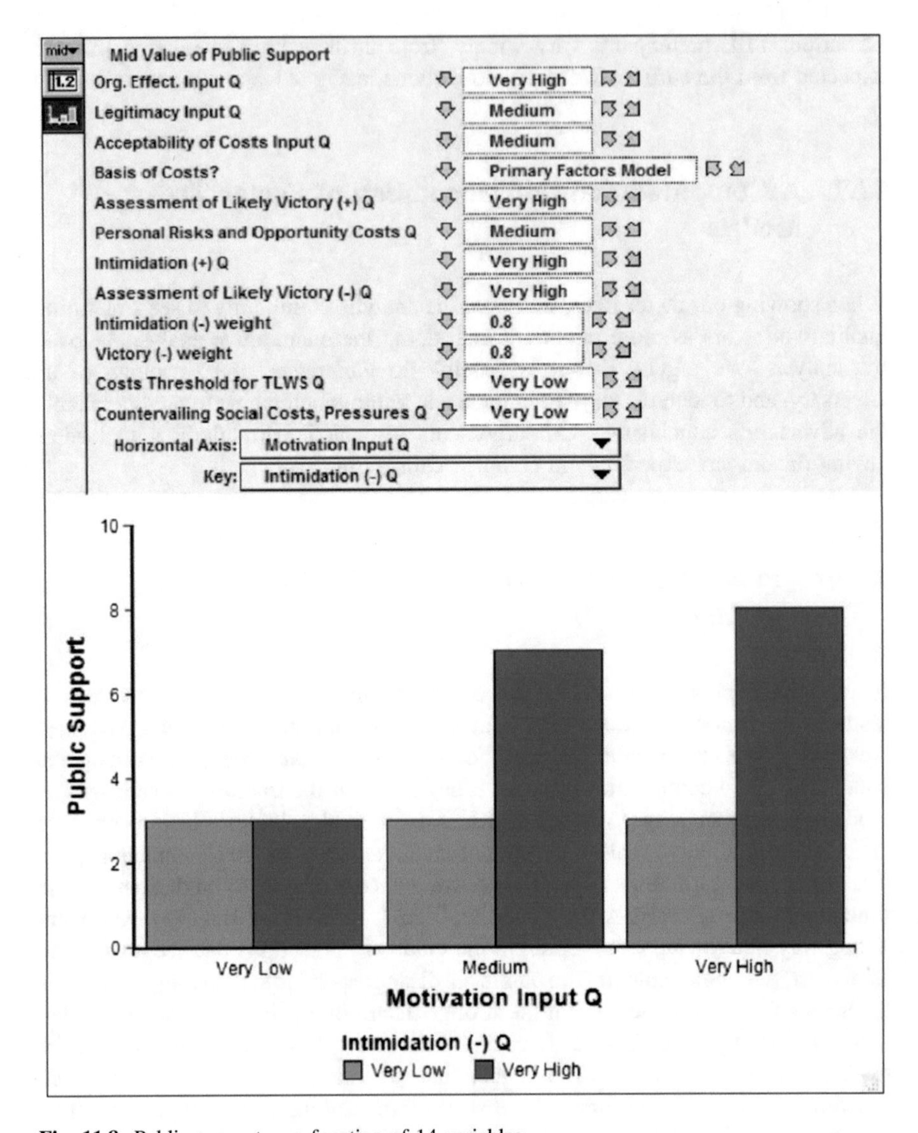

Fig. 11.8 Public support as a function of 14 variables

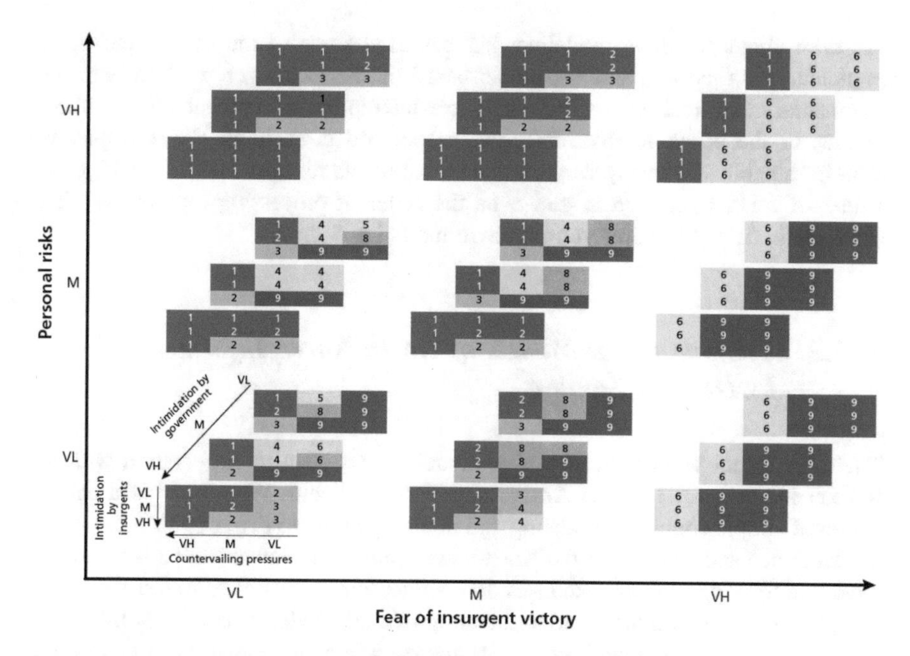

Fig. 11.9 Public support as a function of 5 variables. *Note* The numbers 1, 3, 5, 7, 9 correspond to very low (VL), low (L), medium (M), high (H), and Very High (VH). The ratings are from the perspective of the counterinsurgent side. Thus, 9 (red) is very adverse for it, but very good for the insurgency

and—shifting to the list of variables at the top of the screen, the insurgency's organization effectiveness is very high, etc.[3]

Figure 11.9 illustrates how effects of a great deal of uncertainty-related analysis can be represented [56]. The figure shows the extent of public support (indicated by cell number or color) as a function of five variables. Although briefing such a display must be done slowly and in layers, audiences can understand the results and appreciate just how many what-if questions are being addressed in one fell swoop.

11.7.2 Extending Uncertainty Analysis in Simple Computational Social-Science Models

Very recently, Walter Perry, John Hollywood, David Manheim and I have studied concepts and methods for heterogeneous information fusion in the context of detecting terrorist threats (and also exonerating those falsely suspected). This has

[3]The model was developed in *Analytica*, sold by *Lumina*. It was originally developed at Carnegie Mellon University and features visual programming and "smart arrays," which are powerful for uncertainty analysis.

not been about cognitive modeling, but has used some of the same methods. In particular, we used a qualitative model based on the factor-tree work to help fuse fragmentary information. In that study we attempted to confront *all* the uncertainties of the problem, structural, parametric, and procedural. For example, we built in the ability to easily vary the functional forms for how factors combine, the values of related parameters, and even the order of processing reports [57]. The same methods would apply for cognitive modeling.

11.7.3 Computational Models of North Korea Informed by Human Gaming

The most recent work with cognitive modeling has been in cooperation with the Korean Institute for Defense Analyses (KIDA) in Seoul. We are using a combination of simple cognitive modeling and human gaming to better understand issues of deterrence and stability in the Korean peninsula. A major objective is to use the modeling to design human exercises, to conduct human exercises to test the model and point out shortcomings, and to then improve the models accordingly [58]. This is by analogy with the model-test-model approach in many domains of research and development.

Personality and culture again matter. With respect to culture, what matters is less "Korean" culture than the "Kim Dynasty culture" in which the ruling Kim despot is treated as God-like and people are barraged throughout life with propaganda about the wonders of the North Korean system and the magnificence of their leaders [59, 60]. Even if we accept the fundamental importance of that culture, would Kim Jong Un behave in crisis stereotypically and fatalistically, or would he reason in a way that an economist would regard as rational, although driven by a "bad" utility function prizing the ability to remain in power? The jury is out, but—based on experience with his father Kim Jong Il and Kim Jong Un's behavior to date—it seems that Kim Jong Un is best understood as violent, ruthless, picking up in the footsteps of his father, but also rational.

An interesting question is whether the artificial environment of the Kim Dynasty will degrade his rationality over time. After all, in such a despotic culture, we would hardly expect advisors to tell him things he doesn't want to hear or to question his beliefs. Further, at some point, a person that has been treated like a God might come to believe some of his own propaganda. Would that potentially mean that he would value a "glorious" death to something that would spare his people?

The spectre of an adversary seeking a "glorious" death has some basis. Adolph Hitler went through a period when he seemed strongly to embrace the concept of the glorious death. In discussing the possibility of dying in battle, he expressed the view that his death would be inspirational:

> We shall not capitulate…no, never. We may be destroyed, but if we are, we shall drag a world with us…a world in flames…we should drag half the world into destruction with us and leave no one to triumph over Germany. There will not be another 1918 (Langer, 1943)

In the end, Hitler just committed suicide, but what if he had had nuclear weapons?

As a final example to remind us of how important non-rational considerations can be, including those due to culture, consider the Japanese decision to attack Pearl Harbor. This was a decision made with the *knowledge* that Japan could not plausibly win a long war with the United States. Recent scholarship based on unprecedented access to Japanese primary materials reinforces the conclusion that the Japanese decision was strongly affected by considerations as (1) avoiding shame, (2) a sense of persecution and wounded pride, (3) a belief in the "Yamoto spirit" referring to a perceived Japanese trait of being unique, resilient, disciplined and hard-working, (4) a willingness to gamble based in part on Japan's success in its 1904–1905 war with Russia, (5) *hope* that the United States, if badly bruised would quickly tire and sue for peace, and (7) a sense of desperation because the U.S.-British embargos were undercutting their rightful ability to expand their empire [58, 61]. Could similar ideas influence a Kim-Dynasty leader at the time of some future crisis?

11.8 Conclusions and Suggestions for Research

From this paper's research involving simple cognitive models and related methods, it seems that the most important points for the present volume are these:

- Cognitive modeling should be undertaken with humility. History is replete with examples of failures to understand the adversary. A more fruitful approach is constructing *alternative* models rather than reflecting only the current best estimate, which is often wrong. Even having two well-chosen alternatives can highlight uncertainties and possibilities, improving the ability to construct an appropriately hedged and adaptive strategy. It is necessary, however, for the alternatives to be taken seriously, rather than as the best estimate plus a token variant.
- The models should have a structure *allowing* for a version of rational-actor decision making: multiple objectives, multiple criteria for their evaluation, a range of options, option comparison by the recognized criteria, evaluation of options based on best-estimate, best-case, and worst-case outcomes, and a net-assessment calculation.
- In estimating each of the elements of that model, however, the analyst should consider alternative models, and their *perceptions* of the elements as affected by

information, cultural and personal biases, and situationally dependent factors such as desperation and related risk-taking propensity. Also, the net-assessment calculation should be model-dependent, and may need to be nonlinear (heuristics introduce edges amounting to nonlinearities).

- Both perceptions and net-assessment logic may be affected by emotional considerations, to fear, desperation, and culturally fueled hatred.
- Such estimates should be guided first by *qualitative* constructs in tune with actual human psychology rather than such analyst abstractions as narrow cost-effectiveness. Factor trees are especially useful for such purposes.
- Computational models should be avoided *except* when simple and routinely uncertainty-sensitive. Enforcing a shift to uncertainty-sensitive models is difficult because of ingrained habits and limitations of common modeling and programming technology. With more appropriate methods and technology, routine exploratory analysis under uncertainty can be straightforward. Modern methods, such as those in data mining, can be valuable in inferring conclusions that are relatively robust to assumptions [34].

References

1. Epstein, J.M., Axtell, R.L.: Growing Artificial Societies: Social Science From the Bottom Up, MIT Press, Cambridge, MA (1996)
2. Carley, K.M.: Computational organization science: a new frontier. PNAS **99**(3), 7257–7262 (2002)
3. Chatuverdi, A.R., Dolk, D.R.: Design principles for virtual worlds. MIS Q. **35**(3), 673–684 (2011)
4. North, M.J., Macal, C.M.: Managing Business Complexity: Discovering Strategic Solutions With Agent-Based Modeling and Simulation. Oxford University Press, USA (2007)
5. Kahneman, D.: Maps of Bounded Rationality: A Perspective on Intuitive Judgment and Choice (*Nobel Prize Lecture*) (2002)
6. Kahneman, D., Slovic, P., Tversky, A.: Judgment Under Uncertainty: Heuristics and Biases. Cambridge University Press, New York (1982)
7. Krugman, P.: How Did Economists Get it So Wrong. New York Times Magazine, MM36. Accessed at http://www.nytimes.com/2009/09/06/magazine/06Economic-t.html?pagewanted=all (2009)
8. Thaler, R.H., Mullainathan, S.: How behavioral economics differs from traditional economics. In: Henderson, D.R. (Ed.) The Concise Encyclopedia of Economics (2nd Edn.), Liberty Fund, unspecified (2005)
9. Simon, H.A.: Rational decision-making in business organizations: Nobel Prize lecture. In: Lindbeck, A. (ed.) Nobel Lectures, Economics, 1969–1980. World Scientific Publishing Co, Singapore (1978)
10. Friedman, M.: Essays in positive economics, books.google.com (1953)
11. Gigerenzer, G., Selten, R.: Bounded Rationality: The Adaptive Toolbox. MIT Press, Cambridge, MA (2002)
12. Klein, G.: Sources of Power: How People Make Decisions. MIT Press, Cambridge, MA (1998)
13. Klein, G.: The fiction of optimization. In: Gigerenzer, G., Selten, R. (eds.) Bounded Rationality: The Adaptive Tookit, pp. 103–121. MIT Press, Cambridge, MA (2001)

14. Gladwell, M.: Blink: The Power of Thinking Without Thinking. Little, Brown, MA (2005)
15. Davis, P.K., Kulick, J., Egner, M.: Implications of Modern Decision Science for Military Decision Support Systems. RAND Corp, Santa Monica, CA (2005)
16. Kahneman, D.: Thinking, Fast and Slow, Farrar, Straus and Giroux, New York (2011)
17. Mintzberg, H.: Rise and Fall of Strategic Planning, Free Press, New York (1994)
18. Howard, N.: The M&A play: using drama theory for mergers and acquisitions. In: Rosenhead, J., Mingers, J. (eds.) Rational Analysis for a Problematic World Revisited: Problem Structuring Methods for Complexity, Uncertainty, and Conflict, pp. 249–265. Wiley, Chichester, England (2001)
19. Davis, P.K., Winnefeld, J.A.: The RAND Corp. Strategy Assessment Center, RAND Corp, Santa Monica, CA (1983)
20. Davis, P.K.: Some Lessons Learned From Building Red Agents in the RAND Strategy Assessment System. RAND Corp, Santa Monica, CA (1989)
21. Davis, P.K., Bankes, S.C., Kahan, James P.: A New Methodology for Modeling National Command Level Decisionmaking in War Games and Simulations. RAND Corp, Santa Monica, CA (1986)
22. National Research Council: U.S. Air Force Strategic Deterrence Analytic Capabilities: An Assessment of Methods, Tools, and Approaches for the 21st Century Security Environment, National Academies Press, Washington, D.C. (2014)
23. Schank, R.C., Abelson, R.P.: Scripts, Plans, Goals, and Undersanding: an Inquiry into Human Knowledge Structures (1977)
24. Steeb, R., Gillogly, J.: Design for an Advanced Red Agent for the RAND Strategy Assessment Center. RAND Corp, Santa Monica, CA (1983)
25. Pipes, R.: Why the Soviet union thinks it could fight and win a Nuclear War. Commentary (1977)
26. Wohlstetter, A.: The political and military aims of offense and defense innovation. In: Hoffman, F.S., Wohlstetter, A., Yost, D.S. (Eds.) Swords and Shields: NATO, the Ussr, and New Choices for Long-Range Offense and Defense. Lexington Books of D. C. Heath and Company, Lexington, MA (1987)
27. Garthoff, R.L.: Mutual deterrence and strategic arms limitation in Soviet policy. Int. Secur. **3**(1), 112–147 (1978)
28. Khrushchev, N.S., Talbot, S (Translator and editor): Khrushchev Remembers, Little Brown & Company, MA (1970)
29. Musgrave Plantation: SALT II and the Growth of Mistrust, Conference Tw of the Carter-Brezhnev Project, Carter-Brezhnev Project (1994)
30. Bracken, P.: The Second Nuclear Age: Strategy, Danger, and the New Power Politics. Times Books, New York (2012)
31. Reagan, R.: An American Life. Simon & Schuster, New York (1990)
32. Davis, P.K.: Studying First-Strike Stability With Knowledge-Based Models of Human Decision Making. RAND Corp, Santa Monica, CA (1989)
33. National Research Council and Naval Studies Board: Post-Cold War Conflict Deterrence. National Academy Press, Washington, D.C. (1996)
34. Davis, P.K.: Some Lessons From RAND's Work on Planning Under Uncertainty for National Security. RAND Corp, Santa Monica CA (2012)
35. Davis, P.K., Arquilla, J.: Deterring or Coercing Opponents in Crisis: Lessons From the War With Saddam Hussein. RAND Corp, Santa Monica, CA (1991)
36. Marla, E.J.: The United States Navy and the Persian Gulf, http://www.history.navy.mil/research/library/online-reading-room/title-list-alphabetically/u/the-united-states-navy-and-the-persian-gulf.html (last accessed) (2015)
37. Mearsheimer, J., Stephen M.: Can Saddam Hussein Be Contained? History Says Yes. Belfer Center for Science and International Affairs, Harvard Kennedy School, Cambridge, MA (2002)
38. Stein, J.G.: Deterrence and compellence in the Gulf, 1990–1991: a failed or impossible task. Int. Secur. **17**(2), 147–179 (1992)

39. Walt, S.M.: WikiLeaks, April Glaspie, and Saddam Hussein. Foreign Policy, January 9 (2011)
40. Post, J.M. (ed.): The Psychological Assessment of Political Leaders. University of Michigan Press, Ann Arbor, MI (2008)
41. Zimmerman, W.: The Last Ambassador: A Memoir of the Collapse of Yugoslavia. Foreign Affairs, March/April (1995)
42. Lankov, A.: The Real North Korea: Life and Politics in the Failed Stalinist Utopia. Oxford University Press, New York (2013)
43. Oh, K., Hassig, R.C.: North Korea Through the Looking Glass. Brookings, Washington, D.C. (2000)
44. Arquilla, J., Davis, Paul K.: Modeling Decisionmaking of Potential Proliferators as Part of Developing Counterproliferation Strategies. RAND Corp, Santa Monica, CA (1994)
45. Sagan, S.D.: Why do states build nuclear weapons? three models in search of a bomb. Int. Secur. **21**(3) (1996)
46. Chinoy, M.: Meltdown: The Inside Story of the North Korean Nuclear Crisis. St. Martin's Press, New York (2008)
47. Davis, P.K., Jenkins, B.M.: Deterrence and Influence in Counterterrorism: A Component in the War on Al Qaeda. RAND Corp, Santa Monica, CA (2002)
48. Davis, P.K., Cragin, K. (eds.): Social Science for Counterterrorism: Putting the Pieces Together, Santa Monica. RAND Corp, Santa Monica, CA (2009)
49. Davis, P.K., et al.: Understanding and Influencing Public Support for Insurgency and Terrorism. RAND Corp, Santa Monica, CA (2012)
50. Sageman, M.: Leaderless Jihad: Terror Networks in the Twenty-First Century. University of Pennsylvania Press, Philadelphia (2008)
51. Wright, L.: The Looming Tower: Al Qaeda and the Road to 9/11 (Vintage). Knopf, New York (2006)
52. Lewis, B.: What Went Wrong? Western Impact and Middle Eastern Response. Oxford University Press, New York (2002)
53. al-Zawahiri, A.: Knights Under the Prophet's Banner, London: FBIS-NES-2002-0108 (2001)
54. Wood, G.: What ISIS Really Wants. The Atlantic, March (2015)
55. Quantum: Understanding Jihadists: In Their Own Words. April 7 (2015)
56. Davis, P.K., O'Mahony, A.: A Computational Model of Public Support for Insurgency and Terrorism: a Prototype for More General Social-Science Modeling. RAND Corp, Santa Monica, CA (2013)
57. Davis, P.K., et al.: Uncertainty-Sensitive Heterogeneous Information Fusion: Assessing Threat With Soft, Uncertain, and Conflicting Evidence. RAND Corp, Santa Monica, CA (2016)
58. Davis, P.K., et al.: Deterrence and stability for the Korean Peninsula. Korean J. Defense Anal. (TBD) (forthcoming)
59. Bowden, M.: Understanding Kim Jong Un, the world's most enigmatic and unpredictable dictator. *Vanity Fair*, March (2015)
60. Oh, K., (Katy): The World of North Korea's Kim Jong-UN, Global Experts: Analysis on Demand, April 4. Accessed at http://www.theglobalexperts.org/comment-analysis/world-north-koreas-kim-jongun (2013)
61. Hotta, E.: Japan 1941: Countdown to Infamy. Alfred A. Knopf, Toronto (2013)
62. Davis, P.K., Arquilla, J.: Thinking About Opponent Behavior in Crisis and Conflict: A Generic Model for Analysis and Group Discussion. RAND Corp, Santa Monica, CA (1991)

Author Biography

Paul K. Davis is a principal researcher at the RAND Corporation and a Professor of Policy Analysis in the Pardee RAND Graduate School. He majored in chemistry (B.S., University of Michigan) before specializing in theoretical chemical physics (Ph.D., MIT). He worked initially in systems analysis and the science of observing rockets (Institute for Defense Analyses) and then moved into the U.S. government, where he worked on strategic nuclear forces, strategic nuclear arms control, and international security strategy. He then moved to the RAND Corporation where he has pursued a number of research streams over the years: strategic planning and resource allocation, deterrence theory, advanced concepts for modeling and simulation, structuring social science knowledge relevant to counterterrorism, and heterogeneous information fusion. He teaches a graduate course in policy analysis for complex problems. Dr. Davis serves on the editorial board or reviews for numerous professional journals and has served on numerous national panels for the National Academy of Sciences abd Defense Science Board.

Part II
Cultural Neuroscience

Chapter 12
Cultural Neuroscience

R. Thora Bjornsdottir and Nicholas O. Rule

Abstract Recently, the fields of cultural psychology and cognitive neuroscience have converged to form the research domain of cultural neuroscience. In this chapter, we provide an overview of the research in this burgeoning field and outline the history of the field and its origins. This specific field encompasses a wide variety of research and provides a unique lens through which to study cultural differences. Notably, research in this field has provided evidence of subtle and nuanced differences across cultures where behavioral evidence alone could not, demonstrating the importance of the neuroscientific approach. The primary focus of the chapter is to review work on the most-studied topics within cultural neuroscience: logical processing, auditory and visual perception, and social cognition. This research illustrates how culture affects how people perceive and interact with the world and the those around them, showing convergent evidence from both behavior and neuroimaging. Overall, cultural neuroscience uniquely improves understanding of cultural differences. We discuss how this discipline can inform programs aiming to promote cultural understanding and effective cross-cultural communication.

Keywords Culture · Neuroscience · Language · Attention · Music
Emotion · Person perception · Theory of mind · Self

12.1 Introduction

Cultural psychology and cognitive neuroscience have recently converged to produce the new field of cultural neuroscience. This specific subdomain encompasses a wide variety of research topics, ranging from simple visual processing to more

R.T. Bjornsdottir · N.O. Rule (✉)
Department of Psychology, University of Toronto, 100 St. George Street,
Toronto, ON M5S 3G3, Canada
e-mail: rule@psych.utoronto.ca

R.T. Bjornsdottir
e-mail: thora.bjornsdottir@mail.utoronto.ca

C. Faucher (ed.), *Advances in Culturally-Aware Intelligent Systems and in Cross-Cultural Psychological Studies*, Intelligent Systems Reference Library 134, https://doi.org/10.1007/978-3-319-67024-9_12

abstract and high-level subjects, such as person perception. The focus, critically remains on cultural differences, however, and cultural neuroscience provides a unique lens through which to study these distinctions. Notably, research in this field has yielded evidence for subtle and nuanced cultural variation, often where behavioral evidence alone could not, demonstrating the importance of a neuroscientific approach. By integrating data from studies of the brain with questions from cultural psychology, cultural neuroscience uniquely contributes to understanding cultural differences, thereby providing helpful tools to promote cultural understanding and effective cross-cultural communication.

In this chapter, we present an overview of the research in this burgeoning field and outline its history and origins by focusing on the most-studied topics within cultural neuroscience: auditory perception (including music and language), visual perception, and social cognition. In doing so, we aim to illustrate how culture affects how people perceive and interact with others through convergent and combined behavioral and neurological evidence. Finally, we discuss the ways in which this evidence, and cultural neuroscience as a discipline, can inform programs aiming to increase intercultural understanding.

12.2 The Origins of Cultural Neuroscience

Before examining the field of cultural neuroscience, specifically, we must first consider its predecessors: cultural psychology and cognitive neuroscience. Each of these fields emerged independently in the early 1990s, bringing with them novel ways to consider human thought and behavior. Cultural psychology first began as an interdisciplinary subfield of social psychology, inspiring decades of work that captured differences between cultural groups. Yet it may have only really come to the foreground of psychological research following publication of the highly influential review paper by Markus and Kitayama [1]. They proposed integrating and applying questions about cultural differences with questions in social cognition, encouraging the discovery of much of what is presently known about cultural variability in processing social information. As one of the most-cited publications in psychology, it remains incredibly influential across the behavioral sciences (and undeniably so within cultural psychology).

Cognitive neuroscience also had its beginnings during this time, starting with advances in functional magnetic resonance imaging (fMRI) technology in the early 1990s [2]. The emergence of fMRI technology allowed for a new way of studying the brain through the tracking of cerebral blood flow. Psychology researchers as well as their university departments embraced this novel approach, and studies employing it soon emerged (e.g., [3, 4]). Today, the number of brain imaging studies (using fMRI as well as other methods, such as electroencephalography, or EEG) is ever-increasing, demonstrating the continued popularity of cognitive neuroscience as a way to understand the human brain. Like cultural psychology, cognitive neuroscience is a highly interdisciplinary field and its impact extends beyond psychology to other natural sciences [5]. Cognitive neuroscience also sprouted a number of subfields—for example, social neuroscience. This particular subfield applies imaging techniques to answer questions about social thought and behavior in humans (see [6]). From this, even more specific subcategories emerged—including cultural neuroscience, a subfield now so influential that it boasts a number of books and special issues in scientific journals (e.g., [7–9]).

In brief, cultural neuroscience employs the tools of cognitive neuroscience to understand cultural differences in thought and behavior, thus addressing the questions of cultural psychology from the perspective of the brain. Like its parent fields, cultural neuroscience is very interdisciplinary in nature. It is thus connected to a broad number of domains but retains a narrow focus on culture, specifically. Additionally, the study of cultural neuroscience encompasses varying levels of analysis and points of focus, ranging from the study of basic cognitions (e.g., [10]) to more abstract social phenomena (e.g., [11]). The field has also gone so far as to expand beyond the brain to include the study of genetics (e.g., [12]).

As a field, cultural neuroscience has its challenges and limitations. As with all neuroscience, brain activity is only suggestive; one cannot oversimplify and conclude that activity in a particular brain region equates to that area of tissue being wholly responsible for a specific phenomenon (see [13]). There are also technological restrictions. Most notably, variations across fMRI machines and their calibrations mean that the technology used must be kept as consistent as possible within and across study sites so as to avoid confounds—on top of the more pedestrian challenges of cross-cultural work, such as assuring that study materials are properly translated across languages and cultures [14–16]. Furthermore, there are limits to the generalizability of cultural neuroscience findings. First, studies employing fMRI are only conducted in nations and institutions wealthy enough to afford such expensive equipment, limiting the representativeness of the data for the more general population and for the myriad cultures in which fMRI technology is unavailable (see [17] as well as [18]). Moreover, the expense of running such studies can fiercely constrain sample sizes, reducing the reliability of the findings [19].

Despite these restrictions, cultural neuroscience boasts a variety of features that allow it to uniquely contribute to understanding human behavior and thought. Most simply, the field provides insight to how culture influences the brain, and, conversely, the role that the brain may play in building and maintaining culture [20]. Additionally, it is an approach that integrates a wide variety of methods, allowing for a broader exploration and fuller understanding of cultural differences [20, 21]. Perhaps most interesting, behavior may sometimes appear the same across cultures but differ in its neural representation in each culture (e.g., [22]). Cultural neuroscience is thus poised to provide a unique contribution to the study of human thought and behavior. Here, we review some of the contributions to auditory processing, visual perception, and social cognition to date, with the aim of demonstrating how the merging of these two eclectic fields may afford new understanding of the processes underlying neural function and their sundry manifestations in people's behavior across cultures.

12.3 Auditory Perception and Logical Processing

12.3.1 Language

Few concepts relate to culture as intimately as language. The languages that people speak are emblems, markers, and often dividers of who they are as groups of people. It is no surprise, then, that variation in language correlates strongly with variation in culture, tying them to each other and to the histories and ideas that they uniquely share [23]. Of course, there are instances in which language transcends cultural lines. For example, people in the United States, Australia, Canada, the United Kingdom, and South Africa all speak English. Similarly, Germans, Austrians, and the Swiss speak German. But, importantly, variations of the same language, such as dialects, often mark cultural differences—German-speakers who speak *Plattdeutsch* (found in parts of northern Germany) differ culturally from those who speak *Schwyzerdütsch* (common in Switzerland). Critically, the relationship between language and culture is therefore not merely correlational: the two reciprocally influence one another, demonstrating how language and culture intertwine. For instance, culture introduces new phrases [23] and language provides the terms needed to express cultural concepts [24, 25].

Culture also impacts how one's brain processes language. Notably, the brain's response to different types of written language varies. Ideographic languages, such as Chinese, are based on symbols that historically began as pictures, whereas phonographic languages, such as English, are based on symbols representing the sounds that make up speech [26]. These different forms of orthography influence how a reader processes written words. For Chinese readers, the visual word-form area of the brain is more active than in Western readers—a difference that can be attributed to the nature of readers' written languages [27].

Culture not only affects the processing of writing, but of spoken language as well. For example, one's age of language acquisition, degree of language mastery, and amount of language exposure promote distinct activations in bilingual individuals' brains when processing their second language [28]. For example, whereas participants who had been bilingual since birth showed no activation differences during grammatical judgment tasks in their first and second languages, late-acquisition bilinguals demonstrated greater activity in Broca's area and subcortical structures (language-related regions of the brain) for second- (vs. first-) language grammatical processing [29]. Such variability aside, there is also some universality in speech processing such that similar brain areas become active when hearing phrases in one's native language versus a new or unfamiliar language [30].

The interplay between culture and language goes beyond the processing of words, extending to abstract concepts, such as time. For example, in English, time is described as if it were horizontal, whereas it is characterized as vertical in Mandarin [31]. This subsequently affects how the speakers of these languages think chronologically. Interestingly, bilinguals' concept of time can depend on the age at which they acquired their second language, demonstrating that language powerfully shapes thoughts about abstract concepts but remains malleable [31].

Another abstract concept influenced by culture and language is mathematics. The behavioral differences across cultures in this domain are well-documented: East Asians tend to outperform Westerners in math [32]. However, there are also pronounced differences in neural activation. Specifically, when completing mathematical tasks, English speakers show more activation in linguistic areas (such as Wernicke's and Broca's areas), but Chinese speakers demonstrate activation in areas associated with visuospatial processing [10]. These activation differences may be due to differences in the written languages in which math is learned—that is, a sound-based phonological writing system versus a visuospatial ideographic writing system (see [33–35]). This, in turn, may explain differences in performance: mathematics may simply be easier to process for speakers whose written languages are visuospatial, as mathematical performance is highly related to visuospatial working memory [36].

12.3.2 Music

Like language, music can be communicated through both sound and writing; and the two accordingly evoke similar brain responses [37]. As with language, music moreover has both universals and cultural variations. For instance, music elicits similar emotions across cultures, such that a sad song from one culture will sound sad to a perceiver from another culture [38, 39]. And although culture tunes one's sensitivity to particular rhythms, pitches, and meters through exposure, infants are sensitive to these elements across cultural boundaries [40].

Culture can create an environment in which one is exposed to particular kinds of music. Thus, early exposure to specific types of music leads to the acquisition of culture-specific musical knowledge—similar to language [41]. For example, as pitch varies across culture, perceivers show an advantage in perceiving mistuned notes from their own culture, with which they are more familiar ([42], see also [43]). Furthermore, musicians show distinct event-related potential (ERP) waveforms (i.e., electrical signals from the accumulated firing of neurons in the brain, recorded using EEG) in response to tones that would generally be unexpected in their own culture's music systems, but not in response to combinations of tones that would be unexpected within culturally unfamiliar music systems [44]. Similarly, perceivers' ERP signals for expectancy violations (that is, deviations from a melody's expected pattern) as well as melody congruence showed an own-culture advantage, such that participants were more sensitive to deviations in melodies from their own culture, as opposed to those from an unfamiliar culture [45]. Culture furthermore affects ERP responses for recognizing the boundaries of musical phrase [46, 47]. One possible explanation for this may be that rhythmic groupings in music tend to correspond to a culture's language accent patterns [48]. For example, the rhythms in English and French classical music parallel the prosody of each of these spoken languages [49]. Familiarity with these prosody patterns could thus extend to more accurate identification of musical phrase boundaries.

12.4 Visual Perception

12.4.1 Visual Focus

Behavioral studies have repeatedly demonstrated cultural differences in visual focus. Namely, when presented with an object within a scene, people from Eastern cultures typically attend to the context in which the object appears, whereas Westerners focus solely on the object itself [50]. Thus, Western perceivers separate objects from their contexts quite easily, but Easterners tend to perceive scenes in a holistic manner. This difference in perception parallels cultural differences in individualism and collectivism. That is, individualistic cultures consider individuals as separate from others, whereas collectivistic cultures consider others as part of an individual's context—analogous to how these cultures process objects and scenes.

These cultural differences in visual focus are not only evident in behavior, but also in neural activity. One study presented East Asian and American participants with images of objects and scenes and recorded the participants' brain activity using fMRI [22]. The results showed that, when viewing the stimuli, American participants showed greater activation in areas associated with object processing, including the bilateral middle temporal gyrus, left superior parietal/angular gyrus, and right superior temporal/supramarginal gyrus, than East Asian participants did. These results are consistent with previous behavioral evidence demonstrating Westerners' greater focus on individual objects.

Another study used European-Americans' and East Asians' brain responses during a line judgment task to evaluate the role of culture in visual perception [51]. The line judgment task, adapted from Witkin and Goodenough's [52] rod and frame task, asks participants to judge absolute and relative length, thereby assessing their focus on object and context. Specifically, processing individual features advantages absolute length judgments, whereas judgments of relative length involve holistic processing. East Asian participants demonstrated stronger activation of the dorsolateral prefrontal cortex (dlPFC)—an area implicated in cognitive control—when making judgments of absolute line length compared to relative line length, suggesting that absolute length judgments may have been more difficult than relative judgments for East Asian participants. Importantly, this difference in dlPFC activation decreased with East Asians' degree of acculturation to the United States (that is, the extent of their identification with American culture). Similarly, other researchers have found that age modulates cultural differences in object-scene processing [53]. Together, these results suggest that cultural exposure may attenuate cultural differences in perception.

12.4.2 Attention

Another oft-studied aspect of visual perception is attention—and this aspect, too, is affected by culture. One study recorded ERP signals while European-American and Asian-American participants tried to detect a particular target image among other images in a series of experimental trials [54]. Although the two groups did not differ in their performance on the task, European-Americans showed a larger N2 ERP waveform (indicative of early orienting of attention and target discrimination) after 200 ms, as well as a slow wave (signaling deliberate attention and elaborative processing) around 700 ms, compared to Asian-Americans. Together, these differences suggest that European-Americans allocated more attention to the target object from the start of stimulus processing than Asian-Americans did.

A further investigation of culture's influence on attention examined the interaction between culture and genes [55]. Korean and European-American participants were genotyped and asked to report their chronic locus of attention (that is, their tendency to attend to a central object or its context). Consistent with previous research, Korean participants paid more attention to context than European-Americans did. Interestingly, however, this cultural difference was moderated by a specific serotonin receptor called 5-HTR1A. Participants possessing the type of allele associated with reduced adaptability were more likely to exhibit a culturally-stereotypical locus of attention, demonstrating an interaction between culture and genetic makeup.

12.4.3 Person Perception

Finally, culture's role in visual attention and focus goes beyond just the perception of objects, extending to perceptions of the most salient and interesting stimuli that humans perceive: other people. For instance, culture influences how one visually processes faces. In a study comparing Caucasian-American and Japanese participants' face processing, Japanese participants processed the configuration of the faces' features more than Caucasian-American participants did [56]. Notably, Japanese participants tended to identify images representing the prototypes of groups of faces by using resemblance (i.e., choosing a morphed combination of the faces) rather than matching individual features (i.e., choosing a mixture of individual features puzzled together) and showed greater sensitivity to changes in the spatial configuration of the eyes and mouth within the face than Americans did. These results suggest that the Japanese participants processed the faces in a more holistic manner than the Americans did, considering the configuration (that is, the context) of the entire face rather than parsing it into its individual features (separate from the context), paralleling the similar (holistic vs. analytic) cultural differences in the visual processing of objects. Although configural processing is generally important for discriminating between faces [57], Miyamoto et al. [56] have suggested that it may be of particular importance in interdependent cultures in which cooperation with others is paramount, and fine-tuned discrimination between individuals may therefore be more important. Furthermore, these results demonstrate culture's robust effect on cognition, as it affects the perception of even the most complex and important visual stimuli (i.e., people). Moreover, Asians living in the United States showed an intermediate response, thereby additionally suggesting that the effect of culture on perception is malleable.

12.5 Social Cognition

12.5.1 Impression Formation

Cultural differences in how we perceive people extend beyond processing their facial features to judgments and inferences about their thoughts and social behaviors as well. Although there is cross-cultural consensus in perceptions of others (for example, in personality judgments made about strangers [58]), culture can affect how one interprets and uses those impressions. For instance, Zebrowitz, Montepare, and Lee [59] found that White American, Black American, and Korean raters showed high agreement in their ratings of targets' attractiveness but differed in how they associated these judgments with impressions of how interpersonally warm the targets were (an intertrait relationship known as a "halo effect"). White raters showed this halo effect regardless of the target's race (perceiving more attractive targets as warmer) whereas Korean and Black raters showed this effect only for White and Black targets, respectively. Similarly, American and Japanese perceivers exhibited high agreement in their ratings of American and Japanese political candidates' warmth and power [60]. Yet, their interpretations of these personality ratings differed according to the norms of leadership in their respective cultures: Japanese participants indicated that they would be more likely to vote for the candidates that they had previously rated as warm, whereas American participants selected the candidates that they had rated as powerful. Importantly, these preferences aligned with how the broader population of voters in each nation actually behaved: warm-looking candidates won elections in Japan, and powerful-looking candidates won elections in the United States. Cultural values about leadership therefore accord with the actual selection of leaders.

Parallel discrepancies in cultural values emerged in a neuroimaging study. When Japanese and American participants viewed images of bodies posed to look dominant or submissive, they agreed about which stimuli looked submissive versus dominant [61]. However, the participants processed this information differently depending on their culture's values. Specifically, American participants showed greater activation in areas of the brain associated with rewards (including the caudate and medial prefrontal cortex) when viewing dominant people, whereas Japanese participants displayed greater activation in these areas in response to submissive individuals. This dissociation aligns with cultural values in which the US tends to emphasize assertion whereas Japan encourages deference.

Of course, people evaluate each other using more than just visual stimuli. Vocal cues also contribute importantly to human interaction and the consequent impressions that people form. The speed and volume with which one speaks lead to distinct impressions between cultures. For example, one study found that loudness corresponded to perceptions of power among American and Korean perceivers, but fast speech corresponded to perceptions of power and competence only for American (and not Korean) perceivers [62]. Interestingly, Koreans living in the US showed intermediate responses, interpreting only competence but not power from fast speech. This implies that cultural exposure can alter one's perceptions of others.

12.5.2 Mental State Inferences

Not only do we infer people's *traits* from our interactions with them, we also infer their mental states [63]. We make these inferences both from what we perceive firsthand and from information provided to us about a person by others. For example, the false-belief task is one method of testing people's theory of mind—that is, their capacity to infer the mental states of others [64]. In this test, the reader must guess the thoughts of a character who has less information about the situation than the reader does. For example, in a false-belief scenario used to test children's theory of mind, a protagonist places an object in one location, leaves the scene, and another character transfers the object to a different location without the protagonist's knowledge. Children are then asked to indicate in which of the two locations the protagonist will look for the object upon returning. Children who believe that the protagonist will look in the new location have not yet developed a theory of mind, as they are unable to distinguish what they know from what the protagonist knows. One study tested the brain's involvement in theory of mind judgments by presenting American and Japanese participants with false-belief tasks while recording their neural responses using fMRI [65]. Although the two groups did not differ in their theory of mind accuracy, their neural activation patterns were distinct. Both groups showed responses in the right medial prefrontal cortex (mPFC), the right anterior cingulate cortex, the right middle frontal gyrus and the dlPFC, but the Americans showed more activation specific to theory of mind in the right insula, bilateral temporo-parietal junction (TPJ), and right mPFC; and the Japanese participants showed greater activity in the right orbito-frontal gyrus and right inferior frontal gyrus. These differences suggest that individuals process others' mental states differently according to their cultural background. Similar results have emerged from studies with American and Japanese children, indicating that the neural correlates of theory of mind may vary depending on one's early cultural and linguistic environment [66].

As noted above, people also infer mental states from direct perceptions of others, and principally do so from attending to their eyes [67]. A common test of mental state reasoning ability, the Reading the Mind in the Eyes (RME) test, thus presents individuals with images of the eye region of people's faces and asks them to choose, from a list of adjectives, what the person is thinking or feeling [68]. Using a cross-cultural version of the RME test with Caucasian and East Asian stimuli, Adams et al. [69] found that Japanese and Caucasian-American participants performed better when judging the eyes of people from their own culture. That is, Japanese participants more accurately identified the mental states of the East Asian targets whereas American participants more accurately inferred the Caucasian targets' mental states, demonstrating an intracultural advantage in theory of mind. Importantly, the participants also showed increased activation in the superior temporal sulcus, an area known to be central to inferring others' intentions [70], when judging own-culture stimuli—thereby providing the suggestion for a neural basis to cross-cultural understanding and misunderstanding (see also [71]). This intracultural advantage in RME performance may also decrease with acculturation such that cross-cultural misunderstandings of others' mental states may be mitigated by cultural exposure [72].

12.5.3 Emotions

Related to inferences of mental states is the interpretation and recognition of emotional states. Overall, people perceive emotions accurately across cultures. There is variability in this accuracy, however, including an ingroup advantage. Analogous to the intracultural advantage in theory of mind, perceivers tend to more accurately recognize the emotions of people from their own culture [73, 74]. Furthermore, the neural response for identifying the emotions of cultural ingroup members tends to be stronger than that for identifying the emotions of cultural outgroup members. Specifically, Chiao et al. [75] found that Caucasian-American and Japanese participants showed stronger amygdala activation (implicated in threat perception) in response to fearful Caucasian and Japanese faces, respectively. Interestingly, other work found this effect to be moderated by the targets' eye gaze, such that own-culture faces with averted eyes and other-culture faces with direct gaze elicited greater amygdala activation [76]. Natural fear expressions tend to feature averted gazes (see [77]), and ingroup members' fear may signal a threat more relevant to the self than outgroup members' fear does. In contrast, direct gaze from an outgroup member may actually represent an expression of threat toward oneself. Adams et al. [76] furthermore found that a variety of other regions important to face and eye gaze processing showed differential activation based on the targets' culture, irrespective of the perceivers' culture. Specifically, the bilateral fusiform gyri, bilateral inferior temporal gyri, and bilateral angular gyri responded more to averted- rather than direct-gazing Caucasian faces, but activated more to direct- versus averted-gazing Japanese faces. This appears to indicate that the

American and Japanese participants understood the distinct cultural meanings of the gazes (aligning with American and Japanese cultural values of assertion and deference, respectively), showing heightened activation for the less culturally appropriate gaze for each target group.

Beyond differences in neural responses to ingroup and outgroup members' emotions, there are also culturally-based differences in overall *sensitivity* to emotional displays. For example, when viewing scenes of others in emotional pain, Koreans showed more habitual attention to the needs and perspectives of others and heightened response in the anterior cingulate cortex and insula (part of a network of brain regions that processes pain) compared to Caucasian-Americans [78]. This heightened reactivity to emotional displays has important implications for culture's effect on empathy, indicating that certain cultures may be more empathic than others, due to their chronic attention to others and heightened reactivity to their pain.

Social interactions dictate not only that one must recognize others' emotions, but also that one must often regulate one's own emotions. Here, too, culture plays a role. Different cultures have varying norms for emotion regulation [79]. This, in turn, affects neural responses when individuals are instructed to suppress their emotional responses [80]. ERP data from one study showed that East Asians displayed decreased emotional processing when instructed to suppress their emotions, whereas European-Americans did not. This pattern aligns with cultural norms [81, 82]. Importantly, however, emotion regulation is not only determined by social factors but also biological ones, and evidence suggests that culture interacts with genes when determining a person's emotion regulation style [83]. Particularly, people with one particular genotype of the oxytocin receptor gene OXTR rs53576 (linked to a variety of social behaviors) use culturally-normative emotion regulation styles more than individuals without this variant.

Closely related to emotion is empathy, or the ability to understand and share another person's emotions and feelings. There is evidence that cultural norms and values affect empathy for certain emotions. For example, German and Chinese subjects showed differential brain activation when empathizing with anger, an emotion that disrupts harmony [84]. Consistent with interdependent Chinese cultural values of maintaining harmony, Chinese participants showed emotion regulation during empathy with anger, which was mediated by the left dlPFC, an area involved in emotion inhibition and regulation. However, for Germans (whose culture is independent and more tolerant of the expression of anger), activity was greater in the right inferior temporal gyrus and right superior temporal gyrus (areas involved in understanding others' intentions [85]) the left middle insula (sometimes implicated in perspective-taking [86]), and the right TPJ (associated with theory of mind, as reviewed above).

The effect of culture on empathy goes beyond reactions to specific emotions to also affect *with whom* one empathizes. Cultures vary in their preference for social hierarchy, and this, in turn, affects empathy towards ingroup versus outgroup members [87]. When viewing racial ingroup and outgroup members experiencing emotional pain, Koreans felt greater empathy towards other Koreans than they did towards Caucasian-Americans, and showed greater activity in the bilateral TPJ when viewing ingroup members (similar to the theory of mind findings reviewed above). These differences were associated with preferences for social hierarchy. That is, greater bilateral TPJ activation for racial ingroup members correlated with increased preferences for social hierarchies (as opposed to more egalitarian social structures), possibly due to the link between hierarchies and strict ingroup-outgroup divisions. In contrast, Caucasian-Americans, who reported low social hierarchy preferences, showed no such differences in their empathic reactions toward ingroup and outgroup members, reporting equal empathy for both Koreans and Caucasian-Americans and no differences in right TPJ activation, although left TPJ activity was higher for Koreans' pain. Importantly, however, another study found that differences in empathy towards racial ingroup and outgroup members was attenuated by exposure to the particular outgroup ([88], see also [89]). Specifically, Chinese adults who had spent a large portion of their lives in Western countries had similar neural responses to viewing Caucasians and Asians experiencing pain.

12.5.4 The Self and Others

People in cultures differing in their levels of interdependence versus independence (perhaps unsurprisingly) also differ in how they think about the self and others. Moreover, Americans typically show stronger ventral medial prefrontal cortex (vmPFC) activation when thinking about the self than when thinking about others [90]. In contrast, Chinese individuals show similar patterns of vmPFC activation when thinking about themselves and their mothers [91]—but less activation for their fathers or friends [92]. This suggests that certain close others, such as mothers, represent extensions of the self among people from cultures that encourage more interdependent thinking styles. Additionally, individual differences in the *degree* of one's interdependent self-construal affects self- and mother-driven mPFC and posterior cingulate cortex activations, demonstrating a range in how the self and others are represented by the brain, even within a single culture ([93], see also [94]). Interestingly, bicultural Chinese participants' vmPFC activation for mothers depended on whether they had been primed with Eastern or Western stimuli such that Eastern primes led to similar vmPFC activation for the self and one's mother, whereas Western primes promoted separate activations in the mPFC, thereby demonstrating that self-other representation may be somewhat malleable [95].

12.6 Significance of the Cultural Neuroscientific Approach

What does the perspective of cultural neuroscience provide that its parent fields cannot on their own? Importantly, the two components of cultural neuroscience may complement one another to provide a fuller understanding of human thought and behavior. First, neuroscience offers greater understanding of the underlying processes of cultural differences—even when behavior looks the same. One sees examples of this in topics ranging from visual processing to person perception. For example, Kitayama and Murata [54] found that European-Americans' and Asian-Americans' ERP signals indicated that they attended to target objects differently, even when their behavioral performance on an image-detection task did not differ. Additionally, although Japanese and American perceivers agreed on which bodies were dominant and which were submissive in Freeman et al.'s study [61], their neural responses to them showed evidence of different processes. Furthermore, in one study, American and Japanese participants achieved similar accuracy in a false-belief task, but their brain activation patterns showed pronounced differences in the regions supporting their judgments [65]. Such differences go beyond the topics reviewed in this chapter. For instance, even though members of Eastern and Western cultures may remember objects equally well, the involvement of different neural processes suggests that objects may be seen and processed differently by the two cultural groups [22]. That is, one group sees the objects as part of a context, whereas the other sees them as separate from their contexts—information that would not be evident from their performance on the memory task.

Furthermore, cultural neuroscience demonstrates how culture can serve as a moderating variable in many of the effects found in cognitive neuroscience. Even basic effects are influenced by culture—for example, the particular areas in the occipital cortex that activate in response to visual stimuli [22], or the degree of superior temporal sulcus activation when making mental state inferences of own- and other-culture targets [69]. Last, the integration of cultural psychology and cognitive neuroscience affords a wider array of methods to gain a more nuanced understanding of the phenomena researched by scientists in each of these fields, permitting greater convergence that can lead to more refined and precise accounts of how people think and behave in the context of their cultures [20, 21].

12.7 Applications and Increasing Intercultural Understanding

Despite these benefits, the question of how findings in cultural neuroscience may be applied to everyday cross-cultural interactions still remains. Cultural neuroscience is uniquely able to increase intercultural understanding for a number of reasons. Many of the findings in this field, including those reviewed above, demonstrate that outer similarities in behavior do not always equate to inner similarities in processing. This, in turn, highlights the far-reaching power of culture on processing and perceiving the world. Acknowledging this allows one to understand some of the difficulties in communication across cultures: Because the effect of culture on perception and cognition is so pervasive, it stands to reason that people from varying cultures may not always fully understand one another—they may simply perceive the world differently. This awareness may help individuals to avoid biases in cross-cultural interactions.

Not only does cultural neuroscience promote understanding group differences, it also encourages the detection of similarities. Across the research we have reviewed above, there are multiple examples of how acculturation can attenuate culturally based differences in cognition and perception. This underscores the importance of exposure to, and engagement with, other cultures. To foster cross-cultural understanding, the people constituting those cultures cannot remain separated from one another. Educating people about other cultures and providing them the opportunity to experience and spend time in those cultures thus has the potential to increase intercultural understanding.

12.8 Conclusion

The research reviewed in this chapter provides only a glimpse of the wealth of findings within cultural neuroscience. Nevertheless, it also demonstrates the pervasive influence of culture on cognition and points to this field's unique approach, as well as the benefits of its application. Cultural neuroscience research affords the opportunity to more fully understand the degree to which culture affects how people perceive and process the world, highlighting how even very basic cognitions may differ based on one's culture. This knowledge can accordingly apply to specific intervention programs aimed to increase successful intercultural dialogue, therefore advancing greater cooperation between people from diverse means of thinking and behaving.

References

1. Markus, H.R., Kitayama, S.: Culture and the self: implications for cognition, emotion, and motivation. Psychol. Rev. **98**, 224–253 (1991)
2. Kwong, K.K.: Record of a single fMRI experiment in May of 1991. NeuroImage **62**, 610–612 (2012)
3. Belliveau, J.W., Kennedy Jr., D.N., McKinstry, R.C., Buchbinder, B.R., Weisskoff, R.M., Cohen, M.S., Rosen, B.R., et al.: Functional mapping of the human visual cortex by magnetic resonance imaging. Science **254**, 716–719 (1991)
4. Jaffe, E.: Identity shift. APS Observer **24**, 28–30 (2011)
5. Gazzaniga, M.S.: Handbook of Cognitive Neuroscience. Plenum Press, New York (1984)
6. Lieberman, M.D.: Social cognitive and affective neuroscience: when opposites attract. Soc. Cogn. Affect. Neurosci. **1**, 1–2 (2006)
7. Chiao, J.Y.: At the frontier of cultural neuroscience: introduction to the special issue. Soc. Cogn. Affect. Neurosci. **5**, 109–110 (2010)
8. Han, S., Northoff, G.: Culture-sensitive neural substrates of human cognition: a transcultural neuroimaging approach. Nat. Rev. Neurosci. **9**, 646–654 (2008)

9. Han, S., Pöppel, E.: Culture and neural frames of cognition and communication. Springer, New York (2011)
10. Tang, Y., Zhang, W., Chen, K., Feng, S., Ji, Y., Shen, J., Liu, Y., et al.: Arithmetic processing in the brain shaped by cultures. Proc. Natl. Acad. Sci. USA **103**, 10775–10780 (2006)
11. Rule, N.O., Freeman, J.B., Moran, J.M., Gabrieli, J.D.E., Ambady, N.: Voting behavior is reflected in amygdala response across cultures. Soc. Cogn. Affect. Neurosci. **5**, 349–355 (2010)
12. Way, B.M., Lieberman, M.D.: Is there a genetic contribution to cultural differences? Collectivism, individualism and genetic markers of social sensitivity. Soc. Cogn. Affect. Neurosci. **5**, 203–211 (2010)
13. Horwitz, B.: The elusive concept of brain connectivity. NeuroImage **19**, 466–470 (2003)
14. Sperber, A.D., Devellis, R.F., Boehlecke, B.: Cross-cultural translation: methodology and validation. J. Cross Cult. Psychol. **25**, 501–524 (1994)
15. van de Vijver, F.J.R., Leung, K.: Methods and Data Analysis for Cross-Cultural Research. Sage, Newbury Park, CA (1997)
16. Weeks, A., Swerissen, H., Belfrage, J.: Issues, challenges, and solutions in translating study instruments. Eval. Rev. **31**, 153–165 (2007)
17. Chiao, J.Y., Cheon, B.K.: The weirdest brains in the world. Behav. Brain Sci. **33**, 88–90 (2010)
18. Henrich, J., Heine, S.J., Norenzayan, A.: The weirdest people in the world? Behav. Brain Sci. **33**, 61–83 (2010)
19. Button, K.S., Ioannidis, J.P.A., Mokrysz, C., Nosek, B.A., Flint, J., Robinson, E.S.J., Munafò, M.R.: Power failure: why small sample size undermines the reliability of neuroscience. Nat. Rev. Neurosci. **14**, 365–376 (2013)
20. Kitayama, S., Tompson, S.: Envisioning the future of cultural neuroscience. Asian J. Soc. Psychol. **13**, 92–101 (2010)
21. Ng, B.W., Morris, J.P., Oishi, S.: Cultural neuroscience: the current state of affairs. Psychol. Inq. **24**, 53–57 (2013)
22. Gutchess, A., Welsh, R., Boduroglu, A., Park, D.C.: Cultural differences in neural function associated with object processing. Cogn. Affect. Behav. Neurosci. **6**, 102–109 (2006)
23. Romaine, S.: Language in Society. Oxford University Press, Oxford (1994)
24. Sapir, E.: The status of linguistics as a science. In: Mandelbaum, D.G. (ed.) E. Sapir (1958): Culture, Language and Personality. University of California, Berkeley, CA (1929)
25. Whorf, B.L.: Science and linguistics. Technol. Rev. **42**(229–31), 247–248 (1940)
26. McArthur, T.: Alphabet. In Concise Oxford Companion to the English Language. Oxford University Press (1998). Retrieved 26 June 2015, from http://www.oxfordreference.com/view/10.1093/acref/9780192800619.001.0001/acref-9780192800619-e-53
27. Bolger, D.J., Perfetti, C.A., Schneider, W.: Cross-cultural effect on the brain revisited, universal structures plus writing system variation. Hum. Brain Mapp. **25**, 92–104 (2005)
28. Perani, D., Abutalebi, J.: The neural basis of first and second language processing. Curr. Opin. Neurobiol. **15**, 202–206 (2005)
29. Wartenburger, I., Heekeren, H.R., Abutalebi, J., Cappa, S.F., Villringer, A., Perani, D.: Early setting of grammatical processing in the bilingual brain. Neuron **37**, 159–170 (2003)
30. Pallier, C., Dehaene, S., Poline, J.-B., LeBihan, D., Argenti, A.-M., Dupoux, E., Mehler, J.: Brain imaging of language plasticity in adopted adults: Can a second language replace the first? Cereb. Cortex **13**, 155–161 (2003)
31. Boroditsky, L.: Does language shape thought? Mandarin and English speakers' conceptions of time. Cogn. Psychol. **43**, 1–22 (2001)
32. Stevenson, H.W., Stigler, J.W.: The Learning Gap: Why Our Schools Are Failing and What We Can Learn from Japanese and Chinese Education. Summit Books, New York (1992)
33. Cao, B., Li, F., Li, H.: Notation-dependent processing of numerical magnitude: electrophysiological evidence from Chinese numerals. Biol. Psychol. **83**, 47–55 (2010)

34. Holloway, I.D., Battista, C., Vogel, S.E., Ansari, D.: Semantic and perceptual processing of number symbols: evidence from a cross-linguistic fMRI adaptation study. J. Cogn. Neurosci. **25**, 388–400 (2013)
35. Lin, J.-F.L., Imada, T., Kuhl, P.K.: Mental addition in bilinguals: an fMRI study of task-related and performance-related activation. Cereb. Cortex **22**, 1851–1861 (2012)
36. Dumontheil, I., Klingberg, T.: Brain activity during a visuospatial working memory task predicts arithmetical performance 2 years later. Cereb. Cortex **22**, 1078–1085 (2012)
37. Janata, P., Birk, J.L., Van Horn, J.D., Leman, M., Tillman, B., Bharucha, J.J.: The cortical topography of tonal structures underlying Western music. Science **298**, 2167–2170 (2002)
38. Balkwill, L., Thompson, W.F., Matsunaga, R.: Recognition of emotion in Japanese, Western, and Hindustani music by Japanese listeners. Jpn. Psychol. Res. **46**, 337–349 (2004)
39. Fritz, T., Jentschke, S., Gosselin, N., Sammler, D., Peretz, I., Turner, R., Koelsch, S., et al.: Universal recognition of three basic emotions in music. Curr. Biol. **19**, 573–576 (2009)
40. Hannon, E.E., Trainor, L.J.: Music acquisition: Effects of enculturation and formal training on development. Trends Cogn. Sci. **11**, 466–472 (2007)
41. Trainor, L.J.: Are there critical periods for music development? Dev. Psychobiol. **46**, 343–358 (2005)
42. Lynch, M.P., Eilers, R.E., Oller, K.D., Urbano, R.C., Wilson, P.: Influences of acculturation and musical sophistication on perception of musical interval patterns. J. Exp. Psychol. Hum. Percept. Perform. **17**, 967–975 (1991)
43. Morrison, S.J., Demorest, S.M.: Cultural constraints on music perception and cognition. Prog. Brain Res. **178**, 67–77 (2009)
44. Neuhaus, C.: Perceiving musical scale structures: a cross-cultural event-related brain potentials study. Ann. N. Y. Acad. Sci. **999**, 184–188 (2003)
45. Demorest, S.M., Osterhout, L.: ERP responses to cross-cultural melodic expectancy violations. Ann. N. Y. Acad. Sci. **1252**, 152–157 (2012)
46. Nan, Y., Knosche, T.R., Friederici, A.D.: The perception of musical phrase structure: a cross-cultural ERP study. Brain Res. **1094**, 179–191 (2006)
47. Nan, Y., Knosche, T.R., Friederici, A.D.: Non-musicians' perception of phrase boundaries in music: A cross-cultural ERP study. Biol. Psychiat. **82**, 70–81 (2009)
48. Iversen, J.R., Patel, A.D., Ohgushi, K.: Perception of rhythmic grouping depends on auditory experience. J. Acoust. Soc. Am. **124**, 2263–2271 (2008)
49. Patel, A.D., Daniele, J.R.: An empirical comparison of rhythm in language and music. Cognition **87**, B35–B45 (2003)
50. Kitayama, S., Duffy, S., Kawamura, T., Larsen, J.T.: Perceiving an object and its context in different cultures: a cultural look at New Look. Psychol. Sci. **14**, 201–206 (2003)
51. Hedden, T., Ketay, S., Aron, A., Markus, H.R., Gabrieli, J.D.E.: Cultural influences on neural substrates of attentional control. Psychol. Sci. **19**, 12–17 (2008)
52. Witkin, H.A., Goodenough, D.R.: Field dependence and interpersonal behavior. Psychol. Bull. **84**, 661–689 (1977)
53. Goh, J.O., Chee, M.W., Tan, J.C., Venkatraman, V., Hebrank, A., Leshikar, E., Park, D.C.: Age and culture modulate object processing and object-scene binding in the ventral visual area. Cogn. Affect. Behav. Neurosci. **7**, 44–52 (2007)
54. Kitayama, S., Murata, A.: Culture modulates perceptual attention: An event-related potential study. Soc. Cogn. **31**, 758–769 (2013)
55. Kim, H.S., Sherman, D.K., Taylor, S.E., Sasaki, J.Y., Chu, T.Q., Ryu, C., Xu, J., et al.: Culture, the serotonin receptor polymorphism (5-HTR1A) and locus of attention. Soc. Cogn. Affect. Neurosci. **5**, 212–218 (2010)
56. Miyamoto, Y., Yoshikawa, S., Kitayama, S.: Feature and configuration in face processing: Japanese are more configural than Americans. Cogn. Sci. **35**, 563–574 (2011)
57. Maurer, D., Le Grand, R., Mondloch, C.J.: The many faces of configural processing. Trends Cogn. Sci. **6**, 255–260 (2002)
58. Albright, L., Malloy, T.E., Dong, Q., Kenny, D.A., Fang, X., Winquist, L., Yu, D.: Cross-cultural consensus in personality judgments. J. Pers. Soc. Psychol. **72**, 558–569 (1997)

59. Zebrowitz, L.A., Montepare, J.M., Lee, H.K.: They don't all look alike: individuated impressions of other racial groups. J. Pers. Soc. Psychol. **65**, 85–101 (1993)
60. Rule, N.O., Ambady, N., Adams Jr., R.B., Ozono, H., Nakashima, S., Yoshikawa, S., Watabe, M.: Polling the face: Prediction and consensus across cultures. J. Pers. Soc. Psychol. **98**, 1–15 (2010)
61. Freeman, J.B., Rule, N.O., Adams Jr., R.B., Ambady, N.: Culture shapes a mesolimbic response to signals of dominance and subordination that associates with behavior. NeuroImage **47**, 353–359 (2009)
62. Peng, Y., Zebrowitz, L.A., Lee, H.K.: The impact of cultural background and cross-cultural experience on impressions of American and Korean male speakers. J. Cross Cult. Psychol. **24**, 203–220 (1993)
63. Mitchell, J.P.: Mentalizing and Marr: an information processing approach to the study of social cognition. Brain Res. **1079**, 66–75 (2006)
64. Wimmer, H., Perner, J.: Beliefs about beliefs: representation and constraining function of wrong beliefs in young children's understanding of deception. Cognition **13**, 103–128 (1983)
65. Kobayashi, C., Glover, G.H., Temple, E.: Cultural and linguistic influence on neural bases of "theory of mind": an fMRI study with Japanese bilinguals. Brain Lang. **98**, 210–220 (2006)
66. Kobayashi, C., Glover, G.H., Temple, E.: Cultural and linguistic effects on neural bases of "theory of mind" in American and Japanese children. Brain Res. **1164**, 95–107 (2007)
67. Baron-Cohen, S., Wheelwright, S., Jolliffe, T.: Is there a "language of the eyes"? Evidence from normal adults, and adults with autism or asperger syndrome. Vis. Cogn. **4**, 311–331 (1997)
68. Baron-Cohen, S., Wheelwright, S., Hill, J., Raste, Y., Plumb, I.: The "reading the mind in the eyes" test revised version: A study with normal adults, and adults with Asperger syndrome or high-functioning autism. J. Child Psychol. Psychiatry **42**, 241–251 (2001)
69. Adams Jr., R.B., Rule, N.O., Franklin Jr., R.G., Wang, E., Stevenson, M.T., Yoshikawa, S., Ambady, N., et al.: Cross-cultural reading the mind in the eyes: an fMRI investigation. J. Cogn. Neurosci. **22**, 97–108 (2010)
70. Allison, T., Puce, A., McCarthy, G.: Social perception from visual cues: role of the STS region. Trends Cogn. Sci. **4**, 267–278 (2000)
71. Franklin, R.G., Stevenson, M.T., Ambady, N., Adams, R.B.: Cross-cultural reading the mind in the eyes and its consequences for international relations. In: Warnick, J.E., Landis, D. (eds.) Neuroscience in Intercultural Contexts, pp. 117–141. Springer, New York (2015)
72. Bjornsdottir, R.T., Rule, N.O.: On the relationship between acculturation and intercultural understanding: Insight from the Reading the Mind in the Eyes test. Manuscript submitted for publication (2015)
73. Elfenbein, H.A., Ambady, N.: On the universality and cultural specificity of emotion recognition: a meta-analysis. Psychol. Bull. **128**, 203–235 (2002)
74. Jack, R.E., Garrod, O.G.B., Yu, H., Caldara, R., Schyns, P.G.: Facial expressions of emotion are not culturally universal. Proc. Natl. Acad. Sci. USA **109**, 7241–7244 (2012)
75. Chiao, J.Y., Iidaka, T., Gordon, H.L., Nogawa, J., Bar, M., Aminoff, E., Sadato, N., Ambady, N.: Cultural specificity in amygdala response to fear faces. J. Cogn. Neurosci. **20**, 2167–2174 (2008)
76. Adams Jr., R.B., Franklin Jr., R.G., Rule, N.O., Freeman, J.B., Kveraga, K., Hadjikhani, N., Ambady, N., et al.: Culture, gaze, and the neural processing of fear expressions. Soc. Cogn. Affect. Neurosci. **5**, 340–348 (2010)
77. Adams Jr., R.B., Gordon, H.L., Baird, A.A., Ambady, N., Kleck, R.E.: Effects of gaze on amygdala sensitivity to anger and fear faces. Science **300**, 1536 (2003)
78. Cheon, B.K., Im, D., Harada, T., Kim, J., Mathur, V.A., Scimeca, J.M., Chiao, J.Y., et al.: Cultural modulation of the neural correlates of emotional pain perception: the role of other-focusedness. Neuropsychologia **51**, 1177–1186 (2013)
79. Matsumoto, D., Yoo, S.H., Nakagawa, S.: Culture, emotion regulation, and adjustment. J. Pers. Soc. Psychol. **94**, 925–937 (2008)

80. Murata, A., Moser, J.S., Kitayama, S.: Culture shapes electrocortical responses during emotion suppression. Soc. Cogn. Affect. Neurosci. **8**, 595–601 (2013)
81. Kitayama, S., Mesquita, B., Karasawa, M.: Cultural affordances and emotional ex-perience: socially engaging and disengaging emotions in Japan and the United States. J. Pers. Soc. Psychol. **91**, 890–903 (2006)
82. Tsai, J.L., Knutson, B., Fung, H.H.: Cultural variation in affect valuation. J. Pers. Soc. Psychol. **90**, 288–307 (2006)
83. Kim, H.S., Sasaki, J.Y.: Emotion regulation: the interplay of culture and genes. Soc. Pers. Psychol. Compass **6**, 865–877 (2012)
84. de Greck, M., Shi, Z., Wang, G., Zuo, X., Yang, X., Wang, X., Han, S., et al.: Culture modulates brain activity during empathy with anger. NeuroImage **59**, 2871–2882 (2012)
85. Britton, J.C., Phan, K.L., Taylor, S.F., Welsh, R.C., Berridge, K.C., Liberzon, I.: Neural correlates of social and nonsocial emotions: An fMRI study. NeuroImage **31**, 397–409 (2006)
86. Lamm, C., Batson, C.D., Decety, J.: The neural substrate of human empathy: effects of perspective-taking and cognitive appraisal. J. Cogn. Neurosci. **19**, 42–58 (2007)
87. Cheon, B.K., Im, D., Harada, T., Kim, J., Mathur, V.A., Scimeca, J.M., Chiao, J.Y., et al.: Cultural influences on neural basis of intergroup empathy. NeuroImage **57**, 642–650 (2011)
88. Zuo, X., Han, S.: Cultural experiences reduce racial bias in neural responses to others' suffering. Cult. Brain **1**, 34 (2013)
89. Han, S.: Intergroup relationship and empathy for others' pain: a social neuroscience approach. In: Warnick, J.E., Landis, D. (eds.) Neuroscience in Intercultural Contexts, pp. 31–47. Springer, New York (2015)
90. Kelley, W.T., Macrae, C.N., Wyland, C., Caglar, S., Inati, S., Heatherton, T.F.: Finding the self? An event-related fMRI study. J. Cogn. Neurosci. **14**, 785–794 (2002)
91. Zhu, Y., Zhang, L., Fan, J., Han, S.: Neural basis of cultural influence on self-representation. NeuroImage **34**, 1310–1316 (2007)
92. Wang, G., Mao, L., Ma, Y., Yang, X., Cao, J., Liu, X., Han, S., et al.: Neural representations of close others in collectivistic brains. Soc. Cogn. Affect. Neurosci. **7**, 222–229 (2011)
93. Ray, R.D., Shelton, A.L., Hollon, N.G., Matsumoto, D., Frankel, C.B., Gross, J.J., Gabrieli, J.D.E.: Interdependent self-construal and neural representations of self and mother. Soc. Cogn. Affect. Neurosci. **5**, 318–323 (2010)
94. Chiao, J.Y., Harada, T., Komeda, H., Li, Z., Mano, Y., Saito, D., Parrish, T.B., Iidaka, T., et al.: Dynamic cultural influences on neural representations of the self. J. Cogn. Neurosci. **22**, 1–11 (2010)
95. Ng, S.H., Han, S., Mao, L., Lai, J.C.L.: Dynamic bicultural brains: fMRI study of their flexible neural representation of self and significant others in response to culture primes. Asian J. Soc. Psychol. **13**, 83–91 (2010)

Chapter 13
Cultural Neuroscience and the Military: Applications, Perspectives, Controversies

Kamila Trochowska

> *What seems astonishing is that a mere three-pound object, made of the same atoms that constitute everything else under the sun, is capable of directing virtually everything that humans have done: flying to the moon and hitting seventy home runs, writing Hamlet and building the Taj Mahal – even unlocking the secrets of the brain itself.*
> —Joel Havemann

Abstract As we are entering the golden age of brain research and the Biotechnological Revolution in Military Affairs (BIOTECH RMA), not only civilian entities in research endeavors such as the American *BRAIN* initiative or the European *Human Brain Project*, but also security and defense communities start exploring the cognitive area of human activity. Brain research, however, due to numerous technological and other limitations, does not cover the complexity of the mind, and the cultural variety of the individuals involved. Culturally influenced cognitive, affective and even physical domains embrace emotional antecedents, conditioning, moral reasoning, perception and gaining situational awareness, communication, approach to death, somatic health, aggression, responding to narratives, group relations and many others. However, only a limited amount of other research has been performed and reported on in this aspect. Therefore, the chapter analyzes the existing evidence on the culture-brain nexus and its numerous implications for human functioning in a variety of domains, reviews the existing solutions and projects that leading military institutions already realize in the cognitive field, and in the light of newest findings of cultural neuroscience, proposes new potential solutions and enhancements for the design and conduct of military training and conduct of non-kinetic aspects of military operations. The aim of the research piece is to expand on the added value of the cultural neuroscience research to the field, and to discuss the resulting reservations and controversies of a situation in which winning "hearts and minds" might no longer be a metaphor.

K. Trochowska (✉)
State Security Institute, National Security Faculty,
War Studies University, al. Chruściela 103, Block 25, 00-910 Warsaw, Poland
e-mail: k.trochowska@akademia.mil.pl

C. Faucher (ed.), *Advances in Culturally-Aware Intelligent Systems and in Cross-Cultural Psychological Studies*, Intelligent Systems Reference Library 134, https://doi.org/10.1007/978-3-319-67024-9_13

Keywords Neurobiology · Culture · Neuroanthropology · Military training
Serious games and VR design

13.1 Introduction

Science and war—the debate on how far can this relationship go is almost a cliché one. Is there anything new we can say about this union? Are they a good match? Will there be a happily ever after? Since the 17th century, western militaries have been implementing (and also conceiving) scientific advancements, which created the "scientific way of warfare" characteristic of increasingly technologically developed societies [1]. The 20th century, along with the cybernetic revolution and the dawn of the third-wave information societies, brought us the bloom of research on the information dimension of human functioning, with a parallel military Information Revolution in Military Affairs (INFO RMA). As we are entering the golden age of brain research and cognitive science, not only civilian entities in multi-billion dollar research endeavors such as the American *BRAIN* initiative or the European *Human Brain Project*, but also security and defense communities start exploring the cognitive area of human activity. The upcoming Biotechnological Revolution in Military Affairs (BIOTECH RMA) aims at expediting the efficiency and invincibility of organized state violence on the individual level, through the maximization of the army's human resources potential and impact [2]. The four major drivers of improvement are: (psycho) pharmacologization of warfare (to explain the term through movie analogies: *Drugstore Cowboys* meets *Phenomenon*), cyborgization (*Ex Machina* meets *Full Metal Jacket*), genetic engineering (*Next* meets *Blade Runner*) and nanotechnologies (*Microcosm* meets *Avatar*). It is needless to say that neuroscience is a vast part of the future maths of war, since the human mind is where everything begins and ends (and as Sun Tzu and other military thought classics have claimed for centuries, where all the battles are won). Moreover, the development of increasingly sophisticated and intelligent, ISR, weapon and communication systems, demands highly elevated cognitive abilities of their operators. Since the very beginning of warfare we used to improve human capacities with external enhancements: arms, weapons and instruments, but for the first time in our species history, we can actually change and improve the biological essence of what we are, and through neurobiological, computational and nanotechnological advancements, reach the glorious post-human condition postulated in transhumanism.[1] Raymond Kurzweil, one of the founders of the movement, is also one of the most prominent members of the military trend-setting U.S. Army

[1]Transhumanism is an ideological and philosophical current that promotes radical human evolution aided by technology, in particular in the fields of neurobiology, biotechnology and nanotechnology. It aims at overcoming human physical and mental limitations, improving global social relations and achieving a new stage of human species development. Major representatives of the current are, among others, Raymond Kurzweil, Max More, Nick Bostrom, David Pearce,

Science Advisory Group, so we might expect that soon every private Jones will be his own "Discovery Channel Special".

Let us not get too intimidated by the almighty scientific-military complex yet. Neither neuroscience nor cognitive science, due to numerous technological and methodological limitations, is able to cover the complexity of the mind, and the cultural variety of the individuals involved. First of all, vast majority of empirical studies on neural processes underlying human emotion and cognition that created the contemporary brain functioning paradigms, were performed on non-human animals [3]. Secondly, until recently, neuroscience lacked the technology to gain a comprehensive insight into the brain, and even nowadays, with the introduction of a variety of neuroimaging techniques,[2] we still fail to get the full picture, since even the cutting-edge tools do not provide accurately combined measures in spatial and temporal dimensions. Thirdly, in contemporary cognitive psychology and neuroscience we still have to do with a "Western" bias (with the choice of research samples limited to particular socio-cultural groups, in particular college aged students from the WEIRD demographic – Western Educated Rich and Democratic) [5]. 95% of psychological research samples comes from countries with only 12% of the global population, and 90% of peer-reviewed neuroimaging studies originate from Western countries [6]. And last but not least—we still lack a unified theory of what human mind actually is—and how the mysterious inner mechanisms of consciousness work. Several attempts on appreciating the bidirectional influence of culture and genes to brain and behavior have been made since late 1990s, when the fields of social and cultural neuroscience evolved as domains of cognitive neuroscience. Research results from these fields contributed largely to cognitive neuroscience findings, since culturally influenced cognitive, affective and even physical domains embrace, among others, such crucial for our considerations domains as: memory, emotional antecedents, conditioning, moral reasoning, social cognition, perception and gaining situational awareness, interpersonal communication, approach to death, somatic health, aggression, responding to narratives, group relations and many others.

The aforementioned approaches, however, still base on population-scale samples computational models and laboratory conditioned research, which might not always give us the true insight into how cultural traits such as values, beliefs, practices or

James Hughes Natasha Vita-More or Zoltan Istvan. An official portal that gathers the major thinkers and practitioners of the current is Humanity Plus (H+) http://humanityplus.org/.

[2]The major techniques used to study the brain are: single-unit recording of neuron's activity, event-related potentials (ERPs) that records patterns of brain activity to a stimulus, Positron Emission Tomography (PET) that has great spatial but poor temporal resolution, Functional Magnetic Resonance Imaging (fMRI) and its more advanced event-related version (efMRI), that have superior temporal and spatial resolution but provide indirect measurements through blood oxygenation levels, Magneto-EncephaloGraphy (MEG) that deals with magnetic fields that result from brain activity, and Transcranial Magnetic Stimulation (TMS), and its repeated variation—the rTMS, that uses brief pulses of current that produces a short-lived magnetic field that inhibits neural processing in the area affected [4].

life cycles shape and are shaped by the human mind and underlying biological structures. Thus the field of cultural neurobiology evolved, as an interdisciplinary science that examines the bidirectional influence of culture and brain activity, joining methods from cognitive neuroscience, cultural psychology and neurogenetics.[3] Then, in the mid 1990s, neuroanthropology emerged as a sister science that combines cultural neuroscience and anthropological field work, bringing the analyses down to a more local level through the examination of smaller-scale population samples and confronting them with population-level clinical research. The approach offered by those two sciences enables us not only to understand the interaction of human brain and culture, and hence the implications for mind functioning and behavior determination in a variety of settings and circumstances, determine the role of the nervous system and cognitive processes in creating social structures, constructs and behaviors, but also bring in significant advancements in human science theory [8] and indicate solutions of the most challenging problems in military training and non-kinetic activities.

The upcoming scientific breakthroughs that will enable to crack, model and alter the functioning of the human brain and its expression in cognitive, physical and affective domains of behavior, will also vastly change the way our militaries work. Cultural neurobiology and neuroanthropology, are of particular significance to the armed forces, since they can help explain and enhance solutions for such diverse issues as the cognitive abilities and performance of soldiers' training, Post-Traumatic Stress Disorder (PTSD) management, psychological operations (PSYOPS) or the design of culturally-informed decision-making support systems, to mention a few. The possibilities were partly acknowledged by American defense R&D institutions such as the Defense Advanced Research Projects Agency (DARPA) (*SyNAPSE*, *Narrative Networks*, *N4IA*, *CT2WS*), IARPA (*ICArUS*, *KRNS*, *SHARP*) or Sandia National Laboratories (*HDM*) projects which explore the potential of merging cognitive science, psychometrics, computational neuroscience with social science and humanities to facilitate training, military decision-making process and performance of soldiers. However, only a limited amount of research

[3]It is crucial to draw the distinction between cultural and social neuroscience and neuroanthropology. The subject of interest for all three interdisciplinary branches is the bidirectional culture-nervous system-behavior relation, however they use various approaches, methodology and research tools. Cultural neuroscience explains mental phenomena in terms of a synergic product of mental, neural and genetic events, and analyzes how cultural traits shape brain function, and the human brain gives rise to cultural capacities and their transmission across various timescale. It applies the tools from cultural psychology, neuroscience and neurogenetics. Social neuroscience on the other hand, uses the methodologies and tools developed to measure mental and brain function to study social context of cognition, emotion, and behavior. And neuroanthropology provides the hands-on insight into those matters, as it confronts the neuroscientific findings with field work, joining methods and paradigms of cultural anthropology and neuroscience, and looks at field-based variations by drawing on anthropology, which has over a century of research on human variation. Problem scale (local, intrapopulation level) and of theoretical models (practical and embodied), separate the fields [7].

has been performed and reported on cultural determinants of human cognitive abilities, reasoning and behavior in the security and defense context.
The major research theses that this piece of research work intends to address are:

1. Various brain regions develop and function slightly differently due to cultural influences (including variety in gene expression), which results in cognitive, affective and behavioral differences among individuals.
2. The military can apply findings from cultural and social neuroscience and neuro anthropology to improve its effectiveness in the domains of training, conduct of non-kinetic tasks during military operations and post-operational force sustainment, and upgrade the solutions that are already implemented in the field.
3. Although the field is controversial, scientific ethical standards do not have to be violated during the application of cultural and social neuroscience for security and defense purposes—so that we would not have to welcome the return of MK-ULTRA.[4]

To achieve that, we will review the existing evidence on the culture-brain nexus and its numerous implications for human functioning in a variety of domains, review the existing solutions and projects that leading military institutions already realize in the cognitive field, and in the light of newest findings of cultural neuroscience, propose new potential solutions and enhancements for the design and conduct of military training and conduct of non-kinetic aspects of military operations. The aim of the chapter is to expand on the added value of the cultural neuroscience research to the field, and to discuss the resulting reservations and controversies of a situation in which winning "hearts and minds" might no longer be a metaphor.

13.2 The Culture-Brain Nexus

As Professor Daniel Lende, the pioneer of neuroanthropological research noted, our brains are biosocial. Rather than being exclusively biological, the brain and its neural activity must be considered to be a hybrid of both biological and social influences [8]. Genes and evolution, together with socio-cultural environment," shape the linked variation of brain and behavior. The dimension of individualism/collectivism, for example, is not simply a cultural trait in this model - it has identifiable neural correlates, in part formed by past evolutionary history and in other part by the intersection of culture, development, and neuroplasticity" [8]. Bidirectional relationship between the nervous system and culture can be explained and exemplified in genetic, developmental, physiological, adaptive, environmental

[4]MK ULTRA was the code name of a CIA project realized in the years 1950–1975, that comprised experiments on human subjects. Its aim was to identify and develop drugs and procedures to be used in interrogations and torture, in order to weaken the individual to force confessions through mind control [9].

and evolutionary terms. What is most significant, cultural factors begin their work at the very essence of what human beings are, influencing and changing gene expression, and 70% of genetic information expresses itself in the brain. Although inter-population variations in genome might seem small, reaching around 02–04% [10], it is worth noting that the genetic variation between chimpanzees and humans is measured at 2%—but does it not make an immense difference? Let us then look at examples of cultural influences on genetic diversity among humans.

Evidence from behavioral genetics indicates that the S allele of the serotonin transporter gene (5-HTTLPR),[5] might increase the chance of negative emotional states such as anxiety, fear conditioning and attentional bias to negative information that in the presence of unfavorable environmental factors can lead to major depression. It was found that the allele is significantly more prevalent in East Asian populations (70–80% as compared to the average of 50% in other populations). Another such example might be the 7-repeat allele of the dopamine D4 receptor,[6] which renders individuals prone to risk behaviors and novelty seeking. This combination was discovered to be extremely rare in East Asian populations (less than 1% of carriers), and overwhelmingly prevalent in South American Indian populations (70–80%) [3] and other migratory communities [10]. Ethnicity and culture are drivers of such variations, but they also provide mechanisms that minimize the potential negative influence, with East Asian populations for instance being prevalently collective. Strong social relations and support can eradicate the anxiety and depressive disorders that the representatives of those groups were found to be more prone to. This has significant implications for research, since neurogenetic associations observed in one population might be absent in another one, as it is with the association between amygdala activation and the aforementioned serotonin transporter gene (5-HTTLPR) that exists within Caucasian populations, making them more fit for individual coping with distress, but not within East Asian ones [10].

Another aspect of the gene-culture nexus is the Culture-Gene Co-evolutionary theory (CGC), which asserts that "once cultural traits are adaptive, it is highly likely that genetic selection causes refinement of the cognitive and neural architecture responsible for the storage and transmission of cultural capacities" [10]. CGC began as a branch of theoretical population genetics, which, in addition to modeling the differential transgenerational genetic transmission, incorporates cultural features, such as norms, values, beliefs, cultural dimensions and rituals into further investigations. The two transmission systems cannot be treated independently, both because of individual learning capacity may depend on the genotype, and also

[5]Serotonin (5-hydroxytryptamine, 5-HT) is a chemical mainly found in the brain, bowels and blood platelets. It carries signals along and between nerves - a neurotransmitter. It is considered to be especially active in constricting smooth muscles, transmitting impulses between nerve cells, regulating cyclic body processes and contributing to wellbeing and happiness [11].

[6]Dopamine is also a neurotransmitter, in the brain playing a major role in reward-motivated behavior. Most types of reward increase and a variety of addictive drugs increase dopamine neuronal activity.

because the selection acting on the genetic system may be generated or modified by the spread of a cultural trait (such as the propensity of given cultural group to be more collective or individualistic, feminine or masculine, affective or constrained[7]). To give a very down-to-earth example, the frequency of the sickle cell mutations among populations in West Africa depends on their means of subsistence. Populations that chop down trees to cultivate yams create the conditions where heavy rainfall will leave pools of standing water in which mosquitoes thrive, leading to more intense selection [13].

Recent applications of CGC include an extremely gripping investigation of how the relation influences moral judgment, examining the correlations between the cultural dimensions of Tightness–Looseness (TL), ecological threat, and the aforementioned allelic variants of the serotonin transporter linked polymorphic region (5-HTTLPR) in producing justification of moral behavior [14]. The results of the research conducted across 21 nations proved that the propensity for ecological threat correlates with short (S) allele frequency in the 5-HTTLPR, and allelic frequency in the 5-HTTLPR along with vulnerability to ecological threat both correlate with cultural tightness–looseness cultural dimension that pertains to strength of social norms and tolerance of deviant behavior. It is believed that the distinctive forms of social organization are a form of cultural adaptation to the presence of biological, environmental, and human-made threats. Tight cultural norms are created and maintained to encourage social coordination that facilitates member survival in more threatened areas. Moreover, another bidirectional relation was found —susceptibility to ecological threat predicts tightness–looseness via the mediation of S allele carriers, and frequency of S allele carriers predicts justifiability of morally relevant behavior via tightness–looseness [15]. It is a brilliant example of the research span of the young, but blooming field of cultural neuroscience. Other recent demonstrations of neural plasticity indicate the significance of experience in brain development as non-genetic environmental factors can lead to major differences in gene expressions.

In this place we get to the major feature of the human brain that enables not only the possibility of bidirectional influence of culture and brain, but lays at the very essence of majority of cognitive processes, individual and social development, and creation of culture itself: brain plasticity. Plasticity refers to changes in brain

[7]The dimensions of culture that are most widely used in cultural neuroscientific research are those designated by Geert Hofstede, a social psychologist from Netherlands, who based his research on IBM cross-company studies since 1970s. While investigating diversity of organizational cultures in affiliated companies, he discovered categories that help systematize them through major values that govern them. Those six dimensions of culture are: Masculinity v. Femininity, High v. Low Power Distance, Long v. Short-term Orientation, Individualism v. Collectivism, Uncertainty Avoidance and Indulgence v. Restraint (6th dimension added by Michael Minkov in 2010). For more details see [12]. Despite critical voices on his theory (for instance that the models present static, national cultures which fail to embrace the inner dynamics and heterogeneity of cultures), and the emergence of other cultural dimensions models, such as those of Trompenaars and Hampden-Turner, Hofstede remains a classic in interdisciplinary cross-cultural research, from neurogenetics to marketing.

structure and functioning that depend on experience and affect behavior, at the same time facilitating further learning processes [4]. This particular adaptability of the brain enables its functioning even in critical conditions, since functions of damaged regions of the brain can be overtaken by other regions (and the most awe-inspiring example of that is the ability of people whose one hemisphere is inactive or damaged to function normally[8]). All stages of development of an individual, life script events resulting from a specific cultural environment one is emerged in, social learning and transmission of cultural features from penchant for expensive footwear to religious beliefs and Heisenberg's uncertainty principle, is possible because our brains are able to transform even the most abstract ideas and thoughts to physical processes that change the structure of the organ.

As recent research suggests, however, neural plasticity does not offer infinite response options to experience, and cognitive retooling (finding new ways to use sensimotor stimulations to represent information) will only be possible within the constraints of the brain's ability to adapt its predispositions and pre-existing functions to new but related uses [17]. Culture is a major factor providing the frames for such response, having a profound impact not only on what one perceives through the senses (the environment), but also works at the "how" of everyday cognitive tasks. In other words, while processing the world around us, we operate with cognitive tools that are not biologically determined, but invented and culturally transmitted. What is of crucial importance, they further re-engineer one's cognitive architecture. The piece of neurobiological research on cognitive retooling clinically prove the landmark linguistic relativity hypothesis (also known as the Saphir-Whorf hypothesis) which claims that the language one speaks (a cognitive tool), influences one's perception and interpretation of reality, both in the sense of what is chosen as representation of it, and how it is arranged. For instance, the pioneering research of Whorf compared the different conceptions of time in Hopi and English languages and concluded that it is a function of whether cyclic experiences are classed like ordinary objects—which happens in case of the English language or as recurrent events, as it is in Hopi [18].

There are of course other variables that influence neural processes behind cognitive processing, motivations, emotions and behaviors such as individual features, gender, social status, age, or even situational context. To vast extent, however, those categories are culturally-bound. Let us take gender—not only do the social roles and frames of acceptable behavior differ largely among cultures (and are subject to stark controversies if one compares the British party girls and Pashto women of Afghanistan as two distinct, yet equally socially accepted gender models), but the differences reach biological levels as well. In general, the level of testosterone in females is about twenty times lower than for adult males and is rather constant. Socio-cultural factors create significant exceptions, as women in

[8]The reason of that might be an inborn defect, major hemisphere damage resulting from a serious accident or the very rare neurosurgery procedure of hemispherectomy. More about the cases in [16].

several native societies (such as the matrilineal Khasi tribe of India), are to be competitive, aggressive and assertive, and their usual testosterone levels are significantly higher than in females from other cultures [19]. It is worth to conclude, however, that culture is not the key that opens all doors—since cultures are not homogenous, and even in case of such uniformed culture as the one that the Author origins from, where over 97% of the country's population consists of white Polish Catholics, one finds political, social and ideological divisions so deep that every November there are several conflicting National Independence Day marches that physically fight with each other.

13.3 Military in the Neurobiological Wonderland

Before we move on to the numerous applications of cultural neurobiology to a wide array of non-kinetic military activities, we must first perform a target needs analysis, and review the cognitive military enhancements state of the art.[9] Armed forces nowadays function in a space between pre-modernity and postmodernity. Our armies will be affected by the resulting human advancements, transhumanist ideas and singularity, at the same time having to perform their duties in societies that evolve in a completely different manner, without much regard for technological and civilizational developments. The goal of military activities in the field is to maximize human-system effectiveness where human factor is the key. The contemporary battlefield also brings on extraordinary demands on the soldiers as to their mobility, autonomy, physical performance and cognitive skills [20]. Complex-adaptive threats demand a comprehensive approach and joint and multinational operations, in which the management of training, leadership and conduct of the operations is very complex due to cultural and organizational differences between the entities engaged. The increased cultural diversity both within the armies, the „human domain" and the enemy, implies the growing role of cross-cultural competence and other soft skills. The scheme in Fig. 13.1, presents the relations between the changing nature of contemporary battlefield, the demands it poses for soldiers abilities, and the resulting assumptions of the Biotechnological Revolution in Military Affairs (BIOTECH RMA) that has been developed by the American military for at least two decades now.

As aforementioned, neurobiology is a vital component of the future warfare. As stressed in the landmark document outlining future directions of brain research, the *Opportunities in Neuroscience for Future Army Applications* report of 2009: "Emerging neuroscience opportunities have great potential to improve soldier performance and enable the development of technologies to increase the effectiveness of soldiers on the battlefield. Advances in research and investments by the broader science and medical community promise new insights for future military

[9]We will see to American solutions since U.S. Army has the most extensive experience in the field.

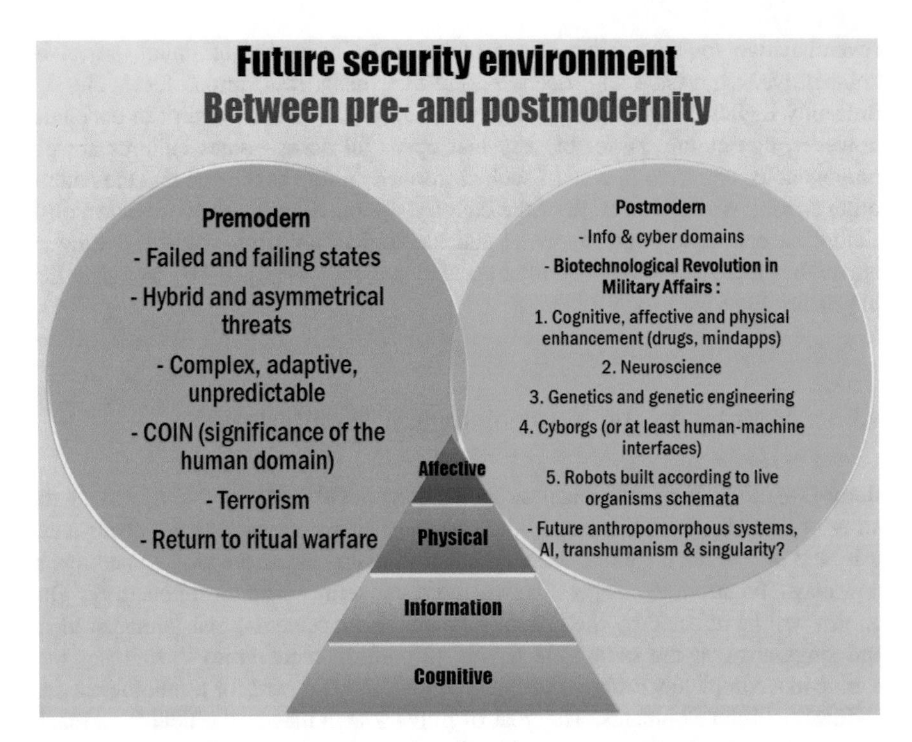

Fig. 13.1 The changing nature of the contemporary battlefield and its implications (*Source* own elaboration)

applications. These include traditional areas of interest to the Army, such as learning, decision making, and performance under stress, as well as new areas, such as cognitive fitness, brain–computer interfaces, and biological markers of neural states" [21]. Leading American military R&D institutions are already working on a robust array of projects that exploit potential applications of neuroscience to the training, conduct of military operations and post-operational force sustainment.

The Defense Advanced Research Projects Agency (DARPA), is the primary breeding ground of military innovations, famous for its visionary interdisciplinary enterprises. One of the major projects in the neurobiological domain, claimed to be one 10 World Changing Ideas according to the *Scientific American Magazine* [22], is the *Systems of Neuromorphic Adaptive Plastic Scalable Electronics* (*SyNAPSE*), which aims at creating electronic systems inspired by the human brain that are able to understand, adapt and respond to information in fundamentally different ways than traditional computers. The program uses a multidisciplinary approach, coordinating aggressive technology development activities in hardware, architecture and simulation. The initial phase of SyNAPSE developed nanometer-scale electronic synaptic components capable of varying connection strength between two neurons in a manner analogous to that seen in biological systems, and tested the utility of

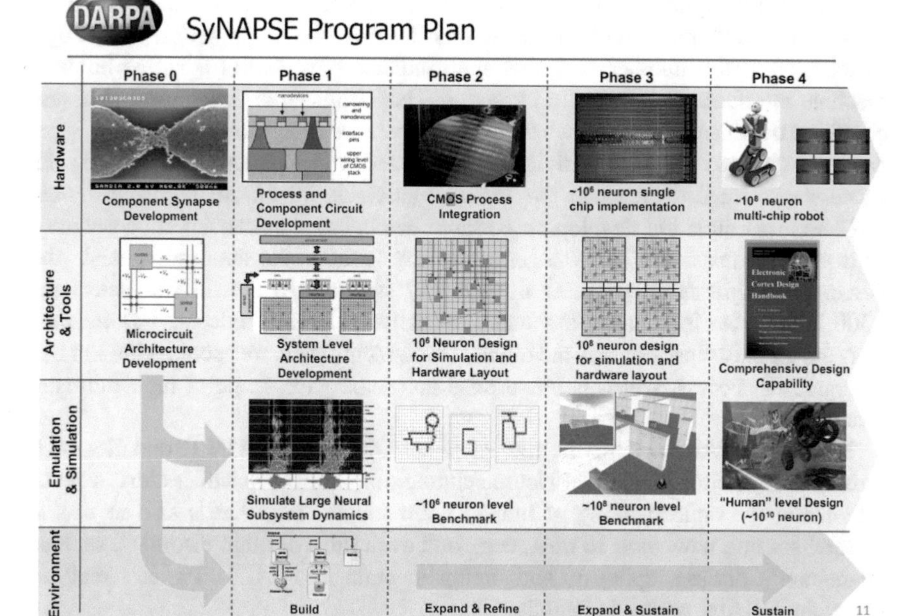

Fig. 13.2 DARPA's SyNAPSE road map (*Source* DARPA's official website, SynaPSE project materials. www.darpa.mil, 20 Aug 2015)

these synaptic components in core microcircuits that supported the overall system architecture [23]. Major further milestones of the project are depicted in Fig. 13.2

Second promising initiative is the *Narrative Networks*, that aims at understanding how narratives influence human cognition and behavior, and apply those findings in international security contexts. The program aims to address the factors that contribute to radicalization, violent social mobilization, insurgency, and terrorism among foreign populations, and to support conflict prevention and resolution, effective communication and innovative post-traumatic stress disorder treatments. *NN* exploits ways in which narratives consolidate memory, shape emotions, cue heuristics and biases in judgment, and influence group distinctions through creating a working theory of narratives, understanding of what role they play in security contexts, and an examination of how to systematically analyze narratives and their psychological and neurobiological impact on individuals [24]. The Project consists of three lines of research, including the development of quantitative analytic tools to study narratives and their effects on human behavior in security contexts, the analysis of neurobiological impact of narratives on hormones and neurotransmitters, reward processing, and emotion-cognition interaction; and the development of models and simulations of narrative influence in social and environmental contexts.

Concerning cognitive enhancements for the warfighter, DARPA launched three projects that addressed the issue. First was the *Augmented Cognition* (*Aug-Cog*), which exploited a number of neural state indicators to control adaptive human–machine interfaces to information systems. Those indicators "were used to assess cognitive overload stress, and when the stress became too great, they would trigger the dynamic load shedding activity of an interface management system" [21]. Its successor to some degree was the *Technology Cognitive Threat Warning System* (*CT2WS*) that aimed at developing portable binoculars that convert subconscious, neurological responses to danger into the user's consciousness, and the *Neuroscience for Intelligence Analysis*(*N4IA*), which used the EEG to detect the P300 brain wave to enhance the analytical skills of IMINT officers.[10] Those projects are already closed and as the official assessments state, met some—but not full—degree of success, which is no surprise given the wide scope of their futuristic premises.

Moreover, DARPA supports the President Obama's BRAIN (Brain Research through Advancing Innovative Neurotechnologies) initiative, which aims at revolutionizing the understanding of human brain functioning, that is said to enable researchers find new ways to treat, cure, and even prevent brain disorders, such as Alzheimer's disease, epilepsy, and traumatic brain injury [25]. Projects realized under the BRAIN umbrella include:

- Electrical Prescriptions (ElectRx) program, which aims to help the human body heal itself through neuromodulation of organ functions using ultraminiaturized devices, approximately the size of individual nerve fibers, which could be delivered through minimally invasive injection.
- Neuro Function, Activity, Structure and Technology (Neuro-FAST), that seeks to enable unprecedented visualization and decoding of brain activity to better characterize and mitigate threats to the human brain, as well as facilitate development of brain-in-the loop systems to accelerate and improve functional behaviors.
- Preventing Violent Explosive Neurologic Trauma (PREVENT), which is comprehensively evaluating the physics of the interaction between explosive blasts and the brain and has identified which blast components are associated with neurologic injury.
- Restoring Active Memory (RAM), that aims to develop and test a wireless, fully implantable neural-interface medical device for human clinical use. The device would facilitate the formation of new memories and retrieval of existing ones in individuals who have lost these capacities as a result of traumatic brain injury or neurological disease.
- Reliable Neural-Interface Technology (RE-NET), that seeks to develop the technologies needed to reliably extract information from the nervous system,

[10]The P300 brainwave is a signal detectable in the brain that is produced 300 ms after a stimulus occurs, only if the visual information is selected by the brain as of utmost importance. What is interesting, it is hardly ever perceived consciously. More on the projects in [2].

and to do so at a scale and rate necessary to control complex machines, such as high-performance prosthetic limbs.

- Development of non-invasive brain-stimulation technologies such as Transcranial Magnetic Stimulation (TMS) which can enhance as variety of neurological functions from mood and social cognition to working memory and learning and Transcranial Direct Current Stimulation (TDCS), which is an even more promising neuromodulation and cognitive enhancement technology [26].

It is crucial to acknowledge that the projects above are an example of "social corporate responsibility" of DARPA—which has always been regarded a controversial R&D institution by the civilian scientific establishment. This subject will be, however, further elaborated on in the part of the chapter devoted to the ethics of merging neuroscience and military.

Although the most (in)famous, DARPA is not the only provider of innovative solutions in the field of neurobiology for security and defense purposes. The Intelligence Advanced Research Projects Agency (IARPA) under the Office of the Director of National Intelligence, introduced even more exotic visionary research projects. The *Integrated Cognitive Neuroscience Architectures for Understanding Sense making* (*ICArUS*) for instance, introduces elements of AI to the research. The main aim of the ICArUS program is to construct integrated computational cognitive neuroscience models of human sense making. The applications will include improved prediction of human-related strengths and failure modes in the intelligence analysis process and are expected to point to new strategies for enhancing analytic tools and methods [27]. Another initiative worth mentioning since it embraces cultural influences on cognition, is the *Knowledge Representation in Neural Systems* (*KRNS*) project. Drawing on findings from neuroscience, linguistics, AI, behavioral science and human factors analysis, the project aims at developing and rigorously evaluating theories that explain how the human brain represents conceptual knowledge. The research bases on the assessment of how concepts can be interpreted from neural activity patterns using algorithms derived from the theories. Moreover, apart from the new theories and algorithms, KRNS features the development of innovative protocols for evoking and measuring concept-related neural activity using neural imaging methods such as, among others, functional magnetic resonance imaging (fMRI) and magnetoencephalography (MEG) [28].

Third and even more brave project, is the *Machine Intelligence from Cortical Networks* (*MICrONS*) which "aims to achieve a quantum leap in machine learning by creating novel machine learning algorithms that use neurally-inspired architectures and mathematical abstractions of the representations, transformations, and learning rules employed by the brain. To guide the construction of these algorithms, performers will conduct targeted neuroscience experiments that interrogate the operation of mesoscale cortical computing circuits, taking advantage of emerging tools for high-resolution structural and functional brain mapping" [29]. Employing research results from theoretical and computational neuroscience, machine learning and connectonomics, the 15-year duration program exceeds in its scope and innovativeness even DARPA's most futuristic projects. Leaving the best to the end, let us move on to the *Strengthening Adaptive Reasoning and Problem-Solving* (*SHARP*) initiative that seeks to advance the science on optimizing human adaptive reasoning and problem-solving capacity in information-rich environments [30]. The program tests and validates a myriad of available methods of cognitive neuroscience, psychometrics, neuroergonomics and other fields (the methods and tools include even transcendental meditation, transcranial stimulation and serious games) to optimize cognitive performance.

Also other American research organizations such as the Office of Naval Research's Human and Bioengineered Systems Division (that welcomes proposals expediting research in among others biometrics and human activity recognition, cognitive sciences, computational neurosciences and biorobotics, human factors, organizational design and decision research, social, cultural and behavioral modeling and training, education and human performance) [31] and Sandia National Laboratories (in among others *The Human Decision Making For National Security* and *Brain-Inspired Computing* projects) [32] are "digging in the brain". Evidently, special units in the fashion of New Earth Army from *The Men Who Stare at Goats* war comedy, are only a step away [33].

Yet, we must not forget about the added value of pharmacologization of warfare which is another important field of development for the BIOTECH RMA, in particular in the neuro-enhancement domain. The aforementioned Committee's on Opportunities in Neuroscience for Future Army Applications Report indicates two fields in which the neuropharmaceutical approach should be investigated: sustainment and improvement of soldiers' cognitive and behavioral performance. Modafinil[11] is already prescribed to pilots, who are tasked to fly prolonged missions to eradicate the negative effects of sleep deprivation. Sertraline hydrochloride (found in popular depression and mood disorder curing drug sold as Zoloft or Lustral) is often prescribed to troops who have sustained repeated combat exposure to reduce the consequences of persistent stress and the risk of depression. The prospective neuropharmacological agents include a nicotinic acetylcholine receptor modulator to improve attention and executive function in attention deficit disorder, N-methyl-D-aspartic acid (NMDA), a receptor-positive modulator to enhance memory consolidation, and a metabotropic glutamate receptor antagonist to treat psychosis [21].

The official list of what already is being used and what is tested is humble as compared to the real pharmacologization of warfare, since the Army has been exploiting a whole arsenal of pharmaceuticals since the foundation of the one nation under God. The American civil war of 1861–1865 welcomed the widespread use of laudanum, opium and morphine, the second World War was ran on amphetamine, in the Korean War of 1950–1953 the soldiers were administered

[11]Modafinil is Food and Drug Administration approved medicine used primarily in treating narcolepsy. In healthy individuals it enables prolonging the waking time by several times, with no side effects yet recoded. It is a popular performance enhancer in the American academic world as well.

dextroamphetamine and methamphetamine (and self-medicated themselves with heroin and speedballs), during the Cold War the Army researched on the potential of, among other hallucynogenes (such as LSD), to reach the most outrageous high in Vietnam (1965–1973), where dextroamphetamine, codein, steroids, tranquilizers and neuroleptics were administered by the army, and soldiers completed the list with amphetamine, heroin, barbiturates, LSD, marihuana, morphine, hashish and opium [9]. The situation improved a little when preposterous side effects of majority of the psychoactive agents used manifested themselves after the conflicts, leaving tens of thousands of veterans addicted, traumatized and incapable of either social or individual functioning. Therefore the wars in Afghanistan (2001–2014) and Iraq (2003–2011) were ran on more "over-the-counter" drugs (as compared to heroin) such as improved dextroamphetamine, tranquilizers (Valium), opoid pain medications and steroids. At least officially.

Nowadays the U.S. military, apart from improving the already tried central nervous system stimulants such as dextroamphetamine and modafinil, is conducting research on several other promising substances. Ampakines are compounds known to enhance attention span and alertness, and facilitate learning and memory, acting selectively on crucial brain regions, improving neuroplasticity through long-term potentation. They were initially researched on as a remedy for Alzheimer, schizophreniaor ADD diseases, but in 2005 the DARPA's *Prevention of Sleep Deprivation* program advanced the research in other directions. The new generation of sedatives is also investigated, together with propanolol, a beta-blocker that might be a blessing for PTSD-affected individuals. The drug however, is subject to many controversies as it desactivates brain regions responsible for emotional link connection of traumatic memories, which might lead to a situation where no reflection on past deeds would follow, and moral reasoning might be later negatively disturbed. Moreover, the decision-making process is also changed under the influence of the substance, with significantly diminished capacity to properly assess the effects of one's actions and the odds of success, which in military context is a real issue and seems like a faustian bargain [2].

The most significant potential, however, is ascribed to oxytocin, popularly called the "cuddle hormone". It is naturally produced in human bodies (in particular during pregnancy, childbirth and lactation in females, during care for the offspring in males, and in various stages of establishing a romantic relationship), responsible for the feelings of security, attachment, trust and an array of other positive emotions. Therefore it is considered the most pro-social neurotransmitter that boosts empathy, altruism, cooperativeness and loyalty towards a social group. The potential of the neurotransmitter was already exploited commercially, since Vero Labs retail oxytocin spray online under the brand name of "Liquid Trust". For only $29,95 per bottle, managers, singles seeking a partner and salesmen (which are said to be the primary targets of the product) can benefit from the instant trust and attachment it is claimed to enhance [2]. It was discovered, however, that once isolated and properly administered, the neurotransmitter reinforces group cohesion, trust and morally proper behaviors, which are the features of a perfect armed formation. As professor Łukasz Kamieński claims in [2], we might soon have the perfectly in-sync, "hormonal bands of brothers" as the basic army organizational units.

Summing, up, as we see from the above projects and applications, the roles of neuroscience in sustaining and enhancing future force as foreseen by the US armed forces are robust and truly promising. Moreover, variability at the neurological level, which by and large is the subject of this chapter, is considered to be a force multiplier. Recommendations presented in the NRC Report on training, decision-making and sustaining and improving cognitive and behavioral performance, stress the significance of individual differences in behavior, cognition, and performance of skilled tasks which are as deeply rooted in the neural structure of individuals as differences in strength, stamina, height, or perceptual acuity are rooted in their physiology. It is noted that such differences have vast consequences for many Army applications, and influence operational readiness and the ability of units to perform assigned tasks optimally. Basing on this logic, recommendation 17 states that "using insights from neuroscience on the sources and characteristics of individual variability, the Army should consider how to take advantage of the variability rather than ignoring it or attempting to eliminate it from a soldier's behavior patterns in performing assigned tasks. The goal should be to seek ways to use individual variability to improve unit readiness and performance" [21]. What is surprising though, hardly any of the existing projects presented above take into account cultural impact on neural functions. Since we know now what to do (incorporate cultural variability into brain research and its applications), let us then investigate the "how" that has not yet been exploited by the military community.

13.4 A Primer for the Neuro-Enhanced Military Training

The effectiveness of a military organization, undeniably has its roots in proper selection of candidates. Popular psychological aptitude and self-assessment tests that are a part of recruitment process, might not always reflect the true potential and predispositions of future cadets and members of armed formations if they do not take into account the cultural variations in self-perception. As indicated by Kitayama and Park in a research devoted to **self-awareness** in the context of social grounding of the brain, structure of the self, varies systematically across cultures at the neurological level. Individuals belonging to more collectivistic cultures, perceive themselves by rule in relation to their group, which is exemplified by East Asians and North Americans with Asian heritage who tend to be more interdependent in the sense that they keep their interpersonal or social self relatively more accessible and value it more, they apply the collectivistic perspective to social perception with consequence to attention biases, their emotional responses are oriented more towards social goals, agendas and issues, which has also implications on their motivation—with prevailing social antecedents and reinforcements [34]. North American and Western European individuals on the other hand, were found to be more individualistic (independent) since they tend to keep their personal self highly accessible, and put more prominence to it, their social perception bases on their individuality, not group point of reference, the emotional life of such individuals is based largely on personal goals, desires and needs, and their internal motivation bases on those individual pursuits.

Initial research on neural influence of religious beliefs performed by Han also brought surprising results as it indicated that they modulate neural responses within one's theory of mind regions and affect neural representations of the self, since different brain regions are involved in **self-evaluation** in atheists and religious individuals'. Neural responses of the first group indicate greater illumination and representation of the self, while "believers" tend to use the brain regions responsible for evaluation through the reference to the higher spiritual instance (however one defines God in numerous denominations) [3]. In the context of HR policy, this distinction has implications in cases where individuals are to assess own abilities, character and performance or perform moral judgment tasks. Moreover, further implications for learning processes, performance self-assessment, moral reasoning, group functioning and social interactions can be drawn from the findings of the cultural variables in self-perception.

Cross-culturally variable components of the learning and skill acquisition process span from such details as representation of number, spatial representations of time through categorization processes, learning feedback and problem-solving strategies. But cultural variables that influence neural underpinnings of learning, begin at the very core level of physical perception. It is possible in case of **visual perception** for instance, since The fundamental architectonic bricks of the primary visual cortex responsible for it, can change in response to experience not only during a limited critical period restricted to the first few months of life as taken for granted in initial neurophysiological research. Recently it was fund that various forms of perceptual learning and cortical plasticity occur throughout the whole human life span [35]. The first evidence that visual perception varies cross-culturally was predicted in a landmark research of Rivers as early as 1905, who investigated the responses of the English, Indian and New Guinean individuals to the popular Muller-Lyer optical illusion.[12] The English tended to assess the difference in length of the lines (which is a result of a perceptual illusion) much more often than the representatives of two previous groups, which was explained by the everyday landscape usual for the representatives of those three cultural groups, with the English being used to more regular angles and buildings that limit the vision perspective, that were absent in the case of India and New Guinea [36]. Also the **representation of 3D objects in two-dimensional space** and their interpretation, are subject to variation due to cross-cultural differences in perception of depth, which has grave implications for the ability to read maps. The very use of maps moreover, which are physical tool that can become internalized under the rule of the

[12]The illusion tests the visual perception of the length of two equally long lines with different endings: >—<, ↔.

aforementioned cognitive retooling theory, to form a basis of a new cognitive tool. Processing spatial navigation information from a survey perspective is different from and involves different brain areas than a route perspective. Therefore, cultural propensity to use maps might not only alter the perception of space but also the very "how" of spatial navigation. It was one of the problems that the Polish Operational Mentor and Liaison Team (OMLT) and Police Operational Mentor and Liaison Team (POMLT) training teams came across while training the Afghan Army during the ISAF operation, as the local trainees had difficulty in using maps, since their spatial overview of the region was based on experience, not 2D representations [37].

Also the way that the environment is perceived visually will vary, either in reality or 2D (images, virtual worlds) or 3D depictions (models, immersive environments) depending on whether one presents holistic **type of perception**, in which case the surroundings of an object of interest are processed with equal attention, or analytic type that creates the full picture by focusing on the features that are most crucial from the observer's point of view. Since visual processing depends on patterns recognition, and perceptions of patterns will definitely differ among cultures, the very content and interpretation of the perceived set of stimuli might be surprising. The differences described above have vast significance not only for producing learning materials (which, additionally, have to characterize with culturally-appropriate and realistic contents regarding for instance taboos, norms of what is and is not acceptable, features that are familiar and evoke desired associations), but also extremely important while creating serious games and immersive environments, and in all sorts of imagery intelligence collection and processing (IMINT).

What has also been found in the cognitive domain that adjusts **emotional responses to visual stimuli**, is the issue of facial emotion recognition. The most frequently quoted research of Chiao et al. from 2008 discovered that both native Japanese and Caucasian Americans exhibited greater brain activation in the amygdala region in response to fear faces expressed by their own cultural group than the other. This indicates that we are more capable to infer mental states from non-verbal cues by turning toward familiar stimuli [3]. Moreover, cultural group one belongs to provides an important means by which people infer mental states of others from non-verbal content, and difference between egalitarian cultures (such as broadly understood Western ones) and hierarchical ones (Middle Eastern) was observed. And since emotions are not just automatic responses to stimuli but are subject to meta-awareness and appraisal, which lead to processes of internal regulation of their outlet and corresponding facial expressions and behavior, the observable non-verbal correlates of emotional states will also differ cross-culturally for the same emotion [38]. This fact is also of significance in designing any type of virtual reality, serious games or tutoring systems, and for officers responsible for contacts with local population—in case of HUMINT officers for instance, the ability to read non-verbal cues outside of their own cultural contents is imperative.

Attention is a next cross-culturally divergent field. In a 2008 neuroimaging study, Hedden and colleagues asserted that European Americans recruit an attention network of the brain when they focus exclusively on the object and its context than while focusing on the object itself. However, Asian participants of the study tended to display an opposite pattern, recruiting the same attention network more in the focused attention than the holistic attention task. The attentional attunement to contextual information contributed to Asian individuals has also been investigated with ERP, and has shown that they display greater sensitivity of incongruence of semantic contents and the inconsistence of an object in a picture and the context. The dissonance was much lower for European Americans [34]. In further line of research that was to test the significance of mnemonic context to learning, it was confirmed that the effect was greater for Japanese than for European Americans. This does not of course imply that context is insignificant in educating the last cultural group, since any learning bases on correlations and contextualization. There are, however, differences in attention bias.

Yet another cross-culturally varied foundation of any type of learning and skills acquisition is **memory**. First of all, the aforementioned cognitive and attention bias serves as a filter of which information is found significant, stored and utilized. Secondly, the cultural variation in holistic versus analytic thinking affect how information is encoded and retrieved. Several studies have proven that Westerners are more likely to encode and retrieve focal objects set in a complex visual scene, while East Asians encode equally the context and the object itself [3]. Further neuroimaging research that compared young and elderly East Asians and Westerners discovered that the processing activity differed in elderly representatives of the two groups, but not among young individuals, asserting that neural regions might exhibit cultural variation as a function of age. Researchers also elaborated on the role of literacy in memory functioning, comparing the ability of students from Ghana and the United States to memorize short stories. Individuals from Ghana, where oral knowledge transmission tradition was at the time prevalent displayed better results than the American counterparts, provided that the stories presented logical whole. Lists of randomly chosen items were memorized with the same effectiveness [26].

Next proof comes from cognitive psychology and pertains to the variables of the reminiscence bump. Elderly people asked to recall personal memories, usually refer to numerous events from adolescence and early adulthood (the reminiscence bump). Empirical evidence from cross-cultural investigation among individuals from America, China, Japan, England and Bangladesh performed by Conway and team in 2005 found that however the bump is present in all five cultures, the Chinese were more likely to recall group and socially-oriented events, whereas Americans exhibited greater tendency to reminisce on events that related directly to them as individuals. The research premises based on the Rubin and Bernstein life script theory of 2003, which claims that life script, namely cultural expectations concerning the nature and order of major life events, guides and organizes the retrieval of autobiographical memories [4]. Moreover, the contents of the memories referred prevalently to the culturally desired and predictable events in their lives, which were different for each researched groups both in their temporal and emotional dimension.

The implications of cultural variability of memory processes are in particular applicable in designing **mnemonic reinforcements**. Obviously, properly adjusted visual or auditory input will result in desired outcome. However, it was found that the affective aspect of the learning process plays even greater role—the more the emotional intensity of the material (positive and negative), the more intensely the amygdala activity is engaged in encoding and consolidation of it [38]. The enhancement of emotional memories happens thanks to a combination of attention-mediated short-term effects on encoding and facilitation of longer term memory consolidation by the amygdala, in a different way than in case of emotionally neutral memories, making them particularly strong and durable. According to Seligman and Brown, rituals which involve physical, sensory and emotional reinforcement at the same time, can be of particular significance to such consolidation, since they "lend themselves to associative learning, and through such learning the social and cultural models they present are imbued with particular significance. This in turn creates particularly strong memories for this information, and links it to powerful emotions that help make it motivationally salient" [38], which might not only expedite the design of Culturally-Aware education strategies, but might also aid reengineering traumatic memories and be used in PTSD treatment. Native American tribes for instance, observe different kinds of ceremonies for returning soldiers. In the Navaho communities, a soldier's family can decide to sponsor a ceremony for him on his return. They contact a spiritual leader, sometimes called a medicineman, who talks to the soldier about his experiences and decides which ceremony would be best. The Enemy Way ceremony, (also known as the Squaw Dance), is an example of Navajo ritual used for soldiers who were in combat, captured or wounded [39]. The effect of those rituals is the re-integration of the soldiers back into the communities and significant decrease of the risk of PTSD development. The significance and mechanisms behind the rituals may be also utilized in non-kinetic aspects of COIN in determining how social ideologies, including those of extremist nature are reinforced through the use of rituals in various cultural groups.

Cultural influences can be observed also in the field of **motivation**, due to the aforementioned significance of individual attributes for independent selves, and collectivistic attributes for interdependent self-perception that influence social motivation. The first body of evidence was presented by Taiwanese researchers who attested that achievement motivation among their fellow nationals is by and large group or socially-oriented. This was confirmed in behavioral studies of Iyengar and Lepper [40] which found that European American children are motivated to perform a task intrinsically, while Asian American children showed greater predisposition towards external motivation provided by an authority (for instance older relative or teacher), which founded a hypothesis that interdependent selves are subject to motivation anchored in social expectations rather than individual pursuit [34]. Once again we have to do with dual potential of the findings, since apart from explaining motivational dynamics in culturally-diverse groups of learners, they can also inform the culturally-bound rationale of individuals to join and actively participate in criminal, terrorist and insurgent organizations.

We must also bear in mind that military education and training is a specific type of adult learning that encompasses equally the cognitive, affective and psychomotor domains of knowledge acquisition. A bunch of evidence on the cross-cultural specificity of adult learners was found also through neuro-anthropological research. The basic andragogy principles that seem universal claim that adults are self-directed and internally motivated, collaborate with instructors to promote their learning and the learning experience itself is grounded in life experience and needs to be applicable – non-practical items will be considered redundant. However, whereas in the "Western" societies adult learning appears to be individualistic and egalitarian, non-western learners, tend to view knowledge as communal. For instance, if a village has no doctor, then the members of the community pull their resources together, and send one of the young individuals to a medical school, so that after he returns, the gap will be filled [16]. The very strong motivation towards learning of such individual and his feeling of duty will be rooted in a very different way than of somebody's who would like to perform the profession of personal passion or for the prestige of it. Moreover, non-Western learners were found to treat education as lifelong, holistic process that does not stop once a person left formal institutions. And educational establishments are not perceived as the sole source of knowledge, since the everyday life experiences are seen as equally legitimate and respected generator of it. Also what it means to be educated varies across cultures, and the level of formal schooling and granted degrees are in some cultures regarded less important than holistic social knowledge, which is of significance both for military education, and in contacts with local population during deployment.

Concerning the **psychomotor domain of learning**, the recent research on neurobiological underpinnings of embodiment, understood as the subjective experience of the body, has significant implications as well. As indicated in a piece of work on emotional embodiment (the global representation of bodily status as somatic mood), the neural correlates of it not only reflect the evolutionary heritage, but are an object of cultural practice as well. In one of the experiments on pain perception, the brain activity of Japanese yoga masters (who claimed to be immune to pain during meditation) was measured, and the activity of insula and anterior cingulated cortex, that are primarily responsible for emotional aspects of pain, was indeed inhibited in the meditative state [41]. Those two brain regions have proven to be not only the primary neural sources of somatic awareness such as pain, but also empathy, racial group identity, social norm violation perception and moral institution, exhibiting thus a link between somatic awareness and social emotions. As the Authors of the research claim, however, further investigations must be performed locally before practical improvements are introduced. And in case of military training, group cohesion, emotional management and kinesthetic aspects of learning may be significantly expedited through the enforcement of bodily awareness and understanding of the emotional embodiment mechanisms.

Last example is inferred from the Author's pre-deployment training (PDT) practice. While designing and providing cross-cultural and other soft skills training to culturally diverse groups of officers,[13] it proved that they differ largely in the **susceptibility to stereotypes**, depending on whether the social environment they originated from was an example of successful multicultural policy (like in Australia, Denmark or the Netherlands) or lack of one (Poland). This resulted in difficulties in instilling flexible, open-minded approaches towards differences and cultural relativism in learners from the latter countries. In consequence, understanding of how the target deployment cultures work, was reduced, therefore all the tasks that aimed at operationalizing certain cultural features and traits for military purposes, had to be facilitated with additional red teaming component to sensitize the groups. Other such consequences were observed in the fields of **perspective plurality**, be it the religious diversity within countries, variety of acceptable gender roles or multiethnic and multinational composition of the environments or institutions that the trainees came from. It is obvious that the greater the plurality, the easier it was to gain the aims of the cross-cultural training. Moreover, the most simple value of respect towards such plurality is not a norm in more conservative and homogenous countries, with numerous ethnocentric approaches prevailing. By rule, in countries like Poland (and several other new-EU countries), with few exceptions, the PDT training in cross-cultural competence is organized locally, in homogenous groups, with instructors coming from the same culture as the trainees. Even if the instructors are well prepared and educated, and can gain a multicultural perspective, they might fail to acknowledge the numerous cognitive biases such dynamics will result in. It is then crucial to take all the above into account during the design and conduct of such training, to explore the whole potential of operationalization of culture.

The application of the findings presented above to serious game design should be also considered, since none of the tactical software used in NATO or EU member states[14] takes into account the cultural variations in the knowledge and skills acquisition domains described above. Additional commercial serious games that train cross-cultural skills as such are being used of course (like *Global Conflicts: Palestine* by Serious Games Interactive, *Tactical Iraqi Language and Culture Training System* used in the US Army or the *Adaptive Thinking and Leadership* simulation game, to mention a few) [43] either in PDT and in overall officers' career development. The analysis of their effectiveness, however, rests on the assertion that all types of learners have similar learning capacity and modes, and that the skill development through the use of the software will be uniform among the trainees, regardless of their cultural background, which as we already know, is not exactly true. Moreover, when we look at the technological limitations of

[13]The Author directs and conducts, among others, the *Cross-Cultural Competence for CSDP Missions and Operations* Course for the European Security and Defence College and the HUMINT officers pre-deployment training for the Polish Armed Forces.

[14]The most common military tactical training and mission rehearsal simulation software is the Virtual Battlespace (VBS) immersive training package from Bohemia Interactive Simulations, with more than 19 NATO nations and nine partner nations as well as three NATO entities are current users of VB [42].

creating realistic immersive environments and modeling human dimension (avatars), the situation is far from perfect. Even the cutting edge VR technologies used in devices like the Oculus Rift head-mounted displays (introduced in 2016), HTC Vive or the expected Sony's Project Morpheus's VR products have numerous limitations. The problem does not lie in the degree of realisticity of the VR, since the virtual reality does not have to be5D for an individual to be able to internalize the experience and fully immerse in it. The perceptible delay between one's movement and the change of the VR image, however, is an issue since even the 20 ms delay causes motion sickness in longer use of HMD equipment. So are the vergence-accomodation conflict or the peripheral and central vision synchronization difficulty [44]. The presented research findings on, among others, various aspects of visual perception, facial emotions recognition and spatial navigation, might contribute to improvement in that field.

13.5 Reservations and Controversies

The review of existing visionary projects and potential cognitive applications that might improve the military status quo is promising indeed. We must not, however, forget that there are certain limitations in applying cultural neuroscience findings to security and defense purposes. First issue pertains to the reception of the cultural neuroscience research—they are simply considered politically incorrect. In the modern world that officially stresses perfect equality of all in all aspects, the claim that people do differ at the very essential, biological level of brain functioning, sounds like a heresy and memories of eugenics cut all potentially constructive discussions on those differences immediately. The role of cultural neuroscience, however, is far from promoting and deepening cultural, racial or ideological divisions. On the contrary, merging the social and natural sciences enables the insight into cultural and behavioral impact on human behavior in novel ways, and inform public policy issues on cultural diversity and interethnic justice by studying the ways in which cultural identity affects the brain and behavior [10]. Moreover, ethnographically driven laboratory studies and the development of ecologically valid experimental protocols enable deeper insight in the complexity of human functioning [38], even in the very limited laboratory settings.

Second reservation is of methodological nature. Cultural neuroscience findings still base on population-scale samples, computational models and laboratory conditioned research, which fails to give us the true insight into how cultural traits and features are shaped by the human mind and underlying biological structures. Neuroanthropological approach must be applied as complimentary, since it gives the hands-on insight into those matters, confronting the neuroscientific findings with field work. It provides local, more exact, intra-population variables examination—as contrasted with the population-scale, cross-cultural neurobiological inquiries, and offers tools for practical utilization of gathered research results in given group. It is worth, however, to begin with the large picture offered by cultural neurobiology, to present major, well-documented differences—since the field of neuroanthropology is a discipline in initial stages of development.

Moreover, numerous ethical concerns are already expressed by the civilian scientific world. Since various national security bodies are exploring neuroscience to advance interrogation methods and detection of deception through for instance fMRI or EEG use for the detection of the aforementioned P300 wave [45], the question whether it does not violate individual integrity and right to privacy must be answered affirmatively. This example is a part of wider group of concerns related to the dual use of research results for purposes that might be considered immoral or harmful. In case of military, an institution that has elimination of other human beings as a permanent point of its agenda, dual use problem is inescapable. Therefore the chapter presented potential solutions for non-kinetic activities only, with full and sad awareness that other applications of neurobiology to military activities do, and will exist, thus any research done in the direction must be performed with extreme ethical scrutiny and responsibility. The issue has yet another aspect—most of the means and technologies of the improvement of human performance begin with military applications, and only after a period of testing by the military or black operations are they unveiled to the general public. In 2008 the JASON group, the Pentagon's top scientific advisers, warned that the U.S. military might face enemies with technologically enhanced abilities, such as brain-machine interfaces and pharmaceutical cognitive enhancements [46]. This is also how the aforementioned MK-ULTRA project started—from a gossip that Soviets possess mind-control tools and techniques. The result was ruthless wave of "neurobiological" research, justified by grave national security concerns.

It is also essential to mention simple technical security concerns. The danger of hacking intricate brain-machine interfaces is one of them. Moreover, the more technologically complex an appliance, the more prone to human or technical errors it will be. Even smallest disturbances in functioning of those high-tech cognitive enhancements might have disastrous consequences both for the individual (with his neurobiological functions disturbed or destroyed in the worst-case scenario) and for the task (imagine the "collateral damage" rate in a combat drone error piloted through a brain-machine interface). The influence of neurobiological re-enchantement of war can have also consequences for the very nature of warfare as such, since, as professor Christopher Coker claims, even such deeply existential aspects of it as courage or devotion of soldiers could be brought down to their chemical components, giving hope that if only we can decode every aspect of our neurobiological and genetic codes, it will be possible to engineer the perfect future warriors [47]. In such situation, conscious human effort towards excellence will become redundant, and since there will be no choice whether one would or would not like to develop the extreme and irreversible warrior capacities, we must ask what will happen with such individuals and their "superhuman" skills, once their service is over.

The very last question that comes to mind is—should we fear the that the "new brave" neuroenhanced social groups, with their cognitive superiority will drift towards an alien, inhuman condition that will change the whole human cultures dynamics? The answer seems to be no—or at least, not so fast, and is rooted in human nature. Regardless of the culture, we are imperfect creatures. We could already have elevated our cognitive abilities, extend life span, optimize social relations and expedite emotional growth with the already available theories and methods. We are all perfectly aware of proper dietary choices, the significance of exercise in healthy lifestyle, constructive methods that enable coping with stress or ways of improving interpersonal relations, to mention a few. We understand the threats of smoking, lying and excessive chocolate consumption. We have produced an infinite number of learning enhancement tools, appliances and techniques. We have created numerous national and international institutions that are to improve our individual and social functioning, and finally bring peace and stability to the world. Moreover, with all the scientific, diplomatic and technological advancements nowadays, we could in theory move war activity entirely to the cyber space. Instead, we still face bloody conflicts that inflict millions of victims. It was already noted by Plato in third c.b.c. that the world we live in is a faint reflection of the perfect world of ideas. This also pertains to the nature of science—ideal models and theories are far from what they were conceived to be when applied. Maybe that is also the reason why in presence of all the perfect solutions, we still find ourselves consuming junk food in front of a TV (with the news channel transmitting yet another war in which hundreds of thousands die and millions suffer). We should then rather not expect that the upcoming cognitive revolution will affect larger groups instantly. Anthropologists, neuroscientists and sociologists will have plenty of time to get ready for the mass neuroenhanced culture.

13.6 Conclusion

As evidenced by cultural neuroscience research, cultural differences play significant role in differentiating basic cognitive and affective processes involved in learning. Both broad and detailed categories like self-awareness, theory of mind, visual perception representation of number, spatial representations of time, problem-solving strategies, attention, memory, motivation and mnemonic reinforcements will be realized and express themselves variously. They have consequences for not only designing and the conduct of military training, but also in recruitment processes, intelligence collection and non-kinetic military activities such as PSYOPS and selected aspects of COIN, to mention a few. A full-spectrum application that will be elaborated on in Author's further research is as follows:

- Responding to cultural diversity of military learners in cognitive, affective and psychomotor domains.
- Projecting and improving virtual training tools and immersive environments.
- Neuroergonomics and neurofeedback.
- Improved leadership and management of multicultural groups and units.
- Better incorporation of cultural factors in operational planning and conduct.

- Support to PSYOPS and other non-kinetic operations.
- Examining the influence of combat conditions on CAP domains in various groups.
- Projecting military mindapps.
- Pharmacologization of warfare.
- Treatment and prevention of PTSD thanks to the improved understanding of the origins, processes and effects in all CAP domains.
- Projecting perceptions, decoding the role of narratives and measuring responses to PTSD treatment.

The very detailed comparative studies presented in the military training section can be for instance applied in the non-kinetic support of operations. Either intelligence gathering and processing (in particular HUMINT), the management of local interactions and contacts with the host nation population, PSYOPS, command and leadership of multicultural teams, the design of cross-cultural decision-making support systems or the introduction of various military mindapps, can benefit from embracing cultural variation at the neurobiological level. Also post-operational force sustainment, in particular the PTSD prevention and treatment will be such fields, since even the biology of fight or flight reaction and emotional antecedents of stress responses differ cross-culturally due to social perceptions of aggression, trauma and acceptable modes and means of emotional regulation.

The Author does realize that the subject matter of the work might be highly controversial, and the clinical research results on cultural influences on variations in brain functioning may lead to overarching generalizations, in particular that cultural neuroscience is a young and still developing branch of knowledge, and neuroanthropology is in its promising, but yet, initial stage of development. The purpose of this research work, however, was not to present a neat "theory of everything", but to indicate current and possible future directions of research in the field.

References

1. Bousquet, Antoine: The Scientific Way of Warfare: Order and Chaos on the Battlefields of Modernity. LSE, London (2007)
2. Kamieński, Łukasz: New Brave Soldier. Biotechnological Revolution and War in the XXI Century. Kraków, WUJ (2014)
3. Chiao, J.: Cultural neuroscience: a once and future discipline. Prog. Brain Res. **178**, 287–304 (2009)
4. Eysenck, Michael, Keane, Mark (eds.): Cognitive Psychology. Psychology Press, London and New York (2015)
5. Adult learning cross cultures: Neuroanthropologyscientific blog. May 30, 2014. http://blogs.plos.org/neuroanthropology/2014/05/30/adult-learning-across-cultures/. 9 June 2014
6. Arnett, J.: The neglected 95%: why American psychology needs to become less American. Am. Psychol. **63**(7), 602–614 (2008)
7. Lende, D.: Advances in Cultural Neuroscience. Neuroanthropology Scientific Blog, March 29, 2013. http://blogs.plos.org/neuroanthropology/2013/03/29/advances-in-cultural-neuroscience/. 10 Aug 2015

8. Lende, Daniel, Downey, Greg: The Encultured Brain. An Introduction to Neuroanthropology. MiT Press, Cambridge (2013)
9. Kamieński, Łukasz: Pharmacologization of Warfare. Kraków, WUJ (2012)
10. Chiao, J. et al.: Theory and methods in cultural neuroscience. Soc. Cogn. Affect. Neurosci. **5**, 356–361 (2010)
11. What is Serotonin: Medical News Today, June 26, 2015. http://www.medicalnewstoday.com/articles/232248.php. 10 Aug 2015
12. Hofstede, G. et al.: Cultures and Organizations: Software of the Mind (2010)
13. Feldman, M. Laland, K.: Gene-culture coevolutionary theory. Trends Ecol. Evol. **11**, 453–457 (1996)
14. Meriam, S., Kim, Y.: Non-Western perspectives on learning and knowing. New Dir. Adult Continuing Educ. **119**, 71–79 (2008)
15. Mrazek, A. et al.: The role of culture–gene coevolution in morality judgment: examining the interplay between tightness–looseness and allelic variation of the serotonin transporter gene. Cult. Brain **1**(2–4), 100–117 (2013)
16. McClelland, S., Maxwell, R.: Hemispherectomy for intractable epilepsy in adults: the first reported series". Ann. Neurol. **61**(4), 372–376 (2007)
17. Wilson, M.: The re-tooled mind: how culture re-engineers cognition. Soc. Cogn. Affect. Neurosci. **5**, 180–187 (2010)
18. Wilson, R., Keil, F. (Ed.): The MIT Encyclopedia of the Cognitive Sciences, pp. 475–476. MIT Press: Cambridge (1999)
19. Cashdan, E.: Hormones and competitive aggression in women. Aggressive Behav. **29** (2003)
20. Parmentola, J.: Strategic Implications of Emerging Technologies. US Army War College SSI, Carlisle (2010)
21. National Research Council: Opportunities in Neuroscience for Future Army Applications. National Academy Press, Washington DC (2009)
22. Fletcher, S.: World Changing Ideas 2014. Scientific American, November 18, 2014. http://www.scientificamerican.com/article/world-changing-ideas-2014/. 15 Aug 2015
23. DARPA. Systems of Neuromorphic Adaptive Plastic Scalable Electronics. http://www.darpa.mil/program/systems-of-neuromorphic-adaptive-plastic-scalable-electronics. 15 Aug 2015
24. DARPA: Narrative Networks. http://www.darpa.mil/program/narrative-networks. 15 Aug 2015
25. The White House: FactSheet: the BRAIN Initiative. February 4, 2013. https://www.whitehouse.gov/the-press-office/2013/04/02/fact-sheet-brain-initiative. 16 Aug 2015
26. DARPA: DARPA and the BRAIN initiative. http://www.darpa.mil/program/our-research/darpa-and-the-brain-initiative. 16 Aug 2015
27. IARPA: Integrated Cognitive-Neuroscience Architectures for Understanding Sense making. http://www.iarpa.gov/index.php/research-programs/icarus/baa. 16 Aug 2015
28. IARPA: Knowledge Representation in Neural Systems. http://www.iarpa.gov/index.php/research-programs/krns. 16 Aug 2015
29. IARPA: Machine Intelligence from Cortical Networks. http://www.iarpa.gov/index.php/research-programs/microns. 16 Aug 2015
30. IARPA: Strengthening Adaptive Reasoning and Problem-Solving. http://www.iarpa.gov/index.php/research-programs/sharp/baa. 16 Aug 2015
31. Office of Naval Research. Human and Bioengeneered Systems Division. http://www.onr.navy.mil/en/Science-Technology/Departments/Code-34/All-Programs/human-bioengineered-systems-341.aspx. 16 Aug 2015
32. Sandia National Laboratories: A multi-scale understanding of decision-making. http://cognitivescience.sandia.gov/capabilities.html. 16 Aug 2015
33. Heslov, G.: The Men Who Stare at Goats. Smokehouse Pictures and BBC Films (2009)
34. Kitayama, S., Jiyoung, P.: Cultural neuroscience of the self: understanding social grounding of the brain. Soc. Cogn. Affect. Neurosci. **5**, 111–129 (2010)
35. Sigman, M. et al.: Neuroscience and education: primetime to build the bridge. Nat. Neurosci. **17**(4), 497–501 (2014)

36. Matsumoto, David, Juang, Linda: Culture and Psychology, 5th edn. Wadsworth, Belmont (2013)
37. Trochowska, K.: Operationalization of Culture in Contemporary Military Operations. Warsaw, AON (2013)
38. Selingman, R., Brown, R. Theory and method at the intersection of anthropology and cultural neuroscience. SCAN **5**, 130–137 (2010)
39. National Museum of the American Indian. Coming Home. Strength Through Culture. http://americanindian.si.edu/education/codetalkers/html/chapter5.html. 26 Aug 2015
40. Iyengar, S.S., Lepper, M.R.: Rethinking the role of choice: a cultural perspective on intrinsic motivation. J. Pers. Soc. Psychol. **76**, 349–366 (1999)
41. Campbell, B., Garcia, J.: Neuroanthropology: evolution and emotional embodiment. Front. Evol. Neurosci. **1**, Article 4 (2009)
42. Immersive Training Spreads Across NATO: Defense News, April 23, 2015. http://www.defensenews.com/story/defense/training-simulation/2015/04/23/training-immersive-itec-iitsec-simulation-nato-virtual/25772997/. 20 Aug 2015
43. Andersen, B. et. al.: The coming revolution in competence development: using serious games to improve cross-cultural skills. TARGET Project website. http://www.reachyourtarget.org/. 20 Aug 2015
44. Zhang, S.: The Neuroscience of Why Virtual Reality Still Sucks. *Gizmodo* online magazine, March 3, 2015. http://gizmodo.com/the-neuroscience-of-why-vr-still-sucks-1691909123. 21 Aug 2015
45. Tennison, M., Moreno, J.: Neuroscience, ethics and national security: the state of the art. PLoS Biol. **10**(3) (2012). www.plosbiology.org. 16 Aug 2015
46. Taylor, D.: The road to singularity: potential annihilation, utopian visions, will liberty prevail?. Old-Thinker News, August 11, 2014. http://www.oldthinkernews.com/2014/08/10/the-road-to-singularity-potential-annihilation-utopian-visions-will-liberty-prevail/. 26 Aug 2015
47. Coker, Christopher: Warrior Geeks: How 21st Century Technology is Changing the Way We Fight and Think about War. CUP, New York (2013)

Part III
Cross-Cultural Psychology

Chapter 14
Cross-Cultural Dimensions, Metaphors, and Paradoxes: An Exploratory Comparative Analysis

Martin J. Gannon and Palash Deb

Abstract We elicit the views of 37 experts who compare three distinctive approaches to the study of cross-cultural understanding: dimensions, cultural metaphors and paradoxes. Underlying this survey, although not openly stated and hopefully invisible to the expert respondents (and confirmed by informal meetings with some of them after they completed the survey), is the assumption that complexity of understanding increases as one moves from dimensions to cultural metaphors and then to paradoxes, with feedback loops connecting them. Prior research supports this progressive perspective based on feedback loops. Also, these three approaches are among the most popular, if not the most popular, methods for describing and analyzing cross-cultural differences, similarities and areas of ambiguity. Indeed, other approaches to cross-cultural similarities and differences can be subsumed in this progressive perspective. This chapter starts with a background discussion of the rationale for focusing on these three approaches, and the justification for analyzing in a comparative manner the major issues that have surfaced about these three approaches relative to their respective strengths and weaknesses. There is then a discussion of our reasons for selecting the 19 survey items, followed by a description of the methodology used, including sample selection and statistical procedures. Since this is an exploratory study of experts, we report only the major findings. However, in the final part of the review we offer suggestions about the manner in which this progression of cross-cultural understanding (via feedback loops) can be applied in the areas of research, teaching and practice, with particular emphasis on modeling human behaviors.

Keywords Cross-cultural understanding · Dimensions · Metaphors Paradoxes · Human behavior modeling · Exploratory analysis

M.J. Gannon (✉)
University of Maryland and Cal State San Marcos,
1248 La Granada Drive, San Marcos, CA 92078, USA
e-mail: mgannon@csusm.edu

P. Deb
Strategic Management Group, Indian Institute of Management Calcutta, K-503, New Academic Block, Diamond Harbour Road, Joka 700104, Kolkata, India
e-mail: pdeb@iimcal.ac.in

C. Faucher (ed.), *Advances in Culturally-Aware Intelligent Systems and in Cross-Cultural Psychological Studies*, Intelligent Systems Reference Library 134, https://doi.org/10.1007/978-3-319-67024-9_14

14.1 Introduction

There are numerous approaches to the study of cross-cultural understanding, almost all of which can be broadly classified as either *etic* or *emic*. Theorists and researchers who employ the dimensional or bipolar approach in the cross-cultural area exemplify the *etic* or culture-general perspective. Outstanding illustrations include the 53-nation study by [18] and the GLOBE study of 62 nations or national societies [19]. Such researchers primarily employ a standardized questionnaire whose items are then used to create dimensions along which these nations can be scored, ranked and compared to one another. By the very nature of this methodology such an approach is general rather than specific.

14.1.1 Cultural Metaphors

By contrast, an *emic* perspective looks at each national culture in depth and simultaneously accepts and attempts to go beyond such broad cultural profiling by exploring the unique and distinctive features of each culture. This perspective employs the idea of a cultural metaphor, which is any institution, activity or phenomenon which members of a given culture consider important and with which they identify cognitively and/or emotionally. Geertz's description of Balinese culture in terms of the metaphor of the cockfight received widespread attention [16]. More recently, [15] have examined 34 national cultures in depth using a distinctive cultural metaphor for each of them. Indeed, a review of the cross-cultural research literature over the last 50 years reveals that while dimensions still represent the dominant approach, metaphors have re-emerged as the most popular *emic* approach [30].

Gannon and Pillai [15] provide several examples of cultural metaphors in their book. Thus, the Swedish *stuga* is a simple, unadorned weekend and vacation home that is found throughout the countryside in this nation. For the Swedish national culture, these distinctive/unique features include the love of untrammeled nature and tradition, individualism through self-development and an emphasis on equality. These authors also provide other interesting examples that help us better understand the concept of cultural metaphors: how the complex rules of American football illustrate the complexity of the many rules and laws of corporate America; how the extraordinary complexity and finesse of French wine capture the intricacies, subtleties and nuances of a historically-rooted, highly evolved and fast-changing culture; how the dance of *Shiva*, a preeminent deity in the Hindu pantheon, represents a cycle of activity that reflects both creation and destruction, and how it shapes the Indian perspective on the cycle of life and reincarnation, and so on.

Popular music, such as the 'samba' in Brazil, the 'tango' in Argentina, or the 'calypso' in the West Indies, can also provide unique examples of cultural metaphors. Similarly, the 'opera' might be uniquely representative of Italian culture, as the pageantry and spectacle of the opera are reflective of the high expressiveness, emotions and animated nature of the average Italian [15]. Further, [5] edited a Special Issue of the *International Journal of Cross-cultural Management* containing five articles, each of which employed a distinctive cultural metaphor to describe either the Caribbean in entirety or one of its national cultures. These metaphors include the *ackee* (the national fruit of Jamaica), the "no ball" concept in cricket, Yoruba proverbs, calypso, and liming (a leisure activity during which members of a group create shared, spontaneous meaning through verbal exchanges reinforced with humor).

14.1.2 Paradoxes

In recent years there has been an increasing interest in a third approach, namely the paradoxical approach [6, 27, 29]. Although there are a few basic types of paradoxes, most cross-cultural experts emphasize one type: a statement seems to be untrue due to the vicious circle created by inconsistent or contradictory elements, when it is in fact true. Operationally, a paradox represents "both-and" thinking (rather than "either-or" thinking) involving inconsistent and/or contradictory elements. Paradoxes are often framed as sentences; an example is the paradox popularized by the Bauhaus school of modern architecture that says, "less is more."

Similarly, [7] analyzes the Chinese negotiating style in terms of a paradox: why do Western negotiators simultaneously consider Chinese negotiators as both very deceptive and very sincere? His answer revolves around three explanations: the long and tortuous history of China, the resulting view of the marketplace as a highly unpredictable and dangerous place similar to a battlefield, and the ideal Confucian gentleman who emphasizes sincerity. Similarly, some cultures see time as involving in a linear progression that goes from past to present to future, while other cultures represent time as only one circle in which there is no distinction between the past, present and future. Both elements of a paradox can exist simultaneously within a single culture in spite of the fact that they are in opposition to one another, particularly in such areas as perception of reality and cross-cultural negotiation. Major world religions follow these divergent paths in trying to explain reality [26]. In the case of Buddhism, there is not even a distinction between past, present and future; one circle rather than three is the Buddhist representation of this concept.

Another study employs the yin-yang perspective (which has traditionally been considered a paradox) as the supposed key to defining culture itself in a dynamic and holistic fashion [8, 9]. Fang's perspective, though debatable, should be pursued, especially in light of the large number of definitions of culture, many of which are inconsistent with one another. Indeed, echoing Fang's perspective, [27] point out that the starting point of a paradoxical methodology for researching groups is that opposition, polarities and conflict are part of the DNA of organizational life. [21] elaborate on these ideas in the following manner: "The idea of change and transformation between two opposite states is the main theme of the *I Ching* … or *Book of Changes*. The book not only discusses change in one direction (from young to old or from small to large) but also discusses changes from one extreme to another extreme. For example, when a moon is full, it starts to wane; when a moon is new, it starts to wax. This is the relationship between yin and yang: when yin reaches its extreme, it becomes yang; when yang reaches its extreme, it becomes yin. … Therefore, yin and yang are dependent upon one another, and transformations between the two occur when one of them becomes extreme." From this paradoxical perspective, it is argued that human beings, organizations and cultures should accept paradoxes to develop in a healthy and mature fashion. Thus, culture is "both-and" rather than "either-or." Similarly the dynamics of yin and yang apply to such paradoxical categories as masculine and feminine, long-term and short-term, individualistic and collectivistic, and so forth.

Another perspective to paradoxes was offered by Gannon [11], who summarized 93 cross-cultural paradoxes by employing the fact that there appears to be three major ways for understanding a paradox and hopefully resolving it. First, we can accept both truths and elements in each paradox, even though they are contrasting and even contradictory. Second, an individual can reframe the situation, which is the method that Bertrand Russell used to understand the famous Liar's Paradox: "all Cretans are liars; I never tell the truth." Russell demonstrates that each of these statements is valid but in different contexts and at different levels of analysis. In the third and final method, the individual accepts the paradox but looks for a higher unifying principle to understand it. Gannon [11] employs this third method, and emphasizes cross-culturally based research to identify a unifying principle for each of the 93 paradoxes, as the following examples drawn directly from the book demonstrate.

For example, one principle that is developed is that of 'value paradox' (e.g., Germans love freedom but feel that too much freedom can lead to disorder) and how it reflects the distinction between the desired and the desirable in life [6]. Thus US advertisements that target supposedly-individualistic Americans usually focus on group activities at home or in a social setting. Another example relates to whether multi-ethnic groups impede or facilitate the formation of national cultures. On the one hand, having several ethnic groups in a culture can lead to conflicts. On the other, countries like the USA, Canada and Australia have benefitted enormously from the contribution of new ethnic groups (e.g., German Jewish professors fleeing Europe before World War 2 contributed to the intellectual growth of US universities). This paradox extends to how immigrant groups can integrate into their adopted society. Thus US society encapsulates the idea of the 'melting pot' in which all groups integrate to form one single culture, while Canadian society is seen as a 'mosaic' wherein each ethnic group can retain its individuality and yet become an integral part of the whole.

Another paradox relates to how languages across the globe are both flourishing and dying. Thus, languages are dying at an alarming rate, from an estimated 20,000 one hundred years ago to 4000 today. However, major language groups such as English and Chinese have flourished. Similarly, globalization, or the increasing integration of national and ethnic cultures, has occurred over the past 200 years, though simultaneously differentiation is occurring. For example, in 1946 there were 76 sovereign nations, while today there are 197. Other examples of paradoxes include how an individually-based need hierarchy can exist in collectivist cultures, how a national culture can value both freedom and dependence, how nations are simultaneously becoming more powerful and less powerful as a result of globalization, how collectivists can also be self-centered and selfish, and so forth. We have presented these examples as illustrations only. However, they serve to underscore the point that paradoxical thinking is especially useful, as it emphasizes a sophisticated understanding that moves beyond mere categorization, e.g., male and female, sincerity and deception, linear and non-linear time, or the integration and differentiation of national and ethnic cultures.

14.1.3 Feedback Loops

Gannon [13] also argued that sophisticated cross-cultural understanding and knowledge proceed from dimensions through cultural metaphors to paradoxes. This sequencing idea also finds empirical support [24, 25]. This sequence also encompasses other recent emphases in the cross-cultural literature, such as the bi-cultural and multi-cultural frames of reference by individuals (e.g., [4]). As these framing mechanisms mature through the acquisition of two or more languages and direct experience in cultures other than the one in which a person is born, sophistication increases. Such results have been reported or at least described at least since the 1950s, and are, at this point in time, well-accepted. Below, we further explain the idea of feedback loop by means of an example.

The United States is consistently ranked as either #1 or in that vicinity in the multi-dimensional studies since Hofstede's original survey of 49 nations and the four territories he treated as the equivalent of nations; in the Hofstede study, the U.S. was #1 in terms of individualism. However, the cultural metaphoric approach provides a more nuanced and deeper understanding through the specification of the particular type of individualism and the distinctive features of the metaphor. As in football, the U.S. is an aggressive, competitive and individualistic culture in which inequality is more acceptable than in egalitarian national cultures, and it is little wonder that the U.S. ranks at or near the top in extrovertic behavior given this type of individualism [23]. At the same time, Americans are taught to work together, even though the reward structures tend to be unequal, with a few players receiving vastly more compensation and acclaim than other players. Also, the focus on the group working together is strengthened by the view that football is war and, by extension, so are other key activities of American life requiring cooperative groups such as business and its "winner take all" mentality.

The weeks-long football training camp prior to the actual season, during which each member of the team must learn his part in each of the complex plays, sometimes numbering at or near 200 plays, reflects this orientation. Football is the only game in the world where all offensive players must synchronize their interdependent motions (when they do not see what other offensive players are doing) if the play is to be successful. Further, the pre-game and half-time lavish entertainment exalting the team's virtues are designed to maximize group effort by team members, and even huddling after each play—the only game in the world that has this distinction after each play—helps to strengthen group effort. Analogs in the business world are the Walmart-influenced daily 10-minute early-morning standing meetings designed as both a pep talk and a clarification of responsibilities, and the periodic meetings during the year at which awards are presented and individual and group efforts are lauded. And, finally, football is treated as religious, even to the extent that devotees term all of the complex activities "the church of football," echoed in comparable talks by politicians and business people justifying the American ideology, e.g., emphasizing a religious rationale to justify specific actions such as a declaration of war or the existence of social inequality.

However, even though cultural metaphors are very complex with many distinctive and unique features, inevitably paradoxes emerge. To go back to the example of US football, individual rewards are emphasized in football but within a rigid group structure. If a highly-valued player's behavior off the field is suspect, his team will dismiss him quickly to ensure that the team or group is not harmed. To provide another example, generosity in sharing with those who have been less fortunate is widespread in the activities of the churches and non-profits in the U.S., contrary to the image that the US only champions individualistic behaviors. Warren Buffet, and Bill Gates and his father, have spearheaded a unique group activity among highly successful families in which a wealthy person bequeaths at least half of her assets to charity, clearly a paradox if one goes by the dimensional notion of the US as an individualistic nation. No other nation in the world has such a group. Thus the initial understanding of cross-cultural behavior that is obtained through the dimensional approach, and deepened through the use of the cultural metaphoric approach, is enriched by incorporating the paradoxical approach. The resultant feedback loops capture the dynamic interactions among these three approaches. In effect then, only the use of all three approaches rather than only one approach has the potential to deepen cultural knowledge and provide the framework to understand human decision processes and behaviors in situations where individuals from different cultural backgrounds may need to interact.

Smith and Berg [27, 28], in their classic work on paradox within small groups, describe some of these feedback loops when discussing paradoxes involving individualism-collectivism within a cross-cultural context. For example, they point out that individual human beings are social animals. As such, they only very rarely live in total isolation from other human beings. An individual wants to feel accepted by the group, at least to some degree. If he is not accepted, he or she will perceive the group in a different manner, and will react negatively to a group that is either too individualistic or collectivistic for him. Hence the feedback loop goes back from the paradox of individuality to culture-general dimensions. Similarly his or her new view of the group tends to create a different narrative or perception of the cultural metaphor that is dominant in the group. Hence the progressive and feedback elements combine to produce cultural knowledge, which is not as limited as that provided individually by cross-cultural dimensions, cultural metaphors or cross-cultural paradoxes.

Thus, given the importance of such sequencing or progression, we decided to undertake a preliminary study of expert perspectives on the major strengths and limitations of each of these three approaches: dimensions, metaphors, and paradoxes. Results are reported below, and implications for research, teaching and practice are developed.

14.2 The Study

It must be emphasized that this exploratory comparative study focused on issues related to these three approaches rather than the testing of a specific theory or theories. Our own experience in teaching managers and students as well as our experiences in navigating across different cultures suggested that each of these three approaches has both strengths and weaknesses, and that cross-cultural understanding is enhanced by the use of all of them. Similarly, Gelfand (in [10]) divided her cross-cultural university class into two groups, one of which argued the case for dimensions and the other did the same for cultural metaphors. The dimensional approach's general strengths included the following: a common metric to compare cultures and a structure to understand an immense amount of detail; quantifiable and verifiable; and amenable to large-scale multi-country studies. This exercise also indicated that: it is hard to keep the Hofstede 5-dimensional model of culture in mind; frequently we look at one dimension separately, yet culture is a complex whole; dimensions can be a-theoretical; one dimension is overwhelmingly emphasized, that is, individualism/collectivism; dimensions are extremely broad and miss important elements; and dimensions can obfuscate within-culture diversity and the dynamics of culture.

Further, Gelfand's students indicated that cultural metaphors afford a rich, detailed and in-depth understanding of a culture, and may include elements not captured in the dimensional approach; provide a dynamic view of culture, which includes actual experiences and vivid images that capture many of the five senses, thus helping to see how people participate in culture; help to create an integrated view of culture that captures the interrelationships among dimensions and how they relate to culture; and are very useful for cross-cultural training and for early-stage research (gaining understanding, for both theory and method). However, Gelfand's students also highlighted some weaknesses of cultural metaphors: They do not easily allow for comparisons; by definition, metaphors highlight some aspects of reality and ignore others; they are more susceptible to stereotyping, and it may be harder to change stereotypes based on cultural metaphors than on dimensions, which can also be stereotypical, because they are vivid and may stick; some metaphorical mappings may be a stretch; and metaphors have been described mainly at the cultural level and not at the individual level. Gelfand's students concluded that the dimensional approach and the cultural metaphoric approach are complementary and need each other to make sense of cross-cultural similarities and differences. As mentioned earlier, by also incorporating paradoxes in addition to dimensions and metaphors, we extend this logic to provide a sequential (or progressive) understanding involving feedback loops, arguing that cross-cultural knowledge becomes more sophisticated as it moves from dimensions to cultural metaphors to paradoxes, with feedback loops tying them together in a dynamic manner.

14.2.1 Methods

In this paper, we report the results of a survey that was completed by 37 cross-cultural experts. We developed 19 items to test the strengths and weaknesses of the three approaches, and grouped these 19 items on an a priori basis into six general categories (discussed in detail below). However, as indicated above, in the survey we simply listed the 19 items in a random fashion without providing any information on the six general categories into which we heuristically placed them before sending out the survey. We e-mailed the short survey comprising the 19 simply-worded items to 58 experts and received 37 usable surveys (63.8% response rate). Note that in the actual survey wording (below), we deliberately altered the sequencing, bringing ‘paradoxes’ before ‘metaphors’. This was done to ensure that respondents’ views were not influenced by the authors’ perspective of the sequencing order as dimensions, metaphors and then paradoxes.

Our e-mail stated: “(The second author) and I would like to ask you for a special favor, namely filling out the attached short survey in Excel that takes five minutes or less to fill out and return to me as an attachment to e-mail. The responses by experts such as yourself will be used to compare the relative advantages and disadvantages of three approaches to national cross-cultural understanding. The first approach, cross-cultural dimensions, is well-known and represented by the work of Hofstede, the GLOBE study, etc. The second, cross-cultural paradoxes, is newer. After reviewing many definitions, we define a paradox as follows: It is a statement, or set of related statements, containing interrelated elements that are opposed to one another or in tension with one another or inconsistent with one another or contradictory to one another (that is, either/or), thus seemingly rendering the paradox untrue when in fact it is true (both/and). The key elements of a paradox are that it:

(a) is a reality that can be expressed in a statement or set of statements;
(b) contains interrelated contradictory or inconsistent elements that are in tension with one another;
(c) leads to the creation of a reality, and any statement or set of statements about this reality or paradox that is seemingly untrue due to the “vicious” circle generated by the contradictory or inconsistent elements is in fact true; and,

(d) is framed or conceptualized as an either-or choice that is better framed as a both-and choice.

The third approach is termed cultural metaphors. A cultural metaphor is any activity, phenomenon, or institution that members of a specific national culture consider important and with which they identify emotionally and/or cognitively, for example, the Japanese garden. The major features of this metaphor are then used to describe a national culture. A person can use a cultural metaphor for an initial understanding of a national culture and can change his understanding as new data and information are processed. That is, the cultural metaphor is a first best guess.

We realize that all of the experts receiving this short survey are busy, but your knowledge of the cross-cultural field will be valuable in comparing the relative advantages and disadvantages of the three approaches. We plan to list the names of the experts who helped us out by filling out the survey and returning it to us in any paper and/or article that we write. If at all possible, we would like the survey returned by e-mail within two weeks. Thank you in advance for your invaluable assistance."

At the top of the survey, we indicated the following:

> Please provide your evaluations of the degree to which each of the 19 items below is attained or occurs using a 1 (low occurrence) to 5 (high occurrence), with the numbers 2, 3 and 4 representing intermediate degrees of occurrence. For each item, please provide evaluations for cross-cultural dimensions, cultural metaphors, and cross-cultural paradoxes.

Originally, we thought of using three separate factor analyses for each of the three approaches. We would have then been able to name factors and look at the individual (raw) ratings for items loading. 6 or above on each factor to obtain insight into the relative strengths and weaknesses of each of the three approaches. However, we had only 37 respondents for 19 items, and factor analysis requires at least a 6 to 1 ratio. As an alternative, we theorized on an a priori basis that the 19 items fall into six broad categories: the perspective on culture; framing culture; theory and related methodology; management education, training and globalization; ease in using each approach; and cognitive complexity. Since the goal of the study was to compare the relative strengths and weaknesses of the three approaches, we felt that the six broad categories and the 19 items within them would provide a basis for coming to some conclusions. However, we did not inform the respondents of this classification and just randomly listed the 19 items. See Table 14.4 for the 19 items subdivided into these six categories.

14.2.2 Explanation of the Six Categories

The six broad categories that we analyzed represent major issues in the cross-cultural area, and we have briefly but only indirectly touched upon them thus far in this chapter, for example, in our discussion of Gelfand's classroom exercise. Category 1, the perspective on culture, includes an item focusing on a detailed, in-depth description of national cultures as perceived by these experts relative to each of the three approaches. As indicated above, the culture-general or dimensional approach is much less in-depth in terms of describing a national culture than the cultural metaphoric approach. Further, the dimensional approach tends to be static and rarely includes measurements at two or more different points in time, whereas the cultural metaphoric approach suggests that a national culture is critically influenced over time by such elements as its birth rate, male-female ratio, rate of prolonged unemployment, population density, religion or religions, and so forth. Cross-cultural paradoxes have the potential of providing fresh insights into dynamic occurrences in a national culture, as some of the examples provided by [11]'s book cited above suggest, and we included an item on this issue in category 1.

The second broad category, framing culture, refers specifically to the manner in which each of the three methods affects cross-cultural experiences. Each of the three approaches has the potential for distortion and inaccurate stereotyping. However, as [1] points out, all of us stereotype and the issue is whether the stereotype is accurate. She indicates that it is acceptable to stereotype provided the stereotype is a first best guess, is based on data and observation, is descriptive and not evaluative, and if the individual is willing to change or even reject the stereotype as new information and experiences become available. Hence we included an item focusing on this issue of distortion and stereotyping. This second broad category has been heavily influenced by Kahneman and Tversky, who have shown that we tend to take more risks when facing an uncertain outcome rather than when facing a guaranteed positive outcome, and that we are influenced much more by stock market losses than the uncertainty of the market itself [22]. This second category also includes an item focusing on how well each of the three approaches enlarges a person's cultural frame of reference or cultural sophistication, and a second item focusing on how well each of the three approaches strengthens attributional abilities/knowledge, which is related to increased cross-cultural knowledge or understanding.

The third category includes six items focused on theory and related methodology. As we have indicated above, a strength of the dimensional approach is the ease of using statistics to test the basic concepts of the approach, while another strength is to use this approach to compare national cultures (since nations are rank-ordered to one another on one or more dimensions). It is also easy for the dimensional approach to show a relationship between a specific dimension and outcome or outcomes. For example, the rank ordering of national cultures on individualism-collectivism has been shown to be significantly related to airline accident rates per nation: the rate among collectivistic national cultures is double that of individualistic national cultures, and if power distance is added, high-power-distant and collectivistic nations exhibit three times the accident rates of low-power-distant and individualistic nations [17]. However, only the metaphorical method has the research potential to build a grounded theory of national cultures, as it emphasizes an in-depth focus. Hence we included an item to that effect. Further, the dimensional approach ignores intra-cultural differences within a specific national culture, whereas the metaphorical approach with its in-depth lens explicitly recognizes such ethnic, religious and even linguistic differences. We created a sixth item focusing on this issue. Given such results and emphases, we included six items in the survey, which is the largest number of items for any of the six categories.

We also wanted to analyze the broad category of management education, training and globalization, which is our fourth category. Research has demonstrated that individualistic national cultures tend to spend more per capita on management education than collectivistic national cultures [17]. Hence we included two items in this category, the first of which touched upon how well management trainees are able to see how to use each of the three approaches. We also wanted to know how suitable each of the three approaches is in a globalizing world.

Further, we created a fifth category examining how easy the experts thought it was to use each of the three approaches. Specifically, we developed three items focused on how easy it is to remember the specifics of each approach, how easy it is to use each of the three approaches in a person's home culture when interacting with those from different national cultures, and how easy it is for a visitor to use each of the three approaches in a host national culture. As noted above, we had only Gelfand's classroom exercise involving undergraduate students as a source of information, and we felt it necessary to supplement it with the opinions of experts.

Finally, since the emphasis of our approach was at least partially to develop a progressive and increasingly in-depth understanding of another culture, we created a sixth category, cognitive complexity, with one item focusing on the degree to which each approach was cognitively complex, and one item focusing on the degree to which each approach required higher-order thinking processes. From the viewpoint of optimizing human decision-making processes, cognitive complexity is positively related to cross-cultural knowledge and understanding.

14.2.3 Sample Selection

We wanted to include experts in this study who represented different viewpoints but who were thoroughly familiar with all three approaches and used them in their teaching, research and consulting. However, it was very difficult to define a population from which the sample was to be drawn. For example, the International Management Division of the Academy of Management is so diverse in membership that we felt it was not appropriate, that is, many of its members would probably not be as thoroughly familiar with the three approaches as would be desirable in such a study. Hence, based on our personal professional knowledge of those working in the cross-cultural area, we decided to send the survey to a large number of such experts. Thus our study is exploratory but, as far as we know, the only one that has been completed. Further, we did test some of our findings using both parametric and non-parametric statistics when appropriate, and report the mean values in the final part of the paper where we offer some suggestions in these six broad categorical areas.

Our sample included prominent cross-cultural psychologists, cross-cultural management educators who primarily teach in business schools rather than psychology departments (and may not be cross-cultural psychologists), and experts from different nations. There were very few non-responses to any item except for item 15 (four non-responses), which was "Reinforcement of the other two approaches." We used the average score for each item within each of the three approaches to complete the statistical analysis when data were missing. The names of the 37 experts and their university affiliations by nation are listed in Table 14.1. As mentioned in the e-mail we sent to these experts requesting their participation, we indicated that we would list their names and university affiliations in a table within the article. Of the 37 respondents, 15 are affiliated with non-US universities and 22 with US universities, and at least 11 teach outside of their country of birth.

14.2.4 Analyses

To assess inter-rater reliability, we chose to use two statistics that statisticians suggest, namely a parametric measure, the ICC (intraclass correlation coefficient) and a non-parametric measure that emphasizes ranked data, Kendall's measure of concordance (W) [20]. ICC values are appropriate for 5-point scales that are assumed to be Likert-type, while W values in this study use the mean values to assess the expert rankings. We added the items together within each of the six categories and used the average means to calculate ICC and W values separately for dimensions, paradoxes and cultural metaphors (see Table 14.3). In the final part of the article we discuss some of the mean values for items in the six broad categories (see Table 14.4). However, we do not engage in statistical testing in this final part but do report some striking results in terms of mean values for each of the three approaches analyzed in terms of the 19 individual items within these six broad categories.

14.3 Results and Implications

As shown in Table 14.2, the pattern is very clear, namely modest but statistically significant agreement among the raters at the.001 level for all 19 items and for the items within each of the three approaches, both for ICC and W values. In reading Table 14.3 horizontally, we can see that the raters agreed with one another across the three different approaches (dimensions, paradoxes and metaphors) only in one category, theory and related methodology. Reading Table 14.3 vertically, we can see that the raters agreed with one another only in one of the six categories when the dimensional approach is analyzed separately, namely theory and related methodology. The strongest agreement was in the area of paradox, where both the ICC and W values indicate that the raters agreed, at least statistically, four out of six times.

Table 14.1 Survey respondents by Country and University

	Name	Country	University
1	Claire Davison	Australia	RMIT University
2	Paul R. Cerotti	Australia	RMIT University
3	Tine Koehler	Australia	University of Melbourne
4	Michael Berry	Finland	University of Turku
5	Michael Hellstern	Germany	University of Kassel
6	Reinhard Huenerberg	Germany	University of Kassel
7	Sonja Sachmann	Germany	University Bw Munich
8	Anne Marie Francesco	Hong Kong	Hong Kong Baptist University
9	Primecz Henriett	Hungary	Corvinus University of Budapest
10	Amit Gupta	India	Indian Institute of Management Bangalore
11	Cormac MacFhionnlaoich	Ireland	University College Dublin
12	Patrick Flood	Ireland	Dublin City University
13	June Poon	Malaysia	Universiti Kebangsaan Malaysia
14	Laurence Romani	Sweden	Stockholm School of Economics
15	Yochanan Altman	UK	London Metropolitan University
16	Amy Kristof-Brown	USA	University of Iowa
17	Asbjorn Osland	USA	San José State University
18	Benjamin Schneider	USA	University of Maryland at College Park
19	Carl Scheraga	USA	Fairfield University
20	Christine Nielsen	USA	University of Baltimore
21	Edwin R. McDaniel	USA	California State University San Marcos
22	Gary Oddou	USA	California State University San Marcos
23	Glen Brodowsky	USA	California State University San Marcos
24	Joyce Osland	USA	San José State University
25	Lawrence Rhyne	USA	San Diego State University
26	Lois Olson	USA	San Diego State University
27	Mark Mendenhall	USA	The University of Tennessee at Chattanooga
28	Michele Gelfand	USA	University of Maryland at College Park
29	Ming-Jer Chen	USA	University of Virginia
30	Nakiye Boyacigiller	USA	San José State University
31	Nancy Napier	USA	Boise State University
32	Paul J. Hanges	USA	University of Maryland at College Park
33	Pino Audia	USA	Dartmouth College
34	Rabi Bhagat	USA	University of Memphis
35	Rajnandini Pillai	USA	California State University San Marcos
36	Stacey R Fitzsimmons	USA	Western Michigan University
37	Walter Lonner	USA	Western Washington University

Note We have survey responses from thirty-seven expert raters. Of these, fifteen are from non-US universities and twenty-two from US universities. At least eleven respondents are teaching at universities outside of their country of birth

Table 14.2 Overall 'ICC' and 'W' values by approaches

Approaches	ICC values	W values
Cross-cultural dimensions	0.22***	0.22***
Cultural metaphors	0.27***	0.26***
Cross-cultural paradoxes	0.29***	0.28***
All approaches combined	0.26***	0.27***

Note ICC = Intra-class correlations; W = Kendall's coefficients of concordance; **p*< 0.05, ***p* <0.01, ****p* <0.001

Table 14.3 Category-wise 'ICC' AND 'W' values for the three approaches

Category descriptions	Dimensions		Metaphors		Paradoxes	
	ICC	W	ICC	W	ICC	W
The perspective on culture	0.04*	0.08	0.01	0.03	0.15***	0.19***
Framing culture	−0.02	0.01	0.19***	0.12*	0.21***	0.14**
Theory and related methodology	0.25***	0.20***	0.33***	0.33***	0.30***	0.26***
Management education, training and globalization	0.01	0.01	0.00	0.00	0.30***	0.29**
Ease in using each approach	0.03*	0.05	0.04*	0.03	0.03	0.06
Cognitive complexity	0.00	0.01	0.17 ***	0.21**	0.06*	0.07

Note ICC = Intra-class correlations; W = Kendall's coefficients of concordance; **p* <0.05, ***p* <0.01, ****p* <0.001

Similarly but less strongly, the raters agreed statistically with one another in three of the six categories for the approach of cultural metaphors.

We next focus on the mean values for each of the three areas (dimensions, metaphors, and paradoxes) described in terms of specific items. As we proceed, we will offer some suggestions in the areas of teaching, research and applications. We feel this is the most appropriate way to proceed rather than testing specific theories and hypotheses, given the methodological issues discussed previously.

In terms of the perspective on cultures, the metaphoric approach is perceived by the respondents as far superior to paradoxes and dimensions: The mean values are 4.08 for metaphors, 3.03 for paradoxes and only 2.62 for dimensions for the item focusing on a detailed, in-depth description of national cultures. This result is not unexpected: when cultural metaphors are used correctly in terms of delineating unique or distinctive features of a particular culture, they provide an in-depth insight not possible when using general cross-cultural dimensions or even paradoxes. Paradoxes dazzle when a reader begins to understand them, but they do not provide the in-depth understanding that metaphors allow. However, paradoxes and metaphors are approximately equal in mean values in terms of a dynamic view of national cultures (3.76 and 3.73). In contrast, the mean value for dimensions is only 2.11. Finally, in terms of an integrated view of national cultures, metaphors clearly are first (3.84) followed by paradoxes (3.03) and then dimensions (2.65).

Based only on these results, one must question why so much research, teaching and even applications are based primarily on dimensions. For the last thirty-five years the emphasis has been on research, where, as we will see, the dimensional approach is highly rated. However, what is generally regarded as substantive issues in cross-cultural understanding emphasizes a detailed, in-depth description of a culture and an integrated view of national cultures, not to mention a paradoxical view of culture. Just reading the New York Times or a similar publication leads inevitably to such a conclusion. Rarely are cross-cultural dimensions the focus of interest. Rather, the focus is on an in-depth description or a paradoxical explanation. Why, then, is this emphasis on researching dimensions significantly different from the results reported directly above?

The answer to this puzzle may possibly be found in the category of theory and related methodology (Category 3). In this category, item 2 shows that it is much easier for dimensional researchers to use statistics to test the basic concepts of this approach (mean value of 4.54) than to use either metaphors (1.89) or paradoxes (2.19). Methodologies have been developed to test both cultural metaphors and cross-cultural paradoxes, but they require much more effort on the part of the researcher, as they cannot rely only on one standardized questionnaire that is employed in numerous nations to test hypotheses. Also, while the respondents felt that both metaphors (3.76) and paradoxes (3.81) emphasize the early stages of grounded theory, the differences between mean values were not as extreme as those reported directly above, as dimensions had a mean value of 2.92 (See Item 1 under Category 3).

These findings bring into focus the largely-unquestioned assumption that dimensions represent the apex of cross-cultural research and understanding. Rather, they may represent only a first step in trying to understand cross-cultural behavior. For example, it is generally accepted that individualism-collectivism, as measured by a standardized survey used in several nations, is the most predictive of the dimensional measures. However, there are so many different types of individualism and collectivism that such a viewpoint is problematic [15]. We need to have culture-general measures in this area of individualism-collectivism supplemented by culture-specific measures and measurements of other dimensions both *etically* and *emically*, and the dimensional approach does not provide such an understanding.

However, what is clear is that the dimensional perspective is clearly superior (mean value of 4.16) to both metaphors (3.35) and paradoxes (2.59) in terms of ease in comparing national cultures (Item 3 under Category 3). Conversely, the dimensional approach is also the highest in terms of ignoring intra-cultural differences within a specific national culture (4.11 versus 3.22 for metaphors and 2.41 for paradoxes) (Item 4 under Category 3). This is a major weakness of the dimensional approach, and even some dimensional researchers have tried to take this weakness into account. For example, Robert House, who initiated and led the well-regarded GLOBE study, hired a cultural anthropologist right at the beginning of this research, but unfortunately he was unable to resolve this problem or dilemma.

Still, there is hope, as the respondents indicated that there are fewer differences in mean values when they judged on the issue of the reinforcement of the other two approaches (3.76, metaphors; 3.30 for paradoxes; and 3.00 for dimensions) (See Item 6 under Category 3). Thus there appears to be recognition of the fact that one approach by itself is insufficient. We need all three approaches to obtain a valid description of an ethnic and/or national culture. Nevertheless, one must question why such an inordinate amount of academic research has been focused on dimensions. If all three approaches reinforce one another, why has there been such limited attention devoted to metaphors and paradoxes, especially in the area of testing results?

Perhaps the major reason is that most researchers seem to be unaware that at least one methodology has been developed to measure cultural metaphors. Specifically, [14] developed three surveys completed by undergraduate students in six nations, one survey for two nations at a time: the US (American football) and India (the Dance of Shiva); Germany (the symphony) and Italy (the opera); and Great Britain (the traditional British home) and Taiwan (the Chinese family altar). College students were asked to respond to a lead-in "Most people in my country" followed by items derived from the specific chapters on each of the six nations found in [15], for example "are honest," "are publicly unexcitable," etc. Each student used an 11-point scale to rate his or her degree of agreement with each item, with 0 indicating "do not agree at all" and 10 indicating total agreement, or he or she could choose any other number between 0 and 10. In addition, the researchers developed two paragraphs for each nation, one of which did not explicitly contain the cultural metaphor while the other did. Each student used the same 11-point rating scale to measure disagreement-agreement relative to the two paragraphs. Appropriate statistical tests were then employed to test whether each cultural metaphor was perceived as reflecting the national culture by these students. There was strong support for at least these six cultural metaphors. The instruments are publicly available in [10], Exercises 4.1 and 4.2, and are also reprinted on http://faculty.csusm.edu/mgannon.

Thus it is possible to statistically test both cross-cultural dimensions at the culture-general level and specific metaphors for each national culture to obtain a more in-depth understanding. Similarly, at least many if not most cross-cultural paradoxes can be tested. Above we have given examples relative to the death of languages while major language groups are flourishing. Similarly we have put forth the paradox relative to time, which can be measured by a standard instrument. Hence we believe that it is possible for human decision-making researchers to test a model of culture that goes far beyond the culture-general dimensional perspective, and even the cultural metaphoric method.

We now turn our attention to another major category, framing culture (Category 2). In this area we do not see the extreme mean value differences reported above.

All three approaches seem to suffer from a susceptibility to distortion and inaccurate stereotyping, at least as judged by mean values (Item 1 under Category 2). Similarly, while cultural metaphors seem to enlarge the frame of reference (4.24), so too do paradoxes (3.92), although it is questionable that dimensions possess this feature to the same extent (3.27), especially when compared to cultural metaphors (Item 2). Further, there appears not to be major differences in the area of strengthening of attritional abilities/knowledge: 3.81, metaphors; 3.78, paradoxes; and 3.14, dimensions (Item 3).

In the category of management education, training, and globalization (Category 4), the differences are not as wide as those reported above. All three approaches appear to be similar within this category, which includes two items (see Table 14.4). Similarly, the mean values in category 5, ease in using each approach, suggest that all three approaches are useful. For the three items, the metaphoric approach has the highest mean value, but the differences in mean values do not seem practically significant. However, in the final category, cognitive complexity, there appears to be significant differences. Paradoxes have a higher mean value (4.22) than either metaphors (3.43) or dimensions (2.81) (Item 2 under Category 6).

This reinforces the concept that cross-cultural understanding progresses through various stages, beginning with dimensions for cultural-general features

Table 14.4 Item means by category for the three approaches

Item description		Dimensions	Metaphors	Paradoxes
I	Category 1: The perspective on culture			
1	Detailed, in-depth description of national cultures	2.62	4.08	3.03
2	A dynamic view of national cultures	2.11	3.73	3.76
3	An integrated view of specific national cultures	2.65	3.84	3.03
II	Category 2: Framing culture			
1	Susceptibility to distortions and inaccurate stereotyping	3.27	3.35	2.97
2	Enlarges the cultural frames of individuals	3.27	4.24	3.92
3	Strengthening of attributional abilities/knowledge	3.14	3.81	3.78
III	Category 3: Theory and related methodology			
1	Research potentiality: early stages of grounded theory	2.92	3.76	3.81
2	Ease of using statistics to test the basic concepts of the approach	4.54	1.89	2.19
3	Ease of comparing national cultures	4.16	3.35	2.59
4	Ignoring intra-cultural differences within a specific national culture	4.11	3.22	2.41
5	Ease of using statistics to relate the approach to other variables	4.32	2.08	2.19
6	Reinforcement of the other two approaches	3.00	3.76	3.30

differentiating national cultures, going through metaphors for specificity, and ending with paradoxes, which provide the sophisticated understanding that is the hallmark of cross-cultural education and training. Admittedly, then, there are feedback loops between these various stages. Still, the natural progression of cross-cultural understanding seems to proceed through a cultural-general phase (dimensions), then through a cultural metaphoric phase (in-depth understanding), and finally to a paradoxical phase that recognizes the importance of both dimensions and metaphors but moves beyond them.

14.4 Conclusion

In this chapter we have put forth a progressive feedback model designed to increase cross-cultural knowledge and understanding that begins with a culture-general approach emphasizing dimensions, moves onto to a more in-depth understanding of a culture through the use of a unique or distinctive cultural metaphor and the specific features of this metaphor, and finally to the third approach of cross-cultural paradoxes that serve as the endpoint of knowledge and understanding. That is, we argue that we can predict human decision-making behaviors and processes in cross-cultural contexts by proceeding sequentially in this manner, and that knowledge created by each of the three approaches feeds back to the other approaches and strengthens cross-cultural knowledge and understanding because of the circularity of the relationships in the model.

For example, [28] agree with many other researchers that individualism-collectivism is a key if not the key dimension in the cross-cultural area. They also argue that there is an inevitable tension between individualism and collectivism, which we see as manifesting itself in the specific type of individualism or collectivism that becomes apparent through the use of cultural metaphors, e.g., the proud and self-sufficient individualism of the Spanish, the interdependent individualism of the Danish, and the competitive individualism of the United States. Inevitably these tensions lead to cross-cultural value paradoxes, e.g., the value paradox such as the high emphasis that Germans place on both freedom and structure.

Further, the respondents generally agreed that all three approaches—dimensions, metaphors, and paradoxes—are useful and add value in trying to understand cross-cultural differences. We likewise believe that all three approaches are useful, especially if tied together in a progressive feedback model maximizing cross-cultural knowledge and understanding. From this perspective, we argue that cross-cultural management education and training should be structured in the manner advocated in this chapter.

Also, given that the dimensional approach's strong suit seems to be the amount of research devoted to it, does that suggest that we need to emphasize research on metaphors and paradoxes more than is currently emphasized if we want to move the field of cross-cultural research beyond where it is today? Will new methodologies need to be developed to test the adequacy of cultural metaphors and paradoxes?

While some methodologies do exist for this purpose, as discussed in this chapter, it appears that much more of an emphasis on methodology in these two areas needs to occur.

There are also other implications that can be derived from our comparison of the three approaches. First, one approach to studying cross-cultural differences and similarities is clearly insufficient. Relatedly, there is clearly interest in the areas of cultural metaphors and paradoxes, judging by conference sessions devoted to them and journal publications. Still, this interest is dwarfed in comparison to that shown to dimensions.

Further, there is an existing body of literature about the manner in which these three approaches can be integrated to improve understanding of human behaviors, cognitive processes and value systems in cross-cultural contexts. [2] describe at length how to use these three approaches in eight different contexts, as described below. As we learn to integrate the three approaches within different contexts, we expect that the field of cross-cultural behavior will flourish and will allow experts to offer suggestions about research, teaching and applications that will tend to positively reinforce one another.

In addition, in support of the proposition that there is nothing as useful as a value-added theory or model, we offer the experiences of various teachers and researchers who have emphasized the use of dimensions, metaphors and paradoxes in their training and educational endeavors. A fuller explanation can be found in [2] and [12]. In the [2] series of eight mini-articles, Nielsen begins by describing how she uses cultural metaphors in her "Leadership Across Cultures" course, focusing on how the overlap between the *fado* metaphor and the Portuguese bullfight provides additional insights into the national culture. Cerotti and Davidson demonstrate how their popular exercise involving posters of cultural metaphors reinforce the idea that cultural knowledge needs to be deep-seated. Scheraga, a well-known economic researcher, shows how he uses dimensions, metaphors and paradoxes to show students how the complexity of culture increases, as described in this article, and how to address the issues raised at each level of analysis through various quantitative methodologies and statistics. Pillai describes a three-hour symposium at the 2009 Academy of Management Conference exploring all three areas of dimensions, metaphors and paradoxes, culminating with a discussion of two metaphors for India, the Dance of Shiva (traditional India) and Kaleidoscopic India (modern India). The Indian Dance Group of Chicago then performed an interpretation of the Dance of Shiva.

Altman goes on to explain how public scandals in France and the USA are viewed, primarily through the prisms of their respective cultural metaphors and their distinctive or unique features (French wine and American football). Rhyne extends the analysis by showing how his student teams incorporate the use of dimensions and cultural metaphors in developing company-specific strategies within a national culture. Köhler and Berry emphasize the role of interpersonal communication using dimensions and cultural metaphors, showing how a well-known 60-minute show on American culture was filled with misunderstandings of Finnish culture, particularly in regard to what silence means (American football and the Finnish sauna). In 1998 the Midwest Academy of Management's International Conference in Istanbul highlighted the use of dimensions and cultural metaphors, asking teams of academics to use the chapters from Gannon's book to understand American football and the Turkish Coffeehouse (plus Chap. 1 which explains the dimensional and metaphoric approaches), prior to interviewing business executives and asking them questions based on these chapters.

Further, [12] describes how he uses all three perspectives (dimensions, metaphors and paradoxes) in various international MBA programs with which he has been intimately involved. First, he presents a short table describing 15 nations in terms of their rankings on the dimensions proposed by Hofstede. He then says something like: the Hofstede perspective is very useful, but what is incomplete about it? A dead silence usually ensues. He then points out that the two major culture-general features of collectivism that political scientists emphasize, paternalistic and authoritarian, are missing completely from the Hofstede framework. He also points out that there are different types of individualism and collectivism both at the culture-general and the individual levels of analysis, as highlighted in [15]. At this point he emphasizes the area of paradoxes, asking the trainees or students to respond to questions that incorporate one of his 93 cross-cultural paradoxes [11].

Gannon ends by suggesting that a broadened approach encompassing dimensions, metaphors and paradoxes creates a situation in which trainees and students are intimately involved in the learning process, rather than the lecturer merely describing each of them. His examples include Maggi Phillips, a professor at Pepperdine University, who has her students read Chap. 1 of the Gannon and Pillai book and the specific chapter devoted to American football, after which the trainees or students are sent to a mall to observe behavior in terms of what they have read. Similarly, Lois Olson of San Diego State University taught in the Program at Sea during a semester in which students visited several nations while living on a ship. She prepared them for each nation just prior to visiting it by asking them to read the appropriate chapter. Such approaches demonstrate the wide range of applicability that a broadened perspective involving dimensions, metaphors and paradoxes emphasizes. This, in effect, is the key thrust of this article. Rather than automatically rejecting alternative viewpoints, the idea is to emphasize the point that there is nothing as useful as a value-added theory or model encompassing multiple viewpoints.

In summary, we are offering a testable framework of culture (based on feedback loops) that seeks to understand human decision-making processes so that cross-cultural knowledge and understanding are enhanced. We have also shown that it is possible to test the effectiveness of the three approaches through methodologies that currently exist (specifically, we offer some thoughts and evidence on ways to test cultural attributes other than dimensions), but we also feel it is probably best if other researchers not heavily identified with these three approaches do such testing, even to the extent that they develop new methodologies for doing so. While testing all aspects of the framework will require a lot of effort, it is possible to do so and to move the cross-cultural area away from an emphasis on specific approaches considered one at a time to a situation in which at least three of the major approaches to culture are integrated with one another to enhance cross-cultural comprehension. As we move towards capturing (by collecting data) and testing (through analysis) the depth and variety of human behaviors in diverse national cultures, these three approaches enhance our cross-cultural intelligence, an essential component of human intelligence itself.

Taken to its logical conclusion, combining the *etic* and *emic* perspectives to culture will enable us to better understand national cultural differences which can then be incorporated in designing interactive systems. As [3] argue, while "the Artificial Intelligence community uses the term 'common sense' to refer to the millions of basic facts and understandings that most people have", changing the culture or environment can change individuals' perceptions of what is common sense, and the "challenge is to try to represent cultural knowledge in the machine, and have interfaces that automatically and dynamically adapt to different cultures". Knowledge bases for different cultures could be developed by incorporating what is common knowledge in different cultures, and software can be developed for comparing these different knowledge bases (this will allow, for example, a US-based teacher to consult the database while developing instructional content for students based in France, or search engines to use that knowledge base to come up with culture-specific search results) [3]. In sum, in this chapter we identify the basis of the multi-dimensional complexity of human behaviors in various cultural settings and suggest that AI experts should be able to capture some of these cross-cultural differences by developing appropriate machine interfaces.

References

1. Adler, N.J., Gundersen, A.: International dimensions of organizational behavior. Cengage Learning (2007)
2. Altman, Y., Berry, M., Cerotti, P., Davison, C., Gannon, M., Nielsen, C., Pillai, R., Rhyne, L., Köhler, T., Scheraga, C.: Applications of cultural metaphors and cross-cultural paradoxes in training and education. In: Friedlmeier, W. (ed.) On-Line Readings of Psychology and Culture (2012). http://scholarworks.gvsu.edu/orpc/
3. Anacleto, J., Lieberman, H., Tsutsumi, M., Neris, V., Carvalho, A., Espinosa, J., Zem-Mascarenhas, S.: Can common sense uncover cultural differences in computer applications? In: Artificial Intelligence in Theory and Practice pp. 1–10 Springer, US (2006)
4. Brannen, M.Y., Thomas, D.C.: Bicultural individuals in organizations: implications and opportunity. Int. J.Cross-Cult. Manag. **10**(1), 5–16 (2010)
5. Corbin, A., Punnett, B.J., Onifa, N.: Special issue: using cultural metaphors to understand management in the Caribbean. Int. J. Cross Cultural Manage. **12**(3), 269–275 (2012)
6. De Mooij, M.: Global marketing and advertising (3rd edn). Sage Publications, Thousand Oaks, CA (2010)
7. Fang, T.: Chinese Business Negotiating Style. Sage Publications, Thousand Oaks, CA (1999)
8. Fang, T.: From "Onion" to "Ocean": Paradox and change in national cultures. Int. Stud. Manag. Org. **35**(4), 71–90 (2005)
9. Fang, T.: Yin Yang: a new perspective on culture. Manag. Org. Rev. **8**(1), 25–50 (2012)
10. Gannon, M.: Working across cultures: applications and exercises. Sage Publications, Thousand Oaks, CA (2001)

11. Gannon, M.: Paradoxes of culture and globalization. Sage Publications, Thousand Oaks, CA (2008)
12. Gannon, M.: Cultural metaphors: their use in management practice as a method for Understanding Cultures. In Friedlmeier, W. (ed.) On-Line Readings of Psychology and Culture (2011). http://scholarworks.gvsu.edu/orpc/. (An earlier version of this paper was published in Walter Lonner (ed.). On-Line Readings in Cross-Cultural Psychology, 2003)
13. Gannon, M.: Sequential cross-cultural learning: from dimensions to cultural metaphor to paradoxes. Ind. Org. Psychol. Perspect. Sci. Practice **5**(2), 239–242
14. Gannon, M., Gupta, A., Audia, P., Kristof-Brown, A.: Cultural metaphors as frames of reference for nations: a six-nation study. Int. Stud. Manage. Org. **35**(4), 4–7 (2005)
15. Gannon, M. Pillai.: Understanding global cultures: Metaphorical journeys through 31 nations, clusters of nations, continents, and diversity (6th edn). Sage Publications, Thousand Oaks, CA (2015)
16. Geertz, C.: The interpretation of culture. Basic Books, New York (1973)
17. Gladwell, M.: Outliers: the story of success. Little, Brown, New York, NY (2008)
18. Hofstede, G.: Culture's consequences (2nd edn). Sage Publications, Thousand Oaks, CA (2001)
19. House, R., Hanges, P., Javidan, M., Dorfman, P., Gupta, V.: Culture, leadership, and organizations: the GLOBE study of 62 societies. Sage Publications, Thousand Oaks, CA (2004)
20. Howell, D.C.: Statistical methods for psychology 7th edition. Wadsworth Publishing, USA (2010)
21. Ji, L., Nisbett, R., Su, Y.: Culture, change, and prediction. Psychol. Sci. **12**(6), 450–456 (2001)
22. Kahneman, D.: Thinking, fast and slow. Macmillan (2011)
23. Keirsay, D.: Please understand me II: temperament, character, intelligence. Prometheus Nemesis Book, Del Mar, CA (1998)
24. Osland, J., Bird, A.: Beyond sophisticated stereotyping: cultural sensemaking in context. Acad. Manag. Exec. **14**(1), 65–87 (2000)
25. Osland, J., Osland, A.: Expatriate Paradoxes and cultural involvement. Int. Stud. Manag. Org. 35(4), 93–116 (2005–2006)
26. Smith, H.: The world's religions: our great wisdom traditions. Harper San Francisco, San Francisco (1991)
27. Smith, K.K., Berg, D.N.: Paradoxes of group life: understanding conflict, paralysis and movement in group dynamics. Jossey-Bass, San Francisco (1987)
28. Smith, K.K., Berg, D.N.: Cross-cultural groups at work. Eur. J. Manag. **15**(1), 8–15 (1997)
29. Smith, W., Lewis, M.: Toward a theory of paradox: a dynamic equilibrium model of organizing. Acad. Manag. Rev. **36**(2), 381–403 (2011)
30. Taras, V., Rowney, J., Steel, P: Half a century of measuring culture: review of approaches, challenges, and limitations based on the analysis of 121 instruments for quantifying culture. J. Int. Manag. **15**(14), 357–373 (2009, December)

Chapter 15
A Model of Culture-Based Communication

Bilyana Martinovski

Abstract Both humans and Virtual Agents interact in intercultural environments. Both humans and Virtual Agents need to behave appropriately according to environment. This paper proposes a dynamic modular model of culture-based communication, which reflects intercultural communication processes and can be used in the design of life-like training scenarios. Culture is defined as a semiotic process and a system, which builds upon Self and Other identities and which is sustained and modified through communication and cognitive-emotive mechanisms such as reciprocal adaptation, interactive alignment and appraisal. Communication is defined as an opportunity for meeting of Otherness. Since culture covers many different aspects of social life, people are practicing intercultural-communication on daily basis and Human-Virtual Agent interaction is seen as a form of intercultural communication.

Keywords Cultural Awareness · Communication · Emotion · Virtual Agents
Intercultural Communication · Adaptation

15.1 Introduction

Simulation of cultural awareness and behavior is a challenge in design of intelligent systems such as Virtual Agents. One of the reasons is the diversity of behaviors and cognitive frameworks in different cultural environments. Another is the unclear relation between emotional states and cultural behaviors. Yet, Virtual Agents often need to behave appropriately according to environment, for instance, when involved in life-like training scenarios. This explains why there has been increasing interest in culturally adaptive agents. The agents are built to relate to particular cultures by exchanges of modules related to behaviors and functions. In most cases,

---

B. Martinovski (✉)
Department of Computer and Systems Sciences (DSV),
Stockholm University, Stockholm, Sweden
e-mail: bilyana@dsv.su.se

C. Faucher (ed.), *Advances in Culturally-Aware Intelligent Systems and in Cross-Cultural Psychological Studies*, Intelligent Systems Reference Library 134, https://doi.org/10.1007/978-3-319-67024-9_15

the agent's designer decides what set of culture-based features to be used in order to generate culturally appropriate behavior [1].

This chapter proposes and motivates a model of culture-based communication for humans and Virtual Agents, which aim is to capture the dynamic relation between culture, communication and emotion. It starts with theoretical foundations and definition of the modeling concepts. Section 15.2 defines culture. Section 15.3 defines communication in culture relevant terms. Section 15.4 offers a description of culture-as-a-process mechanisms such as reciprocal adaptation and interactive alignment. Section 15.5 defines emotion as used in the model of culture-based communication presented in Sect. 15.6.

15.2 Culture as a Process and a System of Interpretation

There are various definitions of culture some emphasizing competence, others—performance, cognitive aspects vs. communicative and activity aspects. Its history of conceptualization is long, starting with cultivation of crops, moving through a dilemma—'culture vs nature', followed by a view of culture as high and low expressions of human talent and insight, and later as 'a historically transmitted pattern of meaning embodied in symbols' ([2]: 89). The currently dominant conceptualization of the term defines culture as 'programming of the mind … interactive aggregate of common characteristics that influence a human group's response to environment' [3]. Although the dominant usage of the culture concept is based on nationality, cultures vary also with respect to ethnicity, age, gender, profession, activity, language, and even species. For instance, Shuter's study [4] of proxemics behavior among Latin American pairs in conversation shows regulations not only according to national culture but in dependence of gender, where women stand closer to each other than to men independent of nationality and Costa Ricans stand closer to each other than Columbians independent of gender. In other words, we participate in different forms of intercultural communication on daily basis and there are measurable behavioral changes depending on cultural context. Therefore, a model of cultural awareness is better grounded on the fundamental relation between Self and Other informed by modular variation of cultural and communicative features.

As cultural values, perception, judgment, emotions and communication are acquired at very early age, cultural awareness becomes a process, which develops from low to high awareness. A child acquires cultural values and habits actively and non-critically, which define his/her sense of Self [5]. Realization of Self is thus contingent on realization of Otherness: if one has no understanding of Self as different from the Other, one has no realization of Self [6, 7]. Without the neural ability to distinguish Self and Other [8], culture-awareness would not be formed. From a neurocognitive point of view, the brain and the body assemble inner and outer information depending on the capacities of the specific species [9]. Except for the sense of smell, which initially does not go through interpretative circuits, the human brain interprets information coming through the outer senses. For instance, when a sound wave reaches the ear receptors there is no interpretation of what that sound is but we cognize that we perceived a sound of a train because we immediately created an interpretation, based on possible already experienced or learned sounds. Thus, what we ordinarily regard as perception is rather the act of interpreting sensory data [9, 10].

This applies also to culture-sensitive phenomena such as proxemics. Through the process of upbringing we learn to 'perceive' certain distance between speakers as right and wrong, comfortable and uncomfortable. During this rather unconscious process, we built our culture-specific model or system of proxemics. From the moment of birth, our environment supplies us with a range of possible interpretations. With time, the possibilities turn into a full system by means of which we conduct all our perceptual transactions. In that sense, culture is not only a full system of preferred interpretations of sensory data but also the very process of acquiring and building that system. It is the process aspect of culture that expresses its dynamics, its ability to change. Both as a system and as a process, Pierce's three basic semiotic signs [11]—icon, index and symbol—are realized in human culture and communication as various modalities [12]. Therefore, the definition of culture adopted in the proposed model is formulated as follows:

Definition of Culture: Culture is a dynamic semiotic process and a semiotic system of interpretation of Self and Otherness by means of which we conduct our perceptual transactions in interaction with environment.

Contemporary research on the functions of mirror-neurons [13] examines the neural mechanisms, which support our understanding of others' and own intentions and states, which, in turn, are used for development of cultural and social cognition [8, 14]. Beliefs about age, gender, language, environment, and so on contribute to cognitive 'models' that individuals form and keep of each other's characteristics and intentions. Baron-Cohen [15] went as far as to claim that the human brain has a particular "mindreading" mechanism, specialized module dedicated to prediction and interpretation of others' intentions. He characterized this as a natural and innate capability, which enables socio-cultural contact to take place: "We mindread all the time, effortlessly, automatically, and mostly, unconsciously" [15]. Communication complements and depends upon "mindreading": "Our mindreading fills in the gaps in communication and holds the dialogue together" (ibid.: 28). This ability, according to Baron-Cohen, is the result of a long process of evolution yet even this form of perception is constrained by cultural interpretations.

15.3 Culture-Relevant Definition of Communication

Culture is tightly related to the concepts of language and communication. This section provides an overview of conceptualization of communication in order to reach a culture-relevant definition operative in the proposed model (see Sect. 15.5).

There are two major views on the relation between language and communication. The more dominant view concludes that

(i) language originated as a tool for communication
(ii) communication is exchange of information.

Within this view, miscommunication is treated as a problem, which can be solved by more communication. Culture-specific concepts are explained with differences in objective reality, e.g. there is no lexicalized word for snow in certain African languages because there is no snow in these regions. However, there are languages that lack specific words for e.g. color, but does that mean that there are no colors in their regions or that the speakers don't perceive color? Universalist studies [16] on perception of color show that the speakers perceive colors although they don't have specific words for color and that their concepts have different semantic fields i.e. they use the word 'blood' to denote even the color of red. These findings suggest another view on the relation between language and communication, namely, that

(i) language originated as an inborn tool for thinking
(ii) communication is not the primary function of language.

This, so called, cognitivist view [17] underlines universal linguistic structures and concepts. Surface lexicalizations and expressions could be culture-specific but not the underlying inborn syntactic and conceptual structures, which define us as species. Miscommunication is not necessarily resolved by more communication because language is not a tool for communication but rather a tool supporting thought processes.

These two views are competitive with respect to origin and function of language but they also express a dichotomy in the modern concept of communication, which Peters [18] sees as a product of 'a dream for instantaneous connection or a nightmare of being locked in a labyrinth of solitude'. In other words, we prefer to define communication and language as means for immediate connection between Self and Other because we are terrified by the idea of being locked in a solitary world of solipsistic thought. In this context, we need to examine the history of the dominant definition of communication as exchange of information.

15.3.1 History of the Concept of Communication

At the end of World War II, Claude Shannon, scientist at Bell Telephone Company, formulated his mathematical theory of signal transmission aiming at maximum telephonic line capacity with minimum distortion. Weaver ([19]: 1) extended this definition of communication as a process of exchange of information even to human communication, beyond its technological origin:

> The word communication will be used here in a very broad sense to include all the procedures by which one mind may affect another. This, of course, involves not only written and oral speech, but also, music, the pictorial arts, the theater, the ballet, and in fact all human behavior. In some connections it may be desirable to use a still broader definition

of communication, namely, one which would include the procedures by means of which one mechanism (say automatic equipment to track an airplane and compute its probable future positions) affect another mechanism (say a guided missile chasing this airplane).

Studies based on Shannon and Weaver's model [20] took two major directions. One applied the engineering principles of transmission and perception within electronic sciences, the other explored interactants' ability or inability to communicate. The communication-as-a-transmission metaphor continues to dominate informatics, interaction design, linguistics, literary theory, art theory, communication studies, ethics, etc. Within cognitive psychology, Johnson-Laird [21] described language communication as a process where a sender intentionally produces a signal to convey information to a receiver. Krebs and Davies's behavior ecology [22] added to that definition intention to modify the behavior of the reactor. Ogden and Richards [23] contributed by qualifying that communication occurs when the intentions are successful. Kimura [24] applied the same model within neuropsychology: interspecies' communication is when one member of the species conveys info to another member of the species.

Peters [18] comments that for the past 60 years the humanities have been dominated by a metaphor of communication borrowed from telephony. In his view, this definition locks us in a dichotomy—instantaneous connection vs. nightmare of solitude—and dislocates our attention from the cultural and ethical nature of communication. Communication becomes either destructive or therapeutic, its main function is to eliminate differences or cure us from our differences [25]. This tendency towards sameness, caused by and resulting in irrational fear of otherness, builds on the temptation to reduplicate the Self, to mirror signal/meaning/ideas. Therefore, Peters (ibid.) revises the definition of inter-human communication as a transmission of information and develops the idea of communication as a manifestation of the ethical, which is described in the next section.

15.3.2 Communication as Tension Between Reproduction of Self and Reconciliation with Alterity

Todorov ([26]: 99) gives a pointed example of the cognitive and communicative mechanisms leading the reproduction of Self to annihilation of otherness. When the Spaniards reached the shores of the New Continent at the end of 1400 AD they shouted from the dock of their ship to the Natives' canoes: "What is the name of this land?" The Natives shouted back: *Ma c'ubah than*. The Spaniards heard "*Yucatan*" and concluded that this is the name of the province. The response of the Natives was taken as an answer already implied in the question, it merely mirrored what was already conceived. What the Spaniards did not suspect is that the meaning of the expression "Ma c'ubah than" in the local language is "We do not understand your language" or "Who are you".

This meeting is a distilled metaphor of communication as the meeting between sameness and otherness. In the initial meeting with the Other, the Spaniards extricated the foreign and placed it in a context furnished by their own knowledge of the world. They did not even suspect the possibility of any other response than an answer intended in their question. With the same strategy they continued taking over and plundering the unknown continent, thus committing a horrific genocide [26]. Unlike many other historical accounts, Todorov's describes with great intensity this specific historical point in which "violence enters the picture under the cloak of innocent ignorance" ([27]: 295). The irresponsible acts of violence were much more frequent and effective i.e. few were killed by a gun, knife or other tools; most were killed by negligence, psychological torture and humiliation, taking over homes, resources and lands, stealing, hitting and raping wives and children, often in front of the humiliated husbands and fathers, spreading of diseases, instigating tribes against each other, etc.

This example illustrates the bond between culture, ethics and communication but it also indicates how ethics emerges through and in language: beyond the contents delivered and the linguistic structure it enforces, language inspires the fundamental response-ability between Self and Other. Responsibility implies both being response-able to the Other's call and being responsible for the Other, reaching out to Alterity, beyond the Self. Communication transpiring between Self and Other is not only exchange of information, or participation in a discursive sphere; it is also a manifestation of the fundamental responsibility to and for the Other. This, according to Martin Buber [6, 7] and Emmanuel Levinas [28–31], is the origin of language and communication: openness, exposure, proximity, and responsibility. Response-ability as a priori agreement is not always representable linguistically, it is the fundamental relation and the condition for any understanding and representation. Locked within a dichotomy in conceptualization of communication "at once a bridge and a chasm" (Peters [18]: 5), telepathy and solipsism, telephony and poesis, missiles and swan lakes, we miss the opportunity to meet the Other and the Self. Instead of viewing human communication as a battle against noise for a clear message or as a therapeutic technique, we can view communication as an ethical process, as an opportunity to meet otherness, outside or inside the Self. As Peters' ([18]: 21) puts it:

> Communication as reduplication of the self or its thoughts in the Other needs to crash for the resulting discovery of the Other (besides knowing and the check on the hubris of the ego) is in essence the way to the distinctness of human beings.

Inspired by Ralph Waldo Emerson, William James, Adorno, and Buber, Peters ([18]: 31) offers a new idea of communication as reconciliation with Alterity:

> The most wonderful thing about our contact with each other is its free dissemination, not its anguished communion….acknowledging the splendid otherness of all creatures that share our world without bemoaning our impotence to tap their interiority.

Based on this analysis of communication, the definition adopted in the proposed model is as follows:

Definition of Communication: Communication is both an opportunity for and a meeting between Self and Other.

This calls for re-evaluation of 'miscommunication'. It is precisely the failure in communication that opens up the very possibility for ethics to manifest. It is in the unsettling moment of incomprehensibility that one is exposed to the Other's otherness with no guidance as to how to respond. Intercultural communication confronts with otherness on daily basis, which is often seen as a challenge, therefore the next section explores the definition of intercultural communication.

15.3.3 Intercultural Communication as Creativity Rather Than a Challenge

The dominant view on language and communication as sender's signal to a receiver dominates the field of intercultural communication yet there is diversity in the evaluation of the effect of intercultural communication. Iles [32] finds that it aggravates relational conflict. "Culture is more often a source of conflict than of synergy. Cultural differences are a nuisance at best and often a disaster." (Hofstede http://www.geert-hofstede.com/). Few emphasize its potential for enhanced performance and creativity [33]. Todorov's account of the history of Yucatan is a clear example of the tragic potential of intercultural communication if realized as linguistic and discursive assumptions of sameness. It is challenging even in peaceful conditions as Peace Corps volunteers return prematurely, large numbers of failed joint ventures in China (the so called China fever), expatriates return with families, etc.

There are not so many studies of intercultural communication. Cross-cultural studies predominate the field. The cross-cultural perspective studies behaviors, values, norms, and perceptions of one culture in mono-cultural conditions and in comparison to those in another culture [34]. Intercultural studies examine behaviors, values, norms, and perceptions of participants representing different cultures in intercultural settings. However, most 'findings' in this field are cross-cultural and oriented towards description of similarities and differences [35], such as:

- In difference from USA, Mexico, Iran prefer leaders responsive to group norms;
- Collective cultures prefer charismatic leaders;
- Individual cultures prefer task-oriented leaders;
- High Power Distance prefer directive leadership;
- In Japan turn-taking is characterized by pauses, in USA by latching and in Brazil —by overlap.

There are only a few pure intercultural negotiation behavior studies. Reeh et al. [36] finds nonverbal reciprocal adaptation [37] in intercultural conditions, which is not observed in mono-cultural conditions. Qui and Wang [38] find verbal reciprocal adaptation in negotiation role-play between Chinese and Swedish women:

- Bargaining—gravitating towards each others' value
- Mirroring, imitation
- Drop learned own cultural behaviors

Whereas cross-cultural studies compare cultures in their intracultural functionality and find or predict an established system of similarities and differences, in intercultural conditions we observe a process of mutual adaptation, of reciprocal reaching towards each other, which along with exposure to otherness produces hybrid cultures. Intercultural communication brings adaptation as a reaction to the exposure to Otherness, which manifests the tension between a desire for fusion, for reduplication of Self and a possibility for reconciliation with Otherness, for openness to the unknown. Exposure to Otherness opens a possibility for communication as manifestation of the ethical. It is therefore important to understand how this adaptation occurs and how it relates to culture-based behavior.

15.4 Reciprocal Adaptation and Alignment

There are interactive mechanisms, which build culture-based frames of reference. Two, relevant for our inquiry, concepts have been introduced, namely reciprocal adaptation and interactive alignment. Gumperz ([37]: 13) definition of reciprocal adaptation is, in short, the following: "the procedure … where each participant gradually learns to adapt and to enter into the other's frame of reference." In his view, reciprocal adaptation is involved in interactive reframing of situations, knowledge and arguments and it does not presuppose conscious or less conscious processing of information. This communicative, cultural and learning mechanism is not only cognitive but also linguistic, emotive and behavioral i.e. speakers adapt to each other on different levels: lexical and semantic choice, syntax, posture, gaze, proxemics, orientation, tone of voice, etc. It is a mechanism behind linguistic phenomena such as creole-like varieties of languages and interactive emotions such as empathy and rapport [39]. Similarly, human users adapt to the speech and behavior of the Virtual Agent [40, 41].

According to Pickering and Garrod [42], communication in discourse is accomplished through an interactive process they call alignment and successful communication is accomplished through good alignment: "the development of similar representations in the interlocutors … interlocutors align situation models during dialogue." (ibid. p. 1). The main claim of their theory is that "automatic processes play a central role and explicit modeling of one's interlocutor is secondary" in communication (ibid.). The alignment involves situation models and non-situational knowledge, such as language knowledge. Interlocutors align their situational knowledge but they also align knowledge of situation and language (for instance, what they think 'right' means with the word 'right'). The situation models include notions, such as "space, time, causality, intentionality, and reference to main individuals under discussion" (ibid. p. 2). Alignment is based on willingness for cooperation and on mechanistic automatic imitation (of lexical choices, syntax, tone of voice, etc.): "Our underlying conceptualization of conversation is collaborative, in that we treat it as a "game of pure cooperation" … in which it is in both interlocutors' interest for it to succeed for both interlocutors" (ibid. p. 22) and "the interactive-alignment account proposes that alignment is primitive. It is a form of imitation and drops out of the functional architecture of the system … In these accounts, imitation is an automatic, non-inferential process and is in some sense the default response. Generally, imitation does not appear to require any decision to act" (ibid. p. 18).

Thus, alignment does not involve building of entire theory of the other as in Theory of Theory of Mind but a primitive turn-to-turn alignment on different linguistic levels of the message: phonetic, syntactic, semantic, etc. Each level is processed and aligned for itself and misalignment on one level enhances alignment on another level. Pickering and Garrod point out also that children can't inhibit alignment, which speaks for the forcefulness of this interactive mechanism. They base their view on situated interaction where participants have to find interactively each other's position in a maze without being able to see it and assume that the same mechanism works on everyday conversation.

"Such models are assumed to capture what people are "thinking about" while they understand a text, and therefore are in some sense within working memory (they can be contrasted with linguistic representations on the one hand and general knowledge on the other). Successful dialogue occurs when interlocutors construct similar situation models to each other." (ibid. p. 1–2).

They point out "that this account differs from Clark [43], who assumes that speakers carefully track their addressees' mental states throughout conversations" (ibid. p. 10) and that "the important point is that effects of partner specificity do not imply that interlocutors need employ complex reasoning whenever they produce an expression. Instead, they have a strong tendency to employ the form that they have just encountered" (ibid. p. 20).

Interactive alignment is similar to the concept of reciprocal adaptation in that they refer to framing in terms of similar discourse processes and do not demand conscious processing during interaction, although the connection has not been made nor explored yet:

Reciprocal adaptation:

> the procedure…where each participant gradually learns to adapt and to enter into the other's frame of reference.

Interactive alignment:

> the development of similar representations in the interlocutors … interlocutors align situation models during dialogue.

However, since alignment is by definition a primitive process, i.e. it does not involve complex reasoning or active long-term memory based monitoring, it is hardly the case that interactants rely only on this process in order to make sense of a situation. Rather, it is more likely to see it as a way of preliminary or first layer framing. In Gumperz' terms, interactants gradually learn to frame their activities through reciprocal adaptation in order to make sense of a situation. It is a more general term than alignment because it does not pose a condition of automaticity and short-term memory basis, but does not exclude it either.

Even more complex cognitive-emotive processes such as some forms of Theory-of-Mind-building produce culture-based behaviors. Martinovski et al. [44] analysis indicates that complex emotions such as empathy involve Theory-of-Mind models. Martinovski [45] found that emotions function as engines in conflict management and involve opposite reciprocal adaptation and that they realize in different manner depending on culture-specific settings. Thus a model on culture-based behavior and awareness needs to involve emotions.

15.5 Culture and Emotion

According to the social and anthropological constructivist theory [46] it is socio-cultural interpretations, which determine emotions and body behaviors. Cornelius [47] gives as an example of emotion attitudes to language variations, such as dialects. However, although attitudes may trigger regularly the same emotions as a result of culture-specific appraisals this does not mean that attitudes are the same as emotions. Test on capacity for interpreting body language, the so called, PONS-tests, Profile of Non-Verbal Sensitivity [48] show that women are generally better then men at recognizing expressions of emotions but worse in interpreting anger. The socio-cultural perspective explains these results by pointing out that in patriarchal, i.e. in most societies, women are prohibited in various ways to express anger, which is part of the subjugation of women [49]. Yet, prosodic comparisons of intended and recognized emotions and attitudes in different unrelated languages found "remarkable similarities" ([50]: 2), which indicates that the relation between content and expression of emotion is not arbitrary i.e. the results supported the universalist perspective on emotion.

In the framework of Darwinism, emotion has a role in adaptation in the course of evolution, i.e. it is universal because expression of emotion is found in other species [47]. In Descartes' era, emotions intertwined with cognition of stimuli. William James [51] introduced the role of the body in the cause and effect chain: the mind perceives the reaction of the body to stimuli, e.g. increased heartbeat; the sensation of the physiological response is a feeling which mental representation is an emotion, e.g. fear.

Contemporary neuroscientists report evidence for the involvement of emotion in so called rational cognitive processing. Neuroscientists such as Von Uexkull [52], Fuster [53], and Arnold Scheibel (personal communication) observe that evolution gave privilege to the limbic system: emotional feedback is present in lower species, but other cortical cognitive feedback is present only in higher species. In that sense, emotion functions in evolution as a coordinator of other cognitive and non-cognitive functions. Damasio [54] suggests that the state of the mind is identical to the state of feeling, which is a reflection of the state of the body. He explores the unusual case of Phineas Gage, a man whose ability to feel emotion was impaired after an accident in which part of his brain was damaged. Damasio finds that, while Gage's intelligence remained intact after the accident, his ability to take rational decisions became severely handicapped because his emotions could no longer be engaged in the process. Based on this case, the neurologist comes to the conclusions that rationality stems from our emotions and that our emotions stem from our bodily senses. Certain body states and postures, e.g. locking of the jaw, would bring about certain feelings, e.g. anger, which in turn will trigger certain thought and interpretations of reality, a thought traced back to William James. Appraisal theory sees emotion as something automatic, non-reflective and immediate and at the same time, cognition leads emotion, i.e. the way we cognize events influences our emotions related to them, not the opposite. In that sense, emotions become and involve coping strategies [55].

Three mutually exclusive theories [56] have been suggested to explain how we understand others i.e. what the cognitive-emotive mechanisms that facilitate communication and learning are:

(i) by imitation (e.g. [8, 32])—ex. we understand what it means to have stomach ache by imitating that state;
(ii) by simulation (e.g. [37, 57, 58])—ex. we understand what it means to have stomach pain by associating it with our state when we ourselves have stomach ache;
(iii) by symbolic representation which does not rely on neither imitation nor simulation (e.g. [3, 59])—we understand when it means to have stomach ache by having a fixed set of features describing how one feels when one has stomach ache and a fixed set of features on how one should react to that state in different contexts.

Martinovski [60] finds indications that all three cognitive-emotive processes are employed in human interaction. Reciprocal adaptation and interactive alignment are directly supported by imitation and simulation. Yet, symbolic representations of states are dependent on simulation. There is correspondence between these cognitive processes and Pierce's semiotic signs: icon (similarity-based relation between signifier and signified), index (indexical associative relation between signifier and signified) and symbol (convention/representation-based relation between signifier and signified). Design of Virtual Agent systems is based on representation-based and simulation-based cognitive-communicative processing which is similar to appraisal theory's view on emotion but even imitation, which is productive in human learning, has proven possible and productive method, especially in culture-based communication training systems [61].

Given the short overview above, the definition adopted here for emotion is as follows:

Definition of Emotion: Emotion is a sensory feedback recognition and a coping strategy, which coordinate decision-making on individual level and in interaction.

15.6 Model of Culture-Based Communication

The model of culture-based communication proposed here relies on definitions motivated by the theoretical overview in the previous sections, namely:

- Culture is a dynamic semiotic process and a semiotic system of interpretation of Self and Otherness by means of which we conduct our perceptual transactions in interaction with environment.
- Communication is both an opportunity for a meeting and a meeting between Self and Other.
- Emotion is a sensory feedback recognition and a coping strategy, which coordinate decision-making on individual level and in interaction.
- Interactive alignment is a primitive cognitive-emotive process based on imitation during which interlocutors develop similar representations and align situation models through dialogue.
- Reciprocal adaptation includes both interactive alignment and complex long-term memory cognition where each participant gradually learns to adapt and to enter into the other's frame of reference.

It suggests a process-like representation of culture-based communication, which builds on connection between different modules, such as modules of culture awareness, culture behavior, communication processing, alignment, appraisal, and revision. Each module has a separate set of features and under-modules. For instance, the module of culture behaviors includes modules on turn-taking, gaze, proxemics, gestures, semantic fields, etc. Culture awareness consists of values related to awareness to each of the behavioral modules as well as other culture related features, such as values, beliefs, and goals. For the purpose of Virtual Agent simulation, they can be represented in XML format as in Jan et al. [1].

In this model (see Fig. 15.1), culture and emotion form and are formed by communication, which is assisted by both primitive and complex cognitive interactive procedures, such as interactive alignment and reciprocal adaptation. Studies suggest (e.g. [62, 63]) that awareness of and reference to co-existence in a larger context facilitate communication and decision-making. Therefore, in this model, each interaction is embedded in a larger existential context, which wraps in all

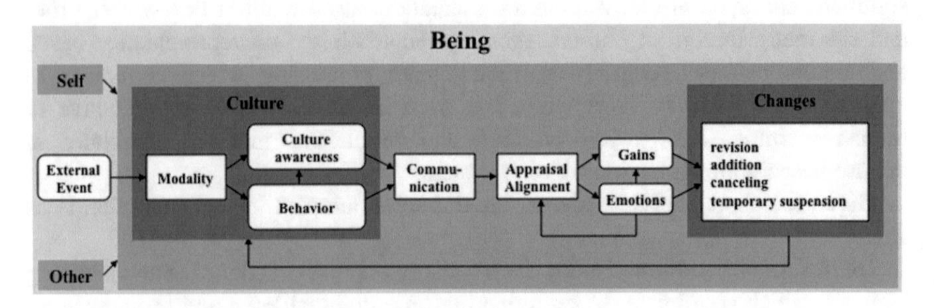

Fig. 15.1 Model of Culture-based Communication

human and other activity. Each interactive situation starts with a basic level of awareness of Self and Other, which is imbedded in a general sense of existence, of being and which can vary from person to person. This existential sense of Self and Other is then processed by degrees of awareness of own and Other culture and expressed in situated behavior in different modalities. Communication is the process through which both non-culture-based and culture-based awareness are realized and have a chance to change. Communication may include activities, such as talking between interlocutors, silent co-presence, reading, singing together, teaching, training, etc.

During communication, beliefs and emotions related to Self and Other are aligned and appraised which brings about coping strategies. These trigger re-evaluation of initial conscious or unconscious models of Self and Other, including goals and beliefs, which may result in need to change cultural models, awareness and behavior. Alignment and appraisal-based states may change without changing culture-based awareness and behavior. Emotions are not static. They are processes on a neurological, biological and expressive level. One and same stimuli can cause a chain of different physiological reactions, emotional sensations and cognitive appraisals, each of which can influence the other in time. That is, a physiological reaction may bring about an emotion, which can influence cognitive appraisal but this appraisal can in turn bring about coping strategies, which generate other emotions.

Reasoning in intercultural situations is based on abduction rather than deduction or induction because the premises of beliefs and goals in the 'awareness' and 'behavior' modules may be incomplete or incorrect. In order to illustrate how the model works we can observe the following scenario: a woman from a culture in which women stand physically far from male acquaintances, especially in public, meets physically and converses on a public street with a male acquaintance who comes from a culture with shorter distance between women and men. Proxemics is usually a low consciousness feature in non-verbal communication, therefore we can define the initial culture-based awareness of Self and Other on both parties i.e. awareness of the exact comfort distance between women and men in own and Other culture to be low. The behavior is defined as expected according to respective proxemics culture. Given their cultural settings of proxemics, the woman may move slightly away from the man as he tries to shorten that distance to fit his cultural habits. Both may experience initial cooperative alignment but may appraise the changes in proxemics as uncomfortable. For example, the woman may feel invaded or in danger and the man may feel unappreciated or avoided. These dynamics will occur in the small loop between appraisal and the module of changes. The reciprocal adaptation mechanism may push both parties towards final acceptance of a distance which is least uncomfortable for both parties but in short term, it may not result in changes of culture-based awareness of own and Other proxemics, nor in changes of culture-based behavior. Only after repeated exposure may culture-based awareness and behavior change and thus yield less emotionally and cognitively loaded interactions. If however, both parties start with high level of culture-awareness regarding the Other but not Self, the process may get stuck in the smaller loop until there is one-sided adaptation or until there is increase in Self culture–awareness. If both parties are highly aware of cultural regulations of proxemics in own and Other culture, the cognitive-emotive load for conscious processing of reciprocal adaptation may be higher i.e. the smaller loop will be shorter, given that there is no special agenda which blocks their will to reciprocal adaptation, but may take more conscious cognitive effort.

15.7 Summary

Based on concept analysis and empirical findings within the research fields of culture and communication, this study concludes that cultural awareness is acquired through communication and filtered through emotion. The proposed dynamic modular model of culture-based communication reflects this process-oriented view of culture and includes emotion as an engine for cultural learning and expression. Culture is defined as a semiotic process and a system, which builds upon Self and Other identities and which is sustained and modified through communication and cognitive-emotive mechanisms, such as reciprocal adaptation, interactive alignment and appraisal. Communication is defined as an opportunity for meeting of Otherness. In that sense, Human-Virtual Agent interaction can be seen as a form of intercultural communication, as a meeting with otherness. Since intercultural communication involves interactive alignment and reciprocal adaptation, these mechanisms can be applied also to simulation of culture-based communication where the adaptation occurs with respect to culturally varied and measurable features and behaviors i.e. the model can be utilized and operationalized in the design of life-like training scenarios.

References

1. Jan, D., Herrera, D., Martinovski, B., Novick, D., Traum, D.: A Computational Model of Culture-Specific Conversational Behavior. In Pelachaud, C. et al. (eds.) IVA 2007, LNAI 4722, pp. 45–56, Springer, Berlin Heidelberg (2007)
2. Geertz, C.: The Interpretation of Cultures. Basic Books (1973)
3. Hofstede, Geert: Cross-Cultural Research and Methodology Series, vol. 5. Sage Publications, Newbury Park, USA (1984)
4. Shuter, R.: Proxemics and tactility in Latin America. Journal of Communication **26**(3), 46–52 (1976)
5. Erikson, E.H.: Identity: youth and crisis. Norton, New York (1968)
6. Buber, M.: Det mellanmänskliga. (Element of the Interhuman). Sweden, Falum: Dualis Förlag AB (1995)
7. Buber, M.: I and Thou. Charles Scribner's sons, New York (2000)
8. Iacoboni, M.: Understanding others: imitation, language, empathy. In: Hurley, S., Chater, N. (eds.) Perspectives on Imitation: From Cognitive Neuroscience to Social Science. MIT Press, Cambridge, MA (2005)

9. Sebeok, T.A.: An Introduction to Semiotics. Pinter, London (1994)
10. Peirce, Ch. S.: Collected Papers In: Hartshorne, Ch Weiss, P. (ed.) Elements of Logic, vol II, pp. 2.228–2.272 (1932)
11. Dewey, John: Pierce's theory of linguistics signs, thought and meaning. J. Philosophy **43**(4), 85–95 (1946)
12. Martinovski, B.: Shifting worlds or deictic signs in WWW. In: Pankow, Ch. (ed.) Indexicality. SSKKII, Göteborg (1995)
13. Rizzolatti, G., Craighero, L.: The mirror-neuron system. Ann. Rev. Neurosci. **27**, 169–192 (2004)
14. Happé, F., Brownell, H., Winner, E.: Acquired 'Theory of Mind' impairments following stroke. Cognition **70**, 211–240 (1999)
15. Baron-Cohen, S.: Mindblindness: An Essay on Autism and ToM. The MIT Press, Cambridge (1995)
16. Berlin, B., Kay, P.: Basic Color Terms: their Universality and Evolution. University of California Press, Berkeley and Los Angeles (1969)
17. Chomsky, Noam: Language and Mind. Cambridge University Press, Cambridge (2006)
18. Peters, John Durham: Speaking into the Air: A History of the Idea of Communication. University of Chicago Press, Chicago (1999)
19. Weaver, W.: (1949) Recent contributions to the mathematical theory of communication. In: Mathematical, The (ed.) Theory of Communication, Claude Shannon and Warren Weaver. The University of Illinois Press, Urbana (1964)
20. Shannon, C., Weaver, W.: The University of Illinois Press. The Mathematical Theory of Communication, Urbana (1964)
21. Johnson-Laird, P.: What is communication? An introduction. In: Mellor, H. (ed.) Communication: Fourth Darwin Lecture Series. Cambridge University Press, Cambridge (1990)
22. Krebs, John, Davies, N.B. (eds.): Behavioural Ecology: An Evolutionary Approach. Blackwell, Oxford (1978)
23. Ogden, C.K., Richards, I.A.: The Meaning of Meaning. Routledge & Kegan Paul, London (1923)
24. Kimura, D.: Neuromotor Mechanisms in Human Communication. Oxford University Press, New York (1993)
25. Derrida, J.: Margins of Philosophy. Trans. A. Bass, University of Chicago Press, Chicago (1982)
26. Todorov, T.: The Conquest of America. The Question of the Other. Harper and Row, New York (1984)
27. Pinchevski, A.: Interruption and Alterity: Dislocating Communication. PhD thesis in Communication, McGill University (2003)
28. Levinas, E.: Totality and Infinity: An Essay on Exteriority. Thans. A. Lingis, Duquesne University Press, Pittsburgh (1969)
29. Levinas, E.: Time and the Other. Duquesne University Press, Pittsburgh (1987)
30. Levinas, E.: The Other in Proust. In: Hand, S. (ed.) The Levinas Reader Blackwell, Oxford (1989)
31. Levinas, E.: Entre Nous. Thinking-of-the-Other. Continuum, New York (2006)
32. Iles, P.: Learning to work with difference. Per. Rev. **24**(6), 44–60 (1995)
33. Enayati, J.: The research: Effective communication and decision-making in diverse groups. In: Hemmati, M. (ed.) Multi-Stakeholder Processes for Governance and Sustainability—Beyond Deadlock and Conflict. England, London (2001)
34. Cultures and Organizations: Software of the Mind: Intercultural Cooperation and Its Importance for Survival. McGraw-Hill, New York (1996)
35. Halverson, C.B., Tirmizi, S.A.: Effective Multicultural Teams. Springer, Amsterdam (2008)
36. Reeh, A., Aydin, S., Moréno, J., García, M.J, Mota, G.R., Martinovski, B.: Body language in intercultural and cross-cultural communication. In: Proceedings of 16th NIC Conference on Intercultural Communication, Borås, Sweden 2009

37. Gumperz, J.J.: Discourse Strategies. Cambridge University Press, Cambridge (1982)
38. Qui, C., Wang, X.: Comparison of business negotiation styles between China and Sweden An experimental role-play according to accommodation theory. Master thesis on Communication supervised by Bilyana Martinovski, Gothenburg University (2010)
39. Martinovski, B.: Emotion in Negotiation. In: Kilgour, M., Eden, C. (eds.) Handbook on Group Decision and Negotiation, Springer, Amsterdam (2010)
40. Bell, L.: Linguistic Adaptation in Spoken Human-Computer Dialogues. Doctoral Thesis, Kungliga Tekniska Högskolan, Stockholm (2003)
41. Martinovski, B., Traum, D.: The Error is the Clue: Breakdown in Human-Machine Interaction. In: Proceedings of ISCA Tutorial and Research Workshop International Speech Communication Association, Switzerland (2003)
42. Pickering, M.J., Garrod, S.: Alignment as the basis for successful communication. Res. Lang. Comput. **4**, 203–228 (2006)
43. Clark, H.H.: Using language. Cambridge University Press, Cambridge (1996)
44. Martinovski, B., Traum D.,. Marsella, S.: Rejection of empathy in negotiation. J. Group Dec. Neg. **16**(1) (2007)
45. Martinovski, B.(ed.): Emotion in Group Decision and Negotiation. Springer, Amsterdam (2015)
46. Averill, J.R.: Anger and Aggression. An Essay on Emotion. Springer, Amsterdam (1982)
47. Cornelius R.R.: Theoretical approaches to emotion. In: Proceedings from ISCA Workshop on Speech and Emotion: A Conceptual Framework for Research 2000
48. Rosenthal, R., Hall, J.A., DiMatteo, M.R., Rogers, P.L., Archer, D.: Sensitivity to Nonverbal Communications: The PONS Test. The Johns Hopkins University Press, Baltimore, MD (1979)
49. Spellman, E.: Anger and insubordination. In: Garry, A., Pearsall, M. (eds.) Women, Knowledge and Reality: Explorations in Feminist Philosophy, pp. 263–273. Unwin Hyman, Boston (1989)
50. Piot, O.: Experimental study of the expression of emotions and attitudes in four languages. In: Proceedings of the 14th International Congress of Phonetic Sciences, San Francisco, pp. 369–370 (1999)
51. James, W.: Principles of Psychology, Dover Publications, Inc., New York (2014/1890)
52. Von Uexkull, J., Kriszat, G.: Streifzuge durch die Umwelten Von Tieren und Menschen Ein Bilderbuch unsichtbarer Welten. Springer, Berlin (1934)
53. Fuster, J.M.: Cortex and Mind: Unifying Cognition. Oxford University Press (2003)
54. Damasio, A.: Descartes' Error: Emotion, Reason, and the Human Brain. Putnam Publishing (1994)
55. Lazarus, R.: Stress and Emotion. Springer Publishing Company Inc, New York (1999)
56. Stich, S., Nichols, S.: Folk psychology: simulation or tacit theory? Mind Lang. **7**, 35–71 (1992)
57. Thagard, P.: The Brain and the Meaning of Life. Princeton University Press, Princeton (2010)
58. Gordon, R.: Folk psychology as simulation. Mind Lang. **1**, 158–170 (1986)
59. Hobbs, J. Gordon, A.: Encoding Knowledge of Commonsense Psychology. In: 7th International Symposium on Logical Formalizations of Commonsense Reasoning. Corfu, Greece, May 22–24, (2005)
60. Martinovski, B.: Empathy and theory of mind and body in evolution. In Ahlsén, E. et al. (eds.) Communication—Action—Meaning, pp. 343–361. A Festschrift to Jens Allwood. University of Gothenburg, Department of Linguistics (2007a)
61. Lhommet, M. Marsella, S.: Gesture with Meaning. In: Intelligent Virtual Agents (2013)
62. Martinovski, B.: Shifting attention as re-contextualization in negotiation, In: Proceedings of GDN, Montreal (2007b)
63. Martinovski, B.: Reciprocal adaptation and problem reframing in group decision and negotiation. J. Group Dec. Neg. **23**(3), 497–514. Springer (2013)

Chapter 16
Dynamic Decision Making Across Cultures

C. Dominik Güss and Elizabeth Teta

Abstract Decision making is a key cognitive process in all aspects of human life, professional as well as private life. The goals of this chapter are threefold. First, the chapter provides a short theoretical background on decision-making research highlighting the need for more comprehensive models of decision making. Whereas decision-making research has focused for a long time on simple choices, recent research has investigated the decision-making process in complex, uncertain, and dynamic situations. Second, the chapter discusses one methodology especially suited for the study of dynamic decision making. The methodology consists of situations that have been simulated on the computer and have been called, for example, microworlds, virtual environments, or serious strategy games. Third, and most importantly, the chapter will discuss new empirical research on how culture influences dynamic decision making both in microworlds and in the real world. Such findings contribute to a more comprehensive theory of decision making and allow for a better understanding of decision-making conflicts. Finally, applications of these findings are discussed, and can be utilized for cultural competence training programs or international work teams.

Keywords Dynamic decision making · Culture · Cross-cultural
Complex problem solving · Microworlds · Virtual environments

C. Dominik Güss (✉) · E. Teta
Department of Psychology, University of North Florida,
1 UNF Drive, Jacksonville FL 32224, USA
e-mail: dguess@unf.edu

E. Teta
e-mail: lizteta@ymail.com

C. Faucher (ed.), *Advances in Culturally-Aware Intelligent Systems and in Cross-Cultural Psychological Studies*, Intelligent Systems Reference Library 134, https://doi.org/10.1007/978-3-319-67024-9_16

16.1 Introduction

Decision making refers to choosing one of several options. Research on judgment and decision making has not only played a profound role in the field of cognitive psychology, but has also influenced various other fields beyond psychology (e.g., Behavioral Finance [15, 45]; Medical Treatments [5]). One indicator of the relevance of studies on decision making for society is the awarding of the Nobel Prize to decision-making researcher Daniel Kahneman in 2002. Together with his colleague Amos Tversky, Daniel Kahneman showed that people often do not follow normative mathematical models to make their decisions, but that people use heuristics when making decisions often times leading to flawed decisions [31, 32, 53]. Both researchers were on the forefront of decision-making research and paved the way for establishing decision-making research in the field of psychology.

16.2 Dynamic Decision Making

16.2.1 Task Characteristics

One way to differentiate research on human decision making is related to the tasks used. The following is an example of a task used by Tverky and Kahneman ([53], p. 1125).

> A certain town is served by two hospitals. In the larger hospital about 45 babies are born each day, and in the smaller hospital about 15 babies are born each day. As you know, about 50 percent of all babies are boys. However, the exact percentage varies from day to day. Sometimes it may be higher than 50 percent, sometimes lower.
>
> For a period of 1 year, each hospital recorded the days on which more than 60% of the babies born were boys. Which hospital do you think recorded more such days?
>
> - The larger hospital (21)
> - The smaller hospital (21)
> - About the same (that is, within 5% of each other) (53)

The numbers in the parentheses show how many undergraduate students chose each answer. The correct answer is "the smaller hospital", because it is more likely for a smaller sample to stray from 50%. Most students, however, chose the last answer option "about the same in each hospital", perhaps because the event was described by the same statistic in both hospitals.

The example stands for a typical procedure to study decision making using well-defined tasks: Describing a problem situation, providing a few possible answer options and asking participants to choose one of them. Decision making in real life, however, is most often not comparable to such tasks, because the situations are mostly ill-defined. This means not all aspects of the situations are known, nor can one predict how these situations will develop or change. Besides, decision makers are often not clear about their own preferences.

To further specify the term "ill-defined", decision situations in daily life can be characterized as being complex, uncertain, and dynamic [8, 16, 17]. Complexity, referring to systems theory, means that a situation consists of a network of variables and their connections. These variables therefore interact with each other. Uncertainty refers to the vagueness and the unpredictability of a given situation. For the decision maker, not all aspects of the situation are known, and it is unclear how the situation will develop. Dynamics refers to time. Situations change over time, either because of events outside of the decision maker's control, intended effects, or side- or long-term effects of previous actions of the decision maker.

Let's clarify the task characteristics referring to an example. A CEO of a company has to deal with a variety of interrelated problems (complexity), such as manufacturing of products, stock, sales, marketing, competitors and/or clients. It is unclear how the market will develop, what competitors will do, and how their products and prizes will develop (uncertainty). The situation of the company and the demands of the market change over time (dynamics). There are seasonal changes related to holidays and seasons of the year, changes in product lifecycle, or unforeseen changes related to market development. The CEO may even face changes within the company as far as management and employees.

Thus dynamic decision making (DDM) can be defined as making decisions in a complex environment that changes over time; these changes are either a result of previous actions of the decision maker or a result of events that are not caused by the decision maker [3, 43].

16.2.2 Demands on the Decision Maker

The special task characteristics pose specific demands for the decision maker. These demands are now discussed sequentially provided with examples referring to a CEO of a company, but one can imagine the decision maker going through these steps recursively and in changing sequences.

- *Goal clarification*: Goals are often not specified and therefore have to be clarified.

  What does it mean, for example to "Increase revenue"? This is a general statement and in order to make decisions, one has to clarify the goal by asking questions like: How can we best cut down on costs? How can we best increase sales? Again, these questions have to be further specified.

- *Information collection*: A next step during dynamic decision making is to collect more information related to the problem situation and related to possible solutions. In uncertain problems, such information collection could take a while.

  What do the competitors do? What does the product development team suggest?

- *Prediction*: This information is then integrated into a mental model of the situation, using strategies such as outlining the causes of the situation and the key problem aspects. This model then allows us to make predictions and to develop hypotheses how the situation will change and further develop. Such predictions are necessary for subsequent steps.

Given the mission, goals, and portfolio of the company and all the information the CEO has collected, it is expected that there is a huge demand on the market for the new product.

- *Planning, decision making, and action*: Based on these predictions, the decision maker can plan several decision alternatives and evaluate them. How easy or difficult is it to implement a certain decision? At what costs? What would be their likely outcomes? Certain alternatives are then pursued and actions have to be taken, so that the situation can be changed.

The top management team has decided to launch the new product. When have we manufactured a sufficient number? Sales and marketing plans are developed and at one point the product is delivered to the stores.

- *Monitoring and Effect control*: At all times, the decision maker has to monitor the process and judge if each step was sufficiently taken care of. It is always necessary to self-reflect and ask: Did I gather enough information? Are the goals reasonably defined? Can I predict relatively well how the situation will further develop? Did the actions really produce the intended outcome?

Did we sell the predicted numbers of our new product? Do the customers reflect the customer profile we expected to buy the product? Which part of the new product can be improved?

16.2.3 Microworlds

DDM has been studied in the real world often referring to case studies, for example investigating Adolf Hitler's decisions [11], career options [23], as well as medical treatment options [5]. One interesting way to study dynamic decision making in complex, uncertain, and dynamic situations, yet maintaining the control of the laboratory, has been the use of so-called microworlds. Microworlds are computer-simulated problems, and have also been called computerized strategy games or virtual environments. All systems require a participant to sit in front of a screen and make decisions using the keyboard. These decisions are automatically implemented, consequences are calculated, and effects of the decisions plus programmed changes in the environment are shown on the screen. Examples of such simulated situations are dealing with forest fires (FIRE CHIEF [39], NEWFIRE [37], WINFIRE [19, 25]), managing companies (SCHOKOFIN/CHOCO FINE [10], TAYLORSHOP [18, 40, 41]), managing a city (LOHAUSEN [13]), managing a water production plant [21] or working as developmental aid assistants in Africa (MORO [14, 49, 50]).

The previous section discussed the demands of complex, uncertain, and dynamic problem situations on the decision maker and the related steps of decision making. We always provided examples for a CEO of a company. All these steps also apply to decision making in microworlds. In CHOCO FINE, for example, goals have to be clarified (How can I get a bigger market share?), information has to be collected (Who are my clients, competitors?), predictions about possible sales have to be made (launching product around holidays), production, marketing and sales have to be planned and specific decisions have to be made (How much bitter chocolate shall I produce this month?), and actions have to be monitored and controlled (Why did I not sell as much as I predicted this month?).

Such microworlds offer various advantages for the study of dynamic decision making [4]. First, they eliminate experimenter effects as the participants work directly on a PC and make decisions using the mouse. These decisions are automatically saved in log files. Second, they allow the analysis of process data such as decision-making strategies or errors over time. Thus they allow the study of multiple interdependent decisions. Third, they also allow the study of decision-making performance over time and of potential variables explaining performance.

16.3 Culture and DDM

Since the decision maker relies on experience to act in such complex, dynamic, and uncertain situations, one can assume that culturally learned knowledge reflected in these experiences strongly influences decision making. More specifically, culturally learned knowledge in this context is understood as explicit and implicit knowledge shared by a specific group of people and transmitted from generation to generation [26, 47]. How specifically does culture influence DDM?

Following the ecocultural framework [1], the ecological context and the sociopolitical context influence adaptation. Individuals make experiences in a specific cultural context and their behaviors adapt and seem functional in this context. Culturally acquired strategies regarded as adaptive and experienced as being successful will most likely be applied to novel situations as well.

Action and sociocultural theories further specify how these cultural experiences take place [48, 54, 55]. Decision making, as many other psychological processes, is learned in social interactions in a specific social, cultural, and historical context [6, 7, 34, 35, 42]. Such social interactions provide opportunities for learning, which Lave ([33], p. 67) called “situated social practice.” Applied to DDM, the learned social conceptions provide frameworks helping to mentally structure a decision problem, to perceive and interpret a problem and to choose specific decision-making strategies. Thus, cross-cultural differences can emerge and strategies that might be adaptive in one cultural environment might not be adaptive in another.

16.3.1 Cultural Values and DDM in Microworlds

The previous section described how people learn culture specific DDM behavior. The following sections discuss specific aspects of culture that can explain cross-cultural differences in DDM. The first cultural dimension of relevance here are cultural values.

Values can be defined as abstract, trans-situational goals [30, 44] that may act as guiding principles for the selection of specific decision-making strategies in complex, dynamic, and uncertain situations. Güss [25] compared DDM strategies and performance in 5 countries: Brazil, Germany, India, Philippines, and the United States. Over 500 participants worked on two different microworlds: Coldstore, which requires cautious decision making, and WinFire, which requires quick decisions/action orientation. Additionally, cultural values for every participant were assessed using a survey [46]. These were horizontal individualism (HI), vertical individualism (VI), horizontal collectivism (HC), and vertical collectivism (VC). HI favors equality and focuses on the self and unique self-identity; VI accepts inequality and focuses on the self and competition; HC favors equality and focuses on the group and caring for the group; VC accepts inequalities and focuses on the group and self-sacrifice for the group.

The four dimensions are not regarded as mutually exclusive and it is possible that participants score high (or low) on several dimensions. Using structural equation modeling, the researcher showed that DDM strategies were related to performance, i.e., low action orientation predicted performance in Coldstore and high action orientation predicted performance in WinFire. Results also showed that, for example, horizontal individualism was positively related to action orientation, and that horizontal collectivism was negatively related to action orientation in WinFire. Thus, these results show that cultural values can trigger specific DDM strategies.

16.3.2 Cultural DDM Strategies in Microworlds

Not only values, but other cultural variables might influence DDM. Predictability of the environmental changes (e.g., economic, political) in one's culture may lead to different planning time frames and require either short-term or long-term planning [24]. During times of extremely high inflation, for example, short-term planning is more adaptive. It is necessary to spend most of the monthly income right away, because money will loose its value when saving it.

Decisions could also be influenced by religious convictions and the extent to which the individual is supposed to make decisions on their own or leave it up to a supernatural being or others to make decisions. At this point, there is not enough empirical evidence to suggest which cultural dimensions affect DDM, but we do know that DDM strategies differ among cultures. Strohschneider and Güss [50] used the complex microworld MORO in their study in India and Germany. MORO puts the participant in the role of a developmental aid assistant who is supposed to help the tribe of the "Moros" in the Sahel zone over a period of several years. The simulation lasted 2 h for every participant in an individual session. The authors created one version that would react positively to high-intensity decisions, and another version that would act favorably to low-intensity decisions. That way, the simulations did not favor one approach over the other. Regardless of the version of the microworld, Indian participants were less active and less control-oriented and committed more errors compared to the German participants. The study showed that DDM strategies differ across cultures and that these preferred DDM strategies are shaped by their learning experiences within a culture. Cultural norms, standards, constraints, and values regulate which strategies are adaptive and desirable and which are not and then result in cultural differences in predictions, information collection, and monitoring [28].

16.4 Empirical Studies on Culture and Dynamic Decision Making in the Real World

16.4.1 Cultural Values and Decisions Regarding Social Media/Internet Usage

Social media is used globally, but it is important to note that social media serves different purposes cross-culturally. The ways that people interact with social media varies in frequency, content, number and types of contacts [51]. Even global networks are utilized differently around the world, such as Facebook, which has a key role in American social networking but is not as popular in other cultures. In Japan and Korea, for example, there is access to Facebook, but people prefer more local networks like Mixi and Cyworld [22].

Goodrich and Mooij [22] attempt to explain the root of cultural differences in social media usage and how these differences affect decision-making, particularly consumer decision-making. The authors first explain that consumers traditionally turned to friends and family for product reviews and insight via word of mouth (WOM). Today, the internet allows us to communicate with people we know personally and impersonally to receive informative WOM from a broader network. This mediated form of WOM is referred to eWOM in this study. Is it possible that WOM and eWOM influence decision-making differently? If so, can cultural values explain how WOM and eWOM influence people differently?

Their study referred to Hofstede's cultural value dimensions [29, 30]. We refer here only to the dimension individualism/collectivism. The authors showed that WOM and eWOM tendencies differed according to dominant cultural values. People in individualistic cultures value independence and tend to look after themselves and their immediate family. Examples of individualistic cultures include

Germany, France, Netherlands, Belgium, Scandinavia, and the United States. When people of predominantly individualistic cultures are faced with making a decision, they tend to seek factual information rather than the opinions of friends or family.

Conversely, members of collectivist cultures identify with their social system and are more conscientious about focusing on their social group as a whole. Collectivist cultures make up about 70% of the world's population and include countries such as Vietnam, China, Thailand, Venezuela, Mexico and Pakistan. Members of collectivist societies have a greater amount of trust and value the opinions of their peers when making decisions. Members of collectivist societies are more likely to use social media to share ideas and opinions with each other about products than to seek information directly from the manufacturer. They also frequently meet and chat with members of their community and are more likely to utilize traditional WOM than individualistic cultures.

Other research has been conducted to identify cultural differences in online decision-making processes in terms of search speed, quantity of information and the types of information used. Li, Takahiko, and Russell [36] sampled Chinese university students in Hong Kong and European Canadian university students in Alberta. Students were asked to imagine that they were looking for an apartment to live in for the upcoming semester. Participants were asked to rate the importance of the apartments attributes on a 6-point Likert scale and then select an apartment they felt was best. The amount of time students spent viewing attributes and the information students opened online were recorded. Students were also randomly assigned to a time-constraint condition or no time-constraint condition for all measures.

There are significant cultural differences between East Asians and North Americans in regards to information search speed and the time spent before making a decision when there were no time constraints, however, these cultural differences were not significant when participants were on a time-constraint. In the no time-constraint condition, Americans took significantly more time to make a final decision than Asians. Also, time constraints had a much greater effect on Americans than Asians. When under a time constraint, Americans took significantly less time to make a decision than when there was no time constraint. Asians showed only a marginal decrease in decision-making time when given a time constraint. Overall, Asians spent less time on decision-making in this apartment scenario.

The second factor of the online decision-making process examined was information search efficiency. Asians searched through information more efficiently than Americans with and without time constraints, showing a significant interaction between culture and quantity of information searched. Interestingly, Americans searched information much more efficiently when there were no time constraints, which is the opposite of how Asians reacted to parsing through information in relation to time. Asians were only marginally less efficient when given a time constraint than without. There were no significant cultural differences in the amount of information searched when both cultures were limited on time.

To summarize, the results of the study suggest significant effects of culture on decision-making in the no-time-constraint condition, but there were no differences in decision-making variables between cultures when time-constraints were in play.

16.4.2 Culture and Team Decisions

Today's business workforce has achieved success incorporating diversity and collaboration more than ever before [2]. Research suggests that culturally diverse teams can increase the quality of decision-making by contributing various points-of-view. Benefits of multicultural teams include a wider range of knowledge and skills beneficial to the task at hand which promotes ingenuity [52]. However, multicultural teams may lack understanding of one another, which in turn can lead to dissatisfaction, negativity, conflict, and counterproductivity.

Takeuchi et al. [52] sought to understand the interaction between cultural identity and teamwork. Participants were college students from the United States and Japan. Teams were either homogeneous (American-American or Asian-Asian) or heterogeneous (American-Asian). All teams worked in a face-to-face environment as well as a video conference environment. Teams were all assigned the same task, to pretend that they were hiring the most qualified person for a job position. Each team member was given a resume that included different information about the job applicant that they were expected to communicate effectively to one another in order to select the most qualified applicant.

Results indicated that Asian-Asian teams performed significantly more poorly than American-Asian teams as well as American-American teams on task cohesion, social cohesion and overall performance. Differences in teamwork may be explained by individualistic and collectivist cultural backgrounds. The benefits of incorporating individualists in teams includes increased comfortability when sharing opinions and even criticisms, which leads to open discussion and fuels creativity. Homogeneous teams feel a greater sense of solidarity which produces positive effects such as friendliness, agreement, reduced tension, increased trust, integration and greater expectations of achievement.

16.4.3 Customer Loyalty Across Cultures

Marshall et al. [38] examined influences on customer loyalty among Singapore and Japan representing the East, and the United State and New Zealand representing the West. Participants were proposed a hypothetical scenario in which they were a customer of either a hairdresser or a bank and were faced with a decision to switch suppliers. The scenario was either positively or negatively framed. In the positively-framed situation, participants had a good relationship with their hairdresser/bank, but had the option to save money by switching to a cheaper supplier. In the negatively-framed situation, participants had a deteriorating relationship with their hairdresser/bank and were given the option to switch services to a different supplier.

All countries differed significantly from each other when examining their willingness to change providers in all given situations, with the United States being the most loyal overall, followed by New Zealand, Singapore and Japan, respectively. Participants were asked to measure their perceived costs and benefits of switching or remaining loyal to their providers. Results suggest that economic costs had a greater influence on their decisions than psychological costs for the United States, New Zealand and Japan. Singapore considered psychological costs of switching providers to be a greater burden than economical costs.

16.4.4 Future-Oriented Decision Making Across Cultures

Research by Gong, Krantz, and Weber [20] was conducted to determine differences in future-oriented decision making among China and America. The study focused on environmental and financial decision making where participants were forced to choose if they would rather the benefits/costs of a gain/loss occur now or in a year. Survey questions were followed by open-ended responses for participants to explain what influenced their decision.

Both cultures discounted losses similarly, and also shared similar reasons. Both cultures considered the weight of potential psychological consequences of dealing with debt, as well as fear of bankruptcy. The common ground of their similarities in discounting future environmental and financial gains was uncertainty of the future.

Both cultures discounted gains significantly more than losses in both environmental and financial categories. However, there were significant differences between cultures when comparing environmental use that affected them directly, such as air quality, with Chinese discounting future gains much more than Americans. Both cultures discounted futures gains equally when considering environmental existence, such as a growing animal population.

When considering finances, Chinese discounted lucky monetary gains, such as winning the lottery, much more than Americans. Chinese participants explained this choice by their lack of trust in the lottery and government systems. Chinese participants discounted self-earned future monetary gains only slightly more than Americans. An example of this is the choice of either receiving a paycheck now, or waiting one year to receive the paycheck along with a bonus. Common explanations for discounting this future financial gain were fear of inflation and labor mobility. Chinese also explained that if they chose to receive a lesser amount of money now, they could invest the money themselves and still possibly gain future income.

These findings show that a decision is made considering cultural context and constraints, such as trust in organizations or possible inflation.

16.5 Integration of Findings: Towards a More Comprehensive Theory of Decision Making

Figure 16.1 summarizes the key points of our previous discussion on theory and empirical results from microworld and real-world studies. The decision maker lives in a specific social, cultural, and historic context and learns DDM strategies that have been promising and adaptive in this environment in the past [27]. These DDM strategies have been passed on from generation to generation. In a concrete DDM situation, the decision maker then relies on these culturally transmitted DDM strategies. These are also partly triggered through dominant cultural values, goals, and other people involved in the decision-making process. In a specific situation, cultural strategies and goals trigger a specific search for information and specific decision making. In DDM research the focus is then on the sequence of decisions and their related outcomes and potential side—and long-term effects. The decision

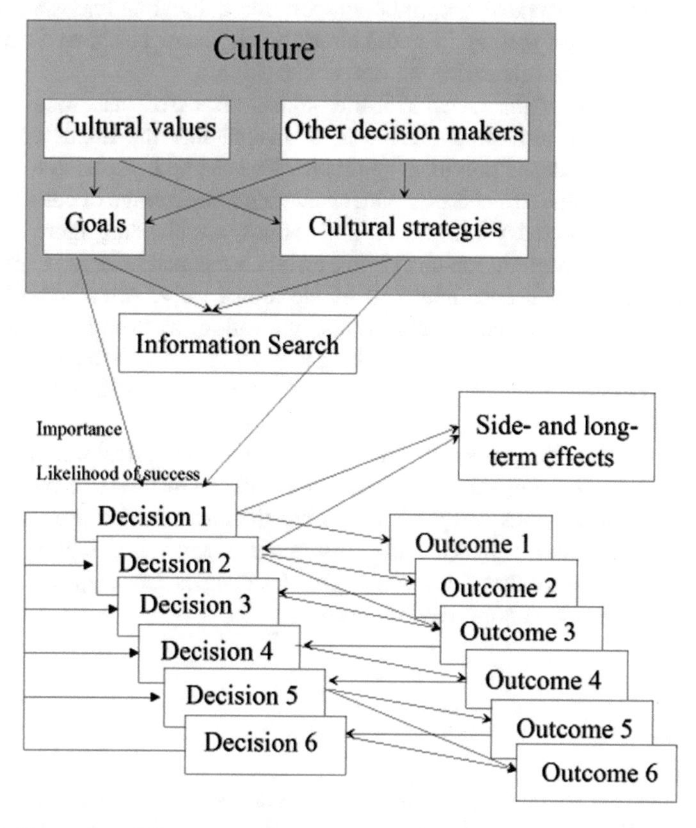

Fig. 16.1 How culture influences DDM

alternatives that are important and show high likelihood of success are selected [9, 12]. This means that ultimately importance and likelihood of success for decision alternatives are culturally determined.

16.6 Conclusion

Although the field of decision making and culture is relatively new [56], and no coherent theory exists that explains the link between culture and decision making, we discussed several empirical findings bringing light to some aspects of this linkage. Referring to empirical studies using microworlds across cultures, we have shown that DDM strategies differ across cultures and how cultural values can influence the selection of DDM strategies. We have then discussed studies on DDM in the real world in the domains of business and technology. Also, these studies showed the influence of cultural values on decision making and information search. Additional studies provided empirical support for cultural differences in group cohesion and decision making in multicultural work teams, customer loyalty, and the influence of future-orientation on decision making.

Although not all of the recent real-life studies discussed here focus on independent DDM, they provide a great deal of insight into the minds of decision makers. With the evidence provided, generalizations can be made as to how certain cultures typically come to a decision. Understanding a wide range of cultures can be especially helpful in today's world as populations are blending more than ever before. A prime example in which cultural awareness is particularly beneficial are business settings, which encompass all of the topics we've discussed—from the benefits of cross-cultural collaboration and innovation, to how individuals seek trusted sources of information before making a decision. Recognizing that decisions stem from cultural roots could lead us to more successful interactions in everyday life.

To summarize, the discussed theories on culture and empirical findings explain how DDM strategies evolved as enduring adaptation to historical, environmental, and sociocultural demands. People learn during the enculturation process through culture-specific tasks, modes of instruction and interaction, specific decision-making approaches, and which DDM strategies are adaptive and functional in their specific cultural environment.

References

1. Berry, J.W.: An ecocultural perspective on the development of competence. In: Sternberg, R.J., Grigorenko, E.L. (eds.) Culture and Competence, pp. 3–22. American Psychological Association, Washington, DC (2004)
2. Boone, L., Kurtz, D.: Essentials of Contemporary Business. Wiley, Hoboken, NJ (2014)

3. Brehmer, B.: Dynamic decision making: human control of complex systems. Acta Physiol. **81**(3), 211–241 (1992). doi:10.1016/0001-6918(92)90019-A
4. Brehmer, B., Dörner, D.: Experiments with computer-simulated microworlds: escaping both the narrow straits of the laboratory and the deep blue sea of the field study. Comput. Hum. Behav. **9**, 171–184 (1993). doi:10.1016/0747-5632(93)90005-D
5. Charles, C., Gafni, A., Whelan, T.: Shared decision-making in the medical encounter: what does it mean? (or it takes at least two to tango). Soc. Sci. Med. **44**, 681–692 (1997). doi:10.1016/S0277-9536(96)00221-3
6. Cole, M.: Cultural Psychology. A Once and Future Discipline. Belknap and Harvard University Press, Cambridge, MA (1996)
7. D'Andrade, R.G.: The cultural part of cognition. Cogn. Sci. **5**, 179–195 (1981)
8. Dörner, D.: The Logic of Failure. Holt, New York (1996)
9. Dörner, D.: Bauplan für eine Seele [Blueprint for a soul]. Rowohlt, Reinbek (1999)
10. Dörner, D., Gerdes, J.: SchokoFin. Computerprogramm [Choco Fine. Computer program]. Institut für Theoretische Psychologie, Universität Bamberg, Germany (2003)
11. Dörner, D., Güss, C.D.: A psychological analysis of Adolf Hitler's decision making as Commander in Chief: Summa confidentia et nimius metus. Rev. Gen. Psychol. **15**(1), 37–49 (2011). doi:10.1037/a0022375
12. Dörner, D., Güss, C.D.: PSI: a computational architecture of the human soul. Rev. Gen. Psychol. **17**, 297–317 (2013). doi:10.1037/a0032947
13. Dörner, D., Kreuzig, H.W., Reither, F., Stäudel, T.: Lohausen. Vom Umgang mit Unbestimmtheit und Komplexität. [Lohausen. On managing uncertainty and complexity]. Hans Huber, Bern (1983)
14. Dörner, D., Stäudel, T., Strohschneider, S.: MORO—Programmdokumentation. Universität, Lehrstuhl Psychologie II, Bamberg, Germany (1986)
15. Fama, E.F.: Market efficiency, long-term returns, and behavioral finance. J. Financ. Econ. **49**, 283–306 (1998). doi:10.1016/S0304-405X(98)00026-9
16. Frensch, P., Funke, J. (eds.): Complex Problem Solving: The European Perspective. Lawrence Erlbaum, Hillsdale, NJ (1995)
17. Funke, J.: Problemlösendes Denken [Problem solving thinking]. Kohlhammer, Stuttgart (2003)
18. Funke, J.: Complex problem solving: a case for complex cognition? Cogn. Process. **11**, 133–142 (2010). doi:10.1007/s10339-009-0345-0
19. Gerdes, J., Dörner, D. Pfeiffer E.: Interaktive Computersimulation "WINFIRE" [The interactive computer simulation "WINFIRE"]. Otto-Friedrich-Universität Bamberg: Lehrstuhl Psychologie II, Bamberg, Germany (1993)
20. Gong, M., Krantz, D.H., Weber, E.U.: Why Chinese discount future financial and environmental gains but not losses more than Americans. J. Risk Uncertainty **49**(2), 103–124 (2014). doi:10.1007/s11166-014-9200-5
21. Gonzalez, C., Lerch, F.J., Lebiere, C.: Instance-based learning in dynamic decision making. Cogn. Sci. **27**, 591–635 (2003). doi:10.1016/S0364-0213(03)00031-4
22. Goodrich, K., de Mooij, M.: How 'social' are social media? A cross-cultural comparison of online and offline purchase decision influences. J. Mark. Commun. **20**, 103–116 (2014). doi:10.1080/13527266.2013.797773
23. Guan, Y., Chen, S.X., Levin, N., … Han, X.: Differences in career decision-making profiles between American and Chinese university students: the relative strength of mediating mechanisms across cultures. J. Cross Cult. Psychol. **46**, 856–872 (2015). doi:10.1177/0022022115585874
24. Güss, C.D.: Planen und Kultur? [Planning and culture?] Lengerich. Pabst Science Publishers, Germany (2000)
25. Güss, C.D.: Fire and ice: testing a model on cultural values and complex problem solving. J. Cross Cult. Psychol. **42**, 1279–1298 (2011). doi:10.1177/0022022110383320

26. Güss, C.D.: Kulturvergleichende Psychologie [Cross-cultural Psychology]. In: Schütz, A., Brand, M., Selg, H., Lautenbacher, S. (eds.) Psychologie. Eine Einführung in ihre Grundlagen und Anwendungsfelder (5., überarbeitete und erweiterte Auflage), pp. 550–562. KohlhammerStuttgart, Germany (2015)
27. Güss, C.D., Robinson, B.: Predicted causality in decision making: the role of culture. Front. Psychol. **5**, 479 (2014). doi:10.3389/fpsyg.2014.00479
28. Güss, C.D., Wiley, B.: Metacognition of problem-solving strategies in Brazil, India, and the United States. J. Cogn. Cult. **7**(1–2), 1–25 (2007). doi:10.1163/156853707X171793
29. Hofstede, G.: Culture's consequences. Sage, Beverly Hills, CA (1980)
30. Hofstede, G.: Culture's Consequences, 2nd edn. Sage, Thousand Oaks, CA (2001)
31. Kahneman, D., Tversky, A.: Prospect Theory: an analysis of decision under risk. Econometrica **47**, 263–291 (1979). doi:10.2307/1914185
32. Kahneman, D., Tversky, A.: On the psychology of prediction. Psychol. Rev. **80**, 237–251 (1973). doi:10.1037/h0034747
33. Lave, J.: Situating learning in communities of practice. In: Resnick, L.B., Levine, J.M., Teasley, S.D. (eds.) Perspectives on Socially Shared Cognition, pp. 63–82. American Psychological Association, Washington, DC (1991)
34. Lave, J., Wenger, E.: Situated Learning: Legitimate Peripheral Participation. Cambridge University Press, Cambridge (1991)
35. Leontiev, A.N.: Activity. Consciousness. Personality. Prentice Hall, Engelwood Cliffs, NJ (1978)
36. Li, L.W., Masuda, T., Russell, M.J.: Culture and decision-making: investigating cultural variations in the East Asian and North American online decision-making processes. Asian J. Soc. Psychol. **18**(3), 183–191 (2015). doi:10.1111/ajsp.12099
37. Løvborg, L. Brehmer, B.: NEWFIRE—A flexible system for running simulated fire fighting experiments. in: Risø National Laboratory, Risø-M-2953, Denmark (1991).
38. Marshall, R., Huan, T.C., Xu, Y., Nam, I.: Extending prospect theory cross-culturally by examining switching behavior in consumer and business-to-business contexts. J. Bus. Res. **64** (8), 871–878 (2011). doi:10.1016/j.jbusres.2010.09.009
39. Omodei, M.M., Wearing, A.J.: Fire Chief (Version 2.2) [Computer program]. University of Melbourne, Department of Psychology, Melbourne, Australia (1993)
40. Putz-Osterloh, W.: Über die Beziehung zwischen Testintelligenzund Problemlöseerfolg. Z. Psychol. **189**, 79–100 (1981)
41. Putz-Osterloh, W.: Strategies for knowledge acquisition and transfer of knowledge in dynamic tasks. In: Strube, G., Wender, K.-F. (eds.) The Cognitive Psychology of Knowledge, pp. 331–335. Elsevier, Amsterdam (1993)
42. Resnick, L.B.: Situated rationalism: Biological and social preparation for learning. In: Hirschfeld, L.A., Gelman, S.A. (eds.) Mapping the Mind. Domain Specificity in Cognition and Culture, pp. 474–493. Cambridge University Press, New York, NY (1994)
43. Schmid, U., Ragni, M., Gonzalez, C., Funke, J.: The challenge of complexity for cognitive systems. Cogn. Syst. Res. **12**, 211–218 (2011). doi:10.1016/j.cogsys.2010.12.007
44. Schwartz, S.H., Boehnke, K.: Evaluating the structure of human values with confirmatory factor analysis. J. Res. Pers. **38**, 230–255 (2004). doi:10.1016/S0092-6566(03)00069-2
45. Shleifer, A.: Inefficient Markets: An Introduction to Behavioral Finance. Oxford University Press, New York, NY (2000)
46. Singelis, T.M., Triandis, H.C., Bhawuk, D.S., Gelfand, M.J.: Horizontal and vertical dimensions of individualism and collectivism: a theoretical and measurement refinement. Cross Cult. Res. **29**, 240–275 (1995). doi:10.1177/106939719502900302
47. Smith, P., Bond, M.H., Kagitcibasi, C.: Understanding Social Psychology Across Cultures: Living and Working in a Changing World. Sage, London (2006)
48. Sperber, D., Hirschfeld, L.A.: The cognitive foundations of cultural stability and diversity. Trends Cogn. Sci. **8**, 40–46 (1999). doi:10.1016/j.tics.2003.11.002
49. Strohschneider, S.: Zur Stabilität und Validität von Handeln in komplexen Realitätsbereichen. Sprache Kognition **5**, 42–48 (1986)

50. Strohschneider, S., Güss, D.: The fate of the Moros: a cross-cultural exploration of strategies in complex and dynamic decision making. Int. J. Psychol. **34**, 235–252 (1999). doi:10.1080/002075999399873
51. Su, N.M., Wang, Y., Mark, G., Aiyelokuin, T., Nakano, T.A.: Bosom buddy afar brings a distant land near: are bloggers a global community? in: Proceedings of the Second National Conference of Communities and Technologies, June 13–16, pp. 171–190. Milano, Italy (2005)
52. Takeuchi, J., Kass, S.J., Schneider, S.K., VanWormer, L.: Virtual and face-to-face teamwork differences in culturally homogeneous and heterogeneous teams. J. Psychol. Issues Organ. Cult. **4**, 17–34 (2013). doi:10.1002/jpoc.21112
53. Tversky, A., Kahneman, D.: Judgment under uncertainty: heuristics and biases. Science **185** (4157), 1124–1131 (1974)
54. Viale, R., Andler, D., Hirschfeld, L.A. (eds.): Biological and Cultural Bases of Human Inference. Lawrence Erlbaum Associates, Mahwah NJ—London (2006)
55. Vygotsky, L.S.: Mind in Society: The Development of Higher Psychological Processes. Harvard University Press, Cambridge, MA (1978)
56. Weber, E.U., Hsee, C.K.: Culture and individual judgment and decision making. Appl. Psychol. Int. Rev. **49**, 32–61 (2000). doi:10.1111/1464-0597.00005

Chapter 17
When Beliefs and Logic Contradict: Issues of Values, Religion and Culture

Vladimíra Čavojová

Abstract In real debates, we often don't think about the validity of the arguments from the strictly logical point of view and we often disagree even before we hear the particular argument. This chapter deals with confirmation bias in reasoning about controversial issues (in this case abortions) and it examines the effect of values (pro-life, pro-choice, neutral), religious and political affiliations on syllogistic reasoning. It shows how our beliefs prevent us from acknowledging the logic to the same type of arguments if they are made by the other side of the dispute. First, evidence of studies showing my-side bias and confirmation bias is presented, together with studies suggesting cultural differences in preference for distinct cognitive style or problem-solving. Then the results from one non-WEIRD (Slovak) sample ($N = 321$, M age = 20.47 years) are analysed. Participants first indicated their attitudes toward abortions in a short questionnaire (6 items from General Social Survey), then they solved 24 syllogisms, which had conclusions either in line with pro-choice or pro-life attitudes and 12 neutral syllogisms. The results showed that people holding opposing beliefs did display confirmation bias, but this confirmation bias was stronger for one side of the dispute, i.e. "pro-lifers". Christian participants performed worse in neutral valid syllogisms, but mainly in all types of invalid syllogisms, where they differed by 10% from the non-religious participants. This chapter shows that when beliefs and evidence clash, it is often belief that wins. It is no surprise that people untrained in critical, scientific thinking resort to beliefs as their compass in navigating through the vast ocean of many conflicting information (claiming their origin in research) and many conflicting values (such as rights of children vs. rights of their mothers).

Keywords My-side bias · Belief bias · Formal reasoning · Motivated reasoning

---

V. Čavojová (✉)
Centre of Social and Psychological Sciences SAS, Dúbravská cesta 9,
84104 Bratislava, Slovak Republic
e-mail: vladimira.cavojova@savba.sk

C. Faucher (ed.), *Advances in Culturally-Aware Intelligent Systems and in Cross-Cultural Psychological Studies*, Intelligent Systems Reference Library 134, https://doi.org/10.1007/978-3-319-67024-9_17

17.1 Introduction

In September 2013 two events that took place during the same weekend, polarised the citizens of Slovakia for much longer time. One of them was Gay Pride in Bratislava, which is the manifestation of equal rights of all citizens regardless their sexual orientation, the event reflecting mainly liberal values; the other one was March for Life, which is the manifestation for the rights of unborn children and the strengthening of the traditional forms of family, the event reflecting mainly conservative Christian values. Although, in principle, one can be against abortions but promote (or at least not degrade) the rights of LGBT people for marriage or adoption, or be homophobic but in favour of liberal abortion laws, in reality (or at least in Slovakia in 2013 until now) people tend to be divided on these two issues reflecting different sets of values—liberal vs. conservative—and these values dictate not only our view on the controversial topics, such as these, but they (sometimes unconsciously) affect also our decision-making and choices we make.

What we believe in, what ideologies we buy into—religious, political, or other—shape what we pay attention to, even how we see things and what we make of them. In this chapter I want to explore, how holding some specific and value-laden beliefs affects the way we reason. It has been already shown that people treat arguments from their camp more favourably than the arguments from the opposing camp [1, 23], and when given a choice we search only for arguments and evidence confirming our views [18]. In this chapter I want to go a step further—not only to show confirmation bias in people on the both sides of a controversial issue, like one dealt in September 2013 in Slovakia, but also to explore whether displaying confirmation bias is moderated by any other demographic variables, such as gender, religion or political affiliation. Holding opposing beliefs is probably related to political preferences, and it often seems (not only in Slovakia) that group (political, religious, alternative) membership creates distinct norms and subcultures. To gain a better insight into the way our "sub-cultural" membership dictates our beliefs and attitudes, and subsequently affect our reasoning (that should ideally be context-free) is important if we are to promote better understanding and communication among more distant cultures.

To understand how culture shapes our reasoning we first have to consider, how culture shapes our values that often have powerful but unconscious influence on our reasoning, which can then be the cause of many misunderstandings and conflicts, in which each party claims to be right and accuses the other party of the lack of logic in their argumentation. The same is true also for the opening example, where "liberals" and "conservatives" form almost distinctive cultures even within one small country. So how can we understand each other in the "global village" when we even do not understand our neighbour who even speaks the same language, although laden with different implicit values? As Nisbett [19] puts it: "If people really do differ profoundly in their systems of thought—their worldviews and cognitive processes—then differences in people's attitudes and beliefs, and even their values and preferences, might not be a matter merely of different inputs and teachings, but rather an inevitable consequence of using different tools to understand the world. And if that's true, then efforts to improve international understanding may be less likely to pay off than one might hope (pp. xvii–xviii)."

The world is getting smaller and we no longer live our whole life in a community of people with the same opinions and set of values. On the one hand, we need to be more open to the other opinions, values, cultures to increase mutual understanding and tolerance, on the other hand, we need to be able to critically evaluate the beliefs of others as well as our own—the ability that is reflected in the concept of critically open-mindedness [13].

One of the core values, in which cultures can differ, is the extent to which they protect the rights of all its citizens, including minorities, women and unborn children, even though the rights of various subgroups can be in conflict. One such controversial issue is the right of a woman to decide about her own body (including ending of unwanted pregnancy) versus right of a child to be born. Therefore exploring the reasoning about such controversial issue as abortions is especially suitable in this context for several reasons. Firstly, this issue is hotly debated in many countries, especially those with strong religious affiliations and conservative values and the people tend to be divided according whose rights they favour more. Secondly, based on their attitudes towards this issue, people tend to be divided into two distinct camps. This division is reflected in naming the stances as "pro-life" (those favouring rights of unborn children) and "pro-choice" (those favouring the right of women to have a choice over their bodies and lives). Thirdly, discussions about controversial issues tends to get very emotional as they usually reflects the inner values of the people, but therefore it is even more important to be able to recognize good arguments on either sides from the basically flawed and logically inconsistent arguments. Fourthly, strong affiliation towards one of the stances makes it possible to examine the extent to which the beliefs affect the reasoning about the arguments of other side versus one's own side—my side bias.

Therefore, in this chapter, I want to explore one form of confirmation bias in reasoning about abortions. My main question was whether the people in opposing camps regarding their attitudes towards abortions will differ in their reasoning about the validity of the conclusions that go against their beliefs. Before I describe the research in detail, I will firstly discuss what is reasoning, which kind of biases influence it and research findings linking the partisanship with biased (motivated) reasoning. Although I want to concentrate on analysing the effect of culture, I will not deal directly with cross-cultural comparison, but will try to focus more on analysing our results from non-WEIRD[1] sample and compare it with similar results from Western samples.

---

[1]WEIRD as acronym for Western, Educated, Industrialised, Rich, & Democratic. Henrich, Heine and Norenzayan [8] suggested that too much research in psychology is done on this kind of WEIRD samples and more research is needed on less typical samples, such as those from non-WEIRD countries.

17.2 Reasoning About "Hot" Issues and Logic: Confirmation Bias, My-Side Bias and Belief Bias

To make responsible and informed choices, we should be able to judge various arguments on the basis of their validity, factual information and logic, not solely on the strength of our beliefs. Ability to decouple our prior beliefs and attitudes from the evaluation of arguments and evidence is one of the fundamental bases of the critical thinking [14]. In the modern and increasingly multicultural world (in which different cultural values can clash together) and with the instant access to any information, ability to reason well, to distinguish the valid information from invalid, and to evaluate the facts and arguments in non-biased and open way becomes an essential skill when dealing with the world around us.

Often we have to draw conclusions which are not based on observable facts or evidence, and in these occasions it is important to follow the rules of correct reasoning, which are manifested in logic. Logic does not deal with descriptions of how people really think, but it states the normative rules for reasoning, which aims for truth and logical validity. It does not deal with the factual truths of the conclusion, but whether they were drawn correctly from the premises.

Sometimes people associate this kind of reasoning with somewhat sophisticated justifications of their own interest at stake and tend to view logic behind this kind of reasoning sceptically as something enabling people to persuade others that white is actually black.[2] Historically, "logic" was often misused for justification of logically unjustifiable attitudes, especially ideological ones [6]. However, regardless whether we consider reason trick, skill or strategy, reasoning is the best way to decide whom and what to believe and it is a hallmark of human species [7, p. 175] and it is one of the best tools against such propaganda.

Reasoning is the process of forming conclusions, judgments, or inferences, from facts and premises, or from evidence and principles [25]. According to Mercier and Sperber [16, p. 57] reasoning is a very special form of inference at the conceptual level, where not only is a new mental representation (or conclusion) consciously produced, but the previously held representations (or premises) that warrant it are also consciously entertained.

According to logic, a conclusion is valid if it necessarily follows from some statements that are accepted as facts. Often, when we are faced with logical problems, we arrive at wrong answers or if we happen to arrive at right answers, it is because of the bad reasons. For instance, paraphrasing Henle's example (cited in [7]), if someone tells us that it is important to talk about sex (or other controversial

---

[2]This notion is probably based on many known sophisms used to trick people by stating premises that everybody agrees to and then drawing conclusion, which is based on premises but is against intuitive logic, e.g. if Diogenes is not Socrates and Socrates is a man, then Diogenes is not that Socrates is, i.e. Diogenes is not a man. In fact, study of formal logic began as a reaction to sophism [6].

issue), because it is important to talk about things that are on our mind (and sex obviously is on our minds), we can rewrite it formally as:

Premise 1: It is important to talk about things that are on our mind.

Premise 2: Sex is on our mind.

Therefore (Conclusion): It is important to talk about sex.

If we think whether the conclusion is logically right, we can come up with several (wrong) answers, reflecting more our actual beliefs or opinions about things, which are important to talk about, such as "No, it is not important to talk about things that do not worry us and if sex does not worry us, it is not important to talk about it", or "No, there are some things that are on our minds that we should not talk about." Even though the answers make perfect sense to us, they are not the answers to the questions, whether the conclusion follows logically from the premises.

On the other hand, a person can give a right answer but for the wrong reasons, such as "Yes, it is important to talk about sex, so that people can avoid disappointments or wrong ideas they can have." Also in this case the person adds her own beliefs concerning the topic (talking about sex) and does not arrive at the answer solely on the information provided, even though her answer happens to be right. Henle [7] has termed this error in reasoning the failure to accept the logical task. Kahneman [11] calls it "substitution"—when we are faced with a difficult problem that would require effortful cognitive processing, we often switch it to the easier problem by answering the question, to which we already have an answer. For example, an executive facing the question, whether he should invest into Ford stock, may easily answer "yes", because the answer came quickly and intuitively into his mind. But this answer could be related to easier question: Do I like Ford cars? Similarly, Stanovich [22] calls this tendency "miserly processing", i.e. we are inclined to invest the least possible amount of cognitive effort into solving any problem.

Prescriptive logic states that we should apply the same rules regardless the context, but it is often not the case. In the real life, when logic and belief collide, people often respond on the basis of their prior knowledge—giving rise to a "*belief-bias*" effect [26]. It means that we tend to evaluate the logical validity of deductive arguments mainly on the basis of our personal beliefs regarding the empirical status of the conclusion [15].

Belief bias is studied most often by syllogistic reasoning paradigm[3] where the validity and the believability of the conclusion are put in conflict [14, 15, 17], such as in the example below.

[3]Belief bias research uses two main paradigms: production tasks (participants are asked to draw conclusions from the presented premises) and evaluation tasks (participants are presented with some premises and a conclusion to be evaluated—as valid if it necessarily follows from the premises and as invalid if it does not).

Premise 1: No religious people are healthy.

Premise 2: Some healthy people are priests.

Therefore (Conclusion): Some priests are not religious.

Belief bias is then defined as greater acceptance of believable than unbelievable conclusions and logical competence can be defined as greater acceptance of valid conclusions than invalid conclusions. However, the ability to reason logically from intuitively false premises (or premises that can turn false) is important for many professions where the premises are false or arbitrary, such as scientists constructing hypotheses about planets without gravity [7].

It should be no surprise that people very often have difficulties in separating truth from validity, especially when the conclusion runs counter to their strong beliefs and values, or when discussing emotional issues. Content influences very much which conclusions we choose as valid (mostly according to which conclusion we already believe in, or which seems intuitively appealing). Moreover, people have strong tendency to look for confirmation of their favoured beliefs and hypothesis, which is vastly documented in research of *confirmation bias* [1, 2, 18]. This inability to decouple one's own beliefs and selecting arguments and conclusions favouring our already held beliefs interferes with good thinking in almost every context [18].

This tendency of people to seek confirmation rather than disconfirmation of what they already believe is usually referred to as confirmation bias [18, 25]. However, the confirmation bias has been used in psychological literature to refer to a variety of phenomena, so it is worthwhile to briefly describe the terminology related to this term. Nickerson [18] in his review of studies relating to confirmation bias uses this term as "generic concept that subsumes several more specific ideas that connote the inappropriate bolstering of hypotheses or beliefs whose truth is in question" (p. 175). More specific types of confirmation bias are belief bias and my-side bias. My-side bias occurs when people evaluate, generate, and test evidence in a manner, which is biased toward their prior attitudes and opinions. Belief bias, on the other hand, occurs when we accept or reject conclusion of the argument on the basis on its un/believability. Belief bias arises from people's factual knowledge about the world in contrast to my-side bias which is reasoning biased toward personal opinions or stances [14]. Belief bias is probably the narrowest of all these concepts, because it deals with conflict between the validity and believability of conclusions and the believability does not need to reflect our inner beliefs or values, but just some factual information about the world (As in example above, even if we are atheists we probably believe that priests are religious). My-side bias deals specifically with our personal beliefs, attitudes and opinions and adds more subjectivity to the studied issues.

Baron [1] studied my-side bias in thinking about early abortion and he found that people tend to give more arguments for the attitude they prefer than for the opposing attitude, but he argues that this is probably caused by preference of one-sided arguments, which we see as stronger than two-sided arguments. Participants in his experiment judged even opposing arguments as more persuasive, if the argument was only one-sided. His results suggest that people's standards (their beliefs about the nature of what comprises good thinking) influence the way they conduct their own thinking. These inner standards can be largely influenced by cultural beliefs, as certain cultural traditions actively discourage people from questioning these particular beliefs and traditions and thus lead to general distrust of open-mindedness [1].

Stanovich and West [24] replicated the results of Baron [1] using slightly modified method of reasoning about abortions to increase the coherence and naturalness of arguments. They used Baron's arguments as well as their own and created four fictional students, who each made one of the following sets of arguments: (1) four anti-abortion reasoning statements, (2) two pro-choice followed by two anti-abortion reasoning statements, (3) four pro-choice reasoning statements, (4) two anti-abortion followed by two pro-choice reasoning statements. Thus each participant evaluated the reasoning of two students whose reasoning statements represented only one side of the abortion issue (one consistently anti-abortion, and one consistently pro-choice) and two students whose reasoning statements represented two sides of the abortion issue. My-side bias was displayed by evaluating better those arguments which were in line with the person's own attitudes, and they also found evidence of one-side bias, i.e. favouring and better evaluating arguments only from one side (even from the opposing side).

Jurkovič [10] studied my-side bias in Slovak sample using materials (agreement with the arguments on studied topics) adapted from Stanovich and West [24] and he found significant my-side bias regarding participants' smoking habits, drinking habits and religion. The level of intelligence did not affect my-side bias and therefore cannot be regarded as protective factor against irrational thinking.

17.3 Reasoning—Universal or Culturally Dependent?

Until only recently it was often believed that humans from all cultures think and reason basically in the same way, which Nisbett [19] summarised in four assumptions: (1) People rely on the same basic cognitive processes for perception, memory, causal analysis, categorization, and inference. (2) If the people from various cultures differ, it must be not because of the different cognitive processes, but because of different experience they were exposed to in life. (3) Higher order processes of reasoning rest on formal rules of logic. (4) Process of reasoning is separate from the content of reasoning.

Nisbett [19] started to question the assumption of universality of human thought on the basis of the results which showed that even the brief training in formal logic can improve the way people think outside laboratory. Together with his colleagues [20] they argue that the considerable social differences that exist among different cultures affect not only their beliefs about the specific aspects of the world but also their basic cognitive processes—the ways by which they know the world. More specifically, the different cultures form different social organisations, which in turn direct the attention to some aspects of the field at the expense of others. What we attend to (e.g., objects vs. relationships) then affects our beliefs about the world and causality. This in turn influences our tacit epistemology, i.e., beliefs about what is important to know and how this knowledge can be obtained. Epistemology then dictates the development and application of some cognitive processes at the expense of others. Social organisation then serves for further promoting the epistemology as well as certain cognitive processes (whether we are explicitly taught to use and value harmony and relationships more than truth and logic).[4]

There are only few studies that examined cultural bases of logical reasoning, though. Norenzayan, Smith, Kim, and Nisbett [21] cite some of the research from Vygotskian tradition conducted among non-Western cultures (e.g., Uzbek farmers, Kpelle of West Africa, Maya in Mexico), which found that participants were unwilling to decouple the content of the deductive problem from its logical structure. For example,

> A Kpelle man was given the following problem: All Kpelle men are rice farmers. Mr. Smith is not a rice farmer. Is he a Kpelle man? The man's response, characteristic of many of these participants, was, "I don't know the man in person. I have not laid eyes on the man himself." (Scribner, 1977, p. 490). Thus, non-Westerners in these studies typically refused to solve deductive problems, on the grounds that the content is unfamiliar or contrary to experience. This suggests that knowledge may counter logic for non-Westerners to a greater extent than for Westerners, especially when the two are in conflict [21].

Norenzayan with his colleagues [21] conducted four experiments in which they put into conflict intuitive and analytic reasoning and compared East Asian (Chinese and Korean), Asian American and European American college students. They theorised that if East Asians favour intuition more than formal rules, they should be less willing to abandon intuition for formal rules and they studied these preferences in four cognitive domains: category learning (Study 1), classification and similarity judgments (Study 2), convincingness of deductive arguments (Study 3), and belief bias in deductive reasoning (Study 4). Their Study 4 was the first investigation into possible cultural differences in belief bias and they expected Asians to show stronger belief bias than Americans and this belief bias to be independent of participants' reasoning ability measured by performance on abstract syllogisms.

The participants had to evaluate 16 syllogisms, 4 in each category: valid/believable, valid/unbelievable, invalid/believable, and invalid/unbelievable and then 24 abstract syllogisms using letters and unfamiliar foreign words. The dependent measure was percentual endorsement of each argumentation type of syllogism, indicating whether participants thought the given conclusion followed logically from the premises. They found that Koreans showed a stronger belief bias than European Americans, though only for valid arguments and this was not due to different reasoning ability, as there were no differences between Americans and Asians in abstract syllogisms. However, Norenzayan et al. [21] found also that

[4]Nisbett [19] observed that Westerners differ from East-Asians in the basic view on problems and how they can be solved. While Westerners (successors of ancient Greek philosophies) believe that world can be categorised and understood by means of simple rules and logic, East Asians despise using logic on complex problems, which plays only minor role in problem solving.

Asians considered believable syllogisms as less plausible than Americans, suggesting that the same believable conclusions, contained also in invalid arguments, were weaker for Koreans than for Americans.

The results concerning lower likelihood of Korean participants of accepting even valid syllogisms led Unsworth and Medin [28] to question the conclusion of Norenzayan et al. [21] that Koreans are less likely to decontextualize an argument's content from its logical structure. Unsworth and Medin [28] argue that the observed results were caused by response bias not belief bias, because Koreans were less likely to indicate that argument is valid than American participants. Therefore, they analysed the hit and correct rejection rates along with the averages of these rates in each condition for each cultural group. Their results suggest that when judging the validity of an argument, European American accuracy is not significantly different from Koreans' accuracy, regardless of the believability of the conclusion. Furthermore, it seems that participants were better able to discriminate between valid and invalid arguments when the arguments were concrete, not abstract, and when they were believable rather than unbelievable. Unsworth and Medin [28] conclude that these two cultures do not differ in the extent to which they are affected by belief bias.

Henrich, Heine, and Norenzayan [8] have recently noted that much of the psychological research is performed on highly restricted sample—that yielded the acronym WEIRD standing for Western, Educated, Industrialised, Rich and Democratic. Henrich, Heine, and Norenzayan [8] reviewed this research and showed that not only people from non-western cultures differ from Western cultures in their perception and reasoning processes, but people from non-US countries differ from US samples, and even within US samples there are still large differences according to socio-economic status, etc. They conclude that members of these WEIRD societies (including small children) are among the least representative populations for generalisations about human thinking.

Talhelm et al. [27] conducted several studies with more than 5000 participants (from United States and Mainland China) and they argue that the liberals are even WEIRDer, so the liberalism should be the sixth attribute of WEIRD cultures (thus becoming WILDER cultures). They hypothesised that liberals would think more analytically because liberal culture is more individualistic, putting priority on self-expression and individual identity over group identity. On the other hand, conservativism is associated with more close-knit communities and interconnected groups, such as churches, fraternities, and the military, where the emphasis is on the group identity and belonging, which can create higher pressure to conform. In this sense, the collectivism associated with conservativism is a "system of tight social ties, but less trust and weaker ties toward strangers—stronger in-group/out-group distinction" [27, p. 252]. In their studies they distinguish between social and economic politics, because they argue that many inconsistent findings about liberals and conservatives were caused by the fact that when we give people only two choices (liberal vs. conservative), libertarians tend to self-identify more like conservatives, even though in their social orientation they are the most individualistic of the political affiliations.

In the study with college sample (Study 1), Talhelm et al. [27] found that liberals made more categorical pairings (similar to the dominant Western style) and conservatives made more relational pairings (more similar to East Asian style) in triad task.[5] In the study with internet sample (Study 2) they found that social conservatives think more holistically and the relationship was true also for non-US participants from the sample. Even controlling for gender, age, socioeconomic status, education, and even cognitive ability (measured by Cognitive Reflection Test, CRT, [5]) still left significant a relationship between thought style and social politics. Participants who scored higher on the CRT also thought more analytically, and the relationship between politics and cultural thought was stronger among people who scored lower on the CRT. Talhelm et al. [27] replicated their findings with Chinese sample in their Study 3, but the relationship between social politics and thought style only emerged among people from more developed areas. They argue that finding that liberals think more analytically even in this more holistic culture gives one piece of evidence that the relationship between politics and thought style is not just an American phenomenon.

In Study 4 and 5 Talhelm et al. manipulated the way people processed information (categorically/analytically or relationally/holistically) on triad task and then measured their responses for political articles favouring liberal program. They found out that the liberal plan would have won after they had people think analytically (64% support) and lost after they had people think holistically (38%). Furthermore, it seemed that training was effective regardless the political affiliation, which was backed up by the lack of interaction between politics and condition. Moreover, the training changed only processing the new information; it did not change self-reported political identity on the subsequent questionnaire.

Generally, they confirmed in five studies across two different cultures that social liberals consistently thought more analytically than social conservatives and these results can be viewed as evidence that social style is connected to cultural thought, as proposed by [19]. This study is important also for our research, as it showed that political differences and divisions are partly cultural divisions and these two sides think about the world as if they really came from different cultures.

17.4 Formal Reasoning About Abortions in Slovak Sample

In the previous sections I firstly reviewed the studies dealing with reasoning about controversial issues showing either my-side bias or belief-bias and then the studies of formal reasoning in various cultural settings that showed that culture possibly

[5]Triad task is used to distinguish between relational and taxonomical categorization. Participants are given target object (e.g. cow) and then are asked to choose which one of the two alternatives (e.g. chicken and grass) is the best associated with the target object. Asians tend to categorise more relationally (i.e. associate cow with a grass, because cow feeds on a grass), while Westerners tend to categorise more taxonomically (i.e. cow and chicken belong to the same category – animals) [27].

influences our preconceptions about the world and our values, and these in turn influence the way we think about the problems.

In this section I want to introduce our approach to studying reasoning about controversial issues, such as abortions, but with modified measuring paradigm and on participants from Eastern Europe, and thus contributing to generalizability of findings about possible cultural differences in reasoning.

In traditional testing paradigm of my-side bias, participants are asked to evaluate the thinking behind reasoning on the various ranging scale and the arguments are constructed as to reflect the natural flow of language to increase the ecological validity of the results. The focus here is on the quality, amount and one-sidedness of arguments provided by fictional students. However, to properly evaluate the effect of one's personal beliefs regarding such difficult issue, we should look also at the belief bias that can affect the way we evaluate validity of the opponent's conclusion and control for the quality of deductive reasoning of these fictional students. Therefore, to better understand psychological processes behind these real-life issues I chose belief-bias paradigm using formal syllogistic reasoning to study my-side bias in reasoning about abortions. The idea was to merge these two paradigms and have participants to evaluate the validity of conclusions that resembled real-life arguments about some actual controversial topic. Also in an everyday speech we often use statements in the form of syllogisms, linear orderings, disjunctions, and if-then statements, but they are embedded in discourse and not labelled by premise and conclusion [7]. However, as we already saw, people are hugely affected by content and in everyday context these statements are used in misleading way either malevolently or by ignorance.

17.4.1 Background

321 of college students (mostly from Pedagogical faculty) participated in the study, the majority of whom were women (N = 259, 80.7%). The mean age of our sample was 20.47 (SD = 1.86). Fifty six reported no religious affiliation, 221 were Roman Catholic, 12 Protestant, 5 Greek Catholic, 12 other Christian, 3 agnostic and others either provided no information or indicated "exotic" affiliation, such as Slavic, Nordic, Pagan or other. 63.9% refused to state their political affiliation, the rest was mainly liberal (22.7%), conservative (17%) or other (middle, democratic without party affiliation, Christian, Stoic, etc.—4.8%). Ninety two percent did not have any experience with formal course of logic, 7% stated that they had some experience with formal course of logic, which was taken into account in the analyses.

The participation in the study was anonymous (on-line) and voluntary—participants were motivated to engage in the study by getting extra credits for a course of Social psychology depending on the number of correctly solved syllogisms. Firstly, they had to indicate their attitudes toward abortions in a short questionnaire on 6-point scale (1 meaning absolutely agree, 6 meaning absolutely disagree). I used 6 items from General Social Survey reported by Jelen and Wilcox [9], which ask respondents whether abortion should be legal for a series of six circumstances: when the mother's health is in danger, when the pregnancy is the result of rape, when the foetus is severely defective, when the family is too poor for additional children, when a single pregnant woman does not want to marry, and when a married couple wants no more children. According to Jelen and Wilcox [9], these items have been successfully used since 1972.

After indicating their attitude in these 6 items, participants had to indicate their answers on 36 syllogisms and they were asked about several demographic variables, such as age, gender, school they attend, whether they attended any formal course of logic, their religion and political affiliation. For each question it was possible to choose the "Prefer not to answer" option.

17.4.2 Anatomy of Used Syllogisms

We constructed[6] two sets of value-laden conclusions for our study: (1) in favour of women's right for deciding for abortion (pro-choice) and (2) in favour of unborn child's right to life (pro-life) and one neutral set (using animal content). In each set, 5 conclusions were valid (they followed from the two preceding premises) and 7 invalid (conclusions did not necessarily follow from the two preceding premises).[7] Moreover, we constructed one additional set of neutral concrete syllogisms. Examples of each type are in Fig. 17.1.

I gave one point for each correct answer, i.e. accepting valid syllogism and rejecting invalid syllogism, although usually researchers calculate acceptance rates. However, as this can give rise to response bias [28], I decided to use a measure of "hits" instead.

In table A in the Appendix, I give the comparison of difficulty of each syllogism across each type. Difficulty was calculated as a percentage of the people who correctly answered a particular syllogism.

17.4.3 Confirming the Confirmation Bias

I compared how people are influenced by their prior attitudes regarding abortions when evaluating the value-laden syllogisms.

[6]Syllogisms were constructed with the help of my student, Yeon Joo Lee (MeiCogSci program in Vienna).

[7]Syllogism No. 25 (see Appendix) was later eliminated from all analyses due to a mistake in wording, which made it invalid. Thus, in pro-choice category there were four valid and seen invalid syllogisms. However, elimination of syllogism No. 25 did not substantially change the results.

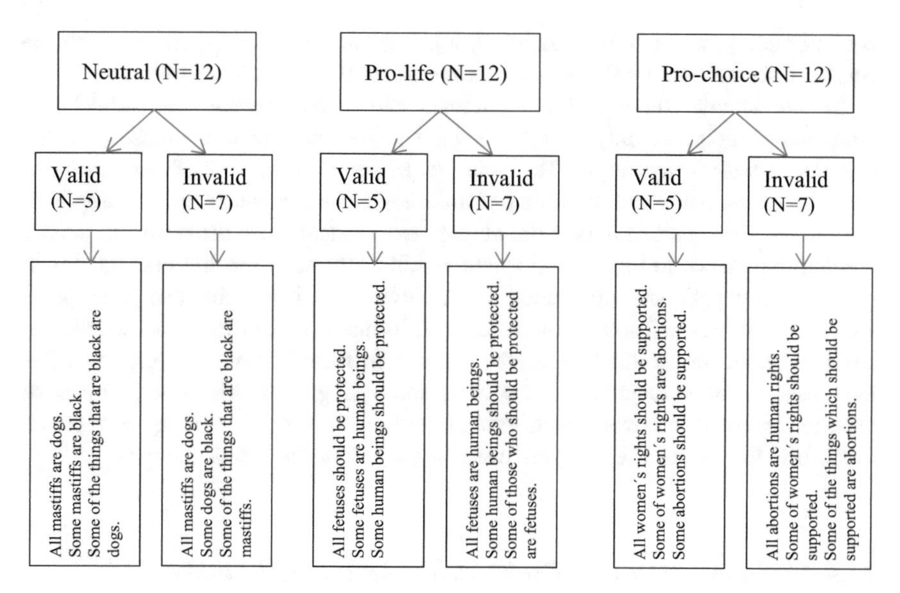

Fig. 17.1 Structure and examples of three different sets of syllogisms

Table 17.1 Descriptive statistics for Slovak sample

Sample	Attitudes- extreme groups	*N*	%	*M*	*SD*
Slovak (N = 314)	Pro-choice	73	23.2	10.59	3.00
	Neutral	151	48.1	22.50	3.17
	Pro-life	90	28.7	32.62	2.52
	Total	314	100	22.63	8.44

I divided the participants into three (extreme) groups according to their attitude towards abortions. Those scoring in the lowest 25% (below 15 in attitude toward abortions questionnaire) were assigned into "Pro-Choice" group, those scoring in the upper 25% (above 29 in attitude toward abortions questionnaire) were assigned into "Pro-Life" group, and 50% scoring in between were assigned into "Neutral" group.

Below is the table with descriptive statistics (Table 17.1):

There was no significant effect of gender on the attitudes toward abortions ($\chi^2 = 1.413$, $p = 0.493$).

Because there was an uneven number of valid (5) vs. invalid (7) syllogisms in each category (neutral pro-life, pro-choice), I calculated the proportions of correctly solved syllogisms in each category: neutral valid (NV), pro-life valid (PLV), pro-choice valid (PChV), neutral invalid (NI), pro-life invalid (PLI), pro-choice invalid (PChI) and these six scores were treated as main dependant variables.

Before I started to analyse the differences between the groups of participants, I checked whether there are any differences in the abilities of Slovak participants to correctly solve the various kinds of syllogisms and how the experience with the course of formal logic influences reasoning about the controversial topics.

In our sample, there was a significant difference between the trained and untrained participants only in pro-life valid syllogisms favouring those with the experience with formal logic. The more surprising was the lack of any other differences (participants with training in formal logic were no better in reasoning with syllogisms than participants without any experience), but furthermore, people without experience in formal logic were significantly more pro-life orientated than people with experience with formal logic. However, it is possible that participants indicated that they had some experience with formal course of logic, but we did not check whether they actually received training in formal reasoning (e.g., whether they used Venn's diagrams for solving the syllogisms). Moreover, any such experience could be from high school and in fact had no long-term effect. Therefore, the rest of the analyses were performed on the whole sample.

17.4.4 Interaction of Effects: Prior Attitudes, Validity and Direction of Conclusion

The proportions of correctly solved syllogisms in each category were analysed with a split-plot ANOVA with three attitude groups (pro-life, pro-choice & neutral) as the between-participants factor and validity (valid vs. invalid) and direction of the conclusion (pro-life, pro-choice & neutral) as within-participants factors. This analysis revealed that main effect of validity was significant (F (1,311) = 252.372, $p < 0.001$, $\eta^2 = 0.448$), and main effect of direction of conclusion was marginally significant (F (2,622) = 2.784, $p = 0.063$, $\eta^2 = 0.009$). In other words, solving valid syllogisms was generally easier for participants than invalid syllogisms and syllogisms with pro-life conclusions generally easier for participants to solve correctly. When we look at Fig. 17.2 we can compare confidence intervals (*CI*) of the similar colours—greenish for neutral syllogisms, reddish for pro-life syllogisms and bluish for pro-choice syllogisms and we can clearly see that valid syllogisms of each type had higher success rates than invalid. For example, success rates for pro-life valid syllogisms was in all three attitude groups over 70%, as compared to less than 50% success rates for pro-life invalid syllogisms. The only exception of this trend was proportion of pro-choice valid syllogisms in the group of pro-lifers, where the success rates were more similar to invalid than valid syllogisms.

Furthermore, I found significant interactions between validity and direction of conclusion (F (2,622) = 5.797, $p = 0.003$, $\eta^2 = 0.018$), attitudes toward abortions and validity (F (2,311) = 4.180, $p = 0.016$, $\eta^2 = 0.026$), and finally between all factors—attitudes, validity and direction of conclusion (F (4,622) = 3.055, $p = 0.016$, $\eta^2 = 0.02$). In other words, differences in successfully solving syllogisms were caused not only by validity (valid syllogisms as easier) and direction of conclusion (pro-choice syllogisms were more difficult), but it interacted also with

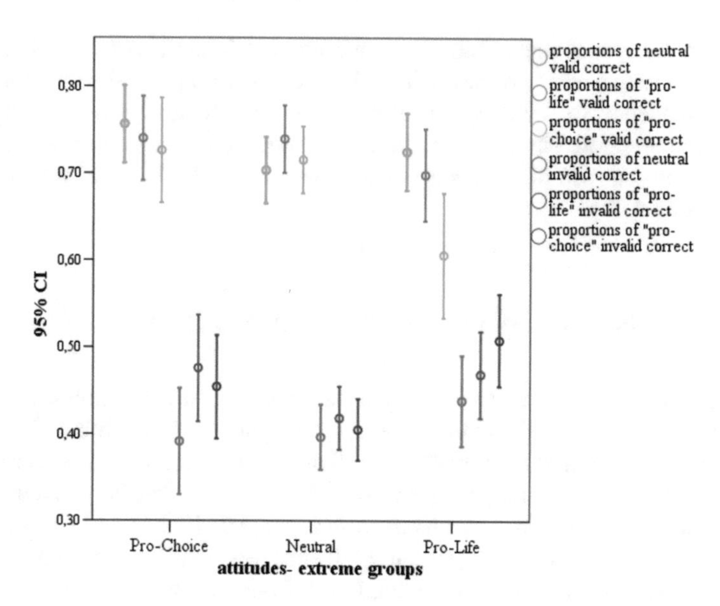

Fig. 17.2 Confidence intervals and interaction of attitudes, validity and direction of conclusion

the participants' prior attitudes. It can be best seen in the group of pro-lifers who rejected significantly more valid pro-choice syllogisms than pro-life and neutral syllogisms of exactly the same kind, while there were no significant differences between three attitude groups in accepting neutral and pro-life valid syllogisms.

There are several interpretations we can draw from these results. Because pro-life valid conclusions were accepted more than pro-choice valid conclusions even in the group of pro-choicers, it can serve as an argument (often overlooked by the camp of pro-life activists) that people with more "pro-choice" attitude are not for abortions, but for freedom of choice. There is generally more agreement on the part of pro-life arguments, such as abortion destroys unborn life (although there is much disagreement in what stage of intrauterine life we start to consider foetus as human being)—the attitude that even many pro-choicers hold. Preference for pro-life arguments can be also caused because they are more vivid—it is generally easier to imagine unborn babies being killed than more abstract freedom of choice, or long-term consequences for the unwanted babies and their families, which are implied in the pro-choice arguments. Lastly, our results indicate the highest incidence of motivated reasoning in people with pro-life attitude—their prior attitudes to the topic do not allow them to see the same logical structure in the syllogisms opposing their attitudes. Generally, it highlights why it is often so difficult to engage in discussion with people holding strong attitudes to the value-laden topic and that "logical" argumentation often fails. Similarly, Klaczynski [12] found in-group biases favouring one's own religious group and he showed how we use double-standards when we evaluate evidence either confirming or contradicting our theories. (We are much more critical toward evidence opposing our theories and engage in analytical reasoning, while we are much more benevolent and use intuitive heuristics when we deal with evidence confirming our theories.) It seems that this tendency for motivated reasoning and confirmation bias is stronger for people who are more "invested" in the topic, therefore in the next section I briefly examine how the pro-life attitude is connected with self-identified religious and political affiliations.

17.5 Effects of Religious and Political Affiliations

I also examined possible effect of religious and political affiliations on confirmation bias. In our demographic questionnaire, participants self-identified themselves freely with any religion or political affiliation. For our analyses, I collapsed participants in two religion groups: no religion (all those who explicitely said they have no religion or regard themselves as agnostics; $N = 80$, 19.2%) and Christian religion (all those who explicitely identify themselves with any of the Christian churches—Roman Catholics, Protestant, Greek Catholic, other Christian; $N = 338$, 80.2%). Similarly, I treated political affiliation, but here I collapsed participants into three groups: no political affiliation or those who preferred not to indicate their political affiliation ($N = 216$, 51.9%), liberal ($N = 93$, 22.4%), and conservative ($N = 19$, 4.6%) (Table 17.2).

I found no differences between liberal, neutral and conservative students, although it can be due to the fact that half of the students refused to identify with any political affiliation whatsoever.

17.6 Discussion

The aim of this chapter was to show that discussions about value-related controversial topic are often aimless due to confirmation bias displayed at both sides of the dispute. I wanted to show how holding strong beliefs regarding any issue related to one's personal and moral values often prevents us from seeing logic also in other party's arguments. In real debates, we often don't think about the validity of the arguments from the strictly logical point of view and we often disagree even before we hear the particular argument. In this chapter, I wanted to explore how our beliefs prevent us from acknowledging the logic to the same type of arguments if they are made by the other side of the dispute.

Moreover, I wanted to explore the notion according to which people holding different beliefs form somewhat distinct "cultures" within one common culture, such as nationality. It seems that these values that are shared within a group play a more important role than the citizenship, and thus it makes sense to talk about "cultures" even within one country.

Table 17.2 Effect of religion on confirmation bias

	Religion	*N*	*M*	*SD*	*t*	*p*	Cohen's *d*[a]
NV	no religion	61	0.74	0.17	0.762	0.447	0.098
	Christian	253	0.72	0.23			
PLV	no religion	61	0.74	0.25	0.378	0.706	0.052
	Christian	253	0.72	0.23			
PChV	no religion	61	0.65	0.27	1.204	0.230	0.174
	Christian	253	0.70	0.28			
NI	no religion	61	0.46	0.26	1.937	0.054	0.268
	Christian	253	0.39	0.24			
PLI	no religion	61	0.52	0.25	2.809	0.005	0.392
	Christian	253	0.43	0.23			
PChI	no religion	61	0.51	0.25	2.343	0.020	0.328
	Christian	253	0.43	0.24			

Legend Legend, *NV* neutral valid, *NI* neutral invalid, *PLV* pro-life valid, *PLI* pro-life invalid, *PChV* pro-choice valid, *PChI* pro-choice invalid

[a]Cohen's *d* is a measure of effect size and it tells as about magnitude of difference between the two groups. With a Cohen's *d* of 0.2 (lowest value for the significant findings in this study), 58% of the no religion group will be above the mean of the Christian group, 92% of the two groups will overlap, and there is a 56% chance that a person picked at random from the no religion group will have a higher score than a person picked at random from the Christian group (probability of superiority). With a Cohen's *d* of 0.4 (highest value for the significant findings in this study), 66% of the no religion group will be above the mean of the Christian group, 84% of the two groups will overlap, and there is a 61% chance that a person picked at random from the no religion group will have a higher score than a person picked at random from the Christian group (probability of superiority). (These interpretations of Cohen's *d* were based on interactive visualisations at http://rpsychologist.com/d3/cohend/)

The results showed that people holding opposing beliefs did display confirmation bias, but this confirmation bias was stronger for one side of the dispute, i.e. "pro-lifers". Because holding more pro-life attitude is strongly mediated by religious affiliation, I explored the differences between people who identified themselves as belonging to any of the Christian religion and those who regard themselves as atheists or agnostics. Christian participants performed worse in neutral valid syllogisms, but mainly in all types of invalid syllogisms, where they differed by 10% from the non-religious participants.

However, this result has to be taken with caution from two main reasons. Firstly, the biggest limitation of this approach was that the religious affiliation was a nominal variable and it did not take into account the strength of the participants' religious beliefs. In Slovakia, the majority of people identify themselves as Catholics, even though the percentage of the practicing Christians is lower. Secondly, this results point to the poorer logical ability of people identified as Christians, but it could be as well caused by the fact that people with higher logical ability seeing more discrepancies in the beliefs they were brought up with tend to abandon their religion. No causal effect of religion can be inferred from these data.

The results also show that the difficulty in correctly identifying syllogisms as valid or invalid does not rest solely on the content of the syllogisms, i.e. whether they are in line with our beliefs or not/whether they are pro-choice or pro-life, but also in the form of the syllogisms. Invalid syllogisms of either kind were much more difficult for all participants. According to Halpern [7], negations make the problems more difficult due to the basic cognitive principle: basic cognitive principle: negative information (no, not) is more difficult to process than positive information, in part because it seems to place additional demands on the working memory. It also explains why it is so difficult to treat our beliefs even as possibly invalid and why the confirmation bias is probably best overcome in argumentative setting [16].

Another limitation of the study could be in the syllogisms themselves. I tried to avoid the problem that people typically face in belief-bias paradigm—they have to accept premises as true to correctly infer whether the conclusion follows from them. People find it hard to abandon what they know about the world, therefore I constructed the syllogisms in such a way that premises were acceptable to work with for most of the people, such as "Human beings should not be murdered", "(Some) foetuses are human beings", "(All) women's rights should be supported", and "Abortions are women's right". These premises could still be disagreeable for some participants (especially those holding strong beliefs), who could find it harder to reason from the premises they did not agree with (e.g., that foetuses are already human beings, or that all women's rights should be supported), although this was the point of the study—to examine the ability not to let one's strong beliefs influence one's reasoning. To reason from the premises that go against their beliefs is a core ability of scientists, if they want to fight their own confirmation biases and remain open to the possibility of new discoveries. All our knowledge is provisional and science aims at expanding our understanding of the world around us, which is only possible if we remain sensitive to our own fallibility.

17.7 Conclusion

Culture shapes our values and can have subtle or more evident influences on our opinion on many controversial issues. Knowing these cultural influences could help to promote debates over sensitive topics that arise in multicultural settings, with implications to education and health care. For instance, one of the currently hotly debated issues in our country concerns women's right for abortion, obligatory sexual education in schools or gays' right to marriage.

Culture can influence critical thinking and reasoning in several ways. It is not only explicit educational system that may or may not promote critical thinking directly through curriculum, but there are also more subtle ways—through preferred thinking style or values that it endorses. Cultures can differ much in their beliefs and values, but it seems that there are not many cultural differences in our tendency to engage more effortful thinking, when we already have strong intuitive answer for some problem [3].

One of the core values, in which cultures can differ, is the extent to which they protect the rights of all its citizens, including women and unborn children, even though the rights of various subgroups can be in conflict. One such controversial issue is the right of a woman to decide about her own body (including ending of unwanted pregnancy) versus right of yet unborn child to be born. Reasoning about this topic is strongly affected by religious beliefs, although this study does not allow for causal inferences. However, this study showed that when beliefs and evidence clash, it is often belief that wins. It is no surprise that people untrained in critical, scientific thinking resort to beliefs as their compass in navigating through the vast ocean of many conflicting information (claiming their origin in research) and many conflicting values (such as rights of children vs. rights of their mothers). However, how strong these beliefs and values are can be seen in many examples of highly trained scientists, who decided to prefer beliefs over any scientific evidence (not to mention logical evidence).[8] One of the most publicized examples is Dr. Kurt Wise [29], with his PhD. in palaeontology from Harvard University, who admitted that "if all the evidence in the universe turns against creationism, I would be the first to admit it, but I would still be a creationist because that is what the Word of God seems to indicate. Here I must stand" (https://scepsis.net/eng/articles/id_2.php).

Richard Dawkins [4] disbelievingly calls this example of a scientist disregarding the evidence as "doublethink". This kind of thinking—ability to hold two opposing views as both true—is typical for more holistic cultures. Disregarding logic in case of very personal and deep values may be even at heart of many misunderstandings and hot disputes about controversial topics among people from different cultures and religions. Even favouring logic and analytic thinking is, at least partly, culturally determined and based on our beliefs on how the rational thinking should look like. The understanding that many people from different cultures (or even from our culture) may not share the view that logic and analytic thinking should be normative in personal and public discourse can help us move in discussion further than just keeping banging on their closed door with our logic.

Acknowledgements I would like to thank Yeon Joo Lee for her help in constructing the syllogisms and Jakub Šrol for his invaluable help with statistical analyses. This work was supported by Slovak grant agency VEGA 2/0085/17 "Cognitive limits of effective information processing and communication".

Appendices

[8]Even great philosopher Plato advised us that when senses contradict the reason/logic, we should go with the reason.

Table A Comparison of difficulty of individual syllogisms

Type	Syllogism No./Type/Content			Difficulty	Syllogism No./Type/Content			Difficulty	Syllogism No./Type/Content			Difficulty
MaP	1	NV	All dogs are mammals.	0.807	13	PLV	All human beings should be protected.	0.707	25*	PChV	All women's rights should be supported.	0.55
SiM			Some carnivores are dogs.				Some foetuses are human beings.				Some abortions are women's rights.	
SiP			Some carnivores are mammals.				Some foetuses should be protected.				Some abortions should be supported.	
PaM	2	NI	All dogs are mammals.	0.315	14	PLI	All human beings should be protected.	0.464	26	PChI	All women's rights should be supported.	0.401
SiM			Some carnivores are mammals.				Some foetuses should be protected.				Some abortions should be supported.	
SiP			Some carnivores are dogs.				Some foetuses are human beings.				Some abortions are women's rights.	
MaP	3	NV	All mastiffs are dogs.	0.627	15	PLV	All foetuses should be protected.	0.678	27	PChV	All women's rights should be supported.	0.7
MiS			Some mastiffs are black.				Some foetuses are human beings.				Some of women's rights are abortions.	
SiP			Some of things which are black are dogs.				Some human beings should be protected.				Some abortions should be supported.	
PaM	4	NI	All mastiffs are dogs.	0.471	16	PLI	All foetuses are human beings.	0.341	28	PChI	All abortions are women's rights.	0.406
MiS			Some dogs are black.				Some human beings should be protected.				Some of women's rights should be supported.	

(continued)

(continued)

Type	Syllogism No./Type/Content			Difficulty	Syllogism No./Type/Content			Difficulty	Syllogism No./Type/Content			Difficulty
SiP			Some of things which are black are mastiffs.				Some of those who should be protected are foetuses.				Some of things which should be supported are abortions.	
MiP	5	NI	Some dogs are black.	0.267	17	PLI	Some human beings should be protected.	0.38	29	PChI	Some of women's rights should be supported.	0.387
SaM			All mastiffs are dogs.				All foetuses are human beings.				All abortions are women's rights.	
SiP			Some mastiffs are black.				Some foetuses should be protected.				Some abortions should be supported.	
PiM	6	NI	Some mammals are dogs.	0.517	18	PLI	Some of things which should be protected are human beings.	0.397	30	PChI	Some of things which should be supported are women's rights.	0.413
SaM			All mastiffs are dogs.				All foetuses are human beings.				All abortions are women's rights.	
SiP			Some mastiffs are mammals.				Some foetuses should be protected.				Some of abortions should be supported.	
MiP	7	NV	Some mastiffs are black.	0.856	19	PLV	Some foetuses should be protected.	0.663	31	PChV	Some abortions should be supported.	0.726
MaS			All mastiffs are dogs.				All foetuses are human beings.				All abortions are women's rights.	
SiP			Some dogs are black.				Some human beings should be protected.				Some of women's rights should be supported.	
PiM	8	NV	Some mammals are mastiffs.	0.526	20	PLV	Some foetuses are human beings.	0.788	32	PChV	Some of abortions are women's rights.	0.668
MaS			All mastiffs are dogs.				All human beings should be protected.				All women's rights should be supported.	

(continued)

(continued)

Type	Syllogism No./Type/Content			Difficulty	Syllogism No./Type/Content			Difficulty	Syllogism No./Type/Content			Difficulty
SiP			Some dogs are mammals.				Some of those who should be protected are foetuses.				Some of things which should be supported are abortions.	
MoP	9	NI	Some mammals are not mastiffs.	0.293	21	PLI	Some of women's rights should not be supported.	0.392	33	PChI	Some human beings should not be protected.	0.399
SaM			All dogs are mammals.				All abortions are women's rights.				All foetuses are human beings.	
SoP			Some dogs are not mastiffs.				Some abortions should not be supported.				Some foetuses should not be protected.	
PoM	10	NI	Some mammals are not dogs.	0.661	22	PLI	Some of women's rights should not be supported.	0.784	34	PChI	Some human beings should not be protected.	0.714
SaM			All mastiffs are dogs.				All abortions should be supported.				All foetuses should be protected.	
SoP			Some mastiffs are not mammals.				Some abortions are not women's rights.				Some foetuses are not human beings.	
MoP	11	NV	Some dogs are not mastiffs.	0.738	23	PLV	Some abortions should not be supported.	0.712	35	PChV	Some foetuses should not be protected.	0.615
MaS			All dogs are mammals.				All abortions are women's rights.				All foetuses are human beings.	
SoP			Some mammals are not mastiffs.				Some of women's rights should not be supported.				Some human beings should not be protected.	

(continued)

(continued)

Type	Syllogism No./Type/Content			Difficulty	Syllogism No./Type/Content			Difficulty	Syllogism No./Type/Content			Difficulty
PoM	12	NI	Some mammals are not mastiffs.	0.651	24	PLI	Some abortions are not women's rights.	0.433	36	PChI	Some foetuses are not human beings.	0.495
MaS			All mastiffs are dogs.				All women's rights should be supported.				All human beings should be protected.	
SoP			Some dogs are not mammals.				Some of things which should be supported are not abortions.				Some of those who should be protected are not foetuses.	

Legend: NV = neutral valid, NI = neutral invalid, PLV = pro-life valid, PLI = pro-life invalid, PChV = pro-choice valid, PChI = pro-choice invalid
Syllogism No. 25 was eliminated from the analyses due to mistake in wording in Slovak version

References

1. Baron, J.: Myside bias in thinking about abortion. Think Reason **1**, 221–235 (1995)
2. Betsch, T., Haberstroh, S. (eds.): Routines of Decision Making. Lawrence Erlbaum Associates, Publishers, London (2005)
3. Čavojová, V., Hanák, R.: Culture´s influences on cognitive reflection. in: Cohn, J., Schatz, S., Freeman, H., Combs, D. (eds.) Modeling Sociocultural Influences on Decision Making: Understanding Conflict, Enabling Stability, pp. 85–102. CRC Group, Taylor & Francis Group, Boca Raton, FL (2016)
4. Dawkins, R.: Sadly, an Honest Creationist. Free Inq. Mag. **21**(4) (2001)
5. Frederick, S.: Cognitive reflection and decision making. J. Econ. Perspect. **19**(4), 25–42 (2005)
6. Gahér, F.: Logika pre každého, 3. vydanie. Iris, Bratislava (2003)
7. Halpern, D.F.: Thought and Knowledge: An Introduction to Critical Thinking. 5th edn. Psychology Press, New York (2014)
8. Henrich, J., Heine, S.J., Norenzayan, A.: The weirdest people in the world? Behav. Brain Sci, vol. 33, pp. 61–83–135 (2010)
9. Jelen, T.G., Wilcox, C.: Causes and consequences of public attitudes toward abortion: a review and research agenda. Polit. Res. Q. **56**(4), 489–500 (2003)
10. Jurkovič, M.: Miera konfirmačnej zaujatosti vo vzťahu k osobne relevantným tvrdeniam a inteligencii, in: Psychologica XLII, pp. 247–254 (2014)
11. Kahneman, D.: Thinking Fast and Slow. Farrar, Straus and Giroux, New York (2011)
12. Klaczynski, P.A.: Motivated scientific reasoning biases, epistemological beliefs, and theory polarization: a two-process approach to adolescent cognition. Child Dev. **71**(5), 1347–1366 (2000)
13. Lambie, J.: How to be Critically Open-minded: A Psychological and Historical Analysis. Palgrave Macmillan, London (2014)
14. Macpherson, R., Stanovich, K.E.: Cognitive ability, thinking dispositions, and instructional set as predictors of critical thinking. Learn. Individ. Differ. **17**(2), 115–127 (2007)
15. Markovits, H., Nantel, G.: The belief-bias effect in the production and evaluation of logical conclusions. Mem. Cognit. **17**(1), 11–17 (1989)
16. Mercier, H., Sperber, D.: Why do humans reason? Arguments for an argumentative theory. Behav. Brain Sci., vol. 34, pp. 57–74–111 (2011)
17. Morley, N.J., Evans, J.S.B.T., Handley, S.J.: Belief bias and figural bias in syllogistic reasoning. Q. J. Exp. Psychol. A **57**(4), 666–692 (2004)
18. Nickerson, R.S.: Confirmation bias: a ubiquitous phenomenon in many guises. Rev. Gen. Psychol. **2**(2), 175–220 (1998)
19. Nisbett, R.E.: The Geography of Thought. Nicholas Brealey Publishing, Boston (2003)
20. Nisbett, R.E., Choi, I., Peng, K., Norenzayan, A.: Culture and systems of thought: holistic versus analytic cognition. Psychol. Rev. **108**(2), 291–310 (2001)
21. Norenzayan, A., Smith, E.E., Kim, B.J., Nisbett, R.E.: Cultural preferences for formal versus intuitive reasoning. Cogn. Sci. **26**(217), 653–684 (2002)
22. Stanovich, K.E.: Rationality and the Reflective Mind. Oxford University Press, New York (2011)
23. Stanovich, K.E., West, R.F.: Individual differences in rational thought. J. Exp. Psychol. Gen. **127**(2), 161–188 (1998)
24. Stanovich, K.E., West, R.F.: On the failure of cognitive ability to predict myside and one-sided thinking biases. Think. Reason. **14**(2), 129–167 (2008)
25. Sternberg, R.J., Sternberg, K.: Cognitive Psychology. 6th edn. Wadsworth, Cenage Learning, Belmont, CA (2012)
26. Stupple, E.J.N., Ball, L.J., Evans, J.S.B.T., Kamal-Smith, E.: When logic and belief collide: individual differences in reasoning times support a selective processing model. J. Cogn. Psychol. **23**(8), 931–941 (2011)
27. Talhelm, T., Haidt, J., Oishi, S., Zhang, X., Miao, F.F., Chen, S.: Liberals think more analytically (More 'WEIRD') than conservatives. Pers. Soc. Psychol. Bull. **41**(2), 250–267 (2014)
28. Unsworth, S.J., Medin, D.L.: Cultural differences in belief bias associated with deductive reasoning? Cogn. Sci. **29**(4), 525–529 (2005)
29. Wise, K.: Geology. in: Ashton, J.F. (ed.) In Six Days: Why Fifty Scientists Choose to Believe in Creation, pp. 351–355. Master Books (2001)

Chapter 18
Social Influence and Intercultural Differences

Lionel Rodrigues, Jérôme Blondé and Fabien Girandola

Abstract Culture is an important part of what individuals are and can orient their attitudes, beliefs, and behaviors in several contexts. In equivalent situations, people would be likely to report different reactions depending on their cultural background. The effects of cross-cultural differences (individualistic vs. collectivistic cultures) on cognitive dissonance, social influence, and persuasion are discussed. This chapter shows that intra-individual processes, such as reduction dissonance and the processing of persuasive information, are regulated by cultural orientations and cultural aspects of the self (independent vs. interdependent self-construal). Considering these cross-cultural effects, new avenues of research open up on change and resistance to change in many fields such as health, environment, consumption, and radicalization.

Keywords Intercultural differences · Social influence · Cognitive dissonance
Persuasion

18.1 Introduction

A given population can share the same perceptions, beliefs and representations, and communication because of a common culture [9, 111]. Culture can be apprehended across the spectrum of nations, ethnic groups, and the perception of a consensus. According to Guimond [35, pp. 26–27], culture is "a way of thinking, of feeling and of behaving which characterizes the members of a group and distinguishes them

L. Rodrigues (✉) · J. Blondé · F. Girandola
LPS, Aix Marseille University, 29 Avenue Robert Schuman,
13621 Aix-en-Provence, France
e-mail: rodrigues.lionel@me.com

J. Blondé
e-mail: jerome.blonde@gmail.com

F. Girandola
e-mail: fabien.girandola@univ-amu.fr

C. Faucher (ed.), *Advances in Culturally-Aware Intelligent Systems and in Cross-Cultural Psychological Studies*, Intelligent Systems Reference Library 134, https://doi.org/10.1007/978-3-319-67024-9_18

from other groups (…) finally, culture is not simply the characteristic of a group, it is the characteristic which distinguishes one group from another". These cultural aspects have important repercussions on forming and changing attitudes and behaviors, and also in the social psychological theories which take them into account: the cognitive dissonance theory, persuasive communication [116] in various domains such as public health, the environment, consumption and its advertising contents [1, 13, 85], radicalization[1] [81, 82] or even the diffusion of military messages [62]. We are mainly interested in the effects of culture on change and resistance to change.

In the framework of intercultural psychology, Hofstede [44, 46] identified six structural or sociological dimensions: individualism/collectivism,[2] power distance,[3] uncertainty avoidance,[4] masculinity/femininity,[5] long-term orientation/short-term orientation,[6] and indulgence/restraint.[7] Most of the later models of culture propose dimensions relatively close to those of Hofstede (e.g., [52, 122, 130]). For example, Trompenaars [131] distinguished seven dimensions of culture: individualism/collectivism, objectivity/subjectivity, universalism/particularism, diffuse culture/limited culture, attributed status/acquired status, volition/refusal to control nature, sequential time/synchronous time. Overall, while each dimension has been underlined (cf. [46, 124]), the research undertaken mainly confirms the influence of the individualism/collectivism variable on psychological processes (e.g., [32, 35, 101]).

Nevertheless, some authors (e.g., [137]) underline the overly important simplicity of the individualism/ collectivism dichotomy. Leung and Cohen [87] propose, for example, taking an interest in individual differences in the framework of

---

[1]Radicalization is defined as an extreme commitment to a specific cause leading to a form of violent action [81].

[2]Individualism refers to the preference to act as an individual with priority for one's personal interest and for the immediate family (husband, wife and children). On the contrary, collectivism can be defined as a preference for a social framework in which the individual is integrated from birth in a group which will protect him in exchange for his unconditional loyalty [46].

[3]Hofstede [44] defined power distance as "the extent to which the less powerful members of institutions and organizations within a country expect and accept that power is distributed unequally" (p. 28). Power distance regulates the extent to which different members of society command respect and wield influence (cited by [106]).

[4]Uncertainty avoidance refers "to the extent to which people feel threatened by ambiguous situations, and have created beliefs and institutions that try to avoid these" [47].

[5]As a society, masculinity/femininity "refer to the distribution of values between the genders" [46, p. 12].

[6]Long-term vs. short-term orientation distinguishes societies which prefer maintaining traditions rather than privileging societal change [44].

[7]"Indulgence stands for a society that allows relatively free gratification of basic and natural human desires related to enjoying life and having fun. Restraint stands for a society that controls gratifications of needs and regulates it by means of strict social norms" [46, p. 15].

Table 18.1 Characteristics of the independent self and the interdependent self ([35, p. 126]; cf. [89])

Dimensions	Independent self	Interdependent self
Definition	Separated from the context	Linked to the context
Structure	Unitary, stable	Flexible, variable
Content	Private, internal attributes (aptitudes, thoughts, feelings)	Public, external attributes (status, roles, relations)
Objectives	To be unique, to express oneself, to assert oneself (personal)	To belong to a group, to respect one's obligations (collective)

intra and intercultural variations. Markus and Kitayama [89, 90], followed by Markus et al. [92], suggest that cultures have an impact on the self-construal of the individual. Thus, individuals with different cultural membership (individualistic/ collectivistic) have a different self-construal which in turn influences cognitions, emotions, motivations and behaviors [135]. North American and European cultures would be characterized by an independent self-representation, in other words by an autonomous self free from relations with others and the social context. Behavior would be under the individual's control in that it would result from private attitudes, sentiments, motivations and personality traits (cf. Table 18.1). On the other hand, Asian, African, or even Latin American or East European cultures would be characterized by an interdependent or collectivistic self-representation (cf. [77]). Behavior would mainly be governed by interindividual and intra-group relations, consequently, constraints, situational and contextual obligations would not reflect the private attitudes and sentiments of the individual [45].

Trafimow [127] showed that the cognitions linked to an independent self and those linked to an interdependent self are stored in different areas of the memory. According to Heine, Lehman, Markus, and Kitayama [41], the need to maintain a positive self-image would be characteristic of individualistic cultures. Collectivistic cultures would encourage self-criticism and personal dissatisfaction. Markus and Kitayama [89] stated that it is possible to identify individuals with an independent self-construal in collectivistic societies and individuals with an interdependent self-construal in individualistic societies. Some research considers culture as a construct that is chronically accessible in the context of self-construal. Finally, it is easy to activate the independent self or the interdependent self (for example, by asking to do a task: circle the singular or plural pronouns, cf. [10, 30]).

In this chapter, we will explore the effects of an individualistic/collectivistic culture in a particularly important field of contemporary social psychology: changing attitudes and behaviors, and resistance. This field will be especially illustrated by the cognitive dissonance theory [28], the heuristics or strategies of changing behavior [16], and the classic theories of persuasive communication notably in relation to messages containing threats and those focused on consumption.

18.2 Intercultural Aspects and Cognitive Dissonance

The cognitive dissonance theory [28] is a motivational theory that enables one to explain the behaviors of daily life in various situations [19, 29]. An individual is susceptible to be in a state of cognitive dissonance when two cognitions (e.g., attitude, belief, behavior or knowledge) maintain an inconsistent relation. For example, the cognitions "I know that smoking kills" (knowledge) and "I continue to smoke" (behavior) taken together exemplify a cognitive inconsistency. Dissonance is also aroused in certain cases when the individual's self is threatened (cf. [5, 121]). Attacking elements important to self-concept (e.g., moral values) by performing a problematic behavior that is contrary to moral values, arouses dissonance and determines the scale of its reduction. The personal moral values that individuals hold are "culturally determined, widely shared by most of the individuals in a given society or subculture" [126, p. 596]. Dissonance is a state of internal tension defined by Festinger as a state of psychological discomfort. In order to reduce this particularly uncomfortable state of tension, an individual in a state of dissonance is motivated to restore cognitive equilibrium by resorting to a mode of dissonance reduction such as changing attitude or behavior. In our example, a good way to reduce this dissonance is to change behavior by stopping to smoke and thus recover a positive self.

Several situations are likely to arouse dissonance. Like performing a problematic act, a situation of choice (or post-decisional dissonance) can be a source of dissonance particularly when the appeal of the different outcomes is identical [8]. When an individual is asked to choose between two (or several) outcomes, the decision generates dissonance because the choice of one of the two outcomes implies the rejection of the other one. In order to reduce the dissonance, the alternative chosen is considered as more appealing than the one (those) rejected. This paradigm has been particularly used by the authors who study the link between dissonance and culture (e.g., [38, 74, 6]).

The cognitive dissonance theory [28] was first tested among North American subjects. These individuals possess a self constructed on the principles and values of their own culture (e.g., [89, 128]), mainly the independent self. In certain cases, the self is implicated in the process of dissonance when it is threatened or questioned. Several authors have been interested in the effects of intercultural self-differences on the production and reduction of dissonance (e.g., [51]).

18.3 Intercultural Self and Dissonance

The cultural membership of an individual orients several aspects of the self-construal [39–41, 71, 73, 90, 91]. Intercultural self-differences would impact the process of dissonance (e.g., [89]). For example, according to Murphy and Miller [97], the self of individuals born into the North American culture would be particularly based on the principles and ideology of a consumer society. These authors used the paradigm of post-decisional dissonance (e.g., [8]). In their first experiment, participants who were strongly vs. weakly ingrained in the ideology of consumption were asked to rank five magazines by order of preference. The magazines given to the participants represented different categories: news (e.g., Newsweek, Time), women's magazines (e.g., Cosmopolitan), men's magazines (e.g., Gentlemen's Quarterly), and the last two categories conveyed the ideology of a consumer society more than the other ones. The participants were then told that they could leave with the magazine that they themselves rated in third or fourth position. Choosing between two magazines should have been a source of dissonance because selecting one implied rejecting the other one. Moreover, since the two magazines were equally appealing, the decision should have been more difficult to make and the dissonance felt should have been higher. As expected, the participants who had to choose among the popular magazines (i.e., those that conveyed the ideology of a consumer society) experienced the most dissonance. The magazine chosen has a function of reducing dissonance because it is considered as more appealing than the others. Moreover, this effect was stronger among the participants strongly ingrained in the ideology of consumption.

In a second experiment, these authors tested North American versus Finnish culture. The Finns would be less ingrained in the ideology of a consumer society than the Americans. Consequently, they should experience less dissonance than the Americans when choosing between the two magazines conveying this ideology ranked in the third or fourth position. As expected, the Finns experienced less dissonance than the Americans whether they were strongly or weakly ideologically ingrained. The amplitude of the dissonance here depended on the importance of the decision and the relative appeal of each outcome [8]. Choosing between two objects with a particular ideology was more important for individuals who share this ideology, and was then a potential source of greater dissonance. Thus, menacing an aspect of the self based on a cultural orientation, in this case the ideology associated with the consumer society, was a source of dissonance.

18.4 Individualism Versus Collectivism

According to Heine and Lehman [38], individuals born into an individualistic culture (e.g., Canada) would be more sensitive to dissonance than those born into a collectivistic culture (e.g., Japan). By favoring an independent self-construal, individualistic cultures encourage individuals to perceive themselves as responsible for their acts. As a consequence, they would experience dissonance when the attributes of the independent self are threatened. On the contrary, individuals born into collectivistic or interdependent cultures would give more importance to situational constraints and social obligations to explain their behavior. These individuals would experience less or even no dissonance. As expected, several studies have not observed the effects of dissonance in populations with collectivistic or interdependent cultures (e.g., [12, 42, 96]). However, they are not immune to dissonance. Sakai [109], for example, has shown that the Japanese experience dissonance if they are persuaded that their peers are observing their behavior. The importance given to social expectations in collectivistic cultures would impact the process of dissonance, for example when a decision involves the preferences of others (e.g., [50, 54]). According to Kitayama et al. [74], individuals born into a collectivistic culture experience dissonance if an attribute of the interdependent self (e.g., the approval of others) is threatened. These authors tested the effects of the post-decisional dissonance paradigm (or paradigm of choice) among Japanese students. They had to make a decision according to their own preferences or according to the presumed preferences of people close to them. The second situation should make salient social expectations and threaten the interdependent self. They showed that the Japanese experienced dissonance only in situation of reference to others. In a second experiment, participants (Americans vs. Japanese) were put in a situation of post-decisional dissonance: they were shown or not a poster with representations of several schematic faces with direct emotional gazes (cf. Fig. 18.1). The aim of the poster was to make salient social expectations and to threaten the interdependent self. The Japanese experienced dissonance only when they were exposed to the poster while the Americans experienced dissonance only in the standard condition, in other words in the absence of the poster. Moreover, the social eye was a source of dissonance for the individuals whose self is interdependent only if it was made salient at the time of the decision [55]. In short, individuals born into collectivistic cultures are susceptible to experience post-decisional dissonance when an attribute of the interdependent self, such as the approval of others, is made salient.

Hoshino-Browne et al. [51] also conducted a series of experiments to test the intercultural variableness of dissonance. In a first experiment, the participants, all Canadians of Asian origin, deeply ingrained in European or in Asian culture, were asked to make a decision based on their own preferences (vs. based on the presumed preferences of their families). The results showed that the participants with an independent self experienced dissonance when they were led to make a personal decision. On the contrary, the participants ingrained in a collectivistic culture (Asian) experienced dissonance only when the decision involved their families' preferences. In a second experiment, they tested a procedure for affirming the interdependent self (vs. independent) as a mode of reducing post-decisional dissonance among the individuals deeply ingrained in Asian culture. According to the

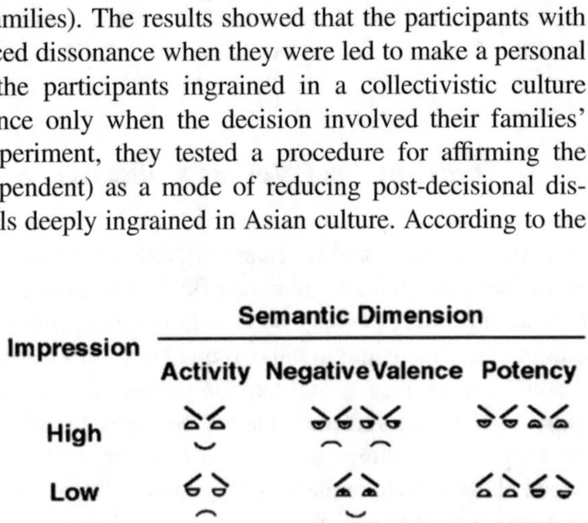

Fig. 18.1 Iconographic representations of faces used to induce the social eye in situation of post-decisional dissonance [75]

theory of self-affirmation (cf. [120]), the self is a resource for reducing dissonance. Giving an individual the opportunity to restore the integrity of the system of the self would allow reducing the dissonance experienced (e.g., [114]). Hoshino-Browne and his collaborators showed that participants who are deeply ingrained in Asian culture reduced dissonance only in the absence of self-affirmation of the interdependent self. Thus, allowing participants of collectivistic cultures to restore the integrity of the system of the interdependent self acts as a mode of reducing dissonance. Finally, Kimel et al. [70] showed that intercultural self-differences can be observed with a physiological measure of dissonance (i.e., "HPA-axis activity"). European-American (vs. Asian) participants expressed stronger physiological arousal when they were led to make a decision in private whereas Asian participants expressed stronger physiological arousal in public.

Thus, the perception of the concept of choice differs from one culture to another (e.g., [65–67]). The individuals of individualistic cultures perceive choice as a manifestation of their personal preferences while those of collectivistic cultures consider it as being dependent on the social context [104]. Choice for the latter has less to do with internal aspects of the self than with the need to satisfy social expectations. According to [55], individuals with an independent self, in other words ingrained in an individualistic culture, would give importance to responsibility for their acts. These individuals would experience post-decisional dissonance only if they have the feeling of being the master of their decision. The social eye is perceived by them as an attempt to influence and it blocks the arousal of dissonance. On the contrary, individuals from a collectivistic or interdependent culture are focused on satisfying social expectations and anticipating the desires of others. A decision made by these individuals when others are watching threatens the interdependent self because it involves public aspects of the self, which is a source of dissonance for them.

18.5 Multiculturalism and Dissonance

An individual's self can be ingrained in several cultures of a different nature [40]. For example, an individual of Asian origin who has been settled in a western country for a long time can assimilate the values of his original culture as well as the values of his host culture. This individual would be multicultural because he would possess a self composed of independent and interdependent attributes. Hoshino-Browne and his collaborators tested self-affirmation of the independent self among multicultural (vs. monocultural) individuals. The results showed that the multicultural participants were able to reduce dissonance by affirming the independent self. In the same vein, Kitayama et al. [72] showed that in the case of multiculturalism, for example when individuals of collectivistic cultures integrate attributes of the independent self, the process of dissonance is similar to the one observed in individualistic cultures. The authors tested the effects of post-decisional dissonance among the residents of Hokkaido, an island north of Japan that has been historically ingrained in North American culture.[8] The participants (Americans vs. residents of Hokkaido vs. Japanese nonresidents of Hokkaido) had to make a decision in the presence (vs. absence) of the social eye. The Japanese nonresidents of Hokkaido experienced the most dissonance in the presence of the social eye, while the Americans and the residents of Hokkaido experienced the most dissonance in the absence of the social eye. Thus, the residents of Hokkaido, ingrained in North American culture, experienced more dissonance when the attributes of the independent self were threatened.

It can then be said that intercultural self-differences impact the arousal of dissonance [49]. On the whole, individuals of a collectivistic culture are more sensitive to situations involving the approval of others (i.e., interdependent self) and those of an individualistic culture to situations of commitment (i.e., independent self). However, this intercultural dichotomy is too restrictive. The process of dissonance would especially depend on the orientation of the self-attributes and their activation in the situation of dissonance [61]. For example, individuals whose self is ingrained in an individualistic culture (the United States) experience dissonance when making a decision for others, but only if the attributes of the interdependent self are activated [69]. Moreover, some individuals can be ingrained in cultures of a different nature. For example, in the case of multiculturalism, individuals born into a collectivistic culture where the self integrates independent aspects, experience dissonance without threatening the attributes of the interdependent self (e.g., [51, 72]).

18.6 Intercultural Aspects and Behavioral Influence

In the domain of social influence, Cialdini [15] presented six strategies or heuristics that are effective for changing behavior: social proof (i.e., tendency to validate one's own behavior in function of the behavior or judgment of others), sympathy (i.e., relations of affinity, similarity or cooperation that individuals have with one another), authority (i.e., the weight of authority requiring obedience), rarity (rare objects seem more appealing), reciprocity (doing a favor implies doing it in return), commitment/consistency (desire to appear consistent in relation to oneself and to others). In a cultural context, Orji [100] used the STPS scale (Susceptibility to Persuasive Strategies Scale, [63]) and showed that the impact of these strategies on behavior differs depending on one's cultural membership (individualistic vs. collectivistic). This scale predicts participant susceptibility to individual strategies and the efficacy

[8]In the middle of the 19th century, the island of Hokkaido was mainly inhabited by indigenes, the Ainu. They underwent a radical change of society in 1867 when the feudal government was returned to the emperor during the Meiji restoration (also called the Meiji revolution). This societal change was accompanied by an openness to commerce and intensive westernization. According to Kitayama and his collaborators, the consequences of these historical particularities would be found in the values shared by the present generation of the island, and the residents are closer to individualist cultures than to collectivist ones.

of the strategies for behavior change in real life. The individuals belonging to a collectivistic culture judged four of the six strategies (i.e., authority, reciprocity, social proof and sympathy) as significantly more effective than those belonging to an individualistic culture [16]. Rarity is the only strategy perceived as being more effective by individuals with an individualistic culture. Commitment/ consistency is judged to be equally effective among both individualistic and collectivistic individuals. However, Cialdini et al. [16] showed that this strategy produced a stronger impact on the behaviors of American individualists than on Polish collectivists. Individuals with an independent self would have a strong need for consistency in both their cognitive world and between the relation attitude-behavior [17, 104]. This need and, as a consequence, the search for consistency would be less important and fulfilling among those with a collectivistic self.

Freedom of choice is also an effective technique to bring about behavioral change. Freedom produces a greater effect on the judgments and behaviors of individuals born into an individualistic culture: Americans are more inclined than Poles to explain their behavior as being based on their personal choice [17], and North American children are more committed to freely chosen activities than their Asian peers [56]. Petrova et al. [104] showed that American subjects who accepted a first request (i.e., answering an online survey), responded significantly more favorably to a final request that was addressed to them (21.6%) than Asian subjects put in the same situation (9.9%). Among all the participants, both Americans and Asians who accepted the first request, those with individualistic orientations were more inclined to accept the final request than those with a collectivistic orientation. Thus the individualistic/collectivistic orientation of each individual mediates the effects of culture on accepting the final request.

In a similar vein, the evoking freedom or "But You Are Free" technique (BYAF; [34]) consists in embellishing a request to another person with the proposal "you are free to…". This technique significantly increases the possibility of acceptance. Guéguen and Pascual [34] asked passers-by to give them money. In the experimental condition, their request ended with the phrase "but you are free to accept or refuse" whereas this phrase was not used in the control condition. They found that 10% of the solicited participants complied with the request in the control condition, whereas 47.5% accepted in the experimental condition. The simple induction of a feeling of freedom can facilitate individual compliance to various types of requests [33]. Pascual et al. [102] showed that this technique would be efficient only in individualistic cultural contexts (i.e., France, Romania). In this cultural context, people are more likely to aspire to a feeling of individual freedom. Inversely, in collectivistic cultures where people are more interdependent, individual liberty has little social value or even meaning, likely rendering the BYAF technique ineffective. This was implied by the results obtained in the three collectivistic countries that were considered: the Ivory Coast, Russia and China. These results were expected since collectivistic individuals are not as easily convinced to partake in an action and are less susceptible to reactance than individualists.

In addition, Salter et al. [112] find that individualists are more sensitive of escalation of commitment than collectivistic individuals. The escalation of commitment describe situation where an individual persists in an unprofitable course of action. For example, this can occur when an individual continues to invest in a financial action whose value is constantly declining. For Sharp et al. [112], the cross-cultural difference of escalation of commitment depend especially on the framing of the decision alternatives and the long-term orientation (cf. [44]).

Using descriptive norms (i.e., "80% of the residents recycle their trash") is also a good way to bring about behavioral change [15]. Lapinski et al. [83] showed that the extent of the effect of descriptive norms is stronger in a collectivistic culture (China) than in an individualistic one (the USA). Exposure to a message advancing descriptive norms or insisting on the prevalence of behaviors in the group produces less favorable attitudes and less behavioral intentions towards the recommended behavior among participants with an individualistic orientation (i.e., individual objectives) than among those with a collectivistic orientation (i.e., group objectives; cf. [68]).

18.7 Intercultural Aspects and Persuasion

Some research on communication aims to study the impact of culture on persuasion, for example in the domain of consumption (e.g., [48, 136] or health [79, 80]). Persuasion is most effective when there is cultural matching or cultural congruity between the message delivered (individualistic vs. collectivistic) and the target of the message (individualistic vs. collectivistic). For example, Han and Shavitt [37] showed that arguments conveying individualistic messages ("The brand Solo cleans with a mildness that you will appreciate") are more persuasive in the United States (predominantly individualistic culture) than in South Korea (predominantly collectivistic culture). In this country, messages focused on collectivistic arguments work best ("The brand Solo cleans with a mildness that your family will appreciate"). Thus attempts at persuasion produce an impact on individuals sharing a collectivistic culture if they are focused, for example, on arguments referring to the history of the society or to a group in general. Symmetrically, they produce an impact on individuals sharing an individualistic culture if they are focused, for example, on personal arguments and private opinions [1, 3, 4, 36]. However, Aaker and Williams [2] showed that a cultural incongruity (e.g., Americans exposed to a "collectivistic" message) can have greater persuasive effectiveness than cultural congruity (e.g., Americans exposed to an "individualistic" message). This incongruity would likely be persuasive because of its innovation or an effect of surprise.

In the framework of persuasive communication, several variables and parameters have been the object of research depending on the intercultural perspective such as the orientation of the message. A positive orientation presents opportunities and the advantages of adopting the recommendations proposed: "If you examine your skin, you will be more likely to notice something abnormal". A negative orientation presents lost opportunities and the drawbacks of rejecting the same recommendations: "If you don't examine your skin, you will be less likely to notice something abnormal". Uskul et al. [133] gave participants belonging to either an individualistic culture (British) or a collectivistic culture (East Asian) a message to read about toothbrushing. The message was either positively (e.g., "If you brush your teeth regularly, you will have healthy teeth and gums" or negatively orientated (e.g., "If you don't brush your teeth, you risk having unhealthy teeth and gums"). The participants born into an individualistic culture expressed a more positive attitude towards brushing and a stronger intention to brush when the orientation was positive rather than when it was negative. The opposite was observed among the participants born into a collectivistic culture. In a later experiment, Uskul and Oyserman [132] made individualism salient among European-American female participants. The participants had to read a message about the relation between coffee consumption and developing fibrocystic diseases with a focus on individual physical consequences (individual orientation: lumps on the breast). The participants judged the message as more persuasive, considered themselves more at risk, and stated that they would adopt prevention behaviors such as drinking decaffeinated coffee. Similar results were observed among Asian participants when collectivism was made salient and they were exposed to a message about fibrocystic diseases that presented relational consequences (relational orientation: e.g., incapacity to take care of one's family).

The question of resistance to persuasion is directly linked to persuasion. For example, Ko and Kim [75] studied resistance to certain health messages according to cultural membership (cf. also [78]). Individuals belonging to an individualistic culture would adopt defenses or resistance when the message underlines the personal risks associated with unprotected sexual practices (e.g., "You risk a lot personally"). However, these defenses or resistance were not observed when the message was focused on collective risks (e.g., "You put your partner's life at risk"). Sherman et al. [115] proposed self-affirmation[9] procedures to enable the reduction of defenses and, as a consequence, to break resistance to persuasion. According to them, self-affirmation must correspond to some aspects of the participants' cultural membership (individualistic self vs. collectivistic self) or more specifically to an independent or an interdependent self [51, 38]. Self-affirmation would be more effective for restoring self-integrity when it is focalized on independent self-attributes for the individualistic participants and interdependent attributes for the collectivistic participants.

Other studies have shown that reasoning about persuasion in binary mode (individualistic/ collectivistic) does not allow taking into account particularities, such as collectivistic countries undergoing an economic transition. This situation is likely to activate different self-construals (interdependent vs. independent) in a certain part

[9]As underlined in the section on culture and dissonance, according to the theory of self-affirmation [119], individuals need to restore their self-integrity. In reaction to a threat to the self that is generated, for example, by reading a preventive message, the individual puts in place a defensive bias which reduces the possibility of changing beliefs or behaviors. Affirming important personal values (self-affirmation) then allows restoring self-integrity and, furthermore, allows accepting the prevention message on both the cognitive and the behavioral level.

of the population. Thus, Zhang and Shavitt [140] showed that China, a country with a traditionally collectivistic culture, uses more and more publicity focused on individualistic values (e.g., independence, individuality, technology) in order to reach those sharing an individualistic culture or an independent self.

Based on the ELM model ("Elaboration Likelihood Model", [105]), Pornpitakpan and Francis [106] showed that cultural membership has an impact on the intensity of processing persuasive information. It must be recalled that according to the ELM model, taking the central route leads individuals to treat or consider each argument contained in a message carefully and quasi-objectively, and, as a consequence, requires a certain cognitive effort. Taking the peripheral route consists of treating or considering persuasive arguments superficially and requires little cognitive effort. A change of attitude in this case is the result of the presence of indices that are peripheral to the message such as the presence of an expert source for example. This can be observed by comparing Thais who are characterized by strong control of uncertainty, a strong power distance, and a collectivistic culture with Canadians who are characterized by weak control of uncertainty, a weak power distance, and an individualistic culture. The results of an identical message suggest that among the individuals moderately implicated by a brand of shampoo, the source's expertise produces a stronger impact on the attitude towards the brand among the Thai participants than among the Canadian participants. The strength of the argument produces a greater impact among the Canadians than among the Thais.

This presentation of the effects of culture on persuasive communication is introductory. It shows the important impact of culture on processing persuasive information which translates a social regulation of cognitive activities [24]. In the following section, we present an important parameter of persuasion: threatening communication and its impact on behaviors.

18.8 Intercultural Aspects and Threatening Communications

Threatening communications are persuasive messages portraying relevant threats to incite individuals to adopt protective behaviors (e.g., stop smoking, eating fruits/vegetables, etc.). Compared with other forms of preventive interventions, that kind of messages still remains extensively used in public health or environmental campaigns in the hopes of increasing risk perceptions and changing unsafe habits. For example, in numerous countries (e.g., Australia, Belgium, Canada, France, New-Zealand, Thailand), shocking and frightening pictorial health warnings have been implemented on the back of all cigarette packs. Inserting vivid pictures of lung tumors or damaged bodies for example, is assumed to make cigarettes less attractive, hold attention on potential risks due to tobacco use, and prevent young people from starting smoking.

Are threatening communications really effective? For more than six decades, a large variety of researches have examined people's reactions when exposed to threatening messages (for reviews, see [7, 113, 59]). Although their effectiveness is sometimes called into question by researchers as well as field practitioners (e.g., [76, 108, 116]), most studies have provided evidence that presenting high threats (i.e., by strengthening personal relevance or by including explicit materials like pictures, videos, testimonials) is an effective lever for action (e.g., [21, 22, 118, 89]). In this line, the recent meta-analysis of Tannenbaum et al. [123], synthetizing more than 250 independent reports, found a linear relationship between threats and persuasiveness of messages. As the threats increase, individuals are more inclined to adopt recommended actions. More specifically, threatening communications are predicted to be persuasive because of the fear that is provoked by the perception of personally relevant threats in one's environment [23, 117]. Fear is a negative emotion, associated with a high level of arousal, which automatically motivates to defend and protect against imminent threats. Thus, when feeling frightened, people's attention is more likely to be drawn to potential risks and, subsequently, on protective means, which facilitates the acceptance of recommendations. However, as argued in the Protection Motivation Theory [107] and the Extended Parallel Process Model [138, 139], threats have beneficial effects only when accompanied with efficacious recommendations that people can easily perform [27, 95]. In cases where recommendations are not effective, people attempt to subdue their anxiety without even considering the risks or recommendations. They are more likely to react defensively, by avoiding threat-related information [64, 99], by denying their own vulnerability or threat severity [21, 22], or by doubting evidence of risks [6, 88].

Despite the fact that the effects of threatening messages have been investigated many times, it remains that studies have been chiefly performed on people from Western countries (i.e., Australia, Canada, US, or the European countries). For instance, when looking at studies that have been included in Tannenbaum et al.'s meta-analysis [123], only 11.7% were run in others countries. Do the beneficial effects of threats are only restricted to Western countries or can be generalized across all countries? Do threatening messages increase the likelihood of adopting appropriate behaviors, regardless the cultural background of message audience? Although few, some researchers have addressed cultural variations.

Accordingly, threatening messages have been found to be affected by individualist-collectivist orientation of message receivers (e.g., [14, 18, 31, 57–59, 98]). Generally, the use of threat produces more impact on collectivist than individualist-oriented people. For example, Chung and Ahn [14] tested the effects of two anti-smoking advertisings varying in threats (low vs. high) in US (individualist culture) and South Korea (collectivist culture). Their results showed a main effect of threat on message acceptance, but only among South Korean participants. No differences were found among participants from the US. In a similar vein, Jacobson et al. [57] have shown that Asian exhibited less defensive reactions than did North-Americans when confronted with a threatening health diagnosis. According to the authors, interdependent self-construal of people from collectivist cultures attenuates defensiveness. To the contrary, as threatening messages emphasize on self-relevant threats and challenge with self-worth, individualist-oriented people, who have an independent self-construal, are more likely to feel threatened and react defensively.

It is worth noting that these cultural variations do not result from national differences but essentially from cultural orientations [58]. For example, within the same country, namely South Africa (which is a country with a great cultural diversity), Terblanche-Smit and Terblanche [125] have noted differences between groups of various cultures. In condition of high threat, colored respondents reported more fear than did white respondents. On the contrary, when threat was low, all groups experienced identical levels of fear. Similarly, Jansen et al. [59] provided evidence of varying responses to threatening HIV risk-reduction messages with respect to individualist-collectivist cultural orientation of South African adolescents only. By contrast, no differences emerged when comparing two countries holding identical cultural orientations (e.g., Spain vs. Netherlands: [60]; France vs. USA: [134]).

Given those findings, does this means that threatening messages exert no influences in individualist countries? Further evidences demonstrated that threatening communications are likely to be equally effective in individualist, as well as in collectivist cultures, depending on which threat is portrayed in messages and how it fits with cultural orientations of receivers. One of the first studies having addressed this issue was that of Murray-Johnson et al. [98], in which two experiments have been conducted to examine how people belonging to varied cultures react to individualist- or collectivist framed HIV/AIDS messages. The first study compared young Mexican immigrants versus young African American adolescents, all of the same age. The Mexican immigrants were assumed to be collectivist while the African Americans were categorized as holding individualist cultural orientation. The second study opposed one group of American students with another group of Taiwanese students, admitting that American were individualist and Taiwanese collectivist (according to Hofstede's categorization). Messages emphasized on HIV/AIDS risks and what can be done to avoid contracting it. It told a story of a young girl suffering from AIDS. Two different messages were created. The individualist-framed message emphasized on individual threats and negative consequences for the girl herself (e.g., "She felt so lonely when people stopped visiting her."). In the collectivist-framed message, threats were group-based and negative consequences were placed on the girl's family (e.g., "Jenny's family suffered the most. They were shunned by their co-workers and friends."). Results showed that African Americans participants reported more fear when exposed to the individualist-framed message, while Mexican participants were more scared after reading the collectivist-framed message. In the second study, authors have measured cultural orientations of people[10] rather than assuming that participants hold

[10]In this study, the scale of individualism and collectivism developed by Hui [53] was used. However, the authors also proposed the use of the individualist-collectivist cultural orientation scale of Triandis et al. [129].

cultural ideologies of groups to which they belong (for a discussion, see [43]). The pattern of results was similar than that of the first study.

In line with Murray-Johnson et al., some other studies examined the effects cultural-frame on message acceptance. For instance, Laroche, Toffoli, Zhang, and Pons [84] tested two kinds of messages. One focused on personal physical pains (individualist) and another on moral pains (collectivist). The results showed that physical pains are more effective among Canadian English targets than Chinese participants. Conversely, moral pain messages are equally impactful for Chinese than Canadian. In another study, Perea and Slater [103] compared reactions to anti-drinking and driving public service announcement of Mexican American with English American. PSAs emphasized risks for family and friends (group-based message) or risks for oneself (self-based message). English-American participants, assumed holding individualist values, responded in a more positive way to the self-based message, while Mexican American participants, who were categorized as more collectivist, reported more appropriate responses to the group-based message. Similar to Chung and Ahn' study, Lee and Park [86] tested the effects of anti-smoking messages among US and South Korean audiences. In addition, they compared an individualist- with a collectivist-framed message. As expected, a significant interaction between cultural orientation and background of participants emerged. Threatening messages are persuasive only when their cultural-frame is congruent with cultural orientation of receivers. In cases of cultural non-congruency, people would be more likely to respond in a defensive manner and reject recommended behaviors (for a similar account, see [75], and also [131]).

Thus, effects of threatening communications are strongly culture-dependent. If using threats may equally promote effective changes across various cultures, not all threats provoke similar effects. Indeed, although each culture is likely to feel threatened and, if so, to react accordingly, not all cultures are jeopardized by the same threats, and not to the same extent. What is threatening varies from culture to culture as a function of norms, values, beliefs, and motivations that are shared in a given group or a society. Threats cannot be assumed to be universal concepts but are reflections of a cultural construction. Group-based threats are real threats for collectivist cultures, while self-based threats are as such for individualist cultures. By extension, one can suggest that what people perceive as an effective recommendation could be also moderated by cultural orientation. Hypothetically, it can be expected that group-based recommendations (e.g., actions that could allow protecting one's family or one's group) would be more persuasive for collectivist rather than individualist cultures and vice versa. Similarly, although Ekman's researches evidenced that facial expression of basic emotions, including fear, are universally recognized (e.g., [25, 26]), high degree of cultural variability in emotion elicitation and regulation has been found [110]. For example, in certain cultures, social beliefs and rules of interaction are instituted to deal with fear and easily regulate it [93, 94]. Cultural variations in people's way of experiencing or controlling fear could also moderate reactions to threatening messages [134]. Regarding practical concerns, it seems thus to be important to account for cultural differences when designing preventive messages and the necessity to carefully adapt message frame to cultural specificities of audience and cultural context in which they are intended to be used. Not all threatening messages, no matter how strong, might have the same impact in every country, although risks can be consistent worldwide For example, this is an important issue in the domain of HIV/AIDS health-risks which implies countries with various cultural backgrounds.

18.9 Conclusion

According to Doise [24], the models and processes of a cognitive nature are insufficient to explain cognitive functioning and reasoning in a concrete situation. It is essential to not isolate the individual's cognitive activities from the social dynamic in which these activities take place. This chapter shows that intra-individual processes, such as the reduction of dissonance and the processing of persuasive information, are regulated socially by individuals who adhere or not to norms, to a culture, and, more specifically, to an independent or interdependent self-construal. This involves sociocognitive processes in which the information processed is endowed with a social status, and processes by which cognitive activity is activated, facilitated, or, on the contrary, inhibited by variables or social structures.

In addition to the conceptual aspect, taking into account the cultural aspect makes it possible to underline an intercultural social psychology of social influence (e.g., [116]) that not only makes it possible to boost the field of social psychology focused on change and resistance, but also to propose numerous possibilities in the domains of public health, sustainable development and consumption. Moreover, cultural aspects can have a role in other domains such as radicalization or the diffusion of military messages as psychological operations. The process of radicalization would depend partly on the cultural milieu in which the individual is integrated [81, 82]. Crettiez [20], for example, has suggested that certain cultural traits are favorable to the socialization of violent radical actors. An individual would have a greater chance of being radicalized if his cultural membership is a martial or bellicose one (e.g., traditional expertise in weapons or war) or one that propagates an obligation of revenge (e.g., lex talionis, "an eye for an eye and a tooth for a tooth"). Finally, taking cultural aspects into account can also be useful in the military domain to develop intercultural competencies that are necessary for certain operations [11], or for diffusing messages adapted to the cultural characteristics of a target population (see the works of Hall, 1980, 1984, in [62]).

References

1. Aaker, J.L., Maheswaran, D.: The effects of cultural orientation on persuasion. J. Cons. Res. **24**, 315–328 (1977)
2. Aaker, J.L., Williams, P.: Empathy versus Pride: The Influence of Emotional Appeals across Cultures. J. Cons. Res. **25**, 241–261 (1998)

3. Agrawal, N.: Culture and persuasion. In: Ng, S., Lee, A.Y. (eds.) Handbook of culture and consumer behavior, pp. 176–192. Oxford University Press, Oxford (2015)
4. Agrawal, N., Maheswaran, D.: The effects of self-construal and commitment on persuasion. J. Cons. Res. **31**, 841–849 (2005)
5. Aronson, E.: The return of the repressed: Dissonance theory makes a comeback. Psychol. Inq. **3**, 303–311 (1992)
6. Beauvois, J.-L., Joule, R.-V.: A radical dissonance theory. Taylor & Francis, London (1996)
7. Berry, J.W., Poortinga, Y.H., Segall, M.H., Dasen, P.R.: Cross-cultural psychology. Cambridge University Press, Cambridge (1992)
8. Block, L.G., Williams, P.: Undoing the effects of seizing and freezing: Decreasing defensive processing of personally relevant messages. J. Appl. Soc. Psychol. **32**, 803–833 (2002)
9. Blondé, J., Girandola, F.: Faire « appel à la peur » pour persuader? Revue de la littérature et perspectives de recherche [Appealing to fear to persuade? Review of literature and research perspectives]. L'Année Psychologique/Topics in Cognitive Psychology **116**(1), 67–103 (2016)
10. Brehm, J.W.: Post decision changes. J. Abnorm. Soc. Psychol. **52**, 384–389 (1956)
11. Brewer, M.B., Chen, Y.R.: Where (who) are collective in collectivism? Toward conceptual clarification of individualism and collectivism. Psychol. Rev. **114**, 133–151 (2007)
12. Brewer, M.B., Gardner, W.: Who is this "We"? Levels of collective identity and self-representations. J. Pers. Soc. Psychol. **71**, 83–93 (1996)
13. Caligiuri, P., Noe, R., Nolan, R., Ryan, A.M., Drasgow, F.: Training, Developing, and Assessing Cross-Cultural Competence in Military Personnel (Technical Report 1284). Retrieved from United States Army Research Institute for the Behavioral and Social Sciences. (2011) http://www.dtic.mil/cgi-bin/GetTRDoc?Location=U2&doc=GetTRDoc.pdf&AD=ADA559500
14. Choi, I., Choi, K.W., Cha, J.-H.: A cross-cultural replication of the Festinger and Carlsmith (1959) study. Unpublished manuscript, Seoul National University, Seoul, Korea (1992)
15. Choi, S.M., Lee, W.-N., Kim, H.-J.: Lessons from the rich and famous: a cross cultural comparison of celebrity endorsement in advertising. J. Advertising **34**, 85–98 (2005)
16. Chung, H., Ahn, E.: The effects of fear appeal: A moderating role of culture and message type. J. Promot. Manag. **19**(4), 452–469 (2013)
17. Cialdini, R.B.: The science of persuasion. Sci. Am. Mind **284**, 76–84 (2004)
18. Cialdini, R.B., Wosinska, W., Barrett, D.W., Butner, J.: The differential impact of two social influence principles on individualists and collectivists in Poland and the United States. In: Wosinska, W., Cialdini, R.B., Barrett, D.W., Reykowski, J. (eds.) Practice of social influence in multiple cultures, pp. 33–51. Lawrence Erlbaum Associates, Mahwah, New Jersey (2001)
19. Cialdini, R., Wosinska, W., Barrett, D., Butner, J., Gornik-Durose, M.: Compliance with a request in two cultures: the differential influence of social proof and commitment/consistency on collectivists and individualists. Pers. Soc. Psychol. Bull. **25**, 1242–1253 (1999)
20. Cochrane, L., Quester, P.: Fear in advertising: The influence of consumers' product involvement and culture. J. Int. Consum. Mark. **17**, 7–32 (2005)
21. Cooper, J.: Cognitive dissonance theory: 50 Years of a Classic Theory. Sage Publications, New York (2007)
22. Crettiez, X.: « High risk activism »: essai sur le processus de radicalisation violente (première partie). Pôle Sud **1**(34), 45–60 (2011)
23. DeHoog, N., Stroebe, W., deWit, J.B.F.: The impact of fear appeals on the processing and acceptance of action recommendations. Pers. Soc. Psychol. Bull. **31**, 24–33 (2005)
24. DeHoog, N., Stroebe, W., deWit, J.B.F.: The processing of fear-arousing communications: How biased processing leads to persuasion. Soc. Influence **3**, 84–113 (2008)
25. Dillard, J.P., Anderson, J.W.: The role of fear in persuasion. Psychol. Mark. **21**, 909–926 (2004)
26. Doise, W.: Logiques sociales dans le raisonnement. Delachaux & Niestlé, Paris (1993)

27. Ekman, P., Friesen, W.V.: Constants across cultures in the face of emotion. J. Pers. Soc. Psychol. **17**, 124–129 (1971)
28. Ekman, P., Friesen, W.V., O'Sullivan, M., Chan, A., Diacoyanni-Tarlatzis, I., Heider, K., Krause, R., Ayhan LeCompte, W., Pitcairn, T., Ricci-Bitti, P.E., Scherer, K., Tomita, M., Tzavaras, A.: Universals and cultural differences in the judgment of facial expressions of emotion. J. Pers. Soc. Psychol. **53**, 712–717 (1987)
29. Eppright, D.R., Hunt, J.B., Tanner, J.B., Franke, G.R.: Fear, coping and information: A pilot study on motivating a healthy response. Health Mark. Q. **20**(1), 51–73 (2002)
30. Festinger, L.: A theory of cognitive dissonance. Stanford University Press, Stanford, CA (1957)
31. Fointiat, V., Girandola, F., Gosling, P.: La dissonance cognitive. Quand les actes changent les idées. Armand Colin, Paris (2013)
32. Fu, J.H., Chiu, C., Morris, M.W., Young, M.Y.: Spontaneous inferences from cultural cues: varying responses of cultural insiders, and outsiders. J. Cross Cult. Psychol. **38**, 58–75 (2007)
33. Green, E.C., Witte, K.: Can fear arousal in public health campaigns contribute to the decline of HIV prevalence? J. Health Commun. **11**(3), 245–259 (2006)
34. Gudykunst, W.B., Matsumoto, Y., Ting-Tooneyk, S., Nishida, T., Kim, K., Heyman, S.: The influence if cultural individualism-collectivism, self-construals, and individual values on communication styles across cultures. Hum. Commun. Res. **22**, 510–543 (1996)
35. Guéguen, N., Joule, R.V., Halimi-Falkowicz, S., Pascual, A., Fischer-Lokou, J., Dufourcq-Brana, M.: I'm free but I'll comply with your request: Generalization and multidimensional effects of the "Evoking Freedom" Technique. J. Appl. Soc. Psychol. **43**, 116–137 (2013)
36. Guéguen, N., Pasual, A.: Evocation of freedom and compliance: The "But you are free of…" Technique. Current Research in Social Psychology **5**(18), 264–270 (2000)
37. Guimond, S.: Psychologie sociale: Perspectives multiculturelle. Mardaga, Wavre (2010)
38. Guimond, S.: Soi, identité et culture. In: Bégue, L. Desrichard, O. (eds.) Traité de psychologie sociale: La science des interactions humaines, pp. 129–146. De Boeck, Bruxelles (2013)
39. Gurhan-Canli, Z., Maheswaran, D.: Cultural variations in country of origin effects. J. Mark. Res. **37**, 309–317 (2000)
40. Han, S.-P., Shavitt, S.: Persuasion and culture: advertising appeals in individualistic and collectivist societies. J. Exp. Soc. Psychol. **30**, 326 (1994)
41. Heine, S.J., Lehman, D.R.: Culture, dissonance, and self-affirmation. Pers. Soc. Psychol. Bull. **23**, 389–400 (1997)
42. Heine, S.J., Lehman, D.R.: The cultural of self-enhancement: An examination of group-serving biaises. J. Pers. Soc. Psychol. **72**(6), 1268–1283 (1997)
43. Heine, S.J., Norenzayan, A.: Toward a psychological science for a cultural species. Perspectives on Psychological Science **1**(3), 251–269 (2006)
44. Heine, S.J., Lehman, D.R., Markus, H.R., Kitayama, S.: Is there a universal need for positive self-regard? Psychological Review, 106, 766–794 (1999)
45. Hirose, Y., Kitada, T.: Actor's and observer's attributions of responsability and attitude change in the forced-compliance situation. Japan. J. Psychol. **56**, 262–268 (1985)
46. Hoeken, H., Korzilius, H.: Conducting experiments on cultural aspects of document design: Why and how? Communications **28**(3), 285–304 (2003)
47. Hofstede, G.: Culture and organizations: Software of the mind. Mc Graw Hill, London (1991)
48. Hofstede, G.: Culture's consequences: Comparing values, behaviors, institutions, and organizations across nations. Sage, Thousand Oaks, CA (2001)
49. Hofstede, G.: Dimensionalizing Cultures: The Hofstede Model in Context. Online Readings in Psychology and Culture, 2(1) (2011)
50. Hofstede, G., Bond, H.: Hofstede's cultural dimensions: an independent validation using Rokeach's value survey. J. Cross Cult. Psychol. **15**, 417–433 (1984)

51. Hornikx, J., O'Keefe, D.J.: Adapting consumer advertising appeals to cultural variables values: A meta-analysis review of effects on persuasiveness and ad liking. Annals of The International Communication Association **33**, 20–31 (2009)
52. Hoshino-Browne, E.: Cultural Variations in Motivation for Cognitive Consistency: Influences of Self-Systems on Cognitive Dissonance. Soc. Pers. Psychol. Compass **6**(2), 126–141 (2012)
53. Hoshino-Browne, E., Zanna, A.S., Spencer, S.J., Zanna, M.P.: Investigating attitudes cross-culturally: A case of cognitive dissonance among East Asians and North Americans. In: Haddock, G., Maio, G.R. (eds.) Contemporary perspectives on the psychology of attitudes, pp. 375–397. Psychology Press, New York (2004)
54. Hoshino-Browne, E., Zanna, A.S., Spencer, S.J., Zanna, M.P., Kitayama, S., Lackenbauer, S.: On the cultural guises of cognitive dissonance: The case of Easterners and Westerners. J. Pers. Soc. Psychol. **89**(3), 294–310 (2005)
55. House, R.J., Hanges, P.J., Javidan, M., Dorfman, P.W., Gupta, V.: Culture, Leadership, and Organizations: The GLOBE Study of 62 Societies. Sage Publications, Thousand Oaks (2004)
56. Hui, C.H.: Measurement of Individualism—Collectivism. J. Res. Pers. **22**(1), 17–36 (1988)
57. Imada, T., Kitayama, S.: Dissonance, self, and eyes of others in Japan and the US. Unpublished manuscript, University of Michigan, Ann Arbor (2005)
58. Imada, T., Kitayama, S.: Social eyes and choce justification: Culture and dissonance revisited. Social Cognition **28**(5), 589–608 (2010)
59. Iyengar, S.S., Brockner, J.: Cultural differences in self and the impact of personal and social influences. In: Wosinska, W., Cialdini, R.B., Barrett, D.W., Reykowski, J. (eds.) Practice of social influence in multiple cultures, pp. 13–28. Lawrence Erlbaum Associates, Mahwah, New Jersey (2001)
60. Iyengar, S.S., Lepper, M.R.: Rethinking the value of choice: A cultural perspective on intrinsic motivation. J. Pers. Soc. Psychol. **76**, 349–366 (1999)
61. Jacobson, J.A., Ji, L.J., Ditto, P.H., Zhang, Z., Sorkin, D.H., Warren, S.K., Legnini, V., Ebel-Lam, A., Roper-Coleman, S.: The effects of culture and self-construal on responses to threatening health information. Psychology & Health **27**(10), 1194–1210 (2012)
62. Jansen, C., Verstappen, R.: Fear appeals in health communication: Should the receivers' nationality or cultural orientation be taken into account? J. Intercul. Comm. Res. **43**(4), 346–368 (2014)
63. Jansen, C., Hoeken, H., Ehlers, D., van der Slik, F.: Cultural differences in the perceptions of fear and efficacy in South Africa. In: Swanepoel, P., Hoeken, H. (eds.) Adapting health communication to cultural needs: Optimizing documents in South African health communication on HIV and AIDS, pp. 107–128. Benjamins, Amsterdam (2008)
64. Jansen, C., van Baal, J., Bouwmans, E.: Culturally-oriented fear appeals in public information documents on HIV/AIDS: An extended replication study. J. Intercul. Comm. **11** (2006).
65. Jia Yan Lee, J., Jeyaraj, S.: Effects of self-construal differences on cognitive dissonance examined by priming the independent and interdependent self. Sage open, 1–9 (2014)
66. Jowett, G.S., O'Donnell, V.: Propaganda and Persuasion Examined. In: Jowett, G.S., O'Donnell, V. (eds.) Propaganda and Persuasion, 6th edn, pp. 179–230. Sage, London (2014)
67. Kaptein, M., De Ruyter, B., Markopoulos, P., Aarts, E.: Adaptive persuasive systems: A study of tailored persuasive text messages to reduce snacking. ACM Trans. Interact. Intell. Syst. **2**, 2 (2012)
68. Kessels, L.T., Ruiter, R.A.C., Jansma, B.M.: Increased attention but more efficient disengagement: Neuroscientific evidence for defensive processing of threatening health information. Health Psychol. **29**, 346–354 (2010)
69. Kim, H.S., Drolet, A.: Choice and Self-Expression: A Cultural Analysis of Variety-Seeking. J. Pers. Soc. Psychol. **85**(2), 373–382 (2003)

70. Kim, H.S., Drolet, A.: Express your social self: Cultural differences in choice of brand-name versus generic products. Pers. Soc. Psychol. Bull. **35**, 1–12 (2009)
71. Kim, H.S., Sherman, D.K.: "Express yourself": Culture and the effect of self-expression on choice. J. Pers. Soc. Psychol. **92**, 1–11 (2007)
72. Kim, H., Markus, H.R.: Uniqueness or deviance, harmony or conformity: A cultural analysis. J. Pers. Soc. Psychol. **77**, 785–800 (1999)
73. Kimel, S.Y., Grossmann, I., Kitayama, S.: When gift-giving produces dissonance: Effects of subliminal affiliation priming on choices for one's self versus close others. J. Exp. Soc. Psychol. **48**, 1221–1224 (2012)
74. Kimel, S.Y., Lopez-Duran, N., Kitayama, S.: Physiological Correlates of Choice-Induced Dissonance: An Exploration of HPA-Axis Responses. J. Behav. Decis. Mak. **28**(4), 309–316 (2014)
75. Kitayama, S., Uchida, Y.: Interdependent agency: An alternative system for action. In: Sorrentino, R., Cohen, D., Ison, J.M., Zanna, M.P. (eds.) Culture and social behaviour: The Ontario symposium, vol. 10, pp. 165–198. Erlbaum, Mahwah, NJ (2005)
76. Kitayama, S., Ishii, K., Imada, T., Takemura, K., Ramaswamy, J.: Voluntary settlement and the spirit of independence: Evidence Japan's "Northern Frontier". J. Pers. Soc. Psychol. **91** (3), 369–384 (2006)
77. Kitayama, S., Markus, H.R., Matsumoto, H., Norasakkunkit, V.: Individual and Collective Processes in the Construction of the Self: Self-Enhancement in the United States and Self-Criticism in Japan. J. Pers. Soc. Psychol. **72**(6), 1245–1267 (1997)
78. Kitayama, S., Snibbe, A.C., Markus, H.R., Suzuki, T.: Is there any free choice? Self and dissonance in two cultures. Psychol. Sci. **15**, 527–533 (2004)
79. Ko, D.M., Kim, H.S.: Message framing and defensive processing: A cultural examination. Health Commun. **25**, 61–68 (2010)
80. Kok, G., Bartholomew, L.K., Parcel, G.S., Gottlieb, N.S., Fernandez, M.E.: Finding-theory and evidence-based alternatives to fear appeals: Intervention mapping. Int. J. Psychol. **49**(2), 98–107 (2014)
81. Kokkoris, M.D., Kühnen, U.: Choice and dissonance in a European cultural context: The case of Western and Eastern Europeans. Int. J. Psychol. **48**(6), 1260–1266 (2013)
82. Kolodziej-Smith R., Friesen, D.P., Yaprak, A.: Does culture affect how people receive and resist persuasive messages? Research proposals about resistance to persuasion in cultural groups. Global Advances in Business and Communication Conference & Journal 2, Article 5 (2013)
83. Kreuter, M.W., Haughton, L.T.: Integrating culture into health information for African American women. Am. Behav. Sci. **49**, 794–811 (2006)
84. Kreuter, M.W., McClure, S.M.: The role of culture in health communication. Annu. Rev. Public Health **25**, 1–17 (2004)
85. Kruglanski, A.W., Gelfand, M.J., Bélanger, J.J., Sheveland, A., Hetiarachchi, M., Gunaratna, R.: The Psychology of Radicalization and Deradicalization: How Significance Quest Impacts Violent Extremism. Advances in Political Psychology **35**(1), 69–93 (2014)
86. Kruglanski, W.A., Webber, D.: The psychology of radicalization. Zeitschrift für Internationale Strafrechtsdogmatik **9**, 379–388 (2014)
87. Lapinski, M.K., Rimal, R.N., DeVries, R., Lee, E.L.: The role of group orientation and descriptive norms on water conservation attitudes and behaviors. Health Commun. **22**, 133–142 (2007)
88. Laroche, M., Toffoli, R., Zhang, Q., Pons, F.: A cross-cultural study of the persuasive effect of fear appeal messages in cigarette advertising: China and Canada. International Journal of Advertising: The Review of Marketing Communications **20**(3), 297–317 (2001)
89. Lee, A.Y., Bradford, T.W.: The effects of self-construal fit on motivation attitudes, and charitable giving. In: Ng, S., Lee, A.Y., (eds.) Handbook of culture and consumer behavior (pp. 193–215). Oxford: Oxford University Press (2015)
90. Lee, C., Green, R.T.: Cross-cultural examination of the Fishbein behavioral intentions. J. Int. Bus. Stud. **22**, 289–305 (1991)

91. Lee, H.S., Park, J.S.: Cultural orientation and the persuasive effects of fear appeals: The case of anti-smoking public service announcements. Journal of Medical Marketing: Device, Diagnostic and Pharmaceutical Marketing **12**(2), 73–80 (2012)
92. Leung, A.K., Cohen, D.: Within- and between- culture variation: individual differences and the cultural logics of honor, face, and dignity cultures. J. Pers. Soc. Psychol. **100**, 507–526 (2011)
93. Liberman, A., Chaiken, S.: Defensive processing of personal relevant health messages. Pers. Soc. Psychol. Bull. **18**, 669–679 (1992)
94. Markus, H.R., Kitayama, S.: Culture and the self: Implications for cognition, emotion, and motivation. Psychol. Rev. **98**, 224–253 (1991)
95. Markus, H.R., Kitayama, S.: A collective fear of the collective: implications for selves and theories of selves. Pers. Soc. Psychol. Bull. **20**, 568–579 (1994)
96. Markus, H.R., Kitayama, S.: Culture and selves: A cycle of mutual constitution. Perspectives on Psychological Science **5**, 420–430 (2010)
97. Markus, H.R., Kitayama, S., Heiman, R.J.: Culture and "basic" psychological principles. In: Higgins, E.T. Kruglanski, A.W. (eds.) Social psychology: handbook of basic principles pp. 857–913. Guildford Press (1996)
98. Matsumoto, D.: Cultural influences on the perception of emotion. J. Cross Cult. Psychol. **20**, 92–105 (1989)
99. Matsumoto, D.: Cultural similarities and differences in display rules. Motivation and Emotion **14**, 195–214 (1990)
100. McMath, B.F., Prentice-Dunn, S.: Protection motivation theory and skin cancer risk: The role of individual differences in responses to persuasive appeals. J. Appl. Soc. Psychol. **35** (3), 621–643 (2005)
101. Monden, K.: Attitude change and attitude recall in a consonant or dissonant: an attempt to integrate cognitive dissonance theory and self-perception theory. Japan. J. Psychol. **51**, 128–135 (1980)
102. Murphy, P.L., Miller, C.T.: Postdecisional dissonance and the commodified self-concept: a cross-cultural examination. Pers. Soc. Psychol. Bull. **23**, 50–62 (1997)
103. Murray-Johnson, L., Witte, K., Liu, W.Y., Hubbell, A.P., Sampson, J., Morrison, K.: Addressing cultural orientations in fear appeals: Promoting AIDS-protective behaviors among Mexican immigrant and African American adolescents and American and Taiwanese college students. J. Health Commun. **6**(4), 335–358 (2001)
104. Nielsen, J., Shapiro, S.: Coping with fear through suppression and avoidance of threatening information. Journal of Experimental Psychology: Applied **15**, 258–274 (2009)
105. Orji, R.: The impact of cultural differences on the persuasiveness of influence strategies. Proceedings of the 11th International Conference on Persuasive Technology, 38–41, (4-7 avril, Salzburg, Autriche) (2016)
106. Oyserman, D., Lee, S.W.S.: Does culture influence what and how we think? Effects of priming individualism and collectivism. Psychol. Bull. **134**, 311–342 (2008)
107. Pascual, A., Oteme, C., Samson, L., Wang, Q., Halimi-Falkowicz, S., Souchet, L., Girandola, F., Guéguen, N., Joule, R.V.: Cross cultural investigation of compliance without pressure: the "you are free to…" technique in France, Ivory Coast, Romania. Russia and China. Cross-Cultural Research **46**, 394–416 (2012)
108. Perea, A., Slater, M.D.: Power distance and collectivist/individualist strategies in alcohol warnings: Effects by gender and ethnicity. J. Health Commun. **4**(4), 295–310 (1999)
109. Petrova, P.K., Cialdini, R.B., Sills, S.J.: Consistency-based compliance across cultures. J. Exp. Soc. Psychol. **43**, 104–111 (2007)
110. Petty, R.E., Cacioppo, J.T.: Communication and persuasion: Central and peripheral routes to attitude change. Springer-Verlag, New York (1986)
111. Pornpitakpan, C., Francis, J.N.P.: The effects of cultural differences, source expertise, and argument strength on persuasion: An experiment with Canadians and Thais. J. Int. Consum. Mark. **13**, 77–101 (2000)

112. Rogers, R.W.: Cognitive and physiological processes in fear appeals and attitude change: A revised theory of protection motivation. In: Cacioppo, J.T., Petty, R.E. (eds.) Social Psychophysiology: A sourcebook, pp. 153–176. Guilford Press, New York (1983)
113. Ruiter, R.A.C., Kessels, L.T.E., Peters, G.J.Y., Kok, G.: Sixty years of fear appeal research: Current state of the evidence. Int. J. Psychol. **49**, 63–70 (2014)
114. Sakai, H.: Induced compliance and opinion change. Jpn. Psychol. Res. **23**, 1–8 (1981)
115. Scherer, K., Wallbott, H.G.: Evidence for universality and cultural variation of differential emotion response patterning. J. Pers. Soc. Psychol. **66**(2), 310–328 (1994)
116. Shavitt, S.: Persuasion and Culture: Advertising Appeals in Individualistic and Collectivistic Societies. J. Exp. Soc. Psychol. **30**, 326–350 (1994)
117. Shavitt, S., Lee, A.Y., Johnson, T.P.: Cross-cultural consumer psychology. In: Haugtvedt, C. P., Herr, P.M., Kardes F.R. (eds.), Handbook of consumer psychology pp. 1103–1131. Lawrence Erlbaum Associates, New York (2008).
118. Salter, S.B., Sharp, D.J., Chen, Y.: The moderating effects of national culture on escalation of commitment. Advances in Accounting, incorporating Advances in International Accounting **29**, 161–169 (2013)
119. Shen, L., Dillard, J.P.: Threat, fear, and persuasion: Review and critique of questions about functional form. Review of Communication Research **2**(1), 94–114 (2014)
120. Sherman, D.K., Nelson, L.D., Steele, C.M.: Do messages about health risks threaten the self? Increasing acceptance of health messages via self-affirmation. Pers. Soc. Psychol. Bull. **26**(9), 1046–1058 (2000)
121. Sherman, D.K., Uskul, A.K., Updegraff, J.A.: The role of the self in responses to health communications: a cultural perspective. Self & Identity **10**, 284–294 (2011)
122. Smith, P.B., Bond, M.H., Kagitcibasi, C.: Understanding social psychology across cultures: Living and working in a changing world. Sage Publications, London (2006)
123. So, J., Kuang, K., Cho, H.: Reexamining fear appeal models from Cognitive Appraisal Theory and Functional Emotion Theory perspectives. Commun. Monogr. **83**(1), 120–144 (2016)
124. Stark, E., Kim, A., Miller, C., Borgida, E.: Effects of including a graphic warning label in advertisements for reduced-exposure products: Implications for persuasion and policy. J. Appl. Soc. Psychol. **38**(2), 281–293 (2008)
125. Steele, C.M.: The psychology of self-affirmation: Sustaining the integrity of the self. In: Berkowitz, L. (ed.) Advances in Experimental Social Psychology, vol. 21, pp. 261–302. Academic Press, New York (1988)
126. Steele, C.M., Spencer, S.J., Lynch, M.: Self-image resilience and dissonance: The role of affirmational resources. J. Pers. Soc. Psychol. **64**, 885–896 (1993)
127. Stone, J., Cooper, J.: A self-standards model of cognitive dissonance. J. Exp. Soc. Psychol. **37**, 228–243 (2001)
128. Schwartz, S.H.: Beyond individualism/collectivism: New cultural dimensions of values. In: Kim, U., Triandis, H.C., Kagitcibasi, C., Choi, S.C., Yoon, G. (eds.) Individualism and Collectivism: Theory, Methods and Applications, pp. 85–119. Sage, London (1994)
129. Tannenbaum, M.B., Helper, J., Zimmerman, R.S., Saul, L., Jacobs, S., Wilson, K., Albarracin, D.: Appealing to fear: A meta-Analysis of fear appeal effectiveness and theories. Psychol. Bull. **141**(6), 1178–1204 (2015)
130. Taras, V., Steel, P., Kirkman, B.L.: Improving national cultural indices using a longitudinal meta-analysis of Hofstede's dimensions. Journal of World Business **47**, 329–341 (2012)
131. Terblanche-Smit, M., Terblanche, N.S.: HIV/AIDS marketing communication and the role of fear, efficacy, and cultural characteristics in promoting social change. J. Public Aff. **11**(4), 279–286 (2011)
132. Thibodeau, R., Aronson, E.: Taking a closer look: reasserting the role of the self-concept in dissonance theory. Pers. Soc. Psychol. Bull. **18**, 591–602 (1992)
133. Trafimow, D.: A theory of attitudes, subjectives norms, and private versus collective self-concepts. In: Terry, D.J., Hogg, M.A. (eds.) Attitudes, behavior, and social context: the role of norms and group membership, pp. 47–65. Lawrence Erlbaum, Mahwa, NJ (2000)

134. Triandis, H.C.: The psychological measurement of cultural syndromes. Am. Psychol. **51**, 407–415 (1996)
135. Triandis, H.C., Chen, X.P., Chan, D.K.: Scenarios for the measurement of collectivism and individualism. J. Cross Cult. Psychol. **29**(2), 275–289 (1998)
136. Trompenaars, F.: Riding the Waves of Culture: Understanding Diversity in Global Business. Irwin Professional Publishing, Chicago (1993)
137. Uskul, A.K., Oyserman, D.: When message-frame fits salient cultural-frame, messages feel more persuasive. Psychology & Health **25**(3), 321–337 (2010)
138. Uskul, A.K., Oyserman, D.: When message-frame fits salient cultural-frame, messages feel more persuasive. Psychology & Health **25**, 321–337 (2010)
139. Uskul, A.K., Sherman, D., Fitzgibbon, J.: The cultural congruency effect: culture, regulatory focus, and the effectiveness of gain- vs. loss-framed health messages. J. Exp. Soc. Psychol. **45**, 535–541 (2009)
140. Vincent, A.-M., Dubinsky, A.J.: Impact of fear appeals in a cross-cultural context. The Marketing Management Journal **15**(1), 17–32 (2005)
141. Wang, C.C.L.: Right appeals for the 'right self': connectedness-separateness self-schema and cross-cultural persuasion. J. Mark. Commun. **6**, 205–217 (2000)
142. Wang, C.L., Mowen, J.C.: The separateness-connectedness self-schema: Scale development and application to message construction. Psychol. Mark. **14**, 185–207 (1997)
143. Wilson, S., Cai, D., Campbell, D., Donohue, W., Drake, L.: Cultural and commutation processes in international business negotiations. In: Nicotera, A. (ed.) Conflict in organizations: Communicative processes pp. 169–188. State University of New York Press, Albany (1994)
144. Witte, K.: Putting the fear back into fear appeals: The Extended Parallel Process Model. Commun. Monogr. **59**, 329–349 (1992)
145. Witte, K., Allen, M.: A meta-analysis of fear appeals: Implications for effective public health campaigns. Health Educ. Behav. **27**, 591–616 (2000)
146. Zhang, J., Shavitt, S.: Cultural values in advertisement to the Chinese X-generation: promoting modernity and individualism. Journal of Advertising **25**, 29–46 (2003)
147. Zou, X., Tam, K.-P., Morris, M.W., Lee, S.-L., Lau, I.Y.-M., Chiu, C.-Y.: Culture as common sense: Perceived consensus versus personal beliefs as mechanisms of cultural influence. J Pers Soc Psychol **97**(4), 579–597 (2009)

Chapter 19
The Influence of Emotion and Culture on Language Representation and Processing

Dana M. Basnight-Brown and Jeanette Altarriba

Abstract Research focused on the study of emotion, specifically how it is mentally represented in the human memory system, is of great importance within the study of cognition. The current chapter will examine the factors that make emotion words unique, as compared to other word types (e.g., concrete and abstract words) that have traditionally been of interest. In particular, key findings from studies where cognitive paradigms were used to explore emotion are emphasized (e.g., Stroop tasks, priming, implicit memory tests, eye tracking, etc.). This chapter will describe the factors that influence how those who know and use more than one language process *and* express emotion, and the role that language selection plays on the level of emotion that is activated and displayed. Finally, cross-cultural differences in emotion are examined, primarily as they relate to differences in individualistic and collectivistic contexts.

Keywords Bilingual · Emotion · Emotion laden · Emotion stroop
Cross-cultural · Artificial intelligence

19.1 Introduction

Researchers in the field of emotion have approached emotion representation from a myriad of perspectives—physiological, linguistic, developmental, psychological—just to name a few [2]. There is no doubt that the processing of emotional stimuli influences mental health and well-being and that knowledge of how emotions are

---

D.M. Basnight-Brown (✉)
United States International University - Africa, PO Box 60875
City Square 00200, Nairobi, Kenya
e-mail: dana.basnightbrown@gmail.com

J. Altarriba
Department of Psychology, Social Science 399, University at Albany,
State University of New York, Albany, NY 12222, USA
e-mail: jaltarriba@albany.edu

C. Faucher (ed.), *Advances in Culturally-Aware Intelligent Systems and in Cross-Cultural Psychological Studies*, Intelligent Systems Reference Library 134, https://doi.org/10.1007/978-3-319-67024-9_19

displayed, represented, and revealed by others allows us to respond in appropriate ways across diverse settings [29]. In fact, we know that emotional language and emotion concepts are some of the earliest concepts learned, and that they are learned through contexts in which they are first experienced and first practiced [1]. If one considers the ways in which language helps to code emotional expression, then it becomes even more interesting to examine the ways in which multiple languages are used to express emotions in a multilingual speaker [14, 18, 35]. While concrete objects or concepts often have a clear one-to-one correspondence across languages, it is known that words used to express emotion may be unique in a given language or at the very least, difficult to translate into a single concept or word, in a different language. Thus, languages carry a richness in emotional content that makes it difficult to engender that emotion into a new learner or even an artificial mode of intelligence [42]. Culture also affects the ways in which languages are used as a vehicle to express emotion [57]. For example, particular words that can be a reflection of one's past experiences, say in a particular phrase dealing with romantic encounters may be viewed as "taboo" in one culture, while being acceptable in a different culture. Therefore, language often moderates the expression of emotion from culture to culture. Thus, it is clear that both language and culture are intertwined and should be considered when discussing everyday cognitive processing and basic communication.

The aims of the current chapter are to introduce the reader to recent work on the processing of emotion words in language and communication, the ways in which culture influences how emotions are learned, stored, and retrieved, and the ways in which emotional concepts may be represented differently in memory by speakers of more than one language [28]. Research in the domains of cognitive psychology, cognitive science, linguistics, and psycholinguistics will be reviewed in order to discuss how researchers have approached these topics from a scientific perspective and what has been learned to date, with regards to these areas of inquiry. An important question that guides this chapter pertains to the ways in which research on emotion word representation, in particular, informs theories of thinking and information processing that readily lead to actual human behaviors [47, 48]. The section below delimits the ways in which emotion words are represented and how they may be distinguished from other units in language, both within and between languages.

19.2 Emotion Words as a Distinct Word Type

Systematic studies of the lexical properties that characterize emotion words (e.g., *love, fear*) reveal several linguistic differences. In an analysis of 1033 words used in 32 published studies on emotion processing, Larsen, Mercer, and Balota [40] reported that negative and positive emotion words tend to be longer in length, have lower word frequencies, and smaller orthographic neighborhood sizes. In addition, concrete words (e.g., *table, pencil*) are typically characterized as being high in concreteness and imagery, while abstract words (e.g., *liberty, myth*) are characterized as being low in concreteness and imagery. Ratings collected in which concrete, abstract, and emotion words were compared, revealed that emotion words are rated higher in imageability and lower in concreteness as compared to abstract words [6, 7]. These findings indicate that emotion words have lexical characteristics that differ from both concrete and abstract words, which suggest that emotional stimuli are distinct. In terms of semantic representation, it has been suggested that emotion words are connected to more words in the lexicon, and have more synonyms as compared to abstract words. For example, Altarriba and Bauer [6] explain that the word *happy* has synonyms such as *delighted, pleased, ecstatic,* and *glad*, while the abstract word *hour* does not have many synonyms (p. 406). The idea that emotion words are more highly connected to each other and have more connections, in general, is further supported by data from a free association task in which participants generated more associations to emotion words as compared to both concrete and abstract words [7].

Emotion words (e.g., *sad* or *happy*) typically label a state of mind that can be directly experienced. For example, one can say, "I feel happy." This is a state of being that can be described by the use of a single word. In fact, this particular emotion word is one that can be universally understood and is translatable across virtually all known languages. Some concepts, however, are associated with an emotional state, and these concepts have come to be connected with certain emotions or feelings, such that those reactions are elicited through the semantic activation of those words (e.g., *death* or *cancer*). These words are known as emotion-laden words, and recent research examining processing mechanisms of emotion and emotion-laden words reveals that they differ on a variety of characteristics [4, 36–38]. Knickerbocker and Altarriba [37] for example, examined the ways in which emotion and emotion-laden words affect responses in a perception-based task that produces an effect known as Repetition Blindness (RB). Research has indicated that when a given word is repeated within a phrase that has been presented word-by-word in quick succession (e.g., 117 ms per word), individuals tend to omit the second representation of that word in their recall of the sentence. Thus, the sentence, "I like steak but this steak tastes awful," might result in a recall response of, "I like steak but it tastes awful." The presentation of the second item is not perceived as a separate event, when both repetitions are presented in close proximity. When emotion and emotion-laden words are used as repetitions, the data indicate that emotion words show a greater RB effect as compared to emotion-laden words. That is, the processing of an emotion word created a strong enough memory trace, based on its valence, that a second instance was less likely to be perceived as a new or novel instance [see related work on the role of valence, 38]. Emotion words appear to have a stronger arousal component and in some situations, can be said to be a "purer" representation of an emotion, as compared to emotion-laden words, where the emotional state is mediated via the words [43, for a discussion on the role of arousal in the representation of emotional words]. Again, this is a comparison between a direct representation of an emotional component and an indirect representation, as in the case of an emotion-laden word. Indeed, most of the extant research on the ways in which emotion words are represented in memory indicates that they exert a stronger influence on linguistic processing and behavior than do emotion-laden words. Most likely, the distinction arises from the fact that emotion words have an early age of acquisition, and as a result are attached to a life experience that is also highly arousing, such as when a small child is repeatedly told, "No!" as he or she reaches for an item they cannot have. In contrast, emotion-laden words require more time and a greater number of instances until they take on the nuance of being positive or negative, and the emotional intensity that comes with experiencing that item over varied contexts and across time begins to increase. Furthermore, research has indicated that the intensity accorded emotion words may be stronger in the native language than in the second language for bilingual speakers [1]. This is likely due to the fact that these first instances of connecting a word to an emotional experience occurred in the native language, often in situations that were highly arousing, physiologically, so as to bind that particular memory to that language, and in particular to a given word or phrase [30]. There is a long line of research confirming that these memories appear to be durable and perhaps more poignant when used and reused in the language in which they were first encountered.

The importance of distinguishing emotion words from emotion-laden words cannot be underestimated, as a review of many published works focusing on emotion word stimuli, reveals that the two types of words have been unsystematically intermixed. Research indicates that this may have influenced the reported effects, as indeed, these two word types affect behavior in measurably different ways. For example, in a recent paper, Knickerbocker et al. [38] asked participants to read sentences that contained neutral target words (e.g., *table*), positive emotion words (e.g., *happy*) or negative emotion words (e.g., *distressed*) while their eye movements were recorded. They found that both emotion word types were easier to process (e.g., as measured by shorter first fixations) than neutral words. These findings were in contrast to those of Scott, O'Donnell, and Ṣereno [60] in which only positive emotion words were found to facilitate eye movements. In this latter study, both emotion and emotion-laden words were randomly intermixed, providing a situation in which the mixing of word types may have influenced reading behavior trial-to-trial and diminished the effects that were viewed in Knickerbocker et al., particularly with regards to negative emotion words. Several other demonstrations have provided ample data to suggest that emotion words are powerful in terms of their ability to draw and maintain attention [63] moderate memory effects within and between languages [3], and influence the ways in which concepts are translated into actual products or objects [41]. In the following section, the manner in which culture interacts with emotion and emotional language will be explored with an eye towards the ways in which emotional concepts themselves may be supra-linguistic in nature.

19.3 Cross-Cultural Differences in the Processing of Emotions

It is evident that the processing of emotion is an extremely important component in the human cognitive and information processing system, and that the different types of emotional stimuli must be carefully examined if one is going to consider differences in how emotional content is encoded and represented in memory [5]. In addition to the study of emotion, we also know that culture plays an important role in how individuals store and process information, and in how they interact with the world. The study of culture has been the focus of thousands of scientific papers, spanning a diverse set of academic disciplines, therefore, we know that culture plays an important role in defining how people view the world, everyday behaviors, what they eat, wear, social interactions, etc. However, it is not only these more overt distinctions that have been observed, research has even pointed to cognitive and neurological differences across individuals as a result of varied cultural backgrounds [53, for a review]. Naturally, out of the study of culture and emotion in separate contexts, researchers gained interest in the intersection between emotion and culture, specifically in terms of whether emotion is processed similarly or differently across cultures.

The study of how individuals from different cultures process emotion extends back several decades to Ekman's eminent work on the universality of emotions [25, for a detailed review of this body of literature]. Early on in his research, he noted the more obvious finding that almost all human beings are able to read facial expressions and to determine how people are feeling based on this process. However, this led Ekman to become most interested in whether the emotional expressions that one perceives were the *same* across cultures. As a result, much of his influential research focused on important questions surrounding human communication: (a) is this process universal? and (b) can we interpret the same emotions from people of all cultures?

In some of the earliest work examining the interaction between emotion and culture, Ekman observed that individuals from various countries (e.g., Argentina, Brazil, Chile, Japan, United States) all appeared to perceive emotional expressions in a similar way [25]. However, there was the concern that these outcomes were influenced by the fact that these cultures had all been exposed to Western media through movies, news, TV, etc., which may have been responsible for the similarities observed. As a result, Ekman and Friesen [26] traveled to New Guinea to examine the processing of emotion in the Fore people, a group who had supposedly not been exposed to Westerners (or to Western type media). In this early, yet still highly influential study, the authors examined the processing of emotion in both children and adult non-English speaking populations. They reported that there were no differences between the Westernized and Nonwesternized groups, and perhaps more importantly, that the Fore seemed to make the same 6 facial expressions they had observed in other cultures. These 6 emotions included: happiness, sadness, anger, surprise, disgust, and fear, all of which came to be known as the classic emotion categories used in future research in the cognitive science domain. Finally, in order to determine the extent of the "universal emotion effect", Ekman and Friesen reported that when university students from the United States were shown facial expressions of the Fore, the students were able to correctly identify these 6 emotions, suggesting that this finding was bi-directional and providing even more support that individuals from different cultures were able to process emotion in a similar manner.

Although many studies exploring the relationship between culture and emotion have revealed strong similarities across cultures, there are more recent studies which point to differences in the way in which individuals from different cultures respond, particularly studies which examine processing at the cognitive and neurological level. For example, recent research using fMRI data has shown that Chinese participants activate the same brain areas (e.g., medial prefrontal cortex) when describing themselves, as well as when describing family members (i.e., in terms of personal characteristics or emotions displayed). Western participants, in contrast, did not show this pattern of activation when describing family members [68]. The authors suggest that even at a neurological level, the values of each culture emerge, as the manner in which the Chinese tend to perceive themselves is based on their relationship with others, while Westerners tend to be more focused on independence and self-sufficiency [51, for an earlier review]. Interestingly, this even extends to the way in which emotions are expressed across cultures (as compared to just perceived). Murata, Moser, and Kitayama [49] examined whether Asians, which they described as being "culturally trained to down regulate emotional processing" in order to fit with characteristics that are valued more heavily within their culture, may be led to express emotion differently [51]. Murata et al. [49] suggest that there are many studies which have revealed that within Asian cultures, low arousal emotions tend to be valued, such that individuals are trained to remain calm in stressful situations. In contrast, many Westerners are often encouraged to show their emotions and to express what they are truly feeling. In order to examine the influence of culture on emotion at a neurological level, Murata et al. [49] used event-related brain potentials (ERPs) to measure the amount of emotional processing that occurred when participants viewed pictures that were high, low, or neutral in arousal. In one condition (i.e., attend condition), American and Asian participants were asked to react normally to each picture, while in the second condition (i.e., suppress condition), they were asked to suppress any emotions they felt when viewing the pictures. As expected, the results indicated clear differences across the two groups, such that the Asian participants showed decreased neurological activity in the suppression condition, suggesting that not only does this "down regulation of emotions" emerge overtly, but that it can be measured at a neurological level as well.

In addition to neurological differences across cultures that have been captured in emotional processing research, social factors relating to "in group" status within a community can also affect how emotional content is perceived. For example, Elfenbein and Ambady [27] reported that emotions are easier to perceive and recognize (i.e., as measured by greater accuracy) when individuals are from the same cultural group (e.g., *in group*), as compared to being from a different cultural group. Therefore, evidence from some of the earliest studies designed to examine the role of culture on the processing of emotion suggest that there are elements that are universal, specifically the key emotions that every culture seems to be able to express and perceive in a similar manner. However, it is evident from some of the more recent research that there are certain differences between people, a finding that is influencing people from simply examining emotion and culture in terms of similarities and differences (e.g., nature vs. nurture focus), to viewing the issue in terms of how universal processes may be *affected* by culture. This has been confirmed across a variety of experimental paradigms and with stimuli that describe both positive and negative emotions [47]. Finally, in terms of emotional *expression*, those in Western cultures with lower social status have been reported to express more anger. Park et al. [52] explored this in both Eastern (i.e., Japanese) and Western populations, and observed a dissociation between the motivation behind the anger that individuals expressed. In American samples, anger was expressed as a result of being frustrated, while for the Japanese, those with higher social status expressed more anger, as higher social status provided them with greater freedom which they felt allowed them to express their anger. These cultural factors that affect emotional expression are further complicated by additional research which has shown that some bilinguals prefer using their L1 to express anger toward themselves, family members, and other individuals, while the L2 is preferred when expressing anger in written formats [21].

When speaking of cultural differences, we must weigh them with caution, as there are additional factors that also mediate these effects or interact with them in unique ways. Several studies have pointed to additional components that should be considered when examining the relationship between emotion and culture, in light of the findings to date. One very important issue is the distinction between accurately perceiving the *same* emotion versus categorizing emotion, which may not actually tap into the same cognitive processes. As Russell [58] so aptly pointed out in his extensive review on culture and its influence on the categorization of emotions, many of the early studies which required participants to determine which emotions were being expressed on faces, used a forced choice facial recognition task. As a result, this type of task only requires individuals to select which emotion best fits that seen on the faces, but is limited in that it cannot show the equivalence of emotion concepts that may or may not exist in different cultures. Russell presents the following example where an individual shown a picture of a woman with a bright smile is asked to select the emotion she is feeling. He suggests that most would likely select *happy*, but if that word is not present, perhaps one would be forced to pick *elated,* as ultimately the person does not have the flexibility to choose the emotion that they truly feel is represented. If *happy* was replaced with another positively valenced word, perhaps that word would be selected instead [65, p. 435]. This is an important distinction that Russell highlights, and one which needs to be considered when interpreting the findings from these studies, as a judgment task is likely not sensitive enough to differences between certain emotions. Therefore, although results obtained from these types of studies are interesting and informative at some level, it may be more accurate to conclude from that body of research that people from different cultures *interpret* facial expressions in a similar way [58].

This issue, concerning differences in how emotions extend across cultures and their degree of overlap in meaning across cultures introduces the important factor of language into the study of emotion and culture. As noted before in Russell's example, *happy* and *elated* do not express the same emotion, yet the subtle differences in this more global positive emotion only emerge when one examines the language used (i.e. choosing to describe someone as *happy* vs. *elated*). This distinction in how emotional concepts translate to other languages and cultures will be discussed in more depth in one of the following sections. However, the role of language and how it intersects with culture is a crucial component in the study of the perception and expression of emotion. In an interesting study of languages around the world, Dodds et al. [22] examined whether there is a universal positivity bias in human communication (referred to as a *Pollyanna Hypothesis*). For the languages examined, they focused on the language used in various news, social media, web data, TV, music lyrics, etc. and reported a positivity bias in human communication, so much so that data on Twitter measurements correlated with well-being polls for some languages and populations.

Finally, several other components of interest in the study of emotion and culture which have recently been reported in the literature focus on the development of new emotions that might not have been examined before, and the role of social factors, like social status on emotional expression. In a study designed to examine the extent to whether the 6 basic emotions outlined earlier still hold, Du et al. [23] suggest that in addition to these 6 basic emotions, humans also produce and perceive compound emotions, which are described as combinations of these 6 key emotions (e.g., being *fearfully surprised*). In their research, they report that there are 15 of these combinations (or "new emotions" as they refer to them). They suggest that this gives cognitive researchers the ability to study how the brain's emotional system processes 21 emotions, as compared to only 6. The practical implications of this are wide reaching, in that it has been suggested that this information can be "used to create better human-computer interaction systems" [67]. As noted earlier, there is a strong interrelationship between emotion, language, and culture; thus, emotional concepts can be expressed in more than one language for a multilingual speaker. The next section explores the ways in which emotion knowledge is represented differentially across more than one language for bilingual and multi-lingual populations.

19.4 How Do Those Who Know and Use More than One Language Process Emotion?

In recent years, interest in the processing of emotional stimuli for individuals who know and use two or more languages has increased considerably. Research in this domain has focused on how emotion words are recognized, their influence on memory, whether they capture and hold attention more than neutral items, and characteristics that separate them from other word classes (such as abstract and concrete words). The emotion Stroop paradigm has been used in dozens of studies with monolingual participants to examine whether emotional stimuli capture attention automatically [see 74 for a review of the emotion Stroop literature]. In this task, neutral (i.e., concrete words) and emotion words are presented in colors to which participants are instructed to name the ink color or make a key press decision denoting the ink color of each item presented. Like the original Stroop task [62], this paradigm rests on the assumption that when a word appears on the computer screen, the semantic meaning of the item is automatically activated. Moreover, emotion words have been associated to specific colors, and that association can drive performance in these tasks [64]. Results from the emotional Stroop task consistently reveal that individuals are slower to name the ink color of emotion words (e.g., specifically negative emotion words—*sad, angry*, etc.) [43, 50] as compared to neutral words (e.g., *dog, lion*, etc.). This *interference effect* is thought to occur because negative emotional stimuli trigger a defense mechanism that "freezes activity" in the presence of threatening information, resulting in a processing delay. In contrast, several studies have revealed that both positive and negative words are processed *faster* than neutral words when the task involved lexical decision [39, 66]. This outcome provides support for theories based on motivated attention, which propose that positive and negative stimuli are processed faster, facilitating goal obtainment [39]. Finally, other studies focused on the cognitive processes surrounding memory for emotional information have reported better recall for emotion words and for emotional narratives in monolingual populations [see 6].

Interestingly, the study of emotion word processing has been extended to bilingual populations, in an effort to determine whether emotional stimuli are represented in memory in a similar way for each language. Researchers have suggested that emotion words are encoded and expressed more deeply in a bilingual's L1, as compared to their L2 [16, 20, 53]. For example, Dewaele [17] reported that the phrase *I love you* was perceived with more emotional weight (felt strongest in) the bilingual's L1 as compared to their L2. It has also been reported that bilingual individuals use fewer emotion words during L2 discourse, as compared to L1 discourse [55]. Additional support for differences in emotion representation between the L1 and L2 have even been observed in marketing research, such that advertising slogans presented in the L1 are viewed as more emotional than those that appear in the L2 [56]. In contrast, the L2 is often described as being the more emotionally distant language, such that bilinguals will switch to this language when they want to distance themselves emotionally from an event [8].

The processing of emotional stimuli in bilinguals has focused on lower-level cognitive processes (e.g., emotion Stroop tasks) and biological responses, as well as higher order, semantic processes that influence the memory system. Physiological measures of emotional influence, such as skin conductance responses (SCRs), reveal stronger SCRs in one's L1 as compared to their L2, suggesting that the emotional representation of a word in the bilingual's first language exerts a stronger biological response [30, 32]. In line with this finding, Colbeck and Bowers [13] observed that it was easier for bilinguals to ignore emotional stimuli in the L2, suggesting that emotion words in a second language do not always activate the same level of attention.

Further demonstrations of the difference in L1/L2 memory representation come from the emotion Stroop task, introduced previously, which has been used to explore how emotion words are processed in the L1 and L2. Sutton, Altarriba, Gianico, and Basnight-Brown [65] investigated emotional Stroop effects in Spanish-English bilinguals. They examined response latencies to negative and neutral color words that appeared in both Spanish and English. Significant interference effects (i.e., slower response times to emotion words) were observed in both languages. Specifically, the size of the interference effects did not differ across languages, which indicates that emotion words are capable of capturing attention in a bilinguals' two languages (i.e., at least for early bilinguals who were highly proficient in both languages). In a second demonstration of emotional Stroop processes in bilinguals, emotion word processing was tested in Finnish-English bilinguals, who also exhibited high levels of proficiency in their two languages [24]. Eilola et al., like Sutton et al., observed significant interference effects in both L1 and L2.

The influence of emotional content and bilingualism on memory has provided several additional results that shed light on the cognitive system of bilingual speakers. For example, free recall revealed that emotion words in a bilingual's L1 were remembered better than in the L2 [9]. Further, retrieval of autobiographical memories was shown to be dependent on the language and context present during encoding (e.g., *Language Dependent Memory Hypothesis*). Marian and Neisser [45] reported that Russian-English bilinguals accessed more autobiographical memories when the language of "questioning" (i.e., interview language) matched that used when the event occurred. In a second study, ratings conducted on the emotional content of autobiographical memories retrieved during an interview session indicated that memories in which encoding and retrieval were matched showed stronger emotional intensity and emotional arousal [44]. These findings clearly show that knowing and using a second language has the ability to influence general cognitive systems, such as memory retrieval and perception of personal events.

In summary, the study of emotion in bilingual populations indicates that emotional concepts are represented differently in each of a bilingual's languages, a finding that is most likely due to the different contexts in which bilinguals learn each of their languages [1, 16, 53, 54]. Context seems to be the key component here, as compared to proficiency of a language, which may be a stronger determinant of second language processing in cognitive tasks that do not explicitly measure the processing of emotion. As Caldwell-Harris so aptly suggests, it is important "not to emphasize proficiency or frequency as root causes, but to propose that words and phrases accrue emotional resonances when they have been learned and used in emotional contexts. This explains why bilinguals could use the same language with similar levels of proficiency and frequency, but experience different levels of emotionality" [12, p. 216]. Building up this important distinction that occurs in multi-language emotional development, Sheikh and Titone [61] hypothesized that negative emotions are enhanced in one's native language, but that positive emotions are enhanced in a foreign language because social interactions as an adult tend to be more positive. This was tested using eye tracking methodology to study natural reading processes in a bilingual population, whereby Sheikh and Titone [61] observed that the processing of positive and negative information was influenced by linguistic, emotional, and sensorimotor information during early stages of activation. While these studies speak to the representation of emotional language in the brain and in memory, an interesting question revolves around the ways in which those who know more than one language actually express emotion in their varied languages in day-to-day communication. This is the topic of the next section below.

19.5 How Do Those Who Know and Use More than One Language Express Emotion?

As noted earlier, Dewaele [17, 18] set out to explore the ways in which the phrase "I love you," was weighted differently across different languages for multilingual speakers. This type of work speaks to the pragmatics of actual emotion word usage when communicating common emotional phrases, thus a closer look at this study is in order. This work consisted of participants completing an online questionnaire with open and closed questions regarding language behavior and emotions. Participants actually rated the strength with which they perceived emotionality in the above phrase in either their first language or a foreign language. Across various analyses with items in various domains, the data indicated that a majority of speakers felt the phrase, "I love you" was strongest in their first or native language [see 75, for differing results that rest on a closer examination of gender and pragmatics]. These speakers represented 77 different first languages in a sample of close to 1500 participants. These findings closely parallel those discussed earlier in which context plays a distinct role in the encoding and learning of emotional language, and when done so in a native language, the co-occurring physiological reactions tend to create a very long-lasting memory trace that also activates the physical responses that co-occurred when that emotion was first labeled [1]. Note, however, that when a given language equates emotional terms with taboo topics, it then becomes the case that their perceived intensity in the native language is a deterrent to using that term because of the strength of that representation [31]. Thus, valence is an important factor moderating the ways in which emotional language, however intense, is used or avoided in basic conversation.

The relationship between emotion and communication is clearly influenced by culture and society, particularly in the case of taboo words. A speaker typically needs to finesse the ways in which pragmatics and sociolinguistic mores work into a computation as to which linguistic sequences to produce and when to produce them. Factors such as gender, race, ethnicity, and word knowledge influence the decision to utter a particular word or phrase and moderate word selection particularly in situations that give rise to emotional language [33, 34]. Dewaele [19] investigated the ways in which speakers of American and British English self-reported their frequency of swearing and their level of perceived offensiveness of 30 negative words extracted from the British National Corpus (e.g., *jerk*). While no significant differences in frequency of swearing arose from the close to 1000 participants overall, in this study, it was the case that the British participants reported a better understanding of the words (though they were known to both language populations) yet reported frequent usage of only about one-third of the words. High frequency of usage for the speakers of American English, in contrast, was reported for about half of the words. Thus, speakers of American English, in this sample, used a larger vocabulary set that was described as offensive or taboo, as compared to speakers of British English for whom this vocabulary appeared to be better understood and more familiar. This suggests that variables that are influenced by culture and current societal norms seem to moderate the pragmatics concerning the actual use of this type of language.

The use of phrases or words that denote emotional states rests on the ways in which those words are organized in memory. For example, as mentioned earlier, Altarriba and Bauer [6] argued that emotion words may have a broader array of associations in memory, as compared to concrete and abstract words. Specifically, when asked to generate words that are directly associated to an emotion word, a larger group of distinct words is produced—most of them emotional—as compared to the same instructions for concrete or abstract words. As a result, one can consider that the mental network of emotion words is broad and interconnected such that activation of a given word in memory may automatically activate an array of words from the same category with similar characteristics or features. Given this is the case, a natural question that emerges is to what extent do these words have multiple translations across languages? That is, is it the case that emotion words also have many more translations across languages as compared to concrete and abstract words? Basnight-Brown and Altarriba [10] explored the ways in which Chinese-English and Spanish-English bilinguals translate concrete, abstract, and emotion words. This work represents the first empirical study focused on how emotion words, in particular, are translated across languages. Prior research had uncovered the notion that abstract words appeared to have far more translations than concrete words, though emotion words had never been investigated in this kind of context. In the present study, participants were presented with a list of 150 concrete words (e.g., *apple*), 151 abstract words (e.g., *thought*), and 43 emotion words (e.g., *happy*). Each English word was paired with the dominant translation in Spanish or in Mandarin. Participants were presented with one or the other word from the translation pair, depending on their known languages, and asked to translate into the other language. When the number of translations was based on the total number of different translations produced across participants, emotion words yielded a significantly higher number of translations than concrete and abstract words in both directions, for Spanish-English bilinguals. For Chinese-English participants, the same findings were reported, in both translation directions. These findings parallel those reported by Altarriba and Bauer's [6] observation that single emotion words produced a higher number of associations within language (e.g., English), as compared to concrete and abstract words, for monolingual English speakers. Therefore, when one examines cross-language access of emotion concepts, it appears that words in a given language elicit more translations across languages, as compared to non-emotional word stimuli [11]. Thus, not only are emotion words highly interconnected within languages, they are also connected to many words *between* languages.

19.6 Applications and Conclusions

The current chapter explored the ways in which language that expresses emotion is represented within and between languages, how those emotional representations are structured in memory, and the pragmatics and cultural factors that affect its use in basic communication. Clearly, the current empirical literature provides a strong case that words that label emotional states (e.g., *happy*, *sad*, *afraid*) are stored in richly interconnected networks within and between languages, carry a strong intensity in terms of their ability to influence human behavior and language use, and are often distinct in terms of their pragmatic use across different language communities and cultures. It is also the case that knowing how emotional language is stored and used can play important roles in everyday life, as given a particular context, a speaker may decide to swear, for example, in a native versus a second language (or vice versa) or hide their emotions behind more neutral language, if this is perceived to be more acceptable in a given societal context [12]. It is also the case that language has been shown to have a direct role in how individuals make moral decisions, and that language is used as a vehicle to make more logical versus more emotional decisions when presented with moral dilemmas [15]. Finally, knowledge of how emotion is coded in a native versus a second language can inform methods of interviewing individuals, either those who serve as eyewitnesses to a crime, for example, or those who are seeking to communicate issues surrounding their mental health in therapeutic settings [46, 59].

Work on emotion has many implications for the fields of business, education, product design, and computer science/robotics. In the field of industrial design, for example, Leblebiçi-Basar and Altarriba [41] were interested in discerning the principles whereby designers translate a concept into an actual form. They asked 12 Turkish industrial designers to consider three different concepts; one concrete (chestnut), one abstract (loyalty), and one emotion (grief) and to then use these concepts to design a perfume bottle. Each designer who was experienced in their training and expertise, was provided with a design brief asking them to design the product with a specific concept in mind. They were asked to provide sketches of the objects and think-aloud protocols that they used to describe their thoughts as they moved from the concept to their design. It was quite clear that designs that focused on the emotion concept included many more associations in preparation for sketching and accrued to human faces for inspiration in their drawings (see Figs. 19.1, 19.2 and 19.3).

Concrete concept designs were drawn more quickly, from beginning to end, than the other two types of concepts; however, participants produced a first representation more quickly for the emotion concept, as compared to the other two.

Fig. 19.1 Example of emotion concept sketch (grief) [41]

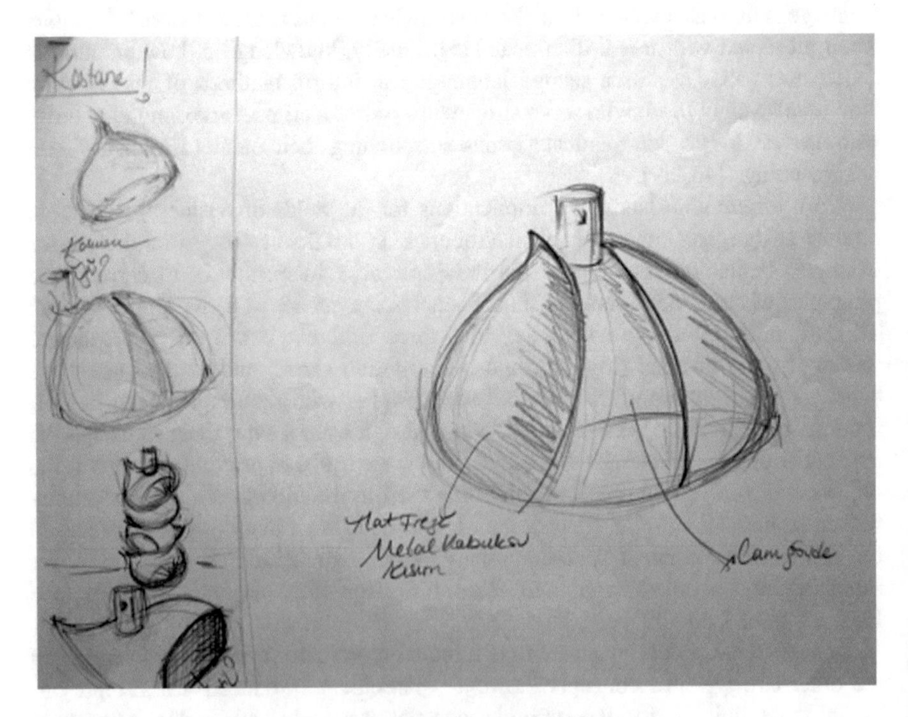

Fig. 19.2 Example of concrete concept sketch (chestnut) [41]

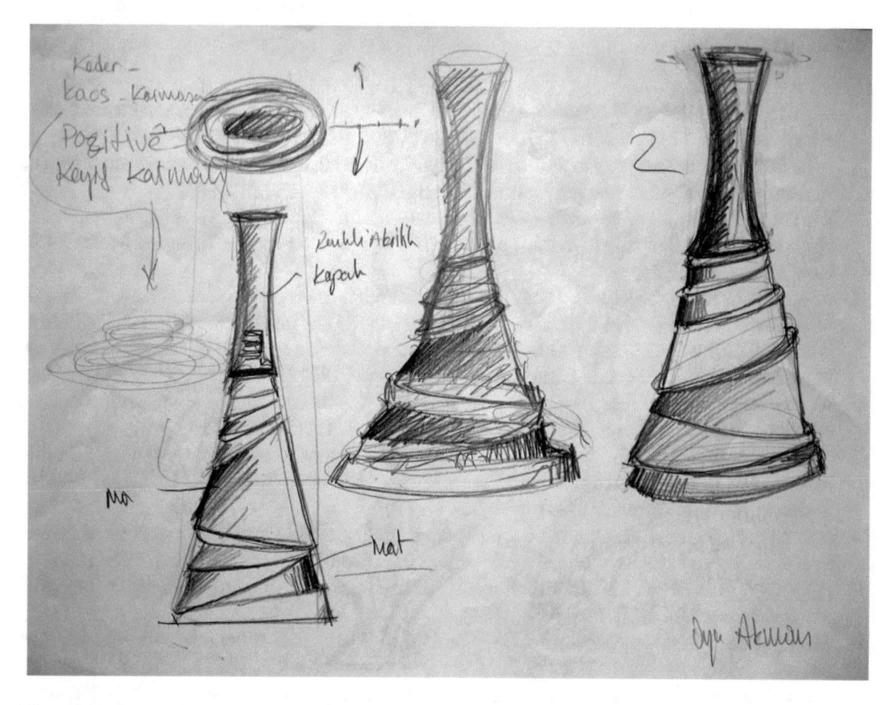

Fig. 19.3 Example of abstract concept sketch (loyalty) [41]

Emotional associations and referents came to mind much more quickly than those for the other two types of concepts. Additionally, there was remarkable consistency across designs and approaches even though the designers were trained in very different areas, did not know each other, and were exposed to the concepts independently of each other in their own respective studios and work sites. This is one application of emotion research that can provide clues to the ways in which training can occur for industrial designers who readily have to design products with varying emotional components. If it takes longer to derive a product with an emotional concept, then modes of training could be developed to make the act of designing more efficient, in these particular cases.

In the field of artificial intelligence, it has been argued that human emotion should be engendered into computers if the goal is to devise an intelligent machine or system that accurately emulates human behavior [42]. Martínez-Miranda and Aldea argue convincingly that computers should show emotion and affection and should communicate emotional states if they are to be true simulations of human behavior. In order to do so effectively, more research needs to be conducted to understand the role of a given language structure on emotional expression, how topics, sensitive or otherwise, are navigated in human communication systems, and how culture affects how emotional language is derived and used. Current cognitive research is aimed at further uncovering the ways in which language, culture, and emotion interact to influence human behavior.

References

1. Altarriba, J.: Does *cariño* equal "liking"? A theoretical approach to conceptual nonequivalence between languages. Int. J. Bilingualism **7**, 305–322 (2003)
2. Altarriba, J.: Emotion and mood: over 120 years of contemplation and exploration in The American Journal of Psychology. Am. J. Psych. **125**, 409–422 (2012)
3. Altarriba, J.: Emotion, memory, and bilingualism. In: Heredia, R.R., Altarriba, J. (eds.) Foundations of bilingual memory, pp. 185–203. Springer Science + Business Media LLC, New York, NY (2014)
4. Altarriba, J., Basnight-Brown, D.M.: The representation of emotion vs. emotion-laden words in English and Spanish in the affective Simon task. Int. J. Bilingualism **15**, 310–328 (2011)
5. Altarriba, J., Basnight-Brown, D.M.: The acquisition of concrete, abstract, and emotion words in a second language. Int. J. Bilingualism **16**, 446–452 (2012)
6. Altarriba, J., Bauer, L.M.: The distinctiveness of emotion concepts: a comparison between emotion, abstract, and concrete words. Am. J. Psychol. **117**(3), 389–410 (2004)
7. Altarriba, J., Bauer, L.M., Benvenuto, C.: Concreteness, context availability, and imageability ratings and word associations for abstract, concrete, and emotion words. Behavior Res. Methods Instrum. Comput. **31**(4), 578–602 (1999)
8. Altarriba, J., Santiago-Rivera, A.: Current perspectives on using linguistic and cultural factors in counseling the Hispanic client. Professional Psych.: Res. Practice **25**, 388–397 (1994)
9. Anooshian, L.J., Hertel, P.T.: Emotionality in free recall: language specificity in bilingual memory. Cogn. Emot. **8**, 503–514 (1994)
10. Basnight-Brown, D.M., Altarriba, J.: Number of translation differences in Spanish and Chinese bilinguals: The difficulty in finding a direct translation for emotion words. In: Cooper, S., Ratele, K. (eds.) Psychology serving humanity, vol. II, pp. 240–251. Taylor & Francis, New York, NY (2014)
11. Basnight-Brown, D.M., Altarriba, J.: Multiple translations in bilingual memory: processing differences across concrete, abstract, and emotion words. J. Psycholinguist. Res. **45**, 1219–1245 (2016)
12. Caldwell-Harris, C.L.: Emotionality differences between a native and foreign language: implications for everyday life. Curr. Dir. Psychol. Sci. **24**, 214–219 (2015)
13. Colbeck, K.L., Bowers, J.S.: Blinded by taboo words in L1 but not L2. Emotion **12**, 217–222 (2012)
14. Comanaru, R.S., Dewaele, J.-M.: A bright future for interdisciplinary multilingualism research. Int. J. Mutlilingualism **12**, 404–418 (2015)
15. Costa, A., Foucart, A., Hayakawa, S., Aparici, M., Apesteguia, J., Heafner, J., Keysar, B.: Your morals depend on language. Proc. Natl. Acad. Sci. **9**, 1–7 (2014)
16. Dewaele, J.-M.: The emotional force of swear words and taboo words in the speech of multilinguals. J. Multilingual Multicultural Dev. **25**, 204–222 (2004)
17. Dewaele, J.-M.: The emotional weight of I love you in multilinguals' languages. J. Pragmat. **40**, 1753–1780 (2008)
18. Dewaele, J.-M.: From obscure echo to language of the heart: Multilinguals' language choices for emotional inner speech. J. Pragmat. **87**, 1–17 (2015)
19. Dewaele, J.-M.: British 'bollocks' versus American 'jerk': do native British English speakers swear more—or differently—compared to American English speakers? Appl. Linguistics Rev. **6**, 309–339 (2015)
20. Dewaele, J.-M., Pavlenko, A.: Emotion vocabulary in interlanguage. Lang. Learn. **52**, 263–322 (2002)
21. Dewaele, J.-M., Qaddourah, I.: Language choice in expressing anger among Arab-English Londoners. Russ. J. Linguist. **19**, 82–100 (2015)
22. Dodds, P.S. et al.: Human language reveals a universal positivity bias. Proc. Natl Acad. Sci. **1073**, 1–6 (2015)

23. Du, S., Tao, Y., Martinez, A.M.: Compound facial expressions of emotion. Proc. Natl. Acad. Sci., 1454–1462 (2014)
24. Eilola, T.M., Havelka, J., Sharma, D.: Emotional activation in the first and second language. Cogn. Emot. **21**, 1064–1076 (2007)
25. Ekman, P.: Emotions revealed: recognizing faces and feelings to improve communication and emotional life. Times Books, New York (2003)
26. Ekman, P., Friesen, W.V.: Constants across cultures in the face and emotion. J. Pers. Soc. Psychol. **17**, 124–129 (1971)
27. Elfenbein, H.A., Ambady, N.: On the universality and cultural specificity of emotion recognition: a meta-analysis. Psychol. Bull. **128**, 203–235 (2002)
28. Fan, S.P., Liberman, Z., Keysar, B., Kinzler, K.D.: The exposure advantage: early exposure to a multilingual environment promotes effective communication. Psych. Sci., 1–10 (2015)
29. Goleman, D.: Emotional intelligence. Bantam Books, New York (1995)
30. Harris, C.: Bilingual speakers in the lab: psychophysiological measures of emotional reactivity. J. Multilingual Multicultural Dev. **25**, 223–247 (2004)
31. Harris, C.L., Ayçiçeği, A., Gleason, J.B.: Taboo words and reprimands elicit greater autonomic reactivity in a first than in a second language. Appl. Psycholinguist. **4**, 561–578 (2003)
32. Harris, C.L., Gleason, J.B., Ayçiçeği, A.: When is a first language more emotional? Psychophysiological evidence from bilingual speakers. In: Pavlenko, A. (ed.) Bilingual minds: emotional experience, expression, and representation, pp. 257–283. Multilingual Matters, Clevedon, UK (2005)
33. Jacobi, L.L.: Perceptions of profanity: How race, gender, and expletive choice affect perceived offensiveness. North Am. J. Psych. **16**, 261–276 (2014)
34. Jia, G., Aaronson, D., Wu, Y.: Long-term language attainment of bilingual immigrants: predictive variables and language group differences. Appl. Psycholinguist. **23**, 599–621 (2002)
35. Kazanas, S.A., Altarriba, J.: Emotion word processing: Effects of word type and valence in Spanish-English bilinguals. J. Psycholinguist. Res. **45**, 395–406 (2016)
36. Knickerbocker, H., Altarriba, J.: Bilingualism and the impact of emotion: the role of experience, memory, and sociolinguistic factors. In: Cook, V., Bassetti, B. (eds.) Language and bilingual cognition, pp. 453–477. Psychology Press/Taylor & Francis, New York (2011)
37. Knickerbocker, H., Altarriba, J.: Differential repetition blindness with emotion and emotion-laden word types. Visual Cognition **21**, 599–627 (2013)
38. Knickerbocker, H., Johnson, R.L., Altarriba, J.: Emotion effects during reading: influence of an emotion target word on eye movements and processing. Cogn. Emot. **29**, 784–806 (2015)
39. Kousta, S.T., Vinson, D.P., Vigliocco, G.: Emotion words, regardless of polarity, have a processing advantage over neutral words. Cognition **112**, 473–481 (2009)
40. Larsen, R.J., Mercer, K.A., Balota, D.A.: Lexical characteristics of words used in Stroop experiments. Emotion **6**, 62–72 (2006)
41. Leblebiçi-Basar, D., Altarriba, J.: The role of imagery and emotion in the translation of concepts into product form. Des. J. **16**, 295–314 (2013) doi: 10.2752/175630613X13660502571787
42. Martinez-Miranda, J., Aldea, A.: Emotions in human and artificial intelligence. Comput. Hum. Behav. **21**, 323–341 (2005)
43. McKenna, F.P., Sharma, D.: Intrusive cognitions: an investigation of the emotional stroop task. J. Exp. Psychol. Learn. Mem. Cogn. **21**, 1595–1607 (1995)
44. Marian, V., Kaushanskaya, M.: Self-construal and emotion in bicultural bilinguals. J. Mem. Lang. **51**, 190–201 (2004)
45. Marian, V., Neisser, U.: Language-dependent recall of autobiographical memories. J. Exp. Psychol. Gen. **129**, 361–368 (2000)
46. Marmolejo, G., Diliberto-Macaluso, K.A., Altarriba, J.: False memory in bilinguals: does switching languages increase false memories? Am. J. Psychol. **122**, 1–16 (2009)

47. Matsumoto, D., Yoo, S.H., Fontaine, J., Altarriba, J., et al.: Mapping expressive differences around the world: the relationship between emotional display rules and individualism versus collectivism. J. Cross Cult. Psychol. **39**, 55–74 (2008)
48. Matsumoto, D., Yoo, S.H., Fontaine, J., Alexandre, J., Altarriba, J., et al.: Hypocrisy or maturity? culture and context differentiation. Eur. J. Pers. **23**, 251–264 (2009)
49. Murata, A., Moser, J.S., Kitayama, S.: Culture shapes electrocortical responses during emotion suppression. Social Cognitive & Affective Neurosci **8**, 595–601 (2013)
50. Nasrallah, M., Carmel, D., Lavie, N.: Murder, she wrote: enhanced sensitivity to negative word valence. Emotion **9**, 609–618 (2009)
51. Ohira, H., Nomura, M., Ichikawa, N., et al.: Association of neural and physiological responses during voluntary emotion suppression. NeuroImage **29**, 721–733 (2006)
52. Park, J., Kitayama, S., Markus, H.R.…et al.: Social status and anger expression: the cultural moderation hypothesis. Emotion **13**, 1122–1131 (2013)
53. Pavlenko, A.: Emotions and multilingualism. Cambridge University Press, Cambridge, MA (2005)
54. Pavlenko, A.: Emotion and emotion-laden words in the bilingual lexicon. Bilingualism: Language and Cognition **11**, 147–164 (2008)
55. Pavlenko, A., Driagina, V.: Russian emotion vocabulary in American learners' narratives. Mod. Language J. **91**, 213–234 (2007)
56. Puntoni, S., de Langhe, B., van Osselaer, S.: Bilingualism and the emotional intensity of advertising language. J. Consumer Res. **35**, 1012–1025 (2009)
57. Robinson, C.J., Altarriba, J.: Culture and language processing. In: Sharifian, F. (ed.) The Routledge handbook of language and culture, pp. 240–252. Routledge, New York, NY (2015)
58. Russell, J.A.: Culture and the categorization of emotions. Psychol. Bull. **110**, 426–450 (1991)
59. Santiago-Rivera, A., Altarriba, J., Poll, N., Gonzalez-Miller, N., Cragun, C.: Therapists' views on working with bilingual Spanish-English speaking clients: a qualitative investigation. Professional Psych.: R. Practice **40**, 436–443 (2009)
60. Scott, G.G., O'Donnell, P.J., Sereno, S.C.: Emotion words affect eye fixations during reading. J. Exp. Psychol. Learn. Mem. Cogn. **38**, 783–792 (2012)
61. Sheikh, N.A., Titone, D.A.: Sensorimotor and linguistic information attenuate emotional word processing benefits: an eye movement study. Emotion **13**, 1107–1121 (2013)
62. Stroop, J.R.: Studies in interference in serial-verbal reactions. J. Exp. Psychol. **18**, 643–662 (1935)
63. Sutton, T.M., Altarriba, J.: The automatic activation and perception of emotion in word processing: evidence from a modified dot probe paradigm. J. Cognitive Psych. **23**, 736–747 (2011)
64. Sutton, T.M., Altarriba, J.: Color associations to emotion and emotion-laden words: a collection of norms for stimulus construction and selection. Behavior Res. Methods (2015). doi: 10.3758/s13428-015-0598-8
65. Sutton, T.M., Altarriba, J., Gianico, J.L., Basnight-Brown, D.M.: Emotional stroop effects in monolingual and bilingual speakers. Cogn. Emot. **21**, 1077–1090 (2007)
66. Williams, J.M.G., Mathews, A., MacLeod, C.: The emotional Stroop task and psychopathology. Psychol. Bull. **120**, 3–24 (1996)
67. Wilson, J.: Happily Disgusted (April, 2014). CNN.com
68. Zhu, Y., Zhang, L., Fan, J., Han, S.: Neural basis of cultural influence on self-representation. NeuroImage **34**, 1310–1316 (2007)

Chapter 20
Creating a Culture of Innovation

Arthur B. Markman

Abstract Innovation—the process of generating and implementing practical new ideas—can be difficult for organizations to do successfully. To make innovation a part of an organization, it is often necessary to change the culture in ways that bring more innovative processes into the workplace. In this chapter, I explore key factors that have to be part of a culture of innovation, including the need to favour innovation over efficiency and to tolerate failure. I also explore the importance of having an ecosystem to support the development of ideas. I illustrate these concepts with an example from the US military.

Keywords Innovation · Innovation ecosystem · Organizational culture Failure

20.1 Introduction

In modern developed economies, innovation is a central part of a strategy for growth. It is hard for the developed world to compete with emerging economies on labor costs, and manufacturing jobs are often located where labor costs are cheapest. Advances in supply-chain management over the past 25 years have made it difficult for firms to compete by offering lower prices for goods that are commodities.

Although the word *innovation* is used in many ways, there is a general consensus that it refers to a process of developing, refining, and commercializing new ideas [1, 2]. Those ideas can lead to new products, but they can also lead to new services, new intellectual property, or new processes within an industry. As a result, innovation can lead to new markets (when a new product is developed) or to new levels of efficiency within an existing market (when a process is refined).

A.B. Markman (✉)
Department of Psychology, University of Texas,
108 E. Dean Keeton St., A8000, Austin, TX 78712, USA
e-mail: markman@utexas.edu

C. Faucher (ed.), *Advances in Culturally-Aware Intelligent Systems and in Cross-Cultural Psychological Studies*, Intelligent Systems Reference Library 134, https://doi.org/10.1007/978-3-319-67024-9_20

Although firms of all sizes talk about the importance of innovation for their success, most are not able to be innovative on a consistent basis. Instead, many new ideas are developed initially within research and development labs owned by companies or organized at universities or government institutions. At companies that thrive on discovery (such as drug companies), there is also an elaborate infrastructure for testing new products and bringing them to market. For other research facilities, new ideas are often commercialized through entrepreneurial ventures that spin off to commercialize a product [3].

This state of affairs leads to a paradox. Firms value innovation and aspire to commercialize new ideas. Yet, they lack an infrastructure that supports innovation. More importantly, the culture at most firms also biases against the development of new ideas.

In this chapter, I begin by thinking about what it would mean for an organization to have a culture. Then, I explore three barriers to a culture of innovation within a firm. First, most organizations are structured around efficiency, which makes innovation difficult. Second, they often punish failure rather than learning from it. Third, large organizations are structured in ways that block the flow of information and ideas. As a result, it is crucial for organizations that want to become more innovative to create structures that allow different groups within the firm to work together. Each of these themes will be addressed in this chapter.

20.2 What Can a Culture of Innovation Be?

The word culture is used in many ways across disciplines. Cultural anthropologists focus on the rituals, social structures, and shared stories of members of groups. They describe cultures and also find dimensions that characterize differences among cultures [4]. These dimensions can be used to predict differences in behavior between members of different cultures [5].

Cultural psychologists then look at how aspects of culture shape the behavior of the members of that culture. For example, Nisbett and his colleagues have demonstrated broad cultural differences in performance on a variety of cognitive tasks between members of Western cultures (which tend to emphasize the importance of individuals) and members of East Asian cultures (which tend to emphasize the importance of the collective) [6, 7].

Cultural psychology also aims to understand the variables that culture can influence that may give rise to these differences in behavior. For example, studies in my lab explored the relationship between fear of isolation and cognitive performance [8]. These studies suggest that members of East Asian cultures have a higher chronic fear of isolation (on average) than members of Western cultures. That difference is correlated with differences in performance on a variety of cognitive tasks like the ones studied by Nisbett and his colleagues. Importantly, inducing a higher level of fear of isolation in members of a Western culture makes their cognitive performance closer to that observed in members of East Asian cultures.

The main idea here is that cultural factors may affect a motivational variable (in this case fear of isolation) that in turn affects cognition. Once culture is seen as an organizational force that influences core motivational factors, we can begin to look for other motivational influences that organizations have on the behavior of the members of that organization. The organizations that create these motivational orientations may be the global culture in which a person is raised, but they may also be the narrower set of social norms imposed by other organizations including religious institutions and workplaces.

This analysis suggests that a core active ingredient in cultures is a set of influences that organizations have on the motivational states of their members. In this paper, I focus on ways that organizations can affect aspects of motivation that influence innovation.

20.3 The Tradeoff Between Efficiency and Innovation

Without realizing it, most companies are structured based on a manufacturing model, even if their core business is intellectual property or service. The core of manufacturing is efficiency. Companies typically focus themselves on efficiency in a number of ways.

First, they try to minimize personnel costs. In many companies, hiring a new employee requires justifying why this position is needed. In some cases, the new hire will bring new skills that enhance the functionality of the organization. More often, though, the existing workload has become too much for the existing employee base to handle, and so the justification of the new hire rests on increasing the capacity of the organization to do business.

In many ways, this mode of hiring parallels the way a manufacturing plant is run. Each machine in the plant should be used to its capacity. Many plants run three shifts a day in order to allow the machinery to be used full-time. New machines are only purchased when they create a necessary new functionality for the plant, or when the current machinery is being fully utilized.

In order to make these determinations, factories and human resource (HR) departments measure employee productivity. In service companies, HR aims to maximize billable hours. In other firms, yearly evaluations lay out a set of goals (or *contributions* to use Peter Drucker's [9] term), and successful evaluation requires reaching these goals.

This strategy is effective for job functions that are focused on execution (which is really the knowledge- and service-work equivalent of manufacturing). When fast-food chains measure time-per-order and call-centers seek to minimize the length of service-calls, they are maximizing the efficiency of execution.

Execution isn't just perfected by focusing on minimizing time, of course. The aviation industry in collaboration with government agencies like the FAA focus on minimizing error [10, 11]. They catalog nearly every error made during routine work and then use those errors to fix procedures that may be leading to those problems. They also analyze all malfunctions and catastrophes to draw lessons to make future air travel safer.

Because errors, time to completion, and billable hours are easy to measure, they become important yardsticks for improving efficiency and evaluating employees. Unfortunately, these evaluations also bias most companies against innovation.

In particular, innovation does not have a straightforward time-course the way manufacturing and execution do. Research and development to create new knowledge is slow and expensive. In addition, there is no clear way to measure how close a project is to a new discovery. Many R&D projects can feel like they are stumbling around creating as many problems as they solve until suddenly the path to a solution is clear.

The same things holds true for other kinds of idea-generation techniques. Often, new ideas emerge as a result of analogies between a current problem and an existing solution in another domain [12, 13]. That means that for employees to have a chance to consistently generate new ideas, they have to continually expand their knowledge base. Furthermore, it is hard to target in advance which knowledge will lead to successful solutions to difficult problems. As a result, employees in creative roles must be given latitude to explore new domains whose applicability to current projects may not be clear. Only in retrospect will it be obvious which aspects of their knowledge were useful.

Even after promising ideas are generated, they must be further refined and ultimately commercialized. At each stage of this process, new problems may emerge that can set back the time-line of a project or lead it to be abandoned. Thus, unlike manufacturing and execution, it is difficult to manage innovative projects efficiently.

On the one hand, firms need to institute procedures that allow them to make decisions about whether to continue or to abandon projects [14]. There is certainly collected wisdom about innovation that can guide firms through the process.

At the same time, firms that want to become more innovative must commit to the expense of overstaffing. The most popular example of this kind of overstaffing is Google's "20% Time," in which (at least early in the company's development) all employees were encouraged to take about one day a week to pursue projects of interest to them, regardless of their other work priorities [15]. Although the need to execute has cut into the amount of flexible time that many employees at Google have [16], this idea reflects a recognition that company-wide innovation means that more employees need to be hired overall than would be required simply to execute existing company priorities.

Thus, the HR process for creating new positions must take into account that a certain amount of over-capacity in each division of an organization is required if the firm is committed to sustained innovation. Yearly evaluations of employees must focus not just on measures of efficiency, but also on engagement with activities that are reliably associated with the creation and commercialization of new ideas.

Finally, within any organization, it is important to set up reward structures that motivate people to act in ways that promote innovation. In every organizational culture, there is what leaders say, what people do, and what the organization rewards. These factors are listed in increasing order of importance.

Regardless of what leaders in the organization claim that they want to do, individuals are sensitive to what they see other people doing. Considerable work on goal contagion demonstrates that people automatically adopt the goals of people they observe. Seeing people in the workplace engaged in activities associated with innovation promotes that same behavior in others.

Rewards at work are particularly important for guiding behavior. Rewards can be pay and promotion, but they can also be praise and opportunity. People like to be recognized for their good work, and so praise influences behavior. In addition, studies of well-being in the workplace suggest that meaningful work is particularly important to people. Providing opportunities for people who engage in desirable behaviors to have more meaningful work opportunities is an excellent way to reward this desirable behavior.

One barrier to rewarding behavior associated with innovation is that existing corporate cultures that are focused on efficiency can create anxiety about work that is not progressing toward a clearly defined goal. One reason why the actions of leaders in the organization is so important is that many employees find it stressful to take time out of the workday to read and improve skills that are not directly related to specific ongoing tasks.

As this section suggests, many organizations (implicitly or explicitly) have adopted a culture of efficiency and execution. This orientation is incompatible with innovation. Without addressing this tradeoff directly, firms will have a hard time motivating employees to engage in more behaviors that lead to innovations.

20.4 Orientations to Failure

James Dyson's invention of a bagless vacuum is often held up as a great example of innovation [13]. Discussions of his achievement often focus on the application of industrial cyclone technology normally used in settings like sawmills to home vacuum cleaners. These discussions do not typically talk about the long years of effort that Dyson put in after having that motivating insight until he perfected a model that functioned well and could be brought to market.

An important lesson from this example is that there are many small failures that are part of even the most successful story of innovation. And most innovative ventures do not succeed, which means that even after running the innovation process flawlessly, there are many factors that can cause the project to fail. Competing projects may cause the market to shift in unexpected ways. An economic downturn can cause potential customers to delay purchases longer than the company can wait for sales.

Because innovation has a number of inherent risks, a company that wants to promote more innovation needs to be willing to tolerate failure. Unfortunately, companies often punish failure in a number of ways.

First, the kinds of employees that companies are likely to hire are often afraid of failure. Success in the modern education environment is defined as getting good grades. Good grades result from high test scores, and that means that the best students are the ones who make the fewest mistakes. Thus, most successful students try to avoid failure rather than embracing it.

Even when particular individuals are comfortable with failure, many firms are not. For example Saxenian [17] explored factors that led Silicon Valley to become the dominant center for high-tech industry in the United States by late 1980's. In the 1940's, it looked like the East Coast of the United States (and particularly the area around Boston, MA) was the best candidate region to excel in high tech. The US Government invested heavily in the research universities in this area (like MIT) and companies (like IBM and Raytheon) during and after World War II.

These large companies had a big technical advantage over West Coast firms in the 1960s, but these larger firms tended to stifle innovation in two ways. First, executive pay was typically tied to success of business units. If an innovative project failed, then profits from that unit lagged. Thus, management often avoided supporting innovative projects.

Second, mid-level managers who were given responsibility for innovative projects were often punished for leading projects that did not succeed. They were passed over for promotion and marginalized within the business.

In contrast to this treatment of failure, firms on the West Coast (many of which were smaller entrepreneurial ventures) were more tolerant of failures. People who worked for startups that failed were seen as having valuable experience that would make them more effective in subsequent companies. There was a general recognition that most of the factors that lead to success or failure of a particular venture are beyond the control of that firm, and so hard-working people with experience are valuable, regardless of whether past ventures succeeded.

This analysis suggests that many companies need to shift their culture to be more tolerant of failures and to treat them as learning experiences. A shorthand recommendation is to suggest that firms should punish negligence rather than failure.

One reason that failure is so important is that it provides valuable information. Scientists focus on disconfirming hypotheses, because a large amount of supporting evidence for a particular theory does not mean that evidence disconfirming it will not ultimately be found [18]. Similarly, it can be difficult to assess which collection of factors led a venture to be successful, but the factors that lead to a failure can be easier to diagnose.

As an example, when working for Apple, Ron Johnson spearheaded the project to develop the Apple Store. The modern design and smart helpful employees were an immediate success and continue to be a popular destination for technology shoppers. Based on this success, Johnson was hired to be the CEO of JC Penney, the mall department store known for inexpensive clothing and housewares. To turn around the struggling retailer, Johnson updated the design of JC Penney stores in an effort to appeal to a younger more stylish crowd [19]. Unfortunately, this strategy was a failure. It alienated existing customers, but failed to attract patrons of chains like H&M that offer trendy designs at low cost.

This example suggests that Johnson had difficulty determining the factors that led to the success of the Apple Store. Consequently, he repeated the formula that worked at Apple without modifying it to suit differences between the customers Apple hoped to attract and the customers that JC Penney was likely to be able to attract. In addition, The Apple Store was a new venture, and so Johnson did not have to worry about an existing customer base. In contrast, JC Penney attracted shoppers who were quite different than those that the redesigned store was target to appeal to.

Because failure creates learning opportunities, many firms need to change their orientation toward failure. The guiding principle is to punish negligence, but not failure. In practice, that means that yearly employee evaluations need to incorporate both employees' accomplishments as well as the process that they go through to achieve those results. This focus on process ensures that members of a firm are working toward positive innovation outcomes. It also recognizes that when employees are working on innovative projects, there are many sources of failure that are beyond the control of an individual or team.

As discussed in the previous section, the execution orientation of many firms leads them to focus on minimizing errors. Thus, tolerance for failure contradicts a central tenet of many corporate cultures. That is why it is crucial to build a focus on innovative activities rather than innovative outcomes into evaluations. People who have been socialized into organizations are likely to feel uncomfortable rewarding someone whose project has not succeeded. Thus, the criteria for evaluation need to be laid out explicitly to help people get beyond their habitual reaction.

20.5 Creating an Ecosystem that Supports Innovation

A critical part of successful innovation in organizations is having a flow of information that allows groups to share ideas and allows the organization to use all of its resources to enable a new idea to succeed. Unfortunately, many large organizations are structured in a way that limits the spread of information.

When organizations are small (particularly when they have less than 25 people), the social structure in the organization need not be that complex. Everyone in the organization is likely to have a clear sense of the roles and knowledge of their colleagues. As a result, information in small organization passes freely. This is one reason why startup companies are often successful as small entrepreneurial ventures.

As organizations get larger, the structure begins to grow. There is some evidence suggesting that when organizations get larger than about 150 people, it is not possible for everyone to really know everyone else [20]. Practically speaking, however, the management structure of an organization begins to get more complex above about 25 people. At that point, groups of people are required to perform specialized functions (such as technical support, customer support, or accounting).

As those groups begin to form, they now contain substructures within the whole organization. The basic psychology of group processes leads these clusters within an organization to form silos that distinguish between those inside the group and those outside [21]. The members within a group work well together, but they are less prone to share information with people outside it. This happens both because members of a group work in proximity to each other and far from others, and because there is a higher level of trust within the group than across groups. As a result, these groups naturally form silos that keep information contained within them and limit the flow of information outward.

Thus, even though members of different silos within an organization are part of the same company overall, the greater group cohesion within a silo than across silos hampers innovation. Members of different silos may not reach out for information or help to the right people. Indeed, as the organization grows later, members of one silo may be unaware of the skills and knowledge contained in different silos.

A related problem involves bureaucratic structures that are developed as organizations get larger. Because the members of one silo are unlikely to be well-acquainted with members of other silos, systems of checks and balances are often put in place to ensure that employees in the organization act honestly and that power is shared. These procedures take the place of the personal relationships that are the basis of trust in smaller organizations.

These structures also limit the flow of information. Members of particular groups often expect information to flow upward to the management of a silo and for knowledge to be transferred across groups by management. Other organizations create procedures for sharing knowledge that involve the creation of reports that are stored in a central database. These reports are meant to be used by teams that want to make use of lessons learned by other segments of the organization.

In practice, these strategies also limit the flow of information that is crucial to innovation. Innovation is most effective when the person who needs the information communicates directly with the person who has that knowledge. Hierarchical structures that pass information place a number of people in between the two people who ultimately need to connect. Those layers of bureaucracy decrease the likelihood that people who need to be in contact will actually connect.

Creating databases of reports is rarely successful as a means of capturing organizational knowledge to support innovation [22]. First, these reports are often written in a cursory way, because people are anxious to move on to the next task rather than capturing what was learned in the task just completed. As a result, the reports themselves do not always provide details of the problem that was solved or the method used to solve it.

Even when the documents are well-written, they need to be found by people who need them for the documents to be useful. In order to retrieve a document in a database, you need to search on terms that are actually in that document. Unfortunately, because many innovations involve knowledge that comes from across disciplines, the search terms are rarely those that were used in previous documents that may have been relevant to solving a new problem [23].

The fundamental difficulty is that search is literal. That is, when people are seeking information in most databases, the documents need to contain specific words that are part of the query. Someone solving a problem about traffic flow is likely to search using terms from that domain like traffic, roads, cars, and driving. That search will yield documents that are also about traffic on the roads.

What we want instead is analogical search [24, 25]. The idea is to find other domains that have the same structure as the problem you are trying to solve. For example, there is a broad analogy between road systems and the circulatory system that might provide productive avenues for exploring a problem.

These analogies preserve relationships among elements in the domain rather than the objects in the domains [26–29]. In the comparison between roads and the circulatory system, there is a parallel between capillaries and arteries, which differ in the amount of blood that can flow through them and side streets and boulevards, which differ in the traffic they can handle. Blood does not look like cars, and veins and arteries do not look like roads. They are analogous just because of these parallel relationships.

To find an analogous domain like this, you could try to use more abstract search terms that might potentially describe both domains. There are two problems with this approach. First, analogies are not really about abstraction, but rather about similarities in relationships. Second, the natural way to write about anything is to use terms relevant to the domain in which you are writing. Thus, the people writing the documents in the database will use specific terms and not the abstract ones. All of these factors clarify why databases of reports are mostly useful for helping people at a company to re-use solutions when the same problem arises again.

Thus, to promote sharing of knowledge, organizations need to find ways to encourage more interactions among people from different silos.

20.5.1 Studying Business Ecosystems

In order to explore ways to encourage more interactions within firms, my colleagues and I looked to an analogous case [30]. We examined how networked technology incubators function.

Technology incubators are organizations that help early-stage technology companies to develop and commercialize technological innovations. In the 1980s, the dominant role of the incubator was to provide inexpensive office space for fledgling companies. Bringing several companies together in the same space was also valuable, because the members of those companies could share expertise.

Over time, the office space became less important than the social interactions. To explore these interactions in more detail, we examined the Austin Technology Incubator (ATI) in great detail. The goal of ATI is to take new companies and help them to get their first round of investor funding. Research assistants followed directors of the incubator, interviewed members of companies and other members of the technology community, and analyzed emails and invitations to incubator events. For comparison, the team also examined less successful incubators in Portugal.

Two important features emerged from these analyses. First, successful incubators focus on companies that have a common goal. When a new venture begins, its owners may want to develop a company that they grow and nurture for life. These *lifestyle companies* often grow slowly and aim to be small to moderate-sized businesses. Other teams want to create companies that will ultimately lead to large exits by IPO or sale that allow the team to reap a large reward. These companies seek significant investment from venture capital and angel investors in order to grow quickly. The cost of taking on these investments is that they have to grow significantly in order for the investors to see a reasonable return.

Successful incubators focus exclusively on one type of company. For example, all of the companies at ATI ultimately plan to have an IPO or to be sold to a larger organization. Unsuccessful incubators bring in many different types of companies with a diverse set of goals.

The reason why a common set of goals is crucial is because the primary value of incubators is that they create an ecosystem for the survival of new companies that functions like a metaphorical *coral reef*. The idea is that the incubator protects new companies. It attracts a broad community of individuals who can help new companies to thrive. Potential investors meet the team. Technology experts help the company to develop its product. Business experts refine the company's business model. Students from nearby universities engage with companies and get experience with entrepreneurship.

The interactions between companies and this community are not structured. Instead, the incubator holds a variety of events that support happenstance interactions among community members and companies in ways that benefit the startups as well as the surrounding business community.

If the companies in an incubator have divergent goals, then community members do not have enough positive interactions when they attend incubator events, and so they stop coming. When there is a critical mass of companies with a common goal, then there is a high probability that a community member who engages with incubated companies will benefit.

It is valuable to understand how these entrepreneurial ecosystems function, because it allows directors of incubators to create more effective ways to nurture a startup community in a region. However, we were also interested in this ecosystem, because a similar structure might be valuable for large organizations to help them break down silo walls.

20.5.2 The Innovation Ecosystem

The same principles for success of networked incubators can be applied to innovation to create an *innovation reef*. The core idea is to create events for people from different silos of a company to come together on a regular basis to exchange ideas and to talk about projects they are working on.

A reef cannot be legislated by the management of an organization. Instead, the individual events need to add value for the participants so that a critical mass of employees engage in them and so that they spur interactions.

It is possible to use some technology to support these interactions. For example, company-wide wikis and bulletin boards where people can ask questions are an excellent support for a reef.

However, there has to be a number of live events for people to attend. The active ingredient in the innovation ecosystem is serendipitous discussions that happen surrounding events rather than targeted queries. So, events within the reef have to include sufficient time for participants to engage in small group discussions. Of course, in order to provide time for people to attend reef events, there has to be enough over-capacity (as discussed earlier) to allow people the flexibility to be part of the reef while still handling their day-to-day responsibilities.

In order to jump start social interactions across silos, it can be helpful for managers to schedule "field trips" in which one group gets together with another for an extended lunch in which key group members give talks to describe what they have been working on and to raise key problems they are addressing. These events help to develop personal relationships among individuals across groups that can be maintained at later events.

In addition, for the innovation ecosystem to succeed, there has to be some clear mechanism for groups that form around innovative projects to get funding to continue their projects. It is also helpful if there is a standardized way for members of groups that are working on innovative projects to apply for more time to focus on those projects when they look promising.

This type of innovation ecosystem contrasts with a more typical way that large companies have tried to spur innovation. Taking a cue from successful design firms like IDEO [31], many companies created in-house design teams that would be staffed with experts in design thinking and idea generation. These teams would often have a fancy off-site location where intact groups from the company would be brought to engage in creativity and innovation exercises.

Unfortunately, many of these centers proved hard to sustain [32]. Sustaining innovative projects requires more than just good design thinking. It requires extensive collaboration across business units throughout the development of a project. Consequently, a more distributed approach to innovation that involves members from different research teams is a more sustainable model for supporting innovation in large organizations.

20.5.3 *Putting the Reef into Practice*

Large organizations are starting to implement this kind of innovation reef. In this section, I focus on two examples, one of which has been more successful to date than the other.

The more successful reef has been developed by USAA, the large insurance and financial services firm [33]. USAA started as an automobile insurance company that catered to members of the United States military, veterans, and their families. Over the years, the company has diversified into a broad-based financial services firm that serves this client base.

In an effort to become more innovative, USAA partnered with the University of Texas to develop training for a broad base of their employees to learn an end-to-end innovation strategy that encompasses idea development, evaluation of new technologies, and implementation of innovative projects. The trainees for this program are selected explicitly from across a variety of segments of the company and from different job functions.

There are three aims to this training. First, it distributes expertise in innovation across business units within the organization. Second, it creates networking opportunities for individuals from different business units who can find common ground and establish relationships with people from other silos. Third, it rewards individuals who are interested in innovation with additional training and opportunities that allow them to advance their career.

A great example of the success of this program is the Pole Cam [33]. USAA provides homeowner's insurance to members. When a policy holder sustains roof damage, an adjuster needs to assess that damage. This job is dangerous, particularly in the aftermath of a significant storm. A cross-disciplinary team at USAA convened to develop a way to minimize the number of roof inspections that required adjusters to get up on the roof to inspect it.

Their first solution involved a portable remote-controlled drone with a camera attached that would fly and hover over damaged roofs. Unfortunately, while the drone worked effectively in preliminary tests indoors, it was easily blown off course by gusts of wind, and so that idea was scrapped. Ultimately, the team developed an extendible fiberglass poll with a digital camera on it controlled by a tablet at the base. This lower-tech solution has been deployed in the field, and has drastically reduced the number of times adjusters need to climb on damaged roofs.

The USAA reef is succeeding, because there is a commitment to continuing to train interested employees in methods of innovation and to provide these employees with time to focus on innovative projects. In this way, the reef is well-aligned with the organization's reward structure. Finally, as the Pole Cam example demonstrates, the teams are willing to learn from their failures and to continue developing innovative projects with the full support of management.

A second example of a reef that was less successful involved United States Special Operations Command (SOCOM). SOCOM, which operates out of MacDill Air Force Base in Tampa, Florida, is the command center for US Special Forces (including the Army Rangers and the Navy Seals). Following the terrorist attacks in the US on September, 11, 2001, Special Operations Command was given a mandate to coordinate defense against terror attacks. As a result, the command structure of SOCOM swelled and came to include military personnel, civilian contractors, and government employees.

In 2013, following a conference on innovation held at the request of Adm. William McRaven (ret.), SOCOM set up their own reef (as evidenced by Twitter account @SOFReef) that aimed to bring together individuals from across the organization interested in innovation. They also developed a number of on-line tools to support communication about problems requiring innovative solutions. One of these, called SOFBox (which had a hashtag #sofbox on Twitter) aimed to be a central bulletin board for the exchange of ideas.

Unfortunately, the reef and the tools like SOFBox were not as successful as the reef at USAA. Indeed, the SOFReef twitter account has not had a new tweet since 2013. A key reason why the reef structure has had a harder time getting established at SOCOM is because of the three types of employees who work there. The HR rules for military personnel, government employees, and civilian contractors are all different. Consequently, it is difficult to create a uniform way to involve people across these groups in innovative projects. The rules governing each of these types of employees has the unintended consequence of maintaining silo walls despite a desire by the upper-level command structure of SOCOM to improve innovation. However, efforts to work with these constraints are under way through the SOCOM iLab (http://www.socom.mil/sofilab/default.aspx).

Looking across these examples, it is clear that a reef can be created that leads to successful innovations. However, it cannot be mandated top-down. Instead, it is important for management to provide support and flexibility to individuals who engage with each other across silos and to tolerate failure. When an organization cannot be flexible in the way that it uses employees, then it is hard to sustain a reef, despite the best of intentions.

20.6 Summary and Future Directions

Most organizational cultures have a variety of elements that tend to perpetuate existing behaviors. In particular, existing reward structures make it easiest for people to continue engaging in behaviors that have brought them success in the past. HR practices that focus on efficiency make it difficult for people to be flexible in their pursuit of innovative ideas. Fear of failure biases individuals away from innovative projects.

A central difficulty that large organizations have when trying to innovate is that it is difficult for them to share information across the silos that inevitably develop when an organization grows larger than about 100 people. To help information flow across silo walls, it is valuable to create an innovation reef that creates events that allow unstructured interactions among people from different groups to share information and to coalesce around new ideas. Organizations interested in creating a reef need to have some flexibility in their HR practices to allow employees to have the time to devote to new projects.

Future work must continue to explore methods for improving information flow in organizations, particularly those that are geographically dispersed. Multi-national companies have a hard time getting employees in distant offices to work together. For one, differences in time zone mean that employees in different countries may only overlap in their workdays for a few hours each day. Even when these individuals synchronize their work schedules, the distance means that conversations must be scheduled and mediated by technology that still does not provide a deep sense of copresence.

In addition, more detailed case studies of organizations that have changed their culture to become more innovative will support the development of new tools. Ultimately, the reef is just one ecosystem metaphor that is useful for thinking about how to improve innovation in organizations.

References

1. Chesbrough, H.W.: Open Innovation: The New Imperative for Creating and Profiting from Technology. Harvard Business Review Press, Cambridge, MA (2003)
2. Markman, A.B., Wood, K.L. (eds.): Tools for Innovation. Oxford University Press, New York (2009)
3. Gibson, D.V., Rogers, E.M.: R&D collaboration on trial. Harvard Business School Press, Boston, MA (1994)
4. Triandis, H.C.: Individualism and collectivism: past, present, and future, In: Matsumoto, D. (ed.) The Handbook of Culture and Psychology, pp. 35–50. Oxford University Press, New York (2001)
5. Hofstede, G., Van Hofstede, G., Minkov, M.: Cultures and Organizations, 3rd edn. New York: McGraw-Hill (2010)
6. Nisbett, R.E., et al.: Culture and systems of thought: Holistic versus analytic cognition. Psychol. Rev. **108**(2), 291–310 (2001)
7. Peng, K.P., Nisbett, R.E.: Culture, dialectics, and reasoning about contradiction. Am. Psychol. **54**(9), 741–754 (1999)
8. Kim, K., Markman, A.B.: Differences in fear of isolation as an explanation of cultural differences: Evidence from memory and reasoning. J. Exp. Soc. Psychol. **42**, 350–364 (2006)
9. Drucker, P.F.: The Practice of Management. HarperCollins Publishers, New York (1954)
10. Helmreich, R.L.: On error management: lessons from aviation. BMJ **320**(7237), 781–785 (2000)
11. Harper, M.L., Helmreich, R.L.: Identifying barriers to the success of a reporting system. In: Henriksen, K., Battles, J.B., Marks, E.S. (eds.) Advances in patient safety: From Research to Implementation (vol. 3: Implementation issues). Agency for Healthcare Research and Quality, Rockville, MD (2005)
12. Basalla, G.: The Evolution of Technology. Cambridge University Press, Cambridge, UK (1988)
13. Markman, A.: Smart Thinking. Perigee Books, New York (2012)
14. Cornwell, B.: Quicklook Commercialization Assessments. R&D Enterprise: Asia Pacific, **1**(1) (1998)
15. Walker, A.: Creativity loves constraints: The paradox of Google's twenty percent time. Ephemera **11**(4), 369–386 (2011)
16. Ross, A.: Why did Google Abandon 20% Time for Innovation. 4/4/16] (2015)
17. Saxenian, A.: Regional Advantage. Harvard University Press, Cambridge, MA (1996)
18. Platt, J.R.: Strong Inference. Science **146**, 347–352 (1964)
19. Lublin, J.S., Mattioli, D., Co, J.P.: Penney CEO Out, Old Boss Back In, in Wall Street Journal (2013)

20. Dunbar, R.I.M.: Neocortex size as a constraint on group size in primates. J. Hum. Evol. **22**(6), 469–493 (1992)
21. Gawronski, B., Bodenhausen, G., Banse, R.: We are, therefore they aren't: Ingroup construal as a standard of comparison for outgroup judgments. J. Exp. Soc. Psychol. **41**, 515–526 (2005)
22. Smith, S.M., et al.: The development and evaluation of tools for creativity. In: Markman, A.B., Wood, K.L. (eds.) Tools for Innovation. Oxford University Press, New York (2009)
23. Linsey, J.S., Wood, K.L., Markman, A.B.: Modality and representation in analogy. Artif. Intell. Eng. Des. Anal. Manuf. **22**(2), 85–100 (2008)
24. Ward, T.B.: ConceptNets for flexible access to knowledge. In: Markman, A.B., Wood, K.L. (eds.) Tools for Innovation. Oxford University Press, New York (2009)
25. Forbus, K.D., Gentner, D., Law, K.: MAC/FAC: a model of similarity-based retrieval. Cognitive Sci. **19**(2), 141–205 (1995)
26. Gentner, D., Markman, A.B.: Structural alignment in analogy and similarity. Am. Psychol. **52**(1), 45–56 (1997)
27. Gentner, D.: Structure-mapping: a theoretical framework for analogy. Cognitive Sci. **7**, 155–170 (1983)
28. Gentner, D.: The mechanisms of analogical learning. In: Vosniadou, S., Ortony, A. (eds.) Similarity and Analogical Reasoning, pp. 199–241. Cambridge University Press, New York (1989)
29. Hummel, J.E., Holyoak, K.J.: Distributed representations of structure: A theory of analogical access and mapping. Psychol. Rev. **104**(3), 427–466 (1997)
30. Pogue, G.P., et al.: Building an innovation coral reef: the austin technology incubator case study. In: Markman, A.B. (ed.) Open Innovation, in press. Oxford University Press, New York
31. Kelley, T., Littman, J.: The art of innovation: Lessons in creativity from IDEO, American's leading design firm. Crown Business, New York (2001)
32. Fisher, W.: Applying the creative problem solving process to open innovation. In: Markman, A.B. (ed.) Open Innovation, in press. Oxford University Press, New York
33. Zintgraff, C., et al.: Better, faster, safer, and cheaper: USAA roof inspections with Pole Cam. In: Markman, A.B. (ed.) Open Innovation, pp. 121–140. Oxford University Press, New York (2016)

Chapter 21
The Wonder of Reason at the Psychological Roots of Violence

Mauro Maldonato

Abstract Aggression, violence and destructiveness have been part of human nature since its origins. Their roots can be traced in unconscious and from an elaboration of mourning that uses division in order to save oneself from anguish and guilt, attributing all good to one's own object of love and all evil to an external enemy—just as happens in the anguish of the stranger, considered dangerous and an enemy, not because he really is, but because onto him the internal enemy is projected. This paper seeks to show how this permanent psychic tension derives from the meeting of opposing, heterogeneous and unpredictable forces and movements which can be neutralized but are never cancelled out. The balance between instinct and rationality can be lost at all times and, on an individual or collective level, it can degenerate into pure violence. But if the life expresses itself through biological functions of a very high complexity, it also does so through history and culture. In other words, a sense of guilt elaborated for the construction of better civilization.

Keywords Unconscious · Rationality · Anxiety · Super-ego · Aggressivity Violence · Psychoanalysis · Destructiveness · Terrorism · Morality

21.1 Introduction

A paradoxical destiny has enveloped history. Utopias turned upside-down and impossible tears and seams provide the context for an experience that is suspended, devoid of direction and permanently discharged by the past. The febrile weariness of the 'laws' of history has given free rein to a mechanism that is indifferent to individual destinies. The difficulty of a coherent representation of history has dramatically impoverished the analysis of the facts, which by now cannot be represented in a complete form. The keys to history have been permanently lost and the effect is a distressing feeling of confusion, that our culture has lost its direction.

M. Maldonato (✉)
Largo Caterina Volpicelli 5, 80136 Naples, Italy
e-mail: mauro.maldonato@unibas.it

C. Faucher (ed.), *Advances in Culturally-Aware Intelligent Systems and in Cross-Cultural Psychological Studies*, Intelligent Systems Reference Library 134, https://doi.org/10.1007/978-3-319-67024-9_21

This discomfort affects, first and foremost, the meaning of life in the contemporary world. Hand in hand with the unprecedented scientific and technological power there is an irredeemable sense of unhappiness. It is as if man's awareness that his will to power cannot be satisfied was more acute and, at the same time, the awareness of one's own fragility, transience and finiteness more dramatic. It is a crisis that is, firstly, a crisis of confidence and hope in one's self and others, and even in the same idea of man and in the possibility of building better forms of co-existence.

Freud maintained that the goal imposed on us by the *pleasure principle* cannot be achieved [17] and that its natural tendency is shattered against an impenetrable wall. Even if the *pleasure principle* tends to continuously assert itself, it is always the *principle of reality that prevails*. Even if art helps us to relieve the stresses of life, it cannot save us from suffering. Beauty is only a feeble defence against the suffering derived from the awareness of the decline of our body and from the negative influence of the external world [18]. The overwhelming force of nature and, at the same time, the fragility of our body remain inescapable. Despite the potent yearning for happiness, man is unable to be happy. The set of rules and institutions which differentiate us from other living beings—whose function is to protect man from nature, governing his relationships within the family, the state and society—appear to be dramatically inadequate [17].

21.2 Aggression, Desire and Lex Naturalis

Freud sees in the behaviour of the masses a strong propensity for submission and the unconditional acquiescence to leaders; a conformist and subordinate form of behaviour tending toward voluntary servitude and a lack of a critical spirit [16]. If culture owes to *eros* phenomenal accomplishments and achievements, it owes to *thanatos* the natural aggressive impulse, the hostility of each person towards everyone and of everyone towards each person and, therefore, the horrors and pains, the grief and destruction that have paved the way of civilisation [18]. The gigantic conflict between *eros* and *thanatos* has as an outcome the preponderance towards war. The desire for control, the feeling of being entitled to everything, pushes men to harm each other, to experience predatory instincts, impulses and desires, which they tend to satisfy with the search for glory and with the use of force to achieve their aims. In a singular convergence with Hobbes [26], the Viennese psychoanalyst maintains that the natural condition is war by all against all. Even if this tendency is difficult to pin down, it is often disguised by cultural and ideological conditioning that pushes us to deny the role of *thanatos* [19]. In what way, therefore, could culture neutralise the aggressive thrust that continuously tends to undermine man's own equilibrium? Hobbes had conceived a system of rational rules (*lex naturalis*) to oppose *natural law*, that is, the right of everyone to everything for their own advantage [26]. The constant and threatening presence of *polemos* imposes a solution that recognises a third party's right to the legitimate use of force: the *Leviathan*.

For the Viennese psychoanalyst the increasing loss of happiness for man is proportional to the fear of the *super-ego* of culture [19]. There are clear differences between processes of civilisation and individual development as well as common elements. If in the individual the pursuit of happiness is always central, in the process of civilisation this always remains in the background: what counts for individuals is to adapt to the demands of society. Each individual is affected by these opposing tendencies, which generate unrest, unhappiness, pain and neurosis. The individual *super-ego* is dominated by the *super-ego* of the culture with its severe ethical demands: penalties, imperatives and prohibitions [17]. The *super-ego* of culture is indifferent to individual psychological needs, in particular to the desire for happiness: indeed, it issues orders and it does not ask itself whether it is possible to carry them out. It takes for granted that the individual *ego* obeys any and every moral prescription, deluding itself that it exercises an unlimited power over its *id*, when on the other hand sovereignty over the *id* is always limited. This is a reason that should lead one to understand why, by demanding increasingly more one produces, on the one hand, anxiety, neurosis and unhappiness [19] and, on the other, the pervasive suffering due to wars, terror, persecution, genocides and so on. After the analysis begun by Freud on the tension between the individual reality which generates desires and civilisation, it is necessary to reflect on the current form of the *super-ego* of culture, on the process of civilisation and on its dramatic tectonic shifts [22]. Nevertheless, today the problem of civilisation is not how to emancipate man from ancient powers that besiege it, but how to deal with that instinctual remnant which can never be entirely integrated into the culture. But to the interminable struggle between the life impulse and the death impulse we will return later.

21.3 The Dark Evil of Culture

When faced with the unbridled totalitarian violence that preceded the Second World War, on 30 July 1932 Einstein wrote to Freud to ask him if it was possible to positively direct the psychic evolution of men and thus put a brake on the impulses of hate and destruction. On the basis of the analysis of dreams, of unconscious slips in healthy people, as well as neurotic symptoms, the Viennese psychoanalyst answered:

> (…) psychoanalysis has drawn the conclusion that the primitive, savage and contemptible impulses of humanity have by no means disappeared, but continue to live, though removed, in the unconscious of each single individual, awaiting the occasion to be able to reactivate themselves. Psychoanalysis has further taught us that our intellect is something fragile and dependent, a trinket and an instrument of our impulses and our affections, and that we are compelled to act sometimes with intelligence and sometimes with stupidity, according to the volition of our personal attitudes and resistances. So, look at what is happening in this war, look at the cruelty and the injustices for which the most civil nations are made responsible, the deceit with which they behave when faced with their own lies and iniquity;

and look finally at how everyone has lost the capacity to judge with rectitude: one must admit that both assertions of psychoanalysis were exact [20].

Aggression is therefore, a 'natural' human characteristic and, as such, cannot be eliminated. One may only try to control its intensity, so that it doesn't mutate into forms of war. But let us not delude ourselves. War reveals the primitive man that is within us: he who transforms the stranger into an enemy who we have to kill, who forces us to be heroes, who prevents us from calmly accepting the idea of death and even from loving our blood brother. Looking at the history of the world, even the famous Christian teaching "love your neighbour as yourself", appears purely to be an ideal: Cane slays Abel, Esau hates Jacob who had deceived him, the brothers want to kill Joseph and they sell him as a slave.

Man is not a kind creature who wishes to be loved or who defends himself only if attacked. Aggression is a significant part of his nature and this has to be taken into account. To this one may add the herd propensity of the majority of men which further highlights the illusion of being able to subdue one's impulses to rationality. Civilisation cannot neutralise man's death instincts [9]. And, in the end, the excessive internalisation of these instincts would not even be desirable, since aggression would explode into tribal, ethnic, ideological and military forms. No ideology, no philosophy, no religion, no theory can delude itself that it can easily rid itself of it. At the basis of the faint critique of power there is the fear of freedom [21]. Of course, man aspires to be free, but he is afraid of freedom. Freedom, in fact, obliges him to make decisions, and these involve risks and the assumption of responsibility. After all, on the basis of what values should one make decisions? Man is used to being told what he should think. Even if from a very early age he is urged to make independent judgements, in reality what is expected of him is clearly determined and established by society. Originality of ideas is intolerable. On the other hand, by submitting to an authority one does not have to worry about what the right thing to do is [28]. Any theory of reform or transformation must take all of this into account.

There is, nevertheless, a further element to consider: the attitude of western culture towards death. On close inspection, it is not very different from that of primitive man. On the one hand he pursues the annihilation of the other, the stranger, the enemy; on the other, he considers death per se as something unreal [20, 5]. The terrifying power of contemporary wars has pushed reason beyond its own limits and there is nothing left for thought but to acknowledge its own radical loss. The presence of death represents a challenge that is impossible to avoid, particularly for the *ontological* life-death link [12]. Compromising with the omnipotence of the ego, the unconscious turns its back on death. More than in its own death, it believes in the death of others, of strangers and enemies. Is it not true perhaps that, in one's own unconscious, everyone aspires to immortality?

The blind frenzy for arms between the first and second world wars drove Freud to reflections that would develop subsequently in *Civilisation and its Discontents* [19]. What is the relation, he asked, between the will to annihilation implemented by governments and the individual conscience which is so disorientated?

The higher the banners of patriotism were raised, the more difficult it was to recognise the identity of one's own nation. Until, all of a sudden, one found one's self lost, and everything appeared mutilated in its essence, in the beauty of its landscapes, disfigured in thought, in art and in any other form of the spirit. Because of all of this is the dissolution of values at the basis of social harmony, of peaceful relations between men in the public sphere: this dissolution is legitimised by the conduct of the same States that made themselves depositories and guarantors of those principles through the law [4]. Those States with a monopoly over the use of violence have denied integration between communities and national identities, pushing peoples to turn on one another with hate and violence [33].

After the tragedies of the twentieth century, non-one imagines that today one would have to witness the return, in the forms of terrorism, of the most ruthless barbarities in the heart of the same western culture. Yet, feelings of disappointment would be unwarranted: it is, instead, the end of an illusion [27]. The revival of conflicts caught unawares those who trusted in the progressive advance of culture, and even more so those who professed certainty about the natural goodness of man. Psychoanalysis exposed the vanity and fallacy of the moral optimism, the ambivalence and the tendency towards conflict of human nature [11]. Love and hate, cruelty and compassion belong to the interplay of instincts which mark the existence of every individual: war brings us back to this truth, one which is entirely impervious to ethical categories and distinctions. Nor, moreover, can war be interpreted as a sort of *regression* [6]. Within culture primitive drives manifest themselves that reveal the fallacy of all social pedagogy. Society began and is sustained thanks to a fiction: education, culture and morality conceal the authentic condition of man, the essence of their inclinations, the passions that obfuscate its rationality [36]. This concealment demands (or rather imposes) the abandonment of the satisfaction of instincts, but it is not able to eradicate them [35]. It cannot eradicate aggression. In fact, when the latter is forced to hide its presence, it merely awaits the opportunity for revenge. Thus, when the vetoes and bans vanish, it takes the upper hand, revealing the violent nucleus of man, to which respect for its own kind is alien.

Intraspecific or interspecific as it may be, war evokes the primal scene; it makes it reappear liberated from the sedimentations of the culture [10]. The same bellicose language, constructed around the obsessive repetition of words such as *enemy* and *hero*, also linguistically brings humanity back to its pre-history: a piece of evidence that *doxa*, scientific rationality and ideology cannot hide [15]. But the reason for violence brings to mind the problem of death. *Polemos*, the feeling of hostility directed outward arises from the defence of life, from the wish to postpone the moment of death: from here arises the struggle of each person against their fellow man through violent appropriation [31]; from here the desire for domination that breaks the cycle of nature. The natural order thus assumes the characteristics of an agonising division between the destiny of the organism that has death as its sole objective and the attempt of the living to outlive itself.

Pre-historic man must have looked on death with some ambivalence: on the one hand there was the *other's* lifeless body (plant, animal or individual of its own species) and on the other, their *own* annihilation, which was unreal and therefore could not be symbolised [8]. What is dead (the prey, the enemy) is before the survivor as a foreign entity. In triumph the winner confirms his own superiority to himself. Only in this alienated form did primitive man take death seriously, considering it the end of life, disavowing and annulling its meaning at the same time [3]. From the beginning, the denial of death marked the progress of humanity as a 'warpath'. Hate, the desire to kill, take the place of the natural tendency towards the dissolution of the organism, prevailing over the drive towards the inorganic state [29]. As is well-known, Freud understood aggression to be an exteriorisation of the death drive that operates in every living being as a primary dynamic factor that tends to lead life back to the state of inanimate matter [11].

In the unconscious impulses of desire the attitude of primitive man to death lives in us almost unchanged. It is only that now the destructive drives move from *factual reality* to *psychic reality*, to the imaginary and the fantasies that are desired [34]. Contemporary man's secret and constant propensity for killing is the dark side of civilisation. We accept death for strangers and enemies and we decree it for them with the same lack of scruples as primitive man [11]. This desire for death, which makes prohibition necessary, is legitimised by war which imposes the killing of the stranger as an enemy [38].

21.4 The Absolute Enemy

As a form of extreme politics, war goes beyond all of the ethical, legal and religious issues. If political identity is founded on belonging, on the difference between individuals and other groups, in war all identities are built on radical contraposition. This contraposition defines the political existence of individuals and groups. Moreover, the separation of their own existence from that of others (who live and act according to different rules, customs and modes of behaviour) could not take place if not through the recognition among counterparts of the same group. The *principium individuationiis* [43] which defines political identity also defines individual identity and belonging as a process of distinction-differentiation.

It is with the birth of the State and conflicts between States, which war changes radically, transforming itself into a technique of annihilation, into a theory of the dehumanisation of the opponent, into the downgrading of the enemy to an inferior race. State and interstate war changes the 'natural' interspecific aggression [10] into civil and ideological conflict, into destructiveness without *pietas*. Before the birth of the modern State, wars were ritualised conflicts that were limited in time and space, which concluded with peace between the victors and the vanquished [32]. Homer and Virgil marvellously tell of the *pietas* that the Greeks and Romans had for the vanquished. The military conquests of the Romans in particular were governed by politics and law in the forms of a *foedus* that granted autonomy to the vanquished cities and peoples and a degree of freedom in their customs and traditions. The frequent wars of the feudal world were also similarly limited. Compared to contemporary wars they look like skirmishes, squabbling between armed groups, which involve only the peoples near the line of combat.

In the Napoleonic era modern war as we know it took form, the absolute war of nation states, with obligatory conscription, industrial technology and economic power [7]. Of course, on the tactical-strategic level, between the Napoleonic wars and the campaigns of Alexander the Great, despite the obvious and profound difference on the social-political landscape, there weren't many differences. Now, war has become an absolute struggle, a mortal conflict that involves all of the citizens. The wonder of Goethe who, reacting to the battle of Valmy (1792), exclaimed "a new era of history has begun!" more from admiration of the genius of Napoleon, was aroused by the power of a new and grandiose historical event: the modern national war. Conversely, the wars of the twentieth century constitute the paradigm of total war. Rummel [39] and Goldhagen [25] cast light on the themes of state violence and genocide in human history. Rummel, a tireless scholar of the violence generated by the modern concentration of power, documented how in the twentieth century alone the policies of contemporary States eliminated almost two hundred million people, in peacetime and during war, sometimes more in the former than the latter: 174,000,000 deaths that, placed end to end, would go around the world four times. All of the water and blood of the 174,000,000 deaths in the twentieth century, flowing over Niagara Falls, would see the passage of water for over 10 h, or the passage of blood equal to almost 43 min [39].

Through the militarisation of the economy and society and the transformation of entire geopolitical spaces into a limitless front, the wars of the twentieth century represented the most direct and coherent expression of the nation-state, with its characteristics of an 'organic body', the unification of national identities, exclusive sovereignty, the total control over an entrenched territory [24]. The step to regimes identified with race and ideology was a short one. Nazism and communism represent a perfect combination of normal and civil war. In reality, each political aggregation always arises against an enemy [42] and the binomial *amicus-hostis*—the political link and the public animosity—represents the essence of every political action: a real and existential opposition, the most intense and extreme of all. In this sense, enemy is not the competitor or the private adversary that hates us for some reason. Enemy is a set of men that opposes another human group of the same kind and an entire people [45]. The enemy is the'*hostis*, not the *inimicus* in a broad sense [42]. If enemy is the political aggregation, the party or the typical partisan formation of civil war, friend on the other hand is the partisan, the soldier of the nation-state, he who lives entirely with reference to the enemy. The *amicus-hostis* dichotomy is the essence of politics, of the State, of *polemos*, of war [37]. The competitor, the adversary in the argument and the private *inimicus* are more suited to representative democracy, economics and civil society. The public friend-enemy battleground is war [40], which as such is never for religious, moral, legal or economic reasons. War, in fact, does not need to be religious, or ethical. When religious, moral and other types of conflict transform into political conflicts, then the decisive conflict is not religious, moral or economic but political [42].

The twentieth century showed how the illusory paradises transformed themselves into real hells, how the pseudo-religion of humanity paved the way for inhuman terror [1]. This lesson imposes on everyone a new awareness of human nature. Man, who the demagogues elected the absolute measure of all things, is not naturally inclined to peace: through terror and annihilation he combats those who don't submit to him. The concept of man embodies a contradictory concept, that of non-man, endowed with tremendous potential, which opens up an abyss of enmities. This is a prelude to another, even more profound distinction between super-man and sub-human. Indeed, the man who treats his fellow man as a non-human distinguishes between superman and sub-human. For the latter no kind of penalty exists: only extermination and annihilation [41].

21.5 The Illusion of Perpetual Peace

Even if today the spectre of another and more extensive nuclear holocaust seems more remote, from the ruins of the old international order of the ideological and colonial empires, ethnic and national realities seem to revive that excite profound cultural mythical and symbolic values [46]. The conforming effect of the 'universalist' ideologies of the twentieth century, which had blunted identities and histories, traditions and cultures, brought religious fundamentalism back to the surface, together with nationalism. Religions, which the expansion of the secular power of the state in the twentieth century had kept in the shadows, now enlist God under their own flags. With the risk of regional, sub-regional and local conflicts and micro-conflicts, the theories on limited war and even on limited nuclear war return.

In other respects, the decline of nation-States swept away the legitimacy of military roles and of permanent armies. After the military hegemony's centuries of endeavours to conquer and to extend frontiers as far as possible, the unitary States with vast territories begin to decline [46]. The decline of the military roles (and of wars between states), paralleled with the weakness of the political function of States and of their internal order, seems by now to be one of the most proven constants of politics [30]. As the history of the last few decades has shown (Northern Ireland, Great Britain, the Basque countries, Italy, the Middle East, the United States, Russia, France, Belgium, etc.), the state security apparatuses—so effective in the control of the individual citizen's daily life—are not able to defend themselves from political terrorism, or to guarantee any peace. Even though powerful on the military level, they appear vulnerable and defenceless. This radically transforms the nature and the operational scenarios of war: from the war between States one passes to regional wars and terrorism. The former that shows the illusory nature of *perpetual peace* and the short range of supranational bodies (United Nations) that attempt to identify forms of international peace and the elimination of conflicts, transforming wars into military police operations: ultimately, making the whole world a great peaceful and protected democracy [23].

The elimination of conflicts from the global scene is little more than an illusion, an abstraction. The idea of a 'common security' is fallacious not only because of the naive vision of human nature that underpins it, but also because no effective security can be common. Its effectiveness, indeed, depends on the will of the states and on their diverging and contingent political interests. Each State acts on the basis of its own national interests and power in order to expand its own prominence wealth and security. The Kantian hope at the basis of the institutionalist paradigms —that is, the idea of an intergovernmental concertation for the restriction of conflicts—comes up against the problems posed by the erosion of sovereignty of the nation-State by sub-national and transnational actors, by the incapacity of the states to mobilise resources previously available, by difficulties in always identifying the aggressor and by the political-strategic criticality of the control and management of operations [2]. All of this makes it difficult to offer a prognosis on the future of peace and on new geopolitical power balances.

Yet, if it is true that war arises from inevitable tensions, conflicts and explosions of violence, it is likewise true that the evolution of trade and competition requires the prevention of the conflicts, non-belligerent scenarios and the limitation of wars [13]. For such a goal one will have to await long and painful transitions, with brief and violent rekindling of hostilities: it is a route that one must follow no longer with the reassuring political-psychological protective screens that, for almost the whole of the twentieth century, guaranteed equilibrium based on hostile interaction between the three ideologies (communism, fascism and democracy) and, after 1945, on the bi-polar balance of power founded on the residual communist and liberal-democratic ideologies. What appears certain is that it is not institutionalist utopias or normativist rhetoric that will make peace a feasible political hypothesis. It can become so, however, within certain limits, in the awareness that it is not enough to reject nationalism, ethnic cultures and diversity. New forms of co-existence will probably be possible on the condition that geo-historical differences and the symbolic dimensions of the different ethoses, spatialities, territories, identities and mythological-symbolic complexes are recognised.

Psychoanalysis has shown how aggression, violence and destructiveness have been part of human nature since its origins [19]. Their roots can be traced in unconscious, terrifying dynamics, with neither face nor words, and from a paranoid elaboration of mourning that uses division in order to save oneself from anguish and guilt, attributing all good to one's own object of love and all evil to an external enemy – just as happens in the anguish of the stranger, considered dangerous and an enemy, not because he really is, but because onto him the internal enemy is projected [14]. It is not implausible to identify here the root of the *amicus-hostis* pattern at the basis of wars, for which man kills without recognising per se the desire to kill, treating death and hate as if they didn't belong to him, as if they were always someone else's, arriving at the great illusion that it is our enemy who wishes to kill us. Alongside the (illusory) function of security directed at an external enemy, war also performs another function that is unconscious and invisible, that hides from the terrifying 'internal enemy'.

This representation of war as a defence from psychotic anguish does not permit pessimism or desperation. Other roads are possible, perhaps by developing a more reparative attitude [14], based on the capacity to live with the pain caused by death and with the growth of individual responsibility. This will be able to take place not beginning from desperation and the burden of guilt for destructiveness, but from the urge to change deriving from the sense of guilt for the idea of killing. A guilt, that is, directed at the future, so that death and destruction do not remain the last words of humanity.

References

1. Arendt, H.: Le origini del totalitarismo. Bompiani, Milano (1978)
2. Attinà, F.: The global political system. Palgrave Macmillan, Houndmills (2011)
3. Bergeret, J.: *La violenza e la vita. La faccia nascosta dell'Edipo*. Trad. it. A cura di E.Cimino, Roma, Borla (1998)
4. Bobbio, N.: Stato, governo, società. Einaudi, Torino (1985)
5. Braudillard, J.: L'illusione dell'immortalità. Armando, Roma (2007)
6. Braudillard, J.: Lo scambio simbolico e la morte. Feltrinelli, Milano (2006)
7. von Clausewitz, C.: Vom Kriege. B. Behrs Verlag, Leipzig (1937)
8. Edwards, I.E.S., Gadd, C.J., Hammond, N.G.L.: *Storia del mondo antico. Preistoria e nascita delle culture in Oriente* (Vol. 1). Milano, Garzanti (1974)
9. Eibl-Eibesfeldt, I.: Liebe und Hass Zur Naturgeschichte elementarer Verhaltensweisen. Piper, München (1970)
10. Eibl-Eibesfeldt, I.: *Krieg und Frieden aus der Sicht der Verhaltensforschung*. München; Zurich: Piper & Co (1975)
11. Einstein, A., Freud, S.: *Warum Krieg?* Zürich, Diogenes (1972)
12. Ellenberger, H.: La scoperta dell'inconscio. Bollati Boringhieri, Torino (1970)
13. Flint, C., Taylor, P.J.: Political Geography: World—Economy, Nation—State and Locality. Pearson Education Limited, Harlow (2007)
14. Fornari, F.: Psicoanalisi della guerra. Feltrinelli, Milano (1988)
15. Fornari, F., Basilica, F., Controzzi, G.: Dell'Acqua GP, (a cura di) La violenza. Firenze, Vallecchi (1978)
16. Freud, S.: Massenpsychologie und Ich-Analyse. Internationaler Psychoanalytischer Verlag, Leipzig (1921)
17. Freud, S.: Das Ich und das Es. Internationaler Psychoanalytischer Verlag, Wien (1923)
18. Freud, S.: Jenseits des Lustprinzips. International psycoanalytischer, Leipzig (1923)
19. Freud, S.: Das Unbehagen in der Kultur. Internationaler psychoanalytischer Verlag, Wien (1930)
20. Freud, S.: Considerazioni attuali sulla guerra e sulla morte. In Opere. Boringhieri, Torino (1976)
21. Fromm, E.: Escape from freedom. Holt, Rinehart and Winston, New York (1961)
22. Fukuyama, F.: La fine della storia e l'ultimo uomo. Rizzoli, Milano (1992)
23. Ikenberry, G.J: Liberal Leviathan. The Origins, Crisis and Transformation of the American World Order. Princeton University Press, Princeton (2011)
24. Galli, C.: Genealogia della politica: Carl Schmitt e la crisi del pensiero politico moderno. Bologna: Il Mulino (1996)
25. Goldhagen, D.J.: Peggio della guerra. Lo sterminio di massa nella storia dell'umanità. Milano, Mondadori (2010)
26. Hobbes, T.: Leviathan. Yale University Press, Yale (ed. or. 1651) (2010)

27. Huntington, S.P.: The Clash of Civilizations? The Debate. Foreign Aff, New York (1996)
28. La Boétie, É.: Discours de la servitude volontaire. Portes de France, Porrentury (1943)
29. Lacan, J.: Écrits. Editions du Seuil, Paris (1966)
30. Lacoste, Y.: Dictionnaire de géopolitique. Flammarion, Paris (1993)
31. Magnani, L.: Filosofia della violenza. Il Melangolo, Genova (2012)
32. Maldonato, M.: Competizione. Enciclopedia del Corpo, Treccani (1999)
33. Maldonato, M.: Dal Sinai alla rivoluzione cibernetica. Guida, Napoli (2002)
34. Maldonato, M.: Psychobiology of conflict. World Futures—Routledge **62**, 392–400 (2006)
35. Nietzsche, F.: Menschliches Allzumenschliches. Kroener, Leipzig (1919)
36. Nietzsche, F.: Zur Genealogie der Moral eine Streitschrift. Reclam, Leipzig (1930)
37. Portinaro, P.: Il realismo politico. Roma; Bari, Laterza (1999)
38. Portinaro, P.: Breviario di politica. Morcelliana, Milano (2009)
39. Rummel, R.J.: Stati assassini. La violenza omicida dei governi. Saggio introduttivo di Alessandro Vitale, Soveria Mannelli, Rubbettino (ed. or. 1994) (2005)
40. Schmitt, C.: Theorie des Partisanen: Zwischenbemerkung zum Begriff des Politischen. Duncker, Humblot, Berlin (1975)
41. Schmitt, C.: Donoso Cortés in gesamteuropäischer Interpretation. Vier Aufsätze, Köln (1950)
42. Schmitt, C.: Der Begriff des Politischen. Akademie Verlag, Berlin (2003)
43. Simondon, G.: L'individu et sa genèse physico-biologique (l'individuation à la lumière des notions de forme et d'information). PUF, Paris (1964)
44. Vitale, A.: Il 'nemico pubblico nella letteratura ideologica d'area slava. In: Amicus (Inimicus) Hostis. Le radici concettuali della conflittualità 'privata' e della conflittualità 'politica'. A cura di G. Miglio, Giuffrè, Milano (1992)
45. Vitale, A.: The re-emergence of historical regions, cities and enclaves in Europe vs. the EU'sintegration concepts, processes and reality. In: "Region & Regionalism" n. 10, vol. I, pp. 29–42. Polish Geographical Society, University Lodz, Silesian Institute in Opole, SilesianInstitute Society. Lodz-Opole (Polonia) (2011)
46. Wade, R.H.: Emerging World Order? From Multipolarity to Multilateralism in the G20, the World Bank, and the IMF. Politics. Soc. **39**(3), 347–378 (2011)

Printed in the United States
By Bookmasters